The SAGE Handbook of
Digital Technology Research

The SAGE Handbook of
Digital Technology
Research

Edited by
Sara Price, Carey Jewitt
and Barry Brown

Los Angeles | London | New Delhi
Singapore | Washington DC

Los Angeles | London | New Delhi
Singapore | Washington DC

SAGE Publications Ltd
1 Oliver's Yard
55 City Road
London EC1Y 1SP

SAGE Publications Inc.
2455 Teller Road
Thousand Oaks, California 91320

SAGE Publications India Pvt Ltd
B 1/I 1 Mohan Cooperative Industrial Area
Mathura Road
New Delhi 110 044

SAGE Publications Asia-Pacific Pte Ltd
3 Church Street
#10-04 Samsung Hub
Singapore 049483

Editor: Jai Seaman
Production editor: Ian Antcliff
Copyeditor: Rosemary Campbell
Proofreader: Michelle Clark
Indexer: Avril Ehrlich
Marketing manager: Ben Griffin-Sherwood
Cover design: Jennifer Crisp
Typeset by: C&M Digitals (P) Ltd, Chennai, India
Printed in Great Britain by Henry Ling Limited,
at the Dorset Press, Dorchester, DT1 1HD

Editorial arrangement and Introduction © Sara Price, Carey Jewitt and Barry Brown 2013

Chapter 1 © Paul Ceruzzi 2013
Chapter 2 © Charles Crook 2013
Chapter 3 © Matt Jones 2013
Chapter 4 © Gary Hsieh & Nicolas Friederici 2013
Chapter 5 © Sonja Baumer 2013
Chapter 6 © Heather Horst & Larissa Hjorth 2013
Chapter 7 © Anna Kouppanou & Paul Standish 2013
Chapter 8 © Sara Grimes & Andrew Feenberg 2013
Chapter 9 © Jeffrey Bardzell 2013
Chapter 10 © Paul Marshall & Eva Hornecker 2013
Chapter 11 © Luigina Ciolfi 2013
Chapter 12 © Kristina Höök 2013
Chapter 13 © Barry Brown 2013
Chapter 14 © Victor Kaptelinin 2013
Chapter 15 © Robert J. Moore 2013
Chapter 16 © Cliff Lampe 2013
Chapter 17 © Carey Jewitt 2013
Chapter 18 © Steven Dow, Wendy Ju and Wendy Mackay 2013
Chapter 19 © Laurel Swan & Kirsten Boehner 2013
Chapter 20 © Sara Price 2013
Chapter 21 © Leah Buechley 2013
Chapter 22 © Eve Hoggan 2013
Chapter 23 © Yvonne Rogers & Nicola Yuill & Paul Marshall 2013
Chapter 24 © Yoosoo Oh and Woontack Woo 2013
Chapter 25 © Ty Hollett & Kevin Leander 2013
Chapter 26 © Niall Winters 2013
Chapter 27 © Catherine Beavis 2013
Chapter 28 © Kirsty Young 2013
Chapter 29 © Kaśka Porayska-Pomsta & Sara Bernardini 2013
Chapter 30 © Lars Erik Holmquist 2013

First published 2013

Library of Congress Control Number: 2012950071

British Library Cataloguing in Publication data

A catalogue record for this book is available from the British Library

ISBN 978-1-4462-0047-6

Contents

Notes on the Editors and Contributors

Jeffrey Bardzell is an Associate Professor of HCI/Design and new media in the School of Informatics and Computing at Indiana University – Bloomington. Having done his doctoral work in Comparative Literature and Philosophy, Bardzell brings a humanist perspective to HCI and is known for developing a theory of interaction criticism. His other HCI specialties include aesthetic interaction, user experience design, amateur multimedia design theory and practice, and digital creativity. Currently, he is using theories from film, fashion, science fiction, and philosophical aesthetics to theorize about users and interaction, especially in the context of user experience design and supporting creativity

Sonja Baumer has a multidisciplinary background in developmental and clinical psychology, as well as in media and communication studies. She has been associated with the Laboratory of Comparative Human Cognition, University of California, San Diego, since 2000. From 2006 to 2008 she was a postdoctoral researcher at the University of California, Berkeley, on the project 'Digital Youth: Kids' Informal Learning with Digital Media'. In addition to studying youth's interactions with digital media, Sonja's research interests also include topics related to the role of play in the development of self-regulation. Sonja is also a practising psychotherapist helping children and youth with atypical development improve self-regulation skills and impulse control.

Catherine Beavis is a Professor of Education in the School of Education and Professional Studies at Griffith University, Australia. She teaches and researches in the area of English Curriculum, Literature and Literacy Education, concerning young people and digital culture, with a particular focus on video or computer games; the changing nature of text; and the nature and implications of young people's engagement with video games for English and literacy education. Her work explores contemporary constructions of English, texts and literacy; the role of game play in young people's lives, games as spaces within which young people play, connections between game play, identity and community, and the critical examination of the in-school use of video games.

Sara Bernardini is a Research Associate in the Planning Group at the Department of Informatics of King's College London. Her research interests are in the area of artificial intelligence and include autonomous intelligent agents, automated planning, technology-enhanced learning, human-computer interaction and knowledge representation and engineering. Before joining King's College, Sara was a Research Fellow at the London Knowledge Lab where she was part of an interdisciplinary team that focused on building an intelligent virtual agent for helping autistic children develop social communication skills. Previously, in 2005–2007, Sara was a research scientist at NASA Ames Research Center (CA, USA) where she worked on developing autonomous planning agents for space mission operations. Sara received her PhD in Artificial Intelligence from the University of Trento, Italy and her Master Degree in Computer Science and Engineering (magna cum laude) from the University of Rome "La Sapienza", Italy.

Kirsten Boehner is a Visiting Fellow at the Interaction Research Studio at Goldsmiths, University of London. Her research focuses on intersections of divergent perspectives and practices in both design and evaluation of digital technology. Kirsten holds a PhD from Cornell University in Communication with a focus on Human Computer Interaction.

Barry Brown has recently moved to become a studio director at the Mobile Life research centre in Stockholm. His recent work has focused on the sociology and design of leisure technologies – computer systems for leisure and pleasure. Recent publications include studies of activities as diverse as games, tourism, museum visiting, the use of maps, television watching and sport spectating. He has also edited books on music consumption (with Kenton O'Hara), and mobile phone use (with Richard Harper and Nicola Green). He was previously an associate professor at the University of California, San Diego, a research fellow on the Equator project at the University of Glasgow and a research scientist at Hewlett-Packard's research labs in Bristol. His qualifications include a degree in computer science from the University of Edinburgh, and a PhD in sociology from the University of Surrey.

Leah Buechley is an Associate Professor at the MIT Media Lab where she directs the High-Low Tech research group. The High-Low Tech group explores the integration of high and low technology from cultural, material, and practical perspectives, with the goal of engaging diverse groups of people in developing their own technologies. She is a well-known expert in the field of electronic textiles (e-textiles), and her work in this area includes developing the LilyPad Arduino toolkit. Her research received a 2011 NSF CAREER award and has been featured in numerous articles in publications including the *New York Times*, *Boston Globe*, *Popular Science*, and the *Taipei Times*. She received PhD and MS degrees in computer science from the University of Colorado at Boulder and a BA in physics from Skidmore College.

Paul E. Ceruzzi is the Chair of the Space History Division at the Smithsonian's National Air and Space Museum in Washington, DC. He is the author of several books on the history of computing and aerospace technology, most recently *A History of Modern Computing* (2003); and *Internet Alley: High Technology in Tysons Corner* (2008). His current research concerns the use of computers for long-range space missions, which is told in a new Smithsonian exhibit on satellite and space navigation, which opened in April 2013.

Luigina Ciolfi is a Reader in Communication in the Communication and Computing Research Centre, Sheffield Hallam University (UK). Her research focuses on the design and evaluation of interactive technologies to support human interaction based on an understanding of the relationship between people, activities and their locales. She has studied heritage sites, urban spaces and work settings through the lens of place as a notion for understanding human interaction (both individual and social) in context. She holds a Laurea (summa cum laude) in Communication Sciences from the University of Siena (Italy) and a PhD in Computer Science/Interaction Design from the University of Limerick (Ireland).

Charles Crook is a developmental psychologist who is currently Professor of Education and Director of the Learning Science Research Institute at the University of Nottingham. He has researched a socio-cultural approach to the adoption of new technologies for learning and recreation.

Steven Dow is an Assistant Professor at the HCI Institute at Carnegie Mellon University where he researches human–computer interaction, innovation, design education, and crowdsourcing methods. He is recipient of a National Science Foundation grant and Stanford's Postdoctoral Research Award, and co-recipient of a Hasso Plattner Design Thinking Research Grant. He received an MS and PhD in Human-Centered Computing from the Georgia Institute of Technology, and a BS in Industrial Engineering from University of Iowa.

Andrew Feenberg is Canada Research Chair in Philosophy of Technology in the School of Communication, Simon Fraser University, where he directs the Applied Communication and Technology Lab. His recent books include *Between Reason and Experience: Essays in Technology and Modernity* (MIT Press, 2010), and *(Re)Inventing the Internet* (Sense Publishers, 2012). His work attempts to inform philosophy of technology with the results of research in STS. His background includes extensive experience with computer networking. He is recognized as an early innovator in the field of online education, a field he helped to create in 1982. His current work includes analysis and critique of the role of technology in Heidegger and the Frankfurt School, and research on the politics of online community.

Nicolas Friederici is an ICT4D & Mobile Innovation Associate at infoDev, a partnership programme at the World Bank. Nicolas was a Fulbright scholar at Michigan State University where he received an MA in Telecommunication, Information Studies and Media. His thesis examined how negative emotions affect online support seeking for health problems. He also worked in a research project on broadband economics and policy and co-taught a class on social computing. In 2010 Nicolas was awarded a Diploma in Media Studies and Media Management from the University of Cologne.

Sara M. Grimes is an Assistant Professor with the Faculty of Information, and Associate Director of the Semaphore Lab, both at the University of Toronto. Her research explores various aspects of children's digital media culture(s), play studies and critical theories of technology, with a focus on digital games. Her previous work has appeared in journals such as *New Media & Society*, *The Information Society*, *The International Journal of Media & Cultural Politics*, and *Communication, Culture & Critique*, and includes discussions of the cultural politics of children's virtual worlds and online communities, examinations of the ethics and politics of online marketing targeted at children, and the articulation of a critical theory of digital game play (co-authored with Andrew Feenberg). Her current research focuses on the democratic rationalization of children's play and creativity within commercial, web 2.0 user-generated content games and platforms.

Larissa Hjorth is an artist, digital ethnographer and Associate Professor in the Games Programs, School of Media & Communication, RMIT University. Since 2000, Hjorth has been researching the gendered and socio-cultural dimensions of mobile, social, and locative media and gaming cultures in the Asia–Pacific. She is particularly interested in the relationship between intimacy, co-presence and place. Her books include *Mobile Media in the Asia-Pacific* (London, Routledge, 2009), *Games & Gaming* (London: Berg, 2010), *Online@AsiaPacific: Mobile, Social and Locative in the Asia–Pacific region* (with Michael Arnold, Routledge, 2013), and *Understanding Social Media* (with Sam Hinton, Sage, 2013). In addition to numerous journal articles, Hjorth has co-edited four Routledge anthologies, *Gaming Cultures and Place in the Asia–Pacific region* (with Dean Chan, 2009), *Mobile technologies: From telecommunication to media* (with Gerard Goggin, 2009), *Studying the iPhone: Cultural technologies, mobile communication, and the iPhone* (with Jean Burgess and Ingrid Richardson, 2012) and *Mobile Media Practices, Presence and Politics: The challenge of being seamlessly mobile* (with Katie Cumsikey, 2013).

Eve Hoggan is an Aalto University Science Fellow and the vice-leader of the Ubiquitous Interaction (UIx) group at the Helsinki Institute for Information Technology, Finland. Hoggan received a PhD in crossmodal audio and tactile interaction from the University of Glasgow, UK in 2010. Prior to obtaining her PhD, Hoggan interned in Barcelona with Telefónica I+D in the multimedia research team and also at Nokia Research Center in Helsinki. Her work involves long-term 'in the wild' studies of multimodal interaction with the audio and haptic modalities, multi-touch surface interaction, and increasing the bandwidth of remote interpersonal communication through the use of novel haptic interaction techniques.

Ty Hollett is a graduate student at Vanderbilt University in the Department of Teaching and Learning, with an emphasis in Language, Literacy and Culture. His research focuses on literacy learning as embodied, mobile and social practice. He is currently exploring the use of mobile devices by adolescents in and out of school settings.

Lars Erik Holmquist leads the Mobile Innovations group at Yahoo! Labs in Sunnyvale, California. Previously, he was Professor in Media Technology at Södertörn University and manager of the Interaction Design and Innovation lab at the Swedish Institute of Computer Science. He was a co-founder and research leader at the Mobile Life Centre, a joint research venture between academia and industry hosted at Stockholm University, with major partners including Ericsson, Microsoft, Nokia, TeliaSonera and the City of Stockholm. He received his MSc in Computer Science in 1996, his PhD in Informatics in 2000, and became an Associate Professor in Applied IT in 2004, all at the University of Gothenburg. In his work he has developed many pioneering interfaces and applications in the areas of ubiquitous computing and mobile services, including location-based devices, handheld games, mobile media sharing, visualization techniques, entertainment robotics, tangible interfaces and ambient displays. All of his work has been carried out in multi-disciplinary settings, mixing technology, design and user studies, often in close collaboration with industrial stakeholders. His first book, *Grounded Innovation: Strategies for Creating Digital Products*, was published by Morgan Kaufman in May 2012.

Kristina Höök is Professor in Interaction Design at KTH, Sweden. She also has a part-time position at SICS. She started the Mobile Life centre in 2007 – a centre that has now grown to include around 50 researchers, working in close contact with industrial partners such as Microsoft Research, Ericsson, Nokia, TeliaSonera, IKEA, ABB and Stockholm City. Höök is most known for her work on social navigation, mobile interactions, affective interactions and design for bodily experiences. She has published in highly rated platforms such as ACM SIGCHI, DIS, NordiCHI, ToCHI, IJHCS, and the Royal Society in the UK.

Eva Hornecker is a Professor of Human-Computer Interaction at the Bauhaus-Universität Weimar in Germany. Previously, she was a lecturer in the Dept of Computer and Information Science at the University of Strathclyde, and held post-doctoral positions at the Open University, the University of Sussex, and the University of Canterbury, NZ. Her research interests focus on the design and user experience of 'beyond the desktop' interaction. This includes multitouch surfaces, tangible interaction, whole-body interaction, mobile devices, physical and physically embedded computing, the support of social/collaborative interactions, and the social/societal implications of technology.

Heather A. Horst is Vice Chancellor's Senior Research Fellow in the School of Media and Communications and Co-Director of the Digital Ethnography Research Centre at RMIT

University, Australia. An anthropologist by training, Horst's research is primarily concerned with the implications of and changing relations to media, technology and material culture in everyday social life across a range of national and transnational contexts. Horst is co-author of *The Cell Phone: An anthropology of communication* (with D. Miller, Berg, 2006) and *Hanging Out, Messing Around and Geeking Out: Living and learning with new media* (with M. Ito et al., MIT Press, 2010) and is currently writing an ethnography on digital media and family life. She has contributed articles on new media, technology and material culture to *Global Networks, International Journal of Communication, International Journal of Cultural Studies, Journal of Material Culture and New Media & Society*, She recently co-edited, with Daniel Miller, *Digital Anthropology* (Berg, 2012).

Gary Hsieh is a joint-appointed Assistant Professor in the Department of Communication and the Department of Telecommunication, Information Studies and Media at Michigan State University. His research focus is on studying, designing and developing technologies to enable people to interact in ways that are efficient and welfare-improving. He has conducted research at a number of industry research labs, including Microsoft, IBM, Intel and Fuji-Xerox. He received his PhD from the Human–Computer Interaction Institute at Carnegie Mellon University and his BS in Electrical Engineering and Computer Science at the University of California, Berkeley.

Carey Jewitt is Professor of Learning and Technology and Head of the Culture, Communication and Media Department at the Institute of Education, University of London. Her research interests are the development of visual and multimodal research theory and methods, video-based research, and researching technology-mediated interaction primarily in educational contexts. She is a founding editor of the Sage journal *Visual Communication*, and Director of MODE (Multimodal Methods for Researching Digital Data and Environments), a National Centre for Research Methods Node funded by the ESRC (http://mode.ioe.ac.uk). Carey's publications include *The Routledge Handbook of Multimodal Analysis* now in its 2nd edition (2009, 2013) and *Technology, Literacy and Learning: A multimodal approach* (Routledge, 2008).

Matt Jones is a Professor of Human Computer Interaction in the Future Interaction Technology Lab (www.fitlab.eu). He has been a Visiting Fellow at Nokia Research and an IBM Faculty Award holder for work with the Spoken Web group at IBM Research Delhi. Over the past 17 years he has worked on mobile interaction research and design projects internationally. He co-authored the book *Mobile Interaction Design* with Gary Marsden (John Wiley; more details are available at: www.undofuture.com).

Wendy Ju is the Executive Director of Interaction Design at the Center for Design Research at Stanford University, and an Assistant Professor in the Graduate Program in Design at the California College of the Arts. Her research focuses on fundamental practices across design disciplines, and on the expressive properties of physical motion in modern-day interaction. Wendy has an MS from the MIT Media Lab, and a BS in Mechanical Engineering from Stanford University.

Victor Kaptelinin is a Professor at the Department of Information Science and Media Studies, University of Bergen, Norway, and the Department of Informatics, Umeå University, Sweden. He has held teaching and/or research positions at the Psychological Institute of the Russian Academy of Education, Moscow Lomonosov University, and the University of California in San Diego, USA. His main research interests are in interaction design, activity theory, and educational use of information technologies. His recent book, co-authored with Bonnie Nardi, is *Activity Theory in HCI: Fundamentals and reflections* (Morgan and Claypool, 2012).

Anna Kouppanou is a primary school educator and also teaches Philosophy of Education at the European University of Cyprus. She holds a BEd and an MA in Cultural Perspectives in Education and Psychology from the University of Cyprus. She is currently pursuing a PhD degree at the Institute of Education, University of London. Her thesis is an investigation of Martin Heidegger's understanding of nearness in relation to technology, space, time, metaphor and imagination. Her research interests include philosophy of education and technology, phenomenology, existentialism, new media and ethics.

Cliff Lampe is an Associate Professor in the School of Information at the University of Michigan. His research focuses on how large, distributed groups are supported through mediation by information and communication technology, and the effects of that mediation on social processes. His work has included studying sites like Slashdot, Wikipedia, and Facebook, as well as creating online communities related to environmental journalism, economic development, water conservation and energy efficiency. His work has been supported by the National Science Foundation, the US Department of Agriculture, the Kellogg Foundation and the Bill and Melinda Gates Foundation, among others. He received his PhD in Information from the University of Michigan in 2006.

Kevin M. Leander is a professor in the Language, Literacy and Culture programme at Vanderbilt University. His work applies and extends spatial theories in the analysis of identity, literacy and learning. Leander has a special interest in changes in literacy as social practice through new media technologies. His most recent research (with Mariette de Haan and Sandra Ponzanesi, Utrecht University) examines the socialization and identity practices of migrant youth through new media. A second project (with Rogers Hall, Vanderbilt University) examines the use of spatial representations and spatial practices in the routine work of professionals.

Wendy Mackay is a Research Director at INRIA Saclay, France, where she heads the In|Situ| research group in Human–Computer Interaction at the University of Paris-Sud. She received her PhD from MIT and created a multi-disciplinary research group at Digital Equipment that produced the world's first commercial interactive video system (IVIS), a pre-Hypercard multimedia authoring language and over 30 multimedia software products in the 1980s. She then created a research group at Xerox PARC's EuroPARC lab that was among the first to explore media spaces, tangible computing and mixed reality interfaces. She is a member of the ACM CHI Academy and has published over a hundred research articles in the area of human–computer interaction. She has served as Chair of ACM/SIGCHI, co-editor-in-chief of the journal *IJHCS* and on the editorial boards of CACM, ACM/TOCHI and RIHM, as well as programme or associate chair of ACM CHI, UIST, CSCW, DIS, IUI and Multimedia. She is chair of CHI'13 in Paris, France. Her research interests include co-adaptive instruments, tangible computing and multi-disciplinary, participatory design methods.

Paul Marshall is a lecturer in interaction design at University College London. His research interests centre on the concept of embodied interaction and how it can be applied to the design and evaluation of technologies that extend and augment individual human capabilities. This has included work on physical interaction and tangible interfaces; on technologies for face-to-face collaboration; on the design of technologies to fit specific physical contexts; and on extended cognition and perception.

Robert J. Moore is a Research Staff Member at IBM Research – Almaden, where he examines work practice, social interaction and human–computer interaction. In the past he has worked as a researcher at Yahoo! Labs and at the Xerox Palo Alto Research Center (PARC) and as a game designer at The Multiverse Network. Moore's past research includes studies of user interaction with search engines using eye-tracking, avatar-mediated interaction in virtual worlds, face-to-face interaction in print shops, work practices in automobile assembly plants and telephone-mediated interaction in survey call centers. He holds a PhD and MS and BA degrees in sociology with a focus on ethnomethodology, conversation analysis and ethnography.

Yoosoo Oh received his BS in the Department of Electronics and Engineering from Kyungpook National University (KNU, Daegu, Korea) in 2002 and his MS in the Department of Information and Communications from Gwangju Institute of Science and Technology (GIST, Gwangju, Korea) in 2003. In 2010, he received his PhD in the School of Information and Mechatronics from GIST. From 2010 to 2012, he was a postdoctoral researcher in the Culture Technology Institute at GIST. In 2012, as a Research Associate, he joined KAIST, Daejeon, Korea. He has been an Assistant Professor with the School of Computer & Communication Engineering at Daegu University (Daegu, Korea) since September 2012,. His research interests include context and activity fusion and reasoning, context-aware middleware, ubiquitous virtual reality in smart space, HCI and ubiquitous computing.

Kaśka Porayska-Pomsta is a senior lecturer in technology-enhanced learning at the London Knowledge Lab, Institute of Education, University of London. She holds a PhD in artificial intelligence from the University of Edinburgh, School of Informatics. She specializes in adaptive technology for learning, including learner modelling in relation to context, affect and motivation and natural language feedback generation. From 2004 to 2006 she was a postdoctoral fellow at the University of Edinburgh, School of Informatics working on the EU-funded (FP6) LeAM project. In 2006 she became the principal investigator of the ESRC/EPSRC-funded ECHOES project, which aimed to design a virtual environment for young children, both typically developing and with autism spectrum conditions, to support them in developing social interaction skills. In the same year she secured an RCUK Academic Fellowship and further funding from ESRC and EPSRC to continue the ECHOES project.

Sara Price is a Senior Lecturer in Technology Enhanced Learning at the London Knowledge Lab, Institute of Education. She has a background in psychology, with extensive experience in HCI. Her research interests focus on the role of digital technologies for learning, approaching technologies as external tools that offer new opportunities for interaction and cognition, in particular the role of embodied forms of interaction on cognition. Much of her work involves the design, development and evaluation of emerging digital technologies (mobile, tangible, sensor, haptic), and exploration of ways in which they can enhance learning through mediating new forms of thinking and reasoning.

Yvonne Rogers is the director of the Interaction Centre at UCL and a professor of Interaction Design. She is internationally renowned for her work in HCI and ubiquitous computing. She is known for her visionary research agenda of user engagement in ubiquitous computing and has pioneered an approach to innovation and ubiquitous learning. Her current research focuses on behavioural change, through augmenting everyday learning and collaborative work activities with interactive technologies. She has published over 200 articles and is a co-author of *Interaction Design: Beyond human-computer interaction* with Helen Sharp and Jenny Preece, the definitive textbook on Interaction Design and HCI, now in its 3rd edition, that has sold over

150,000 copies worldwide. She is a Fellow of the British Computer Society and the CHI Academy. She has been awarded a prestigious EPSRC dream fellowship to rethink the relationship between ageing, computing and creativity.

Paul Standish is Professor and Head of the Centre for Philosophy of Education at the Institute of Education, University of London. His most recent books are *Stanley Cavell and the Education of Grownups* (Fordham University Press, 2012) and *Education and the Kyoto School of Philosophy* (Springer, 2012), both co-edited with Naoko Saito. His sustained interest in technology is reflected in a number of publications, including *Enquiries at the Interface: Philosophical problems of education online* (Wiley-Blackwell, 2000), co-edited with Nigel Blake. He is Associate Editor of the *Journal of Philosophy of Education.*

Laurel Swan is a Research Fellow with Design Interactions Research at The Royal College of Art, London. Her research has examined the mundane practices of everyday family life – from making lists to archiving memories to managing clutter. More recently her research has shifted to focus on the practice of design and the crossover between different design communities. Laurel has various degrees in history, English, psychology and computer science.

Niall Winters is a Senior Lecturer in Learning Technologies for Development at the London Knowledge Lab (LKL), Institute of Education, University of London. He works primarily in the field of Information and Communication Technologies for Development (ICT4D), where his main research interest is in the participatory design of mobile applications and activities for education in developing regions. The current focus of this work is on supporting the training of healthcare professionals in East Africa. Aligned to this, he has an emerging interest in the design of innovative technologies for postgraduate medical education.

Woontack Woo received his BS in Electronics Engineering from Kyungpook National University (KNU, Daegu, Korea) in 1989 and his MS in Electronics and Electrical Engineering from POSTECH (Pohang, Korea) in 1991. In 1998, he was awarded his PhD in Electrical Engineering Systems from the University of Southern California. In 1999, as an invited Researcher, he joined ATR, Kyoto, Japan. From 2001 to 2012, he was a Professor in the Department of Information and Communications and Director of the Culture Technology Institute at Gwangju Institute of Science and Technology (GIST), Gwangju, Korea. Since February 2012 he has been a Professor with the Graduate School of Culture Technology at the Korea Advanced Institute of Science and Technology (KAIST, Daejeon, Korea). The main thrust of his research has been implementing ubiquitous virtual reality in smart space, which includes Context-aware Augmented Reality, 3D Vision, HCI and Culture Technology.

Kirsty Young is a Senior Lecturer in the Faculty of Arts and Social Sciences at the University of Technology, Sydney. Dr Young's research examines the impact of digital technology across the lifespan, with particular emphasis on understanding how the social media permeating daily life can be utilized to facilitate learning in formal educational settings. With a background in Special Education Dr Young also has a keen interest in the use of mobile devices to facilitate literacy learning for individuals with reading delays or disorders. As a member of the UTS Human Research Ethics Committee, Dr Young is also actively engaged in the promotion of ethically sound online research.

Nicola Yuill is the manager of the Children and Technology Lab in the School of Psychology at the University of Sussex. She studies how technology can be used to support collaborative work and play in children with typical and atypical development, based on theories of social and cognitive development. Her PhD work on Theory of Mind led to research in children's language comprehension, followed by work on understanding language ambiguity in riddles and word play, and mothers' conversations with their children. This prompted her interest in technology's role in how children learn through conversation, culminating in setting up the ChaTLab. The underlying theme of this work is the role of social interaction and collaboration in learning and development, and how this can be supported by the design of technology, as informed by developmental theory. How do interactions between peers support learning? How do parents scaffold children's learning at home? Also, how might technology be used creatively and innovatively to support these interactions?

Introduction

Sara Price, Carey Jewitt and Barry Brown

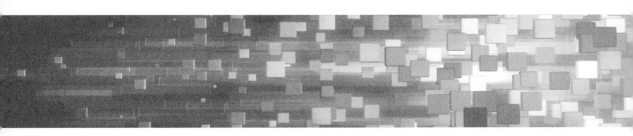

Research on digital technologies is highly significant in the contemporary world, where rapid technological development, social change and the ubiquity of computing technologies are an integral part of people's everyday lives. With rapid and extensive technical development, the digital technology research field is an expanding area of international concern and interest, with a growing number of students, academics and research centres. It is a truly global field, with technology-based courses and research being undertaken across Europe, USA, India and China, and other parts of the world. Researchers across a variety of social science disciplines and beyond are increasingly focused on digital technologies, including sociology, psychology, media, education, visual arts and performance, design and human–computer interaction (HCI)-based fields of computer science.

It is a paradox that while the field of digital technology research is a rapidly changing one (due to the current character of technical development and the continual changes in type of devices, networks and practices that people engage with), the theoretical concerns and approaches of digital technology research are developing at a slower rate, and remain relevant to the progression of digital technology research. This handbook provides an anchor in this changing landscape by addressing the core issues that remain relatively constant and cut across the broad range of digital technologies, including context and location, information access and search, social and personalized media, and ethics. These are addressed alongside expert discussion and debate and empirical cutting-edge research. It provides a comprehensive, up-to-date resource on digital technology research in the twenty-first century. It addresses the key aspects of research within the digital technology field to provide a clear guide for readers wanting to navigate this expanding field of digital technology research.

AIM AND FOCUS

The aim of this handbook is to facilitate digital technology research by comprehensively engaging with current debates and approaches, and exploring the challenges of engaging with cutting-edge emergent technologies and perspectives. It provides an understanding of the 'space' or 'scope' of the research and important ways in which people use technology, as well as challenges such as identifying and engaging with appropriate theoretical and methodological approaches. The handbook provides a comprehensive 'toolkit' of theoretical perspectives, both traditional and emergent, being brought to digital technology research, including theories of embodiment, ethnographic approaches and design-based perspectives, among others. It provides an up-to-date understanding of the current status of research that uses digital technologies by drawing on illustrative examples from a variety of practices and sites, specifically contemporary emergent technologies, such as tangible sensor-based interfaces, wearable technologies and learner-modelled environments.

Importantly, the book brings together authors from a variety of disciplinary backgrounds pertinent to digital technology research in order to provide a rich and comprehensive resource. Contributors are a mix of established academics in the area and emerging early career researchers. This ensures that the handbook reflects established knowledge as well as emergent research within the field of technology research.

This methods-based handbook complements existing handbooks on technology, which tend to focus on specific domains, notably education and psychology. Increasingly, technology features in research both in relation to technology as a topic and as a research tool across the social sciences, and raises many challenges and questions – this handbook aims to provide its readers with the knowledge and the theoretical and methodological resources to address those questions.

WHO IS THIS BOOK FOR?

This handbook is aimed at a social science audience, including postgraduate students at Masters and PhD levels, and researchers and academics undertaking, or interested in research on digital technology or using digital technologies to undertake research.

Digital technologies are increasingly becoming a core part of everyday lives and are thus central to a number of core disciplines. This book is aimed at students, researchers and academics from across the computing sciences to the arts, humanities and all areas of the social sciences, as well as in medical and scientific contexts, and those working across these disciplines. The scope of the handbook thus crosses a broad range of disciplines that include, for example, sociology, psychology, media, education, visual arts and performance, design, humanities, or HCI-based fields of computer science. More specifically, it will be of value to those studying and researching within science and technology, technology-enhanced learning, HCI, mobile and ubiquitous computing, and research, design and evaluation more generally. The handbook is also of interest to practitioners and researchers where digital technology is expanding, for example, educators (including those within museum contexts), designers (particularly those interested in interaction design) and artists (including performance artists).

THE ORGANIZATION AND CONTENT OF THE HANDBOOK

The book is organized into four parts.

Part 1 aims to define the scope of 'digital technology' addressed in the handbook and the emergent trends that inform current research. The chapters in this section provide a foundation for digital technology research that centre on the human uses, effects and challenges of the digital technology that pervades our everyday lives. It begins with a historical perspective (Chapter 1, Ceruzzi) that outlines the development of digital technologies to date. This chapter takes us on a journey from the rise of computing and networking, to the World Wide Web, to the advancing connections between computing technology and social networking, providing an overview of the current position of digital technologies culturally, socially and politically. The following chapter (Chapter 2, Crook) complements this by considering the reconstruction of human activity through digital technologies and the associated impact on social science research, particularly engaging with themes related to developments in theoretical approaches to human cognition, new forms of 'expressive representation', and social communication, as well as generally highlighting the 'rich seam of research opportunity for social scientists'.

Part 2 engages with the key factors and pertinent issues that span digital technology research in the twenty-first century. This set of five chapters begins with an exploration of issues connected with context and location (Chapter 3, Jones). With developments in wireless networking and location-based services the emphasis and nature of the meaning of 'context' and 'location' have changed. This chapter engages with these issues, highlighting important matters from both technical and human meaning-making perspectives. Extending from this, digital development also offers greater information access and dissemination. The next chapter (Chapter 4, Hsieh & Friederici) moves on to explore issues related to information overload, data sharing and questions of authenticity and authorship. This is followed by two chapters that address debates around the social consequences of broadening connection and networking for individuals. One chapter (Chapter 5, Baumer) focuses on debates around social media as 'addiction' and the subsequent psychological effects of social media engagement, offering a critical review of related empirical evidence. The other (Chapter 6, Horst & Hjorth) focuses on the growth of the 'personalization' of technology, and uses two case studies to examine

how personal history and meaning are constructed through different social networking platforms. Part 2 closes with a chapter on ethical considerations around new digital technology and its use (Chapter 7, Kouppanou & Standish). This chapter takes a phenomenological approach to ethical issues, embracing embedded ethical considerations that concern the nature of technology itself and raise new ethical aspects produced by digital ecologies.

The first two parts of the book clearly illustrate the wide-reaching effects of digital technology on social science issues. The breadth and diversity of research that this calls for is highlighted, demanding comprehensive consideration of theoretical and methodological approaches to different research questions and ideas. The next two parts of the book build on this by providing a comprehensive look at key theoretical debates and approaches to digital technology research and methodological considerations. Part 3 presents a wide range of theoretical approaches to digital technology research, while Part 4 provides a broad set of exemplary empirical research that engages with ideas and debates from the earlier sections of the handbook.

Part 3 presents the main theoretical perspectives across 12 chapters, including both established and emergent perspectives that are brought to digital technology research. The origins and key concepts of each of these approaches are outlined, key studies and literature within each perspective are reviewed, and the value and limitations of each perspective are demonstrated through the use of research examples. The first chapter in this section (Chapter 8, Grimes & Feenberg) introduces the background and rationale of the critical theory of technology approach, offering an innovative and practical framework for its application to qualitative and mixed methods research on digital technologies. This chapter is followed by an engagement with critical and cultural approaches to digital technology research (Chapter 9, Bardzell), the author highlighting the value of the coexistence of established scientific and emerging cultural approaches in HCI research. With changes in the nature of interaction through emergent digital technologies, as highlighted in earlier chapters in the handbook, the next three chapters introduce the increasingly prominent theories of embodiment (Chapter 10, Marshall & Hornecker), theories of space and place (Chapter 11, Ciolfi), and experiential approaches (Chapter 12, Höök) to digital technology research. These three chapters collectively foreground notions of 'physicality' to include bodily engagement with the world, and the role of active experience and physical spaces in shaping interaction and cognition – how we perceive, feel and think. Methodological approaches to examining the role of digital technology in the social sciences is the focus of the next set of five chapters, including: ethnographic approaches (Chapter 13, Brown); innovative uses of activity theory (Chapter 14, Kaptelinin); ethnomethodology and conversation analysis (Chapter 15, Moore); analysing online data and communities (Chapter 16, Lampe); and multimodal perspectives (Chapter 17, Jewitt). These chapters draw on existing methodological approaches and consider their application to digital technology research. The theoretical section ends with two chapters addressing design considerations, another key area of work with digital technology, to provide different perspectives on research design and the exploration of 'alternative futures' to gain insight into human needs, desires, emotions and aspirations. The first of these chapters (Chapter 18, Dow, Ju & Mackay) explores the rationale for and methodological implications of 'research through design', and examines the validity of its research contributions. The second (Chapter 19, Swan & Boehner) offers an in-depth description of 'critical design', drawing on Dunne and Raby's work and illustrated through a number of research projects aimed to evoke users' responses to new technological or artefact designs that challenge expectations about their usual use in everyday life.

Part 4 of the handbook links the debates, issues, perspectives, theories and methods discussed throughout the handbook to provide empirical examples of digital technology research from this expanding research field. This collection of chapters presents a range of research that has been undertaken with different kinds of digital technologies and that draw on a variety of sites of practice (including health, education, design). In so doing this section illustrates the diversity of work, as well as the challenges, limitations and advantages of different methodological or theoretical approaches, while providing the foundations for future research through empirical examples. Each chapter provides a review of its area of digital technology research, drawing on key texts and studies, and provides a case study example to demonstrate the particular considerations, potentials and constraints for researching this specific type of digital technology. The chapters focus on an exciting range of technologies at the forefront of digital computing across a range of fields of practice. The section begins with a set of chapters on 'physical' forms of computing: research using tangible technologies is explored here in the context of learning (Chapter 20, Price); while the chapter on material computing (Chapter 21, Buechley) particularly focuses on how new materials can enable interaction design researchers to design and build new technologies; and a chapter on haptic interfaces (Chapter 22, Hoggan) reintroduces a focus on the value of 'touch' through technological advances that enable interactive systems to be enhanced with the haptic modality multi-user displays. The section then moves on to look at multi-user displays (Chapter 23, Rogers, Yuill & Marshall), particularly engaging with a comparison of lab-based and 'in-the-wild' research approaches in this field – raising methodological issues relevant to many digital research contexts. This is followed by a chapter on ubiquitous virtual reality (Chapter 24, Oh & Woo), which describes the primary features of these environments, accompanied by an illustrative example of the development of a ubiquitous virtual reality (UVR) environment. The next two chapters concern location-based technologies. The first (Chapter 25, Hollett & Leander) considers problems of tracking and understanding users, and suggests how sensing technologies extend opportunities to embed environmental and human data within location data, illustrated within both medical and education contexts. The second (Chapter 26, Winters) focuses on the research issues for mobile learning in developing countries. Two further chapters engage with research using online and Internet-based technologies, including gaming (Chapter 27, Beavis) and social networking and (Chapter 28, Young). Our final chapter in this section (Chapter 29, Porayska-Pomsta & Bernardini) moves on to examine learner modelled environments (LMEs), detailing this as a method through which learning can be both supported and studied. The handbook concludes with a chapter (Chapter 30, Holmquist) on the interaction between research and industry, which presents both key challenges for research and new future directions. This chapter highlights the concept of 'grounded innovation' as one future direction for progressing digital technology research.

Together the chapters across these four parts are designed to provide the reader with a clear guide and comprehensive toolkit that will facilitate digital technology research to face the challenges of engaging with cutting-edge emergent technologies and perspectives.

PART 1

An Introduction to the Field of Contemporary Digital Technology Research

This book has been created to give the reader a good solid overview of digital technology research, in particular the methods and approaches that typify contemporary work. Our first two chapters give us an excellent foundation in that task, surveying the field of technology and examining research that focuses on the human uses, effects and challenges of the digital technology that so infuses our modern age.

Our first chapter, by Paul Ceruzzi, provides us with an historical foundation to this task – starting with the coining of the term 'digital' by the Bell Telephone Laboratories mathematician George Stibitz. As Ceruzzi points out, since then this term has come to describe much of the social, economic, and political life of the twenty-first century. Ceruzzi goes on to review how the particular features of computers – machines that stored, computed, and so on – developed out of wartime research. The German mathematician

Zuse took the bold step of using theoretical mathematics to design computers, yet it was Alan Turing who took the opposite step – introducing the concept of a machine into mathematics. On these foundations, Wiener's notions of cybernetics, and Shannon's work in information theory provides the intellectual foundations for computer technology. Indeed, perhaps unusually, computers were theoretically specified before they were physically practical. Yet when they were built they slowly but steadily came to transform not only jobs of computation but also communication. Ceruzzi guides us through the advent of computing and networking, the World Wide Web, and, in turn, the ongoing connections between computing technology and social networking.

In contrast, our second chapter by Charles Crook engages more deeply with the digital technology research tradition, in particular social science research traditions. This lively chapter engages with a large body of work, yet focuses clearly on questions of how human social action is reconstituted in digital form – in particular in virtual worlds and augmented reality. Avoiding the description of digitization as amplification, the chapter engages with how human cognitive activity and social coordination come to the fore when studying digital technologies in use.

The Historical Context

Paul Ceruzzi

INTRODUCTION

In the spring of 1942, the US National Defense Research Committee convened a meeting of high-level scientists and engineers to consider devices to aim and fire anti-aircraft guns. The German *Blitzkrieg* had made the matter urgent. The committee noticed that the proposed designs fell into two broad categories. One directed anti-aircraft fire by constructing a mechanical or electronic analog of the mathematical equations of fire-control, for example by machining a camshaft whose profile followed an equation of motion. The other solved the equations numerically – as with an ordinary adding machine, only with high-speed electrical pulses instead of mechanical counters. One member of the committee, Bell Telephone Laboratories mathematician George Stibitz, felt that the term 'pulse' was not quite right. He suggested another term, which he felt was more descriptive: 'digital' (Williams 1984: 310). The word referred to the method of counting on one's fingers, or, digits. It has since become the adjective that

defines social, economic and political life in the twenty-first century.

It took more than just the coining of a term to create the digital age, but that age does have its origins in secret projects initiated during World War II. But why did those projects have such a far-reaching social impact? The answer has two parts.

The first is that calculating with pulses of electricity was far more than an expeditious way of solving an urgent wartime problem. As the digital technique was further developed, its creators realized that it tapped into fundamental properties of information, which gave the engineers a universal solvent that could dissolve any activity it touched. That property had been hinted at by theoretical mathematicians, going back at least to Alan Turing's mathematical papers of the 1930s; now these engineers saw its embodiment in electronics hardware.

The second part of the answer is that twenty years after the advent of the digital computing age, this universal solvent now dissolved the process of communications.

That began with a network created by the US Defense Department, which in turn unleashed a flood of social change, in which we currently live. Communicating with electricity had a long history going back to the Morse and Wheatstone telegraphs of the nineteenth century (Standage 1998). Although revolutionary, the impact of the telegraph and telephone paled in comparison to the impact of this combination of digital calculation and communication, a century later.

THE COMPUTER

Histories of computing nearly all begin with Charles Babbage, the Englishman who tried, and failed, to build an 'Analytical Engine' in the mid-nineteenth century (Randell 1975). Babbage's design was modern, even if its proposed implementation in gears was not. When we look at it today, however, we make assumptions that need to be challenged. What is a 'computer', and what does its invention have to do with the 'digital age'?

Computing represents a convergence of at least three operations that had already been mechanized. Babbage tried to build a machine that did all three, and the reasons for his failure are still debated (Spicer 2008: 76–77). Calculating is only one of the operations. Mechanical aids to calculation are found in antiquity, when cultures developed aids to counting such as pebbles (Latin calculi), counting boards (from which comes the term 'counter top'), and the abacus. In the seventeenth century, inventors devised ways to add numbers mechanically, in particular to automatically carry a digit from one column to the next. These mechanisms lay dormant until the nineteenth century, when advancing commerce created a demand that commercial manufacturers sought to fill. By 1900 mechanical calculators were marketed in Europe and in the United States, from companies such as Felt, Burroughs, Brunsviga of Germany, and Odhner of Sweden (later Russia).

Besides calculating, computers also store information. The steady increase of capacity in computers, smartphones and laptops, usually described by the shorthand phrase 'Moore's Law', is a major driver of the current social upheaval (Moore's Law will be revisited later). Mechanized storage began in the late nineteenth century, when the American inventor Herman Hollerith developed a method of storing information coded as holes punched into cards. Along with the card itself, Hollerith developed a suite of machines to sort, retrieve, count and perform simple calculations on data punched onto cards. Hollerith founded a company to market his invention, which later became the 'Computing-Tabulating-Recording' Company. In 1924 C-T-R was renamed the International Business Machines Corporation, today's IBM. A competitor, the Powers Accounting Machine Company, was acquired by the Remington Rand Corporation in 1927 and the two rivals dominated business accounting for the next four decades. Punched card technology persisted well into the early electronic computer age.

A third property of computers is the automatic execution of a sequence of operations: whether calculation, storage or routing of information. That was the problem faced by designers of anti-aircraft systems in the midst of World War II: how to get a machine to carry out a sequence of operations quickly and automatically to direct guns to hit fast-moving aircraft. Solving that problem required the use of electronic rather than mechanical components, which introduced a host of new engineering challenges. Prior to the war, the control of machinery had been addressed in a variety of ways. Babbage proposed using punched cards to control his Analytical Engine, an idea he borrowed from the looms invented by the Frenchman Joseph Marie Jacquard (1752–1834), who used cards to control the woven patterns. Whereas Hollerith used cards for storage, Jacquard used cards for control. But for decades, the control function of a Hollerith installation was carried out by *people*: human beings who carried decks of cards from one device to another, setting switches or plugging wires on the devices to perform specific operations, and collecting the results. Jacquard's punched cards were an inside-out version of a device that had been used to control machinery for centuries: a cylinder on which were mounted pegs that tripped levers as it rotated. These had been used in medieval clocks that executed complex movements at the sounding of each hour; they are also found in wind-up toys, and music boxes. Continuous control of many machines, including automobile engines, is effected by cams, which direct the movement of other parts of the machine in a precisely determined way.

COMMUNICATION

Calculation, storage, control: these attributes, when combined and implemented with electronic circuits, make a computer. To them we add one more: communication – the transfer of information by electricity across geographic distances. That attribute was lacking in the early electronic computers built around the time of World War II. It was a mission of the Defense Department's Advanced Research Projects Agency (ARPA), beginning in the 1960s, to reorient the digital computer to be a device for which communication was as important as its other attributes.

Like control, communication was present in early twentieth-century business environments in an ad hoc fashion. People carried decks of punched cards from one machine to another. The telegraph, a nineteenth-century invention, carried information to and from a business. Early adopters of the telegraph were railroads, whose rights-of-way became a natural corridor for the erection of telegraph wires. Railroad operators were proficient at using the Morse code and were proud of their ability to send and receive the dots and dashes accurately and quickly. Their counterparts in early aviation did the same, using the radio or 'wireless' telegraph. Among the many inventions credited to Thomas Edison was a device that printed stock prices on a 'ticker tape', so named because of the sound it made. Around

1914, Edward E. Kleinschmidt combined the keyboard and printing capabilities of a typewriter with the ability to transmit messages over wires (Anon. 1977). In 1928 the company he founded changed its name to the Teletype Corporation. The Teletype had few symbols other than the upper-case letters of the alphabet and the digits 0 through 9, but it provided the communications component to the information processing ensemble. It also entered our culture. Radio newscasters would have a Teletype chattering in the background as they read the news, implying that what they were reading was 'ripped from the wires'. Although it is not certain, legend has it that Jack Kerouac typed the manuscript of his Beat novel *On the Road* on a continuous roll of Teletype paper. Bill Gates and Paul Allen, the founders of Microsoft, marketed their software on rolls of Teletype tape. Among those few extra symbols on a Teletype keyboard was the '@' sign, which in 1972 was adopted as the marker dividing an addressee's e-mail name from the computer system the person was using. Thus to the Teletype we owe the symbol of the Internet age.

The telegraph and Teletype used codes and were precursors to the digital age. By contrast, the telephone, as demonstrated by Alexander Graham Bell in 1876, operated by inducing a continuous variation of current, based on the variations of the sound of a person's voice. In today's terms it was an 'analog' device, as the varying current was analogous to the variations in the sounds of speech. Like 'digital,' the term 'analog' did not come into common use until after World War II. During the first decades of electronic computing, there were debates over the two approaches, but analog devices faded into obscurity.

One could add other antecedents of telecommunications, such as radio, motion pictures, television, hi-fidelity music reproduction, the photocopier, photography, etc. Marshall McLuhan was only the most well-known of many observers who noted the relation of electronic communications to our culture and world-view (McLuhan 1962).

 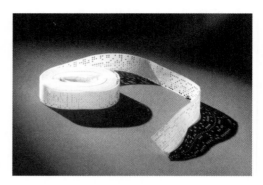

Figure 1.1   (Left) A Teletype Model 'ASR-33', a paper tape reader, which could store coded messages. Ray Tomlinson, a programmer at Bolt Beranek and Newman, says that he chose the '@' sign (Shift-p) to delimit e-mail addresses because it was the only preposition on the keyboard. (Right) Teletype tape containing a version of the BASIC programming language, developed by Bill Gates and Paul Allen, c. 1975 (source: Smithsonian Institution photo).

Others have observed the effects of telecommunications: for example, how the photocopier democratized the process of publishing or how cheap audio cassettes helped revolutionaries overthrow the Shah of Iran in 1979. As mentioned above, nearly all of these antecedents were dissolved and transformed by digital electronics, sadly too late for McLuhan to apply his insights.

DIGITAL ELECTRONIC COMPUTING

Before examining the convergence of communications and computing, we return to the 1930s and 1940s when electronic computing appeared. At the precise moment when the ensemble of data processing and communications equipment was functioning at its peak efficiency, the new paradigm of digital electronics emerged. We have already encountered one reason: a need for higher speeds to process information. That could be achieved by modifying another early twentieth-century invention: the vacuum tube, which had been developed for the radio and telephone. Substituting tubes for mechanical parts introduced a host of new problems, but with 'electronics,' the calculations could approach the speed of light. Solutions emerged simultaneously in

several places in the mid-1930s, and continued during World War II at a faster pace, although under a curtain of secrecy that sometimes worked against the advantages of having funds and human resources made available.

The mechanical systems in place by the 1930s were out of balance. In a punched card installation, human beings had to make a plan for the work that was to be done, and then they had to carry out that plan by operating the machines (Heide 2009). For scientists or engineers, a person operating a calculator could perform arithmetic quite rapidly, but she (such persons typically were women) had to carry out one sequence if interim results were positive, another sequence if negative. The plan would have to be specified in detail in advance, and given to her on paper. Punched card machines likewise had stops built into their operation, signalling to the operator to proceed in a different direction depending on the state of the machine. The human beings who worked in some of these places had the job title 'computer': a definition that was still listed first in dictionaries published into the 1970s. This periodic intrusion of human judgment and action was not synchronized with the speeds and efficiencies of the mechanized parts of the systems.

THE DIGITAL CONCEPT

The many attempts to balance these aspects of computing between 1935 and 1950 make for a fascinating narrative. There were false starts. Some were room-sized arrangements of mechanisms that might have come from a Rube Goldberg cartoon. Others were modest experiments that could fit on a table-top. With the pressures of war, many attacked this problem with brute force. The following are a few representative examples.

In 1934, at Columbia University in New York, Wallace Eckert modified IBM equipment to perform calculations related to his astronomical research (Eckert 1940). He used punched cards for control as well as data storage, combining the methods of Jacquard and Hollerith. At the same time, Howard Aiken, a physics instructor at Harvard University, designed a machine that computed sequences directed by a long strip of perforated paper tape (Aiken 1964). Aiken's design came to fruition as the 'Automatic Sequence Controlled Calculator', unveiled at Harvard in 1944, after years of construction and financial support from the US Navy (Harvard University 1946).

Among those who turned towards electronics was J. V. Atanasoff, a professor at Iowa State College in Ames, Iowa, who was investigating ways to mechanize the solving of large systems of linear algebraic equations. These could be solved by a straightforward sequence of operations, or an *algorithm:* a recipe that, if followed, guaranteed a solution. But the solution of a large system required so many steps that it was impractical for human beings to carry them out. Atanasoff conceived of the idea of using vacuum tubes, and of the idea of using the binary system of arithmetic. In 1940, he proposed building such a machine to Iowa State College, which gave him a modest grant. He completed a prototype that worked, erratically, by 1942. That year Atanasoff left Iowa for the Washington DC region, where he was pressed into service working on urgent wartime problems for the US Navy. He never completed his machine.

The US Army supported a project to aid in the aiming accuracy of large artillery, which led to a machine called the ENIAC: 'Electronic Numerical Integrator and Computer', unveiled to the public in 1946 at the University of Pennsylvania. With its 18,000 vacuum tubes, the ENIAC was a huge step beyond anything done elsewhere. It was designed by John Mauchly and J. Presper Eckert (no relation to Wallace Eckert), and it used vacuum tubes for both storage and calculation. It did not appear *de novo*: Mauchly had visited J. V. Atansoff in Iowa for several days in 1941, where he realized that computing with vacuum tubes was feasible. The ENIAC was programmed by plugging the various elements of it in different configurations. Reprogramming it might require days, but once rewired it could calculate an answer in minutes. Historians are reluctant to call the ENIAC a computer, a term reserved for machines that can be more easily programmed, but the 'C' in the acronym stood for 'computer', a term deliberately chosen by Eckert and Mauchly to evoke the rooms of women operating calculating machines.

If the onset of war hindered Atanasoff's attempts to build a high-speed computer, it had the opposite effect in the UK, where at Bletchley Park, outside London, multiple copies of a device called the Colossus were in operation by 1944. Details remained a closely guarded secret into the 1970s. The Colossus was not a calculator but a machine that aided in the unscrambling of German coded messages. Given the dominance of text on the Internet today, one would assume that the Colossus would be heralded more, but secrecy prevented its taking a more prominent place in history. The Colossus employed vacuum tubes, using base-two circuits that had an ability to follow the rules of symbolic logic.

That simple concept, of designing circuits which had only two instead of 10 states as an ordinary calculator had, was the sword that cut the Gordian knot of complexity (Randell 1980). Because of the secrecy surrounding the Colossus, it is difficult to see how it contributed to this profound conceptual breakthrough. Atanasoff's choice of binary also pointed towards this breakthrough, but his project did not proceed to completion. However, two other efforts have been better documented, and a closer look at them may help us understand the significance of what happened.

In 1937 Konrad Zuse was a 27-year-old mechanical engineer working at the Henschel Aircraft Company in Berlin, where he was occupied with tedious calculations. He began work on a mechanical calculator to automate that process. In June of that year he made several remarkable entries in his diary:

> For about a year now I have been considering the concept of a mechanical brain ... Discovery that there are elementary operations for which all arithmetic and thought processes can be solved ... For every problem to be solved there must be a special purpose brain that solves it as fast as possible ... (Zuse 1962: 44)

Zuse chose the binary system of arithmetic because, as a mechanical engineer, he recognized the advantages of switches or levers that could assume one of only two positions. As he began sketching out a design, he came upon an insight that is fundamental to the digital age that has followed. That was to recognize that the operations of calculation, storage, control and transmission of information, until that time traveling on separate avenues of development, were in fact one and the same. In particular the control function, which had not been mechanized as much as the others in 1937, could be reduced to a matter of (binary) arithmetic. That was the basis for his use of the terms 'mechanical brain' or 'thought processes', which must have sounded outrageous at the time. Zuse realized he could design mechanical devices that could be flexibly rearranged to solve a wide variety of problems: some requiring more calculation, others requiring more storage, each requiring varying degrees of control. In short, he conceived of a universal machine. Once he chose to base his design on the binary system, the rest seemed to flow naturally.

Zuse mentioned his discovery to one of his former mathematics professors, only to be told that what he claimed to have discovered had already been worked out by the famous Göttingen mathematician David Hilbert and his students (an earlier version of the concept had been devised by the Englishman George Boole). That was only partially true: Hilbert worked out a relationship between arithmetic and binary logic, but he did not extend that theory to a design of a computing machine (Reid 1970). The previous year, 1936, the Englishman Alan M. Turing (1912–1954) had done that, in a 36-page paper published in the *Proceedings of the London Mathematical Society* (Turing 1936). So while Zuse took the bold step of introducing theoretical mathematics into the design of a calculator, Turing took the opposite but equally bold step, of introducing the concept of a 'machine' into the pages of a theoretical mathematics journal.

Turing described a theoretical machine to address a mathematical problem. While his paper was admired by mathematicians, it was his construction of this 'machine' that placed Turing among the founders of the digital age (Petzold 2008). The term is in quotation marks because Turing built no hardware; he described a hypothetical device in his paper. One cannot simulate a Turing Machine on a modern computer, since a Turing Machine has a memory tape of unbounded capacity. No matter: the machine he described, and the method by which it was instructed to solve a problem, was the first theoretical description of the fundamental quality of a digital computer, namely that a properly constructed computer can be programmed to perform a limitless range of operations – the limitless 'apps' one finds on a modern smartphone. Turing formalized what Zuse, the engineer, had recognized. A computer, when loaded with a suitable program, becomes 'a special purpose brain', which does whatever the programmer wants it to do.

It took about 15 years before engineers could bring the hardware to a point where these theoretical properties could matter.

Before 1950 it was a major accomplishment to get a digital computer to operate without error for even a few hours. Nonetheless, Turing's insight was significant. By the late 1940s there was a vigorous debate among engineers and mathematicians about computer design, after the first machines began working. From those debates emerged a concept, known as the 'stored program principle', which extended Turing's ideas into the design of practical machinery. The concept is credited to the Hungarian mathematician John von Neumann (1903–1957), who worked closely with Eckert and Mauchly at the University of Pennsylvania (Aspray 1990), and whom Turing visited during the war. Modern computers store their instructions and the data on which those instructions operate, and any new data those instructions generate, in the same physical memory device. Computers do so for practical reasons, and for theoretical reasons: the two types of information are treated the same inside the machine because fundamentally they *are* the same.

Turing's abstract machine, after it had been realized in electronics in the 1950s, had uncanny parallels in other disciplines of knowledge. These are metaphorical, but they all seem to point in the same direction. The notion that a computer consisted of discrete components which moved through successive and discrete 'states' as those components processed binary ones and zeros, had its counterpart in Thomas Kuhn's radical view of the history of science, which for Kuhn did not proceed in smooth increments but rather successive finite states, which he called 'paradigms' (Kuhn 1962). Kuhn's idea was enormously influential on the social sciences, to the extent that the term 'paradigm' is almost forbidden as it has lost its explanatory power (the term's connection to computing permits its use in this chapter, however). Likewise the notion of a computer executing fixed programs from a 'Read-Only Memory' (ROM), which alter the contents of Random Access Memory (RAM), has an uncanny similarity to the mechanism of the Double

Helix as discovered by Watson and Crick in 1953, in which information is transmitted one-way from a fixed DNA code to RNA (Crick 1970). The notion of high-level computer languages (described later), which are compiled by the computer into low-level sequences of bits that the computer executes, likewise suggests the theories of human language acquisition developed by MIT linguist Noam Chomsky in the post-war years (Chomsky 1957). The notion of coding information – any information, not just numbers – in binary form had unforeseen implications.

FIRE-CONTROL

We now return to fire-control, the topic of the secret meetings of the National Defense Research Committee that George Stibitz attended. The committee's chair was Vannevar Bush, a former professor at MIT, where he built an analog computer called the Differential Analyzer, and where he and his students explored a variety of mechanical and electronic devices. In 1938 Bush proposed building a 'Rapid Arithmetical Machine' that would use vacuum tubes. With his move to Washington in 1939, the priorities shifted. Work on the Rapid Arithmetical Machine continued at MIT, but a working system was never completed. Norbert Wiener, one of Bush's colleagues, analyzed the problem of tracking a target in the presence of noise and against an enemy pilot who is taking evasive action. Wiener's mathematics turned out to have an impact, not only on fire-control but also on the general question of the automatic control of machinery. He coined the term 'cybernetics' and later published a book under that title, which became another influential book of the era. Among the many insights found in that book is a discussion of 'Maxwell's Demon': a thought-experiment that purported to violate the Second Law of Thermodynamics. The 'demon' was a fictitious agent who could transfer heat from a cold to a warm body, a clear violation of the

Law. By introducing the notion of 'information' as a physical, not just an abstract quantity, Wiener showed why no violation of the Second Law could occur, and that 'information' was intimately linked to the physical world (Wiener 1961: 56). Thus Wiener, along with his MIT colleague Claude Shannon, laid the foundation for establishing information theory as a science on an equal basis with thermodynamics or physics.

It was out of this ferment of ideas that the theory of information processing emerged, with an articulation of what it meant to be 'digital'. It was also a time when fundamental questions were raised about the proper role of human beings who interacted with complex control machinery. Modern-day tablet computers or other digital devices do not bear a physical resemblance to the anti-aircraft computers of the 1940s, but questions of the human–machine 'interface', as it is now called, are of paramount importance, and those questions do go back to that era.

At the end of the war Vannevar Bush looked again at the world he helped bring into being. He wrote a provocative article for the *Atlantic Monthly*, 'As we may think', in which he warned that a glut of information would swamp science and learning if not controlled (Bush, 1945). He proposed a machine, the 'Memex,' which would address this issue. His description of Memex, and that article, had a direct link to the developers of the graphical computer interface and of the World Wide Web. Likewise, Norbert Wiener's *Cybernetics* did not articulate the digital world in detail, but the term was adopted in 1982 by the science fiction author William Gibson, who coined the term 'cyberspace' – a world of bits.

COMMERCIALIZATION

The ENIAC was a one-of-a-kind, wartime project. It would hardly be remembered, were it not for Eckert and Mauchly's next step. After completing the ENIAC, they sought to build and sell a version that would have commercial applications. That product, the UNIVAC, was conceived and marketed as suitable for any application one could program it for – hence the name, an acronym of 'Universal Automatic Computer'. Eckert designed it conservatively, making the UNIVAC surprisingly reliable. Eckert and Mauchly founded a company – another harbinger of the volatile Silicon Valley culture that followed – which was absorbed by Remington Rand in 1950. The UNIVAC made it clear that the electronic computer was going to replace the machines of an earlier era. That led to a decision by the IBM Corporation to enter the field with electronic computers of its own. By the mid-1950s IBM and Remington Rand were joined by other vendors, while advanced research on memory devices, circuits and, above all, programming was being carried out in US and British universities.

The process of turning an experimental, one-of-a-kind research project into reliable, marketable and useful products took most of the 1950s and 1960s to play out. Those were the decades of the 'mainframe', so-called because of the large metal frames on which circuits were mounted. IBM dominated the industry in both the United States and western Europe as well. Science fiction and popular culture showed the blinking lights of the control panels, but the true characteristic of the mainframe was the banks of spinning reels of magnetic tape, which stored the data and programs that initially came from punched cards. Because these systems were so expensive, a typical user was not allowed to interact directly with the computer – the decks of cards were compiled into batches that were fed into the machine, so the expensive investment was always kept busy. A distaste for batch operations drove computer enthusiasts in later decades to break away from this, at first through interactive terminals, later through personal computers.

Those decades also saw the arrival of 'software', unforeseen by the 1940s pioneers. Computers were harsh taskmasters, demanding to be programmed in binary arithmetic, but beginning in the 1950s researchers developed computer languages that allowed users to write software in more familiar forms. The most popular were COBOL, for business applications, and FORTRAN, for science. These were followed by a Babel of other languages aimed at more specific applications – a trend that continues to the present day, with modern Web applications programmed in Java, C++, Python, and others, many derived from a language called 'C' that was developed at Bell Labs in the 1960s. This was an unforeseen consequence of the stored program principle, although we now see it was implied in Zuse's insight into the 'brain' he sought to build in the 1930s, and in Turing's concept of a universal machine.

Computing technology advanced through the 1950s and 1960s as the devices became more reliable, and as users learned how to write software. The military played a role as well, adapting mainframe computers for air defense with the 'SAGE' system, and supporting very high-performance computers supplied by companies like Cray Research, for classified military research.

Some of the terms introduced at this time give us insight into the digital world that followed. The SAGE system, for example, was designed to operate in 'real time': processing data as fast as it was received by radars tracking enemy aircraft. That implies the entity of time, in the digital world, is no longer a proscenium on which the events of the world are played; it is a variable digital engineers can manipulate and control like any other. It also implies that computers may also deal with forms of time that are less than real: a notion once only found among science fiction writers. Also, in the late 1960s IBM introduced a system that had what it called 'virtual' memory: data were stored on relatively slow disks, but the user had the illusion that the data were in the faster but smaller core memory. The conservative IBM engineers did not realize it, but they highlighted the computer's ability to blur any distinctions between real, virtual, or illusory.

Around 1960 the batch method of operations began to change. The transistor, invented in

the late 1940s at Bell Laboratories, had a long gestation period, but by 1960 mainframe computers were taking advantage of the transistor's reliability, small size, and low power consumption. More significant was the ability now to use transistors to develop a new class of small computers, inexpensive enough to be used in laboratories, industrial settings, and other places where mainframes were impractical. The primary supplier of these so-called minicomputers was the Digital Equipment Corporation (DEC), located in the suburbs of Boston. Minicomputers were impractical for home use, but they set in motion a trend that would bring computers into the home by the 1980s.

Descendants of the minicomputer, first introduced around 1960, are now the laptop computers and tablets in use today. The mainframe was still preferred for heavy duty computing, although its batch method of operation was under attack. Only a handful of people noticed when that attack was first mounted. It had the effect, as it played out, of adding that last crucial component to the digital revolution: the convergence of digital *communications* with the functions of information storage, calculation and control that took place two decades earlier. Once again a war, this time the Cold War, played a crucial role.

TELECOMMUNICATIONS, AGAIN

The convergence began in November, 1962 on a chartered train traveling from the Allegheny Mountains of Virginia to Washington, DC. The passengers were returning from a conference on 'information system sciences', sponsored by the US Air Force and held at a resort in the Warm Springs Valley. Thomas Jefferson had visited and written about these springs, but the conference attendees had little time to take the healing waters. A month earlier, the United States and Soviet Union had gone to the brink of nuclear war over the Soviet's placement of missiles in Cuba. Poor communications, between the two superpowers and the White House, the Pentagon, and commanders of ships at sea, escalated the crisis. Among the conference attendees was J.C.R. Licklider, a psychologist who had just taken on a position at the US Defense Department's Advanced Research Projects Agency (ARPA). For Licklider, the conference had been a disappointment. None of the presenters had recognized the potential of computers to revolutionize military or civilian affairs. The long train ride gave him a chance to reflect on this potential, and then to do something about it.

Licklider was able to act because he had access to Defense Department funds and a free rein to spend them on projects as he saw fit. He also had a vision: he saw the computer as a revolutionary device that could be used to work in symbiosis – his favorite term – with human beings. Upon arrival in Washington, the passengers dispersed

to their respective homes. Two days later, Professor Robert Fano of MIT proposed a project, based on discussions he had with Licklider on the train ride. That led to 'Project MAC': exploring 'machine-aided cognition', by allowing 'multiple-access' to a computer – a dual meaning of the acronym. Licklider arranged for the US Navy to fund the proposal with an initial contract of around $2.2 million (Licklider 1990).

To overcome the impracticality of allowing someone to use an expensive computer, the resource would be *time-shared*: it would spend a fraction of its time with each multiple user, who would not notice that the resource was being shared (note the interesting concept of time). The closest analogy is a grandmaster chess player playing simultaneous games with less capable players. Each user would have the *illusion* (another word they used deliberately) that he or she had a powerful computer at his or her personal beck and call. Time-sharing gave

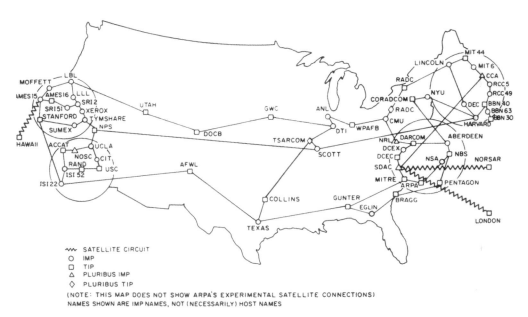

Figure 1.2   ARPANET, ca. 1974. Note the concentration of nodes on the two coasts of the US, with crucial satellite links to London and Norway, the latter for classified Cold War data transfers (source: DARPA).

way to more sophisticated forms of computer networking, but from these efforts came the notion of 'logging on' to a system (borrowed from ship's terminology), and entering a password to gain access (along with the first instances of cracking a password to gain unauthorized access).

ARPA also began funding other research: to interact with a computer using graphics, and to network geographically scattered computers to one another. Time-sharing was the spark that set those other efforts in motion (Kita 2003). A crucial technical step came when ARPA managers learned of the concept of *packet switching*, conceived independently in the UK and US. The technique divided a data transfer into small chunks, called packets, which were separately addressed and sent to their destination, and which could travel over separate channels if necessary. That concept was contrary to all AT&T had developed over the decades, but it offered many advantages over classical methods of communication, and it is the technical backbone of the Internet to this day. The first computers of the 'ARPANET' were linked in 1969; by 1971 there were 15 computers on the network, and the next year it was demonstrated at a computer science conference. ARPANET was a military-sponsored network that lacked the social, political, and economic components which comprise the modern networked world. It did not even have e-mail at first, although that was added relatively quickly. It did demonstrate the feasibility of packet switching. The rules for addressing and routing packets, which ARPA called *protocols*, remain in use.

PERSONAL COMPUTING

The social component of today's digital world came from another arena: personal computing. This was unexpected and not understood by computer scientists and manufacturers. Personal computing was enormously disruptive,

and led to the demise of many established computer companies. It also unleashed a torrent of personal creativity, without which innovations like the ARPANET would never have broken out of its military origins.

The transformation of digital electronics from room-sized ensembles of machinery to hand-held personal devices is a paradox. On the one hand it was the direct result of advances in solid-state electronics. As such it is an illustration of technological determinism: the driving of social change by technology. On the other hand, personal computing was driven by idealistic visions of the 1960s-era counterculture. By that measure, personal computing was the antithesis of technological determinism. Both views are correct. After ten years of transistor development, inventors in Texas and California devised a way of placing multiple transistors and other devices on a single chip of silicon. That led to circuits which could store ever-increasing amounts of data – the storage component of computing described earlier. In 1965 Gordon Moore, working at the California company where the chip was co-invented, noted that the storage capacity of the devices was doubling about every year or so. This became known as 'Moore's Law' – an empirical observation that has persisted into the twenty-first century. Soon after that, Moore co-founded the Intel Corporation, and at about the same time a local journalist christened the region south of San Francisco as 'Silicon Valley'. In 1971, another engineer at Intel, Marcian E. Hoff, led a team that placed all the basic circuits of a computer processor on a chip – another key component of computing – and created what was, next to the airplane, the greatest invention of the twentieth century: the microprocessor.

To the electrical engineers, the microprocessor was trivial: simply by looking at Moore's Law, it was easy to see that by the mid-1970s it would be possible to put on a single chip the same number of circuits that constituted the room-sized UNIVAC of

United States Patent [19]

Hoff, Jr. et al.

[11] **3,821,715**

[45] **June 28, 1974**

[54] **MEMORY SYSTEM FOR A MULTI-CHIP DIGITAL COMPUTER**

[75] Inventors: **Marcian Edward Hoff, Jr.,** Santa Clara; **Stanley Mazor,** Sunnyvale; **Federico Faggin,** Cupertino, all of Calif.

[73] Assignee: **Intel Corporation,** Santa Clara, Calif.

[22] Filed: **Jan. 22, 1973**

[21] Appl. No.: **325,511**

[52] U.S. Cl. .... 340/172.5, 340/173 R, 340/173 SP, 307/238
[51] Int. Cl. G06f 13/00, G11c 11/44
[58] Field of Search .... 340/172.5, 173 SP, 173 R; 307/238, 279

[56]   **References Cited**
          UNITED STATES PATENTS

3,460,094	8/1969	Pryor	340/172.5
3,641,511	2/1972	Cricchi et al.	307/238 X
3,680,061	7/1972	Arbab et al.	340/173 R
3,681,763	8/1972	Meade et al.	340/173 R
3,685,020	8/1972	Meade	340/172.5
3,702,988	11/1972	Haney et al.	340/172.5
3,719,932	3/1973	Cappon	340/173 R
3,731,285	5/1973	Bell	340/172.5
3,735,368	5/1973	Beausoleil	340/173 R
3,737,866	6/1953	Gruner	340/172.5
3,740,723	6/1973	Beausoleil et al.	340/172.5

OTHER PUBLICATIONS

Schuenemann, "Computer Control" in IBM Technical Disclosure Bulletin, Vol. 14, No. 12, May 1972; pp. 3794–3795.

Primary Examiner—Paul J. Henon
Assistant Examiner—Melvin B. Chapnick
Attorney, Agent, or Firm—Spensley, Horn & Lubitz

[57]   **ABSTRACT**

A general purpose digital computer which comprises a plurality of metal-oxide-semiconductor (MOS) chips. Random-access-memories (RAM) and read-only-memories (ROM) used as part of the computer are coupled to common bi-directional data buses to a central processing unit (CPU) with each memory including decoding circuitry to determine which of the plurality of memory chips is being addressed by the CPU. The computer is fabricated using chips mounted on standard 16 pin dual in-line packages allowing additional memory chips to be added to the computer.

17 Claims, 5 Drawing Figures

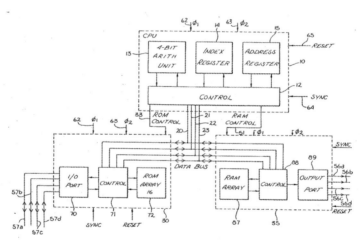

Figure 1.3   Patent for the microprocessor, by Marcian 'Ted' Hoff et al., United States Patent and Trademark Office.

1950. In reality, the invention was anything but trivial. It was the result of careful design and an understanding of the properties of silicon. Intel recognized the power of this invention, but it marketed it to industrial customers and did not imagine that anyone would want to use it to build a computer for personal use. Hobbyists, ham radio operators, and others who were marginally connected to the semiconductor industry thought otherwise. A supplier of circuits for amateur rocket enthusiasts in Albuquerque, New Mexico, was one of the first to design a computer kit

around an Intel microprocessor, and when the company, Micro Instrumentation and Telemetry Systems (MITS), announced their 'Altair' kit on the cover of the January 1975 issue of *Popular Electronics*, the floodgates opened.

The resulting flood was unanticipated by the engineers in Silicon Valley; it was a shock to the ARPA-funded researchers in Cambridge as well. In 1975, Project MAC was well underway on the MIT campus, with a multi-faceted approach towards using large computers. Elsewhere in Cambridge, the

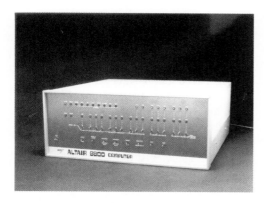

Figure 1.4 The Altair personal computer, the introduction of which on the cover of an issue of *Popular Electronics* in 1975 sparked the personal computer revolution (source: Smithsonian Institution photo).

research firm Bolt Beranek and Newman (BBN) was building the fledgling ARPANET (it was at BBN where the '@' sign was adapted for e-mail). Yet when Paul Allen, a young engineer working at a local electronics company, and Bill Gates, a Harvard undergraduate, saw the *Popular Electronics* article, they both left Cambridge and moved to Albuquerque, to found Microsoft, a company devoted to developing software for the Altair.

The people working on the two coasts at first ignored this phenomenon, but not for long. In Silicon Valley, young computer enthusiasts, many of them children of engineers working at local electronics or defense firms, designed and built personal systems of their own – including the Apple II, built by Steve Jobs and Steve Wozniak in 1977. Among the most fanatical promoters of personal computing were Stewart Brand, editor of the counterculture bible *The Whole Earth Catalog*, who believed that computers would fulfill the broken promise that LSD and other mind-altering drugs was supposed to yield (Turner 2006). The Apple was the best-known, and probably the best-designed, but it was one of literally dozens of competing personal systems, from all over the US and Europe, that used the microprocessor. MITS struggled, but Apple did well, and its financial

success alerted those in Silicon Valley that microprocessors were suitable for more than just embedded or industrial uses.

While Bill Gates and Paul Allen saw the potential for personal computing, others in Massachusetts persisted with research on large mainframes, the power of which would be many times greater than anything one could own personally. Researchers at MIT were working on Artificial Intelligence and advanced programming languages, and they reasoned, correctly, that the small systems were ill-suited for those applications. Local companies were making large profits on minicomputers, and they regarded the personal computers as too weak to threaten their product line. Both groups failed to see how quickly the ever-increasing computer power described by Moore's Law, coupled with the enthusiasm and fanaticism of hobbyists, would find a way around the deficiencies of the early personal computers.

The IBM Corporation was also making profits on its line of mainframes. A small group within the company set out to develop a personal computer based around an Intel microprocessor. The IBM Personal Computer was announced in 1981. It did not have much memory capacity, and it was difficult to network, but in part because of the IBM name, it was a runaway success, penetrating the business world. One reason for the success of both the IBM PC and the Apple II was a wide range of software, which enabled owners of these systems to carry out financial calculations that were cumbersome to do on a mainframe using punched cards. These first programs, VisiCalc and Lotus 1-2-3, came from suppliers in Cambridge, Massachusetts; they were followed by software suppliers from all over the US, with Microsoft soon dominating the industry.

XEROX PARC

Apple's success dominated the news about Silicon Valley in the late 1970s, but equally profound innovations were happening quietly at a nearby laboratory set up by the Xerox

Corporation. The Xerox Palo Alto Research Center opened in 1970, when dissent against the US involvement in the war in Vietnam ended the freewheeling funding for computer research at the Defense Department. Xerox was able to hire many of the top scientists funded by ARPA, and from its laboratory there emerged innovations that defined the digital world: the Windows metaphor, the local area network, the laser printer, the seamless integration of graphics and text on a screen. Xerox engineers developed a computer with icons that the user selected with a mouse – a device invented at Stanford University by Douglas Engelbart, who was inspired by Vannevar Bush's 'As We May Think' article. Xerox was unable to translate those innovations into successful products, but an agreement with Apple's Steve Jobs led to the innovations finding their way to consumers via the Apple Macintosh computer, introduced in 1984, followed by software from Microsoft (Smith and Alexander 1988).

The visionary research at the Xerox lab was far removed from the world of hobbyists tinkering with primative personal computers as they existed in the late 1970s. Stewart Brand, editor of the counterculture bible the *Whole Earth Catalog*, stated that '"Telecommunicating" is our founding domain', but for Brand that was a hope more than a reality (Brand 1984: 139). The first personal computers were used for other things: games, spreadsheets and word processing, which were not practical or permissible on mainframes. One could link a personal computer to a local network over one's home telephone, but larger networks were impractical, as the rate structure of the US telephone system made long-distance links expensive. Commercial and hobby services addressed that problem by interconnecting local networks. The most successful was America Online (AOL), which grew out of a system to allow personal computer users to play games with one another. Hobbyists established local 'bulletin boards' on their home or business computers, exchanging messages with other boards late at night,

when long-distance telephone rates were lower.

The social forces driving AOL and the bulletin boards were the ancestors of the forces driving Facebook, Twitter, and similar programs in the twenty-first century. As with the invention of the personal computer itself, these forces drove networking from the bottom up, while privileged military and academic agencies drove networking from the top down. The two eventually collided, with unexpected results. The growing awareness of the usefulness of networks placed demands on ARPA that, as a military agency, it was unwilling to meet. It responded by establishing a military-only network for internal use and spinning off the rest to the National Science Foundation (NSF). Around 1987 the NSF gave contracts to commercial suppliers to build a network for research and academic use. The NSF found itself unable to keep to its taxpayer-supported mission of restricting the network solely to research or non-commercial use. The dilemma was solved by a revision to the NSF's enabling legislation, signed in 1992, which relaxed the non-commercial restrictions. Within a few years the 'Internet', as it was now being called, was fully open to commercial use (Aspray and Ceruzzi 2008).

THE WORLD WIDE WEB

For many, the Internet is synonymous with a program that runs on it called the World Wide Web, a program developed in the early 1990s. The convoluted story outlined above explains the confusion: the ARPANET was led by military people who wondered at times whether even personal e-mail messages would be permitted over it. Companies like IBM marketed their own proprietary networks. Hobbyists ran networks from their homes. The Internet subsumed all of those, by virtue of its open protocols, lack of proprietary standards, and ability to interconnect existing networks of various designs. The Web, developed by Tim Berners-Lee and

Robert Cailliau at the European Council for Nuclear Research (CERN) near Geneva in 1990, continued this trend by allowing the sharing of diverse kinds of information seamlessly over the Internet (Berners-Lee 1999). Web software was, and remains, free.

The World Wide Web, overlaid on a commercialized but decentralized Internet, dissolved the tangle of incompatible formats, arcane programming languages and impenetrable jargon that were the hallmarks of digital electronics. It was not perfect: the Web had deficiencies that had to be addressed by a variety of approaches in the 1990s and after.

Berners-Lee introduced a program he called a *browser* to view Web pages, but it was a commercial browser from a company called Netscape that transformed the Web. The Netscape browser, introduced in 1994, integrated graphics and text, and the use of the mouse, which made Web 'surfing' painless. Netscape also developed a method of encrypting data, such as credit card numbers, and a more secure Hypertext Transfer Protocol was also developed. Those enabled commercial services like Amazon and eBay – both launched in 1995 and among the largest presences on the Web. Netscape introduced a way of tracking a person's Web session with a piece of data called a 'cookie', probably named after a character from the television program *Sesame Street*. Netscape's public offering of stock in August 1995 triggered an explosion of interest on Wall Street, with the subsequent bubble and its bursting dominating financial news for the rest of the decade.

Berners-Lee hoped that his software would allow people to post information on the Web as easily as they could access it. However, building Web pages required at least a rudimentary facility with programming, even in the simple HTML language. One response was the creation of an application called a 'Web log', soon shortened to *blog*. These began to appear in the mid-1990s and spread quickly. The breakthrough was that the blogger did not have to compile a program or otherwise drop down to HTML programming. The most successful blogs gained large followings and spanned the spectrum from celebrities, journalists and pundits to ordinary folks who had something to say. With the advent of social networking sites like Facebook, blogs lost some of their appeal, although they remain popular.

Another deficiency of the Web resulted from its most endearing feature: its ability to access information directly, whether it was on one's hard drive or on the other side of the planet. But how to navigate through it? As soon as this problem became evident, a number of indexing schemes appeared, variously called portals or search engines. Unlike private networks, which could charge for their built-in indexing schemes, these had to be supported by advertising, similar to the way commercial radio and television evolved in the United States. Some of them, like Yahoo!, combined human indexing with automated tools. The most successful has been Google, which ranks Web pages based on how often others link to them. Those who examine Google's financial success often point out that the opening screen of a Google search is clean, simple, and uncluttered, in contrast to Yahoo!'s or AOL's busy opening screens. The World War II human factors psychologists would have understood.

THE SMARTPHONE

In discussing the significance of what it means to be 'digital', this chapter has proposed that the digital approach has become a Universal Solvent, dissolving any technology that comes into its path. In the new millennium, this phenomenon once again appeared with a vengeance, as the lowly portable telephone, an analog device invented in the early 1970s, was transformed into a general purpose, mainframe, internetworked computer. It is called a 'smartphone', but making phone calls is the least of its capabilities. Young people prefer to use the device not to make calls, but to 'text': in other words, they use it as a portable Teletype. Smartphones incorporate Web access, satellite navigation using the Global Positioning System, on-board accelerometers descended from the US space program, maps, Yellow Pages, an encyclopedia, dictionaries in multiple languages, movies, playback of recorded music, generation of synthesized music, a radio, television, games, and photography. In other words, 'anything' in the sense that was described by Alan Turing.

Currently the dominant supplier of these phones is Apple, but like its Macintosh computer, Apple derived the basic ideas from elsewhere. Engineers had toyed with the idea for a long time as computers kept getting smaller, and had even introduced miniature versions of laptops, with tiny keyboards and displays. However, the human interface was a bottleneck – how to allow people to use computing power comfortably without a standard keyboard and video screen? In 1996 a Silicon Valley firm called Palm developed an ingenious device that replaced all the functions of paper organizers like the Filofax®, using a touch screen. A model introduced in 2002 by a Palm spin-off, Handspring, integrated a telephone. A few years later, Apple's iPhone carried this attention to human factors even farther, driven by the vision of the company's leader, Steven Jobs.

One could end this narrative with the adoption of these smartphones, recognizing that the digital world has evolved more since 2000 than in all the prior decades. The sudden rise of Facebook and Twitter defies understanding and will require a few years to sort out. The reader who has followed this narrative will see that the social forces behind Facebook are not new; they are only in a more accessible form. Early academic networks, for example, had forums where the topics of discussion ranged from the arcana of programming in the Unix operating system to what they openly categorized as 'sex, drugs, and rock and roll' (Salus 1995: 147). Likewise the chat rooms that fueled the growth of America Online were places where people could freely converse who otherwise had trouble socializing or who were engaged in behavior not sanctioned by their communities.

A glut of books, blogs, movies and television programs have appeared to explain what is happening. These analyses of a networked planet have an uncanny parallel to concepts that appeared at the dawn of this age, before the technologies had matured. Marshall McLuhan's 'global village', is getting a second look (McLuhan 1962). Stewart Brand's *Whole Earth Catalog*, inspired by the writings of Buckminster Fuller, likewise seems prescient (Brand 1980). Students of the Facebook phenomenon might want to take a look at the writings of the Jesuit philosopher Pierre Teilhard de Chardin, whose book *The Phenomenon of Man* (1955) introduced a concept of a 'noosphere' of global consciousness. Many, including the Catholic Church, thought his ideas a little too far-fetched, but perhaps he was only a few years ahead of his time.

The desire to use digital technology for social interaction, at the same time as its use as a weapon by the US military, is not a new story. Nor is the tension between a desire to make money, against a desire to share and promote a digital vision of Utopia. What has happened since the turn of the millennium is that the exponential increase in raw digital power, expressed in shorthand as Moore's Law, enabled and made practical the profound implication of the theories of information and computing first uncovered in the 1930s and 1940s. Unless Moore's Law comes to a sudden end – and there is only slight evidence that it will – the coming decades will only bring more astonishing fruits of the 'Digitization of the World Picture' (Dijksterhuis 1961).

REFERENCES

Aiken, Howard H. ([1937] 1964) 'Proposed automatic calculating machine', *Spectrum IEEE*, 1(8): 62–69.
Anon. (1977) 'Edward Kleinschmidt: Teletype inventor', *Datamation* (September): 272–273.

Aspray, W. (1990) *John von Neumann and the Origins of Modern Computing*. Cambridge, MA: MIT Press.

Aspray, William and Paul Ceruzzi (eds) (2008) *The Internet and American Business*. Cambridge, MA: MIT Press.

Berners-Lee, T. (1999) *Weaving the Web: The Original Design and Ultimate Destiny of the World Wide Web*. San Francisco, CA: HarperCollins.

Brand, S. (1980) *The Next Whole Earth Catalog: Access to Tools*. New York: Random House.

Brand, S. (ed.) (1984) *Whole Earth Software Catalog*. New York: Quantum Press/Doubleday.

Bush, V. (1945) 'As we may think', *The Atlantic Monthly*, 176.

Ceruzzi, P.E. (2003) *A History of Modern Computing* (2nd edn). Cambridge, MA: MIT Press.

Chomsky, N. (1957) *Syntactic Structures*. The Hague: Mouton.

Crick, F. (1970) 'Central dogma of molecular biology', *Nature*, 227 (August 8): 561–563.

Dijksterhuis, E.J. (1961) *The Mechanization of the World Picture*. Oxford: Clarendon Press.

Eckert, Wallace J. (1940) *Punched Card Methods in Scientific Computation*. New York: Thomas J. Watson Astronomical Computing Bureau.

Harvard University (1946) *A Manual of Operation for the Automatic Sequence Controlled Calculator*. Cambridge, MA: Harvard University Computation Laboratory.

Heide, L. (2009) *Punched-Card Systems in the Early Information Explosion, 1880–1945*. Baltimore, MD: Johns Hopkins University Press.

Kita, C. (2003) 'J.C.R. Licklider's vision for the IPTO', *IEEE Annals of the History of Computing*, 25(3): 62–77.

Kuhn, T.S. (1962) *The Structure of Scientific Revolutions* (2nd enlarged edn). Chicago, IL: University of Chicago Press.

Licklider, J.C.R. (1990) *In Memoriam: J.C.R. Licklider, 1915–1990*, Palo Alto, CA: Digital Equipment Corporation, Systems Research Center, Report # 61, August 7.

McLuhan, M. (1962) *The Gutenberg Galaxy: The Making of Typographic Man*. Toronto: University of Toronto Press.

Petzold, C. (2008) *The Annotated Turing*. Indianapolis, IN: Wiley.

Randell, Brian (ed.) (1975) *The Origins of Digital Computers: Selected Papers*. Berlin, Heidelberg, New York: Springer-Verlag.

Randell, B. (1980) 'The Colossus', in N. Metropolis, J. Howlett, and Gian-Carlo Rota (eds) *A History of Computing in the Twentieth Century*. New York: Academic Press.

Reid, C. (1970) *Hilbert*. New York: Springer-Verlag

Salus, P. (1995) *Casting the Net: From ARPANET to INTERNET and Beyond...* Reading, MA: Addison-Wesley.

Smith, Douglas and Robert Alexander (1988) *Fumbling the Future: How Xerox Invented, then Ignored, the First Personal Computer*. New York: William Morrow.

Spicer, Dag (2008) 'Computer History Museum Report', *IEEE Annals of the History of Computing*, 30(3): 76–77.

Standage, Tom (1998) *The Victorian Internet: The Remarkable Story of the Telegraph and the Nineteenth Century's On-Line Pioneers*. New York: Walker & Company.

Teilhard de Chardin, Pierre (1955) *The Phenomenon of Man* (English edition). New York: Harper.

Turing, Alan M. (1936) 'On computable numbers, with an application to the Entscheidungsproblem', *Proceedings of the London Mathematical Society* (2nd Series, vol. 42), 230–265.

Turner, Fred (2006) *From Counterculture to Cyberculture: Steward Brand, the Whole Earth Network, and the Rise of Digital Utopianism*. Chicago, IL: University of Chicago Press.

Wiener, Norbert ([1948] 1961) *Cybernetics: or Control and Communication in the Animal and the Machine*. Cambridge, MA: MIT Press.

Williams, Bernard O. (1984) 'Computing with electricity, 1935–1945'. PhD dissertation, Department of History, University of Kansas, Lawrence, Kansas.

Zuse, Konrad (1962) *Der Computer, Mein Lebenswerk*, Berlin: Springer-Verlag. (Translated into English as *The Computer – My Life* in 1993.)

2

The Field of Digital Technology Research

Charles Crook

This chapter will explore the landscape of social activity within which digital technology now plays a significant part. The aim is to understand why social scientists have developed so much interest in our relationship with this technology. The particular ways in which that relationship is played out will be more closely scrutinized by other authors in this handbook. In common with those later chapters, the present overview of the research landscape adopts a social and cultural orientation towards the digital world. This means that, in particular, it will bypass reviewing the emergence of digital technology in engineering and mathematical terms (those matters are addressed in the preceding chapter).

Any such socio-cultural overview must inevitably have a piecemeal quality. Space permits only the sketching of an indicative set of those themes attracting the social scientist towards matters of digital technology. Moreover, no attempt will be made here to offer a unifying theoretical frame for making sense of these themes – that is a challenge for others to address, later in the book. Nevertheless, a scoping exercise such as this one remains useful: it furnishes an organizing bird's-eye view of the territory to be interpreted. To get started on this scene-setting, it will help to reflect a little on the terms 'digital' and 'technology'. This will define and contain the boundaries of our concern and specify a structure for the overview that follows.

THE DIGITAL OBJECT

As I write this, I can hear music from the Miles Davis recording, 'Kind of Blue'. Some readers may own this music in its earliest format: as an 'LP' – a 'record', a 'disc'. In my own case, I also own it as a CD and in the format of an MP3 file. Perhaps first thoughts about the digital world gather around something like this example. That is, a familiar and concrete object, we now see mutating under the influence of digitization. So, there is continuity here, but also discontinuity: similar things but, perhaps, changing engagements with those things. Clearly, there are many examples of fresh engagement around the capture, storing, manipulating and sharing of such digital representations, certainly as music but, also, as all variety (and mix) of image, sound and text. Again, the underlying *practices* of such engagement are surely familiar, yet, at the same time, there is a sense of them being reconfigured.

The mundane but pervasive case of a music recording highlights a technical matter right at the heart of our concerns: namely, something distinctive about the manner of *coding* things. Current enthusiasm for discourses about 'this digital world' can be understood in terms of the radical consequences of a shift towards the widespread digital coding of information. In the case of music, that shift has been away from the traditional analogue method. On the vinyl version of 'Kind of Blue', sound has been captured and stored as a continuous waveform. It is visible as the undulating groves on the surface of a disc. There is an agreeable directness typical of pre-digital representation – a matching of the object and the coding. However, on the CD and MP3 (digital) formats, the waveform of sound has been repeatedly sampled and reconstructed as a long list of successive numerical values: those values being coded as binary numbers. Stark temporal sequences of 1s and 0s are made permanent through the basic electrical states of either being 'on' or 'off'. Such coding is much less easily visible to the naked eye. Yet, compared to the analogue alternative, it offers a highly versatile format. Therefore, a wide variety of input material can be assimilated to the same digital code. In that coded form, material can be readily compressed, manipulated, copied and transmitted. Moreover, given only a small and accessible toolset, it can be *de*-coded and *re*-transmitted. In recent times our appetite for digital representations and our creativity in managing them has flourished.

To exploit the versatility and abundance of items coded digitally, there has evolved an associated infrastructure of access and transport: a public framework for transmission, exchange and participation. Arguably, the *technology* of the digital, in this sense of its mechanisms for access, is one starting point for a deeper consideration of digital artefacts as cultural phenomena. In particular, this might involve considering how the digital infrastructure serves to redefine the time, place and format of our engagements with captured cultural material, thereby altering how it variously enters and interleaves with everyday living. This applies to a wide range of 'information': my newspaper, photo album, city guide, bank statement, unfinished novel, and so on. Moreover, the hosting technologies for these artefacts increasingly invite their owners into an active relationship with the hosted material: interacting, interrogating and manipulating whatever digital objects they can access.

So, digitization shapes how we interact with material artefacts but it also shapes our *social* lives, functioning to offer new points of reference within the interactions we cultivate with others. This may occur at both the interpersonal level ('you and me') and the community level ('us'). For example, I may introduce you to my favourite music by sharing it with you as a copy of some digital file. At the more communal level, a selection from someone's 'Kind of Blue' recording might form the basis of a posting on a website – a space designed to allow a larger (and unknown) audience to share such material. It might, for example, be posted as background to a personal video of a peaceful seascape (see YouTube 2006). Moreover, that same Web service also permits construction of an extended text commentary, whereby large numbers of users can reflect and interact around the posted digital artefact. Of course, all such social exchange does not *demand* digitization but it has been greatly elaborated by that technology.

This modest example illustrates a further sense in which the technology of the digital shapes human activity. Not, in this case, through furnishing an infrastructure of sharing objects, but through offering a technology of tools to act upon them: resources that permit the easy exploration and manipulation of digital representations. The owner of the YouTube posting cited above has *mixed* Miles Davis's music with a video, doing so in a way that offers a novel experience for its audience. Moreover, the growth of such creative activity invites us to explore an interpretative dimension of engaging with digital material. The audience must *understand* the constructed artefact. The users of a digital medium must learn how to 'read' (and, perhaps, admire) its objects. Digital media afford new modes of expression that come with their own syntax and semantics. Thus, in addition to mediating new forms of access and new forms of social interaction, digital technology may demand new strategies of meaning-making: a fresh confidence of interpretation.

To summarize this introduction: the iconic case of a piece of digitally coded music has been worked up as an everyday example of the digital ecology now commonplace within economically developed societies. Implicitly, the example has identified three different senses of 'technology' in the phrase 'digital technology'. First, there is 'technology' as a strategy for digitally *coding* (and compressing) recordable events in the world. Second, there is 'technology' in the form of a digital *infrastructure* designed for transporting those objects and offering engagement with them. Third, there is 'technology' as a set of digital *tools*: resources that permit the creation, representation, manipulation and analysis of the objects so transported.

Now consider the music example in terms of these three senses of digital technology. First, a jazz session may be coded for recording in more than one way. However, the *digital* version imparts an interesting slipperiness to the resulting object, because it is easily replicated, simple to access, and open to be altered with digital tools. Next, an infrastructure of digital transport allows fluent exchange and publication of such objects. Publication creates audience. Audiences create the conditions for acts of interpretation,

debate and commentary. Publication also creates issues of ownership and security. While ownership, in turn, creates questions about authenticity and authority. Ownership may also encourage incorporation of such material into public displays of personal or brand identity. It is not that these patterns of activity are anything new; it is more that digitization has imbued them with a sudden energy – one which is visibly engaging a large sector of society as participants, and one where participation relates to a wide range of cultural practices. It may also be argued that the forms of relationships with the social and material world which emerge from these changes do indeed confront us with a challenging novelty and discontinuity (Caldwell 2000; Weinberger 2002).

All such manipulation and trafficking in digital objects is of natural interest to social scientists, because most of what is happening in such examples does so within the social microstructure of everyday life. However, there is also a societal *macro*structure to consider. Digitization has been illustrated here through the personal and mundane example of musical recording, but this complex of reproduction, manipulation, sharing, publication, commentary and identity management is active in relation to a very wide range of digital material, and, moreover, such material can implicate a very wide range of actors and organizations. So, even the simple acts of individuals in the digital infrastructure may generate products that are wrapped up in the ambitions of industries, institutions and political interests (Castells 1996). Those ambitions may construct for the citizens of a digital world very different conditions of working and living (Harvey 1989). All of this is further reason why the trajectory of digital technology will be of great interest to social scientists.

The discussion so far has dwelt upon digital technology as expressed in the form of digital objects – that is, files or documents broadly understood. The examples above, while familiar, might be judged to convey too narrow a conception of the contexts in which digital technology is encountered. Accordingly, in the next section the nature of the 'digital environment' will be discussed more broadly.

THE DIGITAL ENVIRONMENT: IMMATERIALITY, VIRTUAL WORLDS AND AUGMENTATION

Certainly, one way to characterize the digital environment is in the terms acknowledged immediately above – as a complex of relationships with digital objects – but a striking feature of the digital 'object' is its immaterial character. Although our access to the digital is usually through the physical means of screens, keys, pointers etc., the representations accessed that way are rendered in the bits and bytes of digital code. Accordingly, one strand of social science research has been to understand the consequence of this apparent loss of materiality in the field of action.

This loss is sometimes expressed as a wider nostalgia for ways of acting that ensure continuity between ourselves and the natural world – nostalgia for everyday practices entailing a greater awareness of human agency and a greater sensitivity to the relationships between processes and their products (Watts 1971). Another way to express unease about immateriality, is in terms of how technology underpins the erosion of craftsmanship (Sennett 2008). Traditionally, this concern would have dwelt upon alienation arising from overspecialized forms of labour that disengage the individual from the creative process, but digital technology is also implicated in potentially pulling us apart within the *interpersonal* structures of work, as well as fragmenting and disconnecting labour in terms of material action. Thus the 'information management' perspective cultivated through digital technologies may encourage the enthusiastic proceduralizing or 'engineering' of workplace *processes*. This, in turn, may mean the loss of that informal and lateral communication underpinning the *practice* of work (Zuboff 1984). Relationships

with authority move from a consensual form to an 'informated' form, in which reified categories of analysis replace understanding from traditional forms of social exchange. On the other hand, as Brown and Duguid (2000) illustrate in articulating this concern, digital technology may be recruited to *support* more intimate communication as well as depose it. Clearly, this is a dimension of experience within digital environments that social scientists will want to understand.

The infusion of an immateriality into digitally mediated interactions is often associated with the term 'virtualization'. This defines a second strand of the digital environment. However, 'virtual' is a term that can be used generously – covering all manner of local exchanges with digital artefacts. Yet, in the particular phrasing of 'virtual world', it suggests something more integrated than the rather piecemeal environment of 'digital objects' discussed earlier. In fact, it suggests a more profound version of immateriality: in the shape of comprehensive 'worlds' or wholesale simulations of realistic scenarios. These might range from the goggle-and-glove technology of virtual reality systems to the simpler screen-based designs typical of Second Life and online multiplayer games. Evidently the degree of other-world fidelity that is achieved across this range of virtualizing implementations will vary.

These virtual world contexts have been of interest to social scientists for at least three broad reasons: immersion, embodiment and identity. First, states of personal 'immersion' (strong feelings of presence in some simulation) are claimed for virtual world experience. For example, in the context of games, this is sometimes termed 'GameFlow' (Sweetser and Wyeth 2005). Although the experience of immersion has not been well theorized – perhaps reflecting the wide range of depths and forms around which it is invoked – immersion is often presented as a subset of the more general experience of 'presence', whereby the users of some system have a strong feeling of them (or their avatar) being within a digitally constructed scenario. Understanding the depth of these experiences is of importance because they are increasingly implicated in supporting, amongst other things, risk simulation, skills training, clinical therapies and education.

A second and related research theme around virtual worlds is that of embodiment, or the sense of being disconnected from the physical agency of one's body. Evidently such experiences are related to what has been referred to as 'presence' and its phenomenology. However, a focus on the awareness of one's own body resonates with an established theoretical tradition that considers the role of embodiment in shaping cognition and understanding (Dreyfus 1972; Varela et al. 1991). The pursuit of this topic finds its most extreme form in Moravec's (1988) ambition to download human consciousness into a computer. The complex of digital worlds that virtualization promises has encouraged the tradition of 'cyberculture theorists' (Bell, 2007) who have developed for these worlds a distinct form of theorizing the way participants experience and interact.

A third research theme within the tradition of virtual worlds concerns implications for the construction and expression of personal identity. Turkle (e.g. 1995) has been a leading theorist investigating the way in which ideas about ourselves are influenced by sustained engagement with virtual worlds, or, 'life on the screen'. Since Turkle's early insights around this topic, research interest in the performance of social identity has spread to embrace not just virtual worlds but also the wider space of social networking (boyd, 2010).

The discussion so far has acknowledged 'digital environment' first in terms of pervasive but immaterial objects and the interactions they afford us and, second, in terms of a more all-embracing virtualization of sensory experience. Finally, there is an approach to the construction of 'digital environments' that is less about such wholesale construction of digital worlds and more about a creative interleaving with existing and material worlds. This is often phrased in terms of the 'ubiquity' of digital devices and

the 'augmenting' of reality. Such matters can be understood against an established tradition whereby the typical site of engagement with digital goods has been a circumscribed piece of hardware. Traditionally, the iconic site has been a personal computer. More recently, mobile telephony, laptop or tablet computers, and wireless networking have all made the Internet a 'ubiquitous' experience (Weiser 1994). Users can now find connections on the move and can make them with only minimal hardware. As some describe this pervasive access: ' ... losing it can feel like being stranded. Constant connectivity has changed what it means to participate in life' (Grant et al. 2006). Moreover, this intense connectivity is not only a matter of the person-to-person exchange afforded by a digitization of communication through mobile phones. Digital codes are increasingly embedded in the wider world around us, such that our personal devices can read those embedded codes and connect in yet more novel ways with other people, services and events.

Digitally augmenting an environment typically assumes that a person engaging with it will have a reading device of some kind (this might be a smartphone, for instance). Augmentation may then be simply by location, which might be achieved by GPS estimations or it might be by direct reading of structural features of some object (e.g. a posted picture). This is computationally intensive and so a more common method is through some kind of suitably positioned 'marker' that provides information in standard format, such as a bar code. Such constructions are of research interest because they may enhance people's encounters with specialist environments, such as museums. They are now designed in increasingly dynamic formats, so as to allow more adaptive interaction between individuals and these environments. There has thus evolved a species of augmented reality game – often based in urban settings – and, a developing subset of this species, the augmented reality educational activity.

The digital environment has been characterized here in three broad-brush ways. First, we may consider our environments as variously digitized according to how far interactions with them are shaped by access to and interaction with individual digital objects – music, images, documents, etc. Second an environment may be more dramatically digitized into a self-contained virtual world – offering varying degrees of felt presence and immersion. Finally, environments may be overlayed with digital markers or location 'hotspots' such that space acquires a kind of 'intelligence'. This last case is of special interest where that intelligence is rendered to the actor in a relatively transparent and seamless fashion.

Developments such as those sketched in this section are increasingly familiar to us. What is more intriguing is the challenge of understanding their impact: how they exert an influence on our experience of the world and our cultural practices within it. Later in this chapter some further examples will be explored of how digitization shapes particular aspects of our experience. However, as a preface to that discussion, it is appropriate to consider how such 'digital effects' are best theorized.

CONCEPTUALIZING THE IMPACT OF THE DIGITAL

The highly digitized world has a technical history that has been summarized in Chapter 1 of this handbook and elsewhere (e.g. Gleick 2011). From that history it is tempting to seek key turning points that define 'impact': crucial innovations that might be causally linked to changes in social practice. For instance, primary causes of change might be ascribed to the emergence of the transistor as an electrical switching device or to Shannon's (1948) theorization of information flow using a metric based on binary coding. But such achievements of engineering, logic and mathematics do not simply trigger upheavals that then define 'our new digital world'. Such

achievements must resonate with prevailing socio-cultural forces. Those forces will shape and direct how such technical products are actually adopted: what then evolves reflects the human aspirations and appetites of their times.

From the pressure of such forces there has emerged those 'information goods' (Shapiro and Varian 1999) that define a distinct species of economy: one based upon intangible products. Such products include anything that can be rendered in the bits and bytes of digital coding (music, magazines, invoices, etc.) but they also include digital services (such as search engines and sites for user publishing). Considered together, these may be termed 'digital goods' (Loebbecke 2002). Their dematerialized character allows trading around them to flourish within a particular transmission infrastructure – most notably, the Internet. However, the economic viability of digital goods will often depend on achieving a critical mass of consumer access and engagement; in relation to which, many digital goods (particularly those supporting personal communication) benefit dramatically from 'network effects' (Rohlfs 1974). Under these circumstances, increased levels of adoption act to the advantage of existing users, such that growth in the digital economy thereby will accelerate. That has now happened for many goods. These are developments that have, in turn, stimulated new technologies to further motivate this growth – notably the technologies of faster communications, more intuitive interface designs, and easy-access delivery devices. Caldwell (2000) has argued that within such network effects there can occur 'tipping points' when the pace of adoption and influence manifests a seemingly sudden impact and discontinuity.

The key point is that this growth was not a straightforward or direct response to the technical achievement of digital coding. The growth associated with digitization has been characterized by a generative interplay of technical innovation with cultural practices and preferences. Understanding this interplay has itself been a matter of great interest to economists and researchers of marketing. The digital economy sustains distinctive structures of consumption (Rayna 2008). For instance, on the one hand, the pervasive dynamic of 'network effects' can stimulate innovation and growth. On the other hand, it can be a source of 'lock in' to particular brands and services.

In sum, the current configuration of a digital world has not been brought about by the technology of digital coding exerting some straightforward pattern of direct causal influence or impact. Where we are now has arisen from a complex interplay between technical designs and cultural appetites. Social scientists have been strongly engaged by the challenge of understanding such a dynamic and from that interest has evolved a tradition of theorizing termed 'the social construction of technology' or SCOT (Bijker et al. 1987). One particular caution that such researchers often urge concerns the assumption that digitization brings about dramatic discontinuities – in how we think, act, engage with others, or represent the world (e.g. Weinberger 2002). Current new technologies can be seen as continuous with a long history of technology being incorporated into human construction and craftsmanship (Sennett 2008; Sigfried 1948). New media have always been implicated in shifting patterns of how we think and how we re-present our experience (Friedberg 2006; Olson 1994).

Given that these new technologies are so pervasive in everyday life, it might seem unexceptional to note that social scientists attend to digital matters – but are these digital matters any more than a fine-tuning to the ease of how we live? In practice, some adoptions of digital technology are much more than this and some invite the attention of social science more urgently than others. These may be understood in terms of the manner in which they re-*mediate* human activity in significant ways. In fact, the term 'mediation' deserves careful marking as a valuable one in this context.

Cole and Griffin (1980) contrast the term with a traditional understanding of technological innovations that employs various discourses of efficiency, economy or amplification. They argue that such terminology constrains our vision of change and influence. 'Amplification' (in engineering) implies increasing the strength of some signal – with no change to the basic structure of what is being amplified. In the case of human activity, it may sometimes seem harmless to invoke the amplification metaphor in this 'strengthening' sense. For instance, relative to a bicycle, a motor car amplifies the speed of getting from A to B, but it also radically changes our experience of transport; it reshapes our engagements with the world in all sorts of ways. So, we can say that the internal combustion engine '*remediates*' the cultural practice of travel: motor cars do not simply speed up travelling, they restructure our world – doing so around how they solve the basic need for travel. Similarly, appropriating a digital technology into some cultural practice (say, education, shopping or banking) is not simply 'strengthening' that practice, say in terms of its pace, efficiency or economy, it causes that practice to be executed differently. It remediates the structure of how we act. Social scientists must have a natural interest in these transformations. That scope of that interest – expressed as remediations – will be sketched in the following sections.

Accordingly, in the remainder of this essay, a simple three-part structure will be deployed to organize an overview of some central issues pertaining to digital technology research. Under each heading attention will be given to how this technology remediates forms of human experience and varieties of cultural practice. First, issues concerning how individuals integrate digital technologies with their everyday activity will be considered under the heading of *prosthesis*. Second, a section on digital *representation* will discuss how these media afford new opportunities for expressive activity, and new challenges for the reading or interpretation of such constructions. Finally, a section on *coordination* will address digital media within social interactions, considering the various ways in which this technology supports communication, collaboration and participation with other people.

DIGITAL PROSTHESIS

The term prosthesis implies circumstances in which technologies are incorporated into human action in a manner that creates more elaborate systems of action. Of particular interest here is what may be termed 'cognitive prosthesis' (Clark 2003), meaning the possibility of these technologies extending the way in which we think and reason. Such possibilities are a natural concern of psychology, although that discipline adopted the idea of prosthesis only after having explored other motives for engaging with digital technologies as a resource for understanding mental life.

Arguably, psychology has been the only social science discipline to embrace digitization through appropriating the mathematical and engineering concepts that lie behind it. It was Shannon's (1948) seminal work on information theory that argued the possibility of measuring information and, thus, systematizing the scientific study of communication. His paper opened with the following observation:

> The fundamental problem of communication is that of reproducing at one point either exactly or approximately a message selected at another point. Frequently the messages have meaning; that is they refer to or are correlated according to some system with certain physical or conceptual entities. These semantic aspects of communication are irrelevant to the engineering problem. (1948: 379)

As Gleick comments 'A psychologist could hardly fail to consider the case where the source of the message is the outside world and the receiver is in the mind' (2011: 259). Accordingly, researchers started to conceptualize mental life in computational terms, thinking of the mind as an information processing and transmitting system

(e.g. Attneave 1959; Broadbent 1958). Behaviourism's oppressive language of 'stimulus' and 'response' gave way to a vocabulary that framed human cognition in terms of the transmission and reception of information, with decision making being a matter of reducing (measurable) levels of information uncertainty. Much was made of the limited 'channel capacity' of this human cognitive system, with Miller (1956) stressing our ability to overcome this by strategically recoding incoming information. Indeed the ingenuity with which the cognitive system undertakes this organization of input became the basis of a richer form of cognitive psychology, one richer than that based slavishly on the computational models of information theory. However, an understanding of the neurone as a binary signal device continues to invite theorizing based upon information processing, but, now, more the *pattern*-forming activity of digital systems (e.g. connectionist theories of learning).

Yet this computational metaphor of mind has recently been reconsidered. Since the early 1990s, the work of Vygotsky (1978) and other cultural-historical theorists has stressed the significance of tools as a theoretical resource for understanding human thinking. For example, much has been made of the emergence of writing and, subsequently, printing as technologies that alter human cognition, consciousness and social relations (e.g. Olson 1994). At the same time, empirical studies of human thinking in natural situations of problem solving (e.g. Hutchins 1995) have encouraged a conception of cognition as 'distributed' (Salomon 1993) – meaning that mental activity naturally incorporates the resources of the material and social world into its computations. Subsequently, Clark and others further articulated this idea through the notion of an 'extended mind', radically questioning the idea of the human mind as something bounded by the human skull (Clark 2003; Clark and Chalmers 1998).

This direction of theorizing is often illustrated through a popular quote from the writing of the anthropologist Gregory Bateson. He reflects:

> Suppose I am a blind man, and I use a stick. I go tap, tap, tap. Where do *I* start? Is my mental system bounded at the hand of the stick? Is it bounded by my skin? Does it start halfway up the stick? Does it start at the top of the stick? (Bateson 1972: 459)

The man's stick is a prosthesis. It could be said to be a 'cognitive prosthesis' in that it doubtless incorporates its data (tap, tap, tap) into the man's thinking and reasoning. Evidently the rich information management that is afforded by digital tools will make them a significant component of this 'extended mind'. However, as stressed earlier, this would not be a simple matter of 'amplifying' the mind's capabilities. A recent study by Sparrow, Liu and Wegner (2011) demonstrates how digital extensions for remembering work by changing the structure of that cognitive practice. They show how being aware of the Internet as a memory resource remediates the manner in which we learn new information. So, expecting to have Internet access actually attenuates remembering for the information itself, while enhancing memory for where to find it (online).

This cognitive prosthetic conception of digital technology evidently attracts the research activity of psychologists but it also has implications for pedagogy and the design of schooling. It is, therefore, not surprising to find many manifestos for educational innovation placing digital technology in the foreground of their vision. Yet a recurring concern within the social sciences is the apparent reluctance of educational practitioners to embrace the demands of adapting practice towards a more cognitive prosthetic conception of learning and enquiry (Collins and Halverson 2009).

REPRESENTATION

Central to the cultural evolution of the modern mind has been our ingenuity in capturing experience and then re-presenting it, both for the self and for others (cf. Donald 1991). We naturally think of language as a key representational vehicle for such purposes: first through the voice patterns of oral tradition and then, most powerfully, by means of writing things down. Indeed, much educational practice relates to this: it concentrates on the effort of establishing print literacy in young people (Olson and Torrance 2009).

Written text can, of course, be encountered in a digital format. This observation may seem unexceptional – merely a matter of migrating text from one physical medium (page) to another (screen) – but there is plenty of research that illustrates the demands of engaging comfortably with this re-mediated version of writing (Haas 1996), because these 'migrations' entail formats which may sometimes disorientate the inexperienced user – particularly if they impose new designs on objects known by names inherited from older traditions. For example, a digital newspaper might need to be read differently from the print version and yet it is still presented to the user as 'a newspaper'. Similarly, a student's (digital) essay may exploit presentational possibilities not easily recruited in traditional academic formats and so demand a reading different from a purely textual composition.

Certainly, the impact of digital technologies on the representation of human knowledge goes further than shifting the medium of writing from page to screen. So much so that the phrase 'digital literacy' has become fashionable. Its cultivation is often presented as a modern challenge for educational practice (Cervetti et al. 2006). The term 'literacy' has thereby expanded to embrace much more than its original association with the printed word: extending from 'the ability to read and write' to 'the ability to understand information however presented' (Lanham 1995: 198). Kress (2003) in particular has argued for the increased importance of multimodality or the ability to express ideas across a wide range of representational systems. On this analysis, each medium of communication has its own constraints and affordances.

Digital literacy is about acquiring confidence in 'reading' these systems (Buckingham 1993). Social science research explores both the nature of these new expressive forms as well as the practical challenge of preparing us for the interpretative reading that they demand (Bateman 2011).

However, the above sketch of changing literacy demands needs to be expanded. It needs to acknowledge the wider and institutional reach of these innovations: in particular, the ways in which traditional and public genres of expression or communication are being remediated by digital technology. In the arts, this often reflects the potential of digital media to enhance the interactivity of performance or the participation of audiences (e.g. Ryan 1997), and there is growing interest in 'electronic literature' (Hayles 2008) with its multimodal possibilities. In the context of scholarship, an influence of digital media has been felt less at the level of multimodality and more in terms of a greater inclusion of authorship and new methods of knowledge organization and access. Remarks on each of these topics below should indicate how they would interest social science research.

In relation to the first issue – authorship of knowledge – both (Internet) digitization of publication and the flat structuring of access to the global networks has allowed wider participation in the public conversations of knowledge construction. Anderson's (2006) thesis of the 'long tail' draws attention to how the Internet supports access to more obscure items (i.e. those that would normally be lost on the 'long tail' of a retail demand curve). Today, items that would not survive in the real-world marketplace may more easily find an Internet niche where they might then be discovered. Our ability to find low-popularity books and music is often cited as the reward of protecting (and celebrating) the long tail. However, information of any kind can be long-tail protected in this way, including scholarship. The ubiquitous Web provides powerful tools for a wider constituency of authors (those defining the flat extent of this

tail) to publish their ideas. This in turn raises issues of authority: a concern that has been highlighted by the sceptic Keen (2007), who diagnoses a dangerous 'cult of the amateur', as processes that protect the authority and credibility of scholarship (or news or commentary) are rendered more permeable. One sense in which these become core social science concerns is in relation to the new imperatives that are created for inducting students into thoughtful enquiry in this arena. In a sense this is a further extension of the need to prepare digital literacy – but a form of literacy that includes more than usual attention to the social and political construction of public knowledge.

The same imperative applies to digital influence in the structural organization of knowledge, particularly in relation to designing architectures that permit comfortable enquiry and search. It is inevitable that the growth of digitally coded information increasingly challenges our ability to conduct rational search. Of course, it is also true that the tools of digital search become more sophisticated in parallel with this abundance of information, but the authority and strategies of those tools needs to be interrogated and understood. It may be for the information sciences to articulate those properties (e.g. Morville 2005) but it is for social sciences to design and implement the pedagogic processes that ensure such digital literacy is effectively cultivated.

COORDINATION

This final section considers the mediation of digital technology within the social exchanges in which we take part. It is interesting that this technology – at least in the form of the personal computer – was originally characterized in terms of its potential for socially isolating its users. Early research observers in both work and play settings were fond of documenting a rather compulsive pattern of engagement (Kidder 1981; Turkle 1984). Typical concerns voiced at the time cautioned against technology cultivating within learners 'thought in isolation' (Kreuger et al. 1989: 113), predicting that 'What is learned, then, is passivity and alienation from oneself and others, and that the most fruitful relationships with people will be as passive and impersonal as the solitary interaction with the computer (Kreuger et al. 1989: 114). While such marginalized absorption can still be documented, it is striking how the technology is now seen in terms of its potency for social networking, not social isolation.

The range of issues that could be discussed under this heading is very large. The intention here is merely to summarize them and give a flavour of the challenges currently available to social science research. This summary will be organized through brief consideration of three themes: the digital mediation of personal communication, collaborative relationships, and structures for participation.

Mediating Coordination – Relationships and Interactions

For describing patterns of personal communication that are digitally mediated, the phrase 'social networking' is very familiar. It conveys a positive tone. Perhaps a human concern to be in harmony with others. Yet the designs of social networking are not universally applauded. Benninger (1989) ventures a parallel between the mechanization of labour in the nineteenth century and the current spread of a proceduralizing bureaucracy that now mechanizes personal relations. Social software may contribute by formalizing the informal. As boyd (2007a) observes, networked conventions, such as dichotomizing relationships into friends and non-friends, violate ways of perceiving relationships that have matured over a long period of personal development. A more immediately troubling aspect of the digital social world is the prevalence of online bullying – sometimes termed 'cyberbullying'. Many young people have reported this kind of persecution as an unwanted consequence of Internet participation (Li 2007; Stomfay-Stitz and Wheeler 2007).

However, there is a risk of demonizing digital communication. The opportunity for research to understand the way in which it allows young people to explore their social identity is identified by boyd (2007b), and it is certainly implicated in fostering digital romance (Doring, 2002). Clearly those aspects of social science that address matters of personal communication and relationships have much to explore within digital technology.

Mediating Coordination – Collaborating

A form of social relationship of special interest to social scientists is that in which interacting partners work towards the creation of some shared knowledge. Semantics around this topic can be difficult. So, it is not always clear when a simple 'conversation' should be termed a 'collaboration'. This move is usually made when the conversation has a strong focus; when it is orientated towards constructing a particular product or outcome. In which case it acquires a more organized and directed flavour. It is in respect of managing that organization and direction that digital tools play an interesting role.

A task at the heart of collaborating is creating and updating an external (and therefore shared) representation of what the collaborators know. Crook (1994) has argued that digital media provide a powerful resource for creating and managing joint understandings. The personal computer furnishes an environment characterized by powerful opportunities for joint activity: versatile modes of problem representation, tools for interacting on such representations, and a sustained narrative of what the collaborators have done. The working of such collaborative mediation has become a core concern in the area of 'computer supported collaborative learning' (CSCL).

CSCL is not exclusively about the intimate forms of exchange associated with small group problem solving. Digital tools also have empowered individuals to collaborate while separated by distance, and even in arrangements that do not require their activity to be synchronous. This has largely been made possible by the infrastructure of digital networks. These environments create a structure in which participants can converse but, also, representational tools that express and preserve the evolution of their shared understanding. Researching the ways in which such tools mediate a more distributed form of collaboration is a priority in one particular domain of these designs: namely, 'networked learning' (Goodyear 2002).

Mediating Coordination – Participations

Through the Internet, digital technology offers a striking platform for the individual voice. Moreover, the means available for individual expression are generous. That 'voice' may be encountered in various forms: writing, sound, image, video or as mixtures of these modalities. Accordingly, users of this platform may see it, amongst other things, as an opportunity to be creative, to agitate, or to make knowledge claims. However, the Internet furnishes a platform for multiple voices and, therefore, the possibilities for them to interact. What the individual does may then be termed 'participative' in so far as the expression of individual voice becomes recognized and coordinated with others and in so far as this coordination creates some sense of shared engagement.

However, there are different ways in which such coordination can occur. At its simplest, it may take the form of relatively contained dialogue or conversation. Such exchanges might occur as commentary on the postings of bloggers. Or they might be less intimate, such as might occur within the focused concerns of a text-based discussion forum. Or they might be extremely fragmented, as those that occur in the Twitter stream.

Richer forms of conversation can be achieved when the interactions involved acquire continuity and coherence, and when they are sustained over significant periods of time. It then becomes natural to speak of online 'communities' having been formed – either through deliberate shaping by participants or as the result of a more improvised or spontaneous consolidation. Arguably, the growth of digital communication has encouraged a lazy use of the term 'community'. For instance, Nunberg (2001) notes how it tends to have a status that is inherently positive (it would be a little odd to speak of the 'terrorist community' or the 'paedophile community'). Consequently, any invoking of the term to characterize successful digital coordination might mean the communication is judged rather uncritically, perhaps with limited consideration as to whether 'community'-based argument, decision making or problem solving has actually been optimized. Such uncertainties relating to understanding the workings of digitally empowered communities makes them a natural topic of attention for social science research. This applies whether such communities are exclusively online or they blend online communication with more tradition methods of convening face to face.

The bonding that can occur for online communities illustrates the socially richer end of a continuum of digitally mediated and participative coordination. Meanwhile, at the other extreme is the loosely knit structures often associated with networked communication: that is, structures involving very large numbers of individuals, perhaps exchanging infrequently and perhaps unknown to each other at the personal level. Such groups are sometimes termed 'crowds', particularly when their constituent individuals are polled for evaluations or opinions, the outcome of such polling being valued as the 'wisdom of crowds' (Surowiecki 2004). Some authors advocate efforts to mobilize and structure such networks of loose participation, thereby orchestrating more formal but large-scale collaborative thinking in pursuit of 'mass creativity' (Leadbetter 2008). This notion is not without its critics (e.g. Lanier 2010), and social science research must help the understanding of where it works satisfactorily.

The reality of a crowd wisdom is just one area of participative coordination where there is doubt and criticism. There is a rhetoric in this area that stresses inclusion. Yet unmanaged inclusion must challenge the ease with which confidence about voice and message can be achieved (Keen 2007). Moreover, it may seem churlish to question the success of digital communication in opening up an arena of political participation and agitation. However, some commentators have noted the difference between the fragmented political engagement of digital communications and an earlier form of participation based on the bonding achieved by well-structured and sustainable social groups (Caldwell 2010). Once more it is for social science to make more visible the forces of re-mediation that shape a new digital experience of participation.

CONCLUSION

When significant changes in identity, society and culture are too firmly ascribed to a particular technology we are right to feel uneasy. Adopting a deterministic attitude to digital impact will only serve to conceal from us a more complex and interesting dynamic of influence and causality. However, articulating that dynamic is a far bigger task than can be embraced in one book chapter. Fortunately, later chapters in this volume will go further in both exploring complexity and arousing interest. From the present chapter it is hoped that a rough map has emerged, one that describes the landscape of relevant socio-cultural forces operating around digital technology. So, emphasis has been given to how digital coding has created novel species of artefact and representation and how new cultural practices of creating, sharing and interpretation have evolved around them. Such novel cultural practices are an inevitable interest for social scientists. That interest becomes amplified as engagement with digital artefacts generates quite new structures of experience – such as those encountered in the virtual worlds that may be constructed from digital raw materials. Finally, we have also identified

how a growing awareness of digital coding has stimulated new forms of theory building within the social sciences. In short, this is a rich territory of concern for theorist, designer and practitioner.

REFERENCES

Anderson, C. (2006) *The Long Tail: Why the Future of Business is Selling Less of More*. New York: Hyperion.

Attneave, F. (1959) *Applications of Information Theory to Psychology: A Summary of Basic Concepts, Methods, and Results*. New York: Holt.

Bateman, J. (2011) *Multimodality and Genre: A Foundation for the Systematic Analysis of Multimodal Documents*. Houndmills, Basingstoke: Palgrave Macmillan.

Bateson, G. (1972) *Steps to an Ecology of Mind: Collected Essays in Anthropology, Psychiatry, Evolution, and Epistemology*. London: University of Chicago Press.

Bell, D. (2007) *Cyberculture Theorists*. Abingdon, Oxon: Routledge.

Benninger, J. (1989) *The Control Revolution: Technological and Economic Origins of the Information Society*. Cambridge, MA: Harvard University Press.

Bijker, W.E., Hughes, T.P. and Pinch, T.J. (1987) *The Social Construction of Technological Systems: New Directions in the Sociology and History of Technology*. Cambridge, MA: MIT Press.

boyd, d. (2007a) 'None of this is real', in J. Karaganis (ed.), *Structures of Participation*. New York: Social Science Research Council.

boyd, d. (2007b) 'Why youth (heart) social network sites: The role of networked publics in teenage social life', in D. Buckingham (ed.), *Youth, Identity, and Digital Media*. Cambridge, MA: MIT Press.

boyd, d. (2010) 'Social network sites as networked publics: Affordances, dynamics, and implications', in Z. Papacharissi (ed.), *Networked Self: Identity, Community, and Culture on Social Network Sites*. New York: Routledge, pp. 39–58.

Broadbent, D.E. (1958) *Perception and Communication*. Oxford: Pergamon.

Brown, J.S. and Duguid, P. (2000) *The Social Life of Information*. Boston, MA: Harvard Business School Press.

Buckingham, D. (1993) *Changing Literacies: Media Education and Modern Culture*. London: Tufnell Press.

Castells, M. (1996) *The Information Age: Economy, Society, and Culture. Vol. 1: The Rise of the Network Society*. Malden, MA: Blackwell.

Cervetti, G., Damico, J. and Pearson, P. (2006) 'Multiple literacies, new literacies, and teacher education', *Theory Into Practice*, 45(4): 378.

Clark, A. (2003) *Natural-Born Cyborgs: Minds, Technologies and the Future of Human Intelligence*. New York: Oxford University Press.

Clark, A. and Chalmers, D. (1998) 'The extended mind', *Analysis*, 58(1): 7–19.

Cole, M. and Griffin, P. (1980) 'Cultural amplifiers reconsidered', in D.R. Olson (ed.), *The Social Foundations of Language and Thought*. New York: Norton.

Collins, A. and Halverson, R. (2009) *Rethinking Education in the Age of Technology: The Digital Revolution and Schooling in America*. New York: Teachers College Press.

Crook, C.K. (1994) *Computers and the Collaborative Experience of Learning*. London: Routledge.

Donald, M. (1991) *Origins of the Modern Mind: Three Stages in the Evolution of Culture and Cognition*. Cambridge, MA: Harvard University Press.

Doring, N. (2002) 'Studying online love and cyber romance', in B. Batinic, U-D. Reips and M. Bosnjak (eds), *Online Social Sciences*. Seattle, WA: Hogrefe and Huber, pp. 333–356.

Dreyfus, H. (1972) *What Computers Can't Do: A Critique of Artificial Reason*. New York: Harper & Row.

Friedberg, A. (2006) *The Virtual Window*. Cambridge, MA: MIT Press.

Gladwell, M. (2000) *The Tipping Point*. London: Abacus.

Gladwell, M. (2010) 'Small change: Why the revolution will not be tweeted', *New Yorker*, 4 October.

Gleick, J. (2011) *The Information: A History, a Theory, a Flood*. New York: Pantheon.

Grant, L., Owen, M., Sayers, S. and Facer, K. (2006) *Social Software and Learning*. Opening Education Reports. Bristol: Futurelab. Available at: http://archive.futurelab.org.uk/resources/publications-reports-articles/opening-education-reports/Opening-Education-Report199/ (accessed 9 August 2011).

Goodyear, P.M. (2002) 'Psychological foundations for networked learning', in C. Jones and C. Steeples (eds), *Networked Learning: Perspectives and Issues*. Berlin: Springer-Verlag, pp. 49–75.

Harvey, D. (1989) *The Condition of Postmodernity: An Enquiry into the Origins of Cultural Change*. Oxford: Basil Blackwell.

Haas, C. (1996) *Writing Technology: Studies on the Materiality of Literacy*. Mahwah, NJ: Erlbaum.

Hayles, N.K. (2008) *Electronic Literature: New Horizons for the Literary*. Notre Dame, IA: University of Notre Dame Press.

Hutchins, E. (1995) *Cognition in the Wild*. Cambridge, MA: MIT Press.

Keen, A. (2007) *The Cult of the Amateur: How Today's Internet is Killing our Culture*. New York: Currency.

Kidder, J.T. (1981) *The Soul of a Machine*. Toronto: Little Brown.

Kress, G. (2003) *Literacy in the New Media Age*. Abingdon, Oxon: Routledge.

Kreuger, L.W., Karger, H. and Barwick, K. (1989) 'A critical look at children and microcomputers: Some phenomenological observations', in J. Pardeck and J. Murphy (eds), *Microcomputers in Early Childhood Education*. New York: Gordon and Breach.

Lanham, R. (1995) 'Digital literacy', *Scientific American*, 273(3): 160–161.

Lanier, J. (2010) *You Are Not a Gadget: A Manifesto*. New York: Alfred A. Knopf.

Leadbetter, C. (2008) *We-Think*. London: Profile Books.

Li, Q. (2007) 'New bottle but old wine: A research of cyberbullying in schools', *Computers in Human Behavior*, 23(4): 1777–1791.

Loebbecke, C. (2002) 'Digital goods: An economic perspective', in H. Bidgoli (ed.), *Encyclopedia of Information Systems*. San Diego, CA: Academic Press, pp. 635–647.

Miller, G.A. (1956) 'The magical number seven plus or minus two: Some limits on our capacity for processing information', *Psychological Review*, 63(2): 81–97.

Morville, P. (2005) *Ambient Findability*. Sebastopol, CA: O'Reilly Press.

Moravec, H. (1988) *Mind Children: The Future of Robot and Human Intelligence*. Cambridge, MA: Harvard University Press.

Nunberg, G. (2001) *The Way We Talk Now*. Boston, MA: Houghton Mifflin.

Olson, D.R. (1994) *The World on Paper: The Conceptual and Cognitive Implications of Writing and Reading*. Cambridge: Cambridge University Press.

Olson, D.R. and Torrance N. (2009) *The Cambridge Handbook of Literacy*. New York: Cambridge University Press.

Rayna, T. (2008) 'Understanding the challenges of the digital economy: The nature of digital goods', *Communications & Strategies*, 71(1): 13–36.

Rohlfs, J.H. (1974) 'A theory of interdependent demand for a communication service', *Bell Journal of Economics and Management Science*, 5(1): 16–37.

Ryan, M-L. (1997) 'Interactive drama: narrativity in a highly interactive environment', *Modern Fiction Studies*, 43(3): 677–707.

Salomon, G. (1993) *Distributed Cognitions: Psychological and Educational Considerations*. Cambridge: Cambridge University Press.

Sennett, R. (2008) *The Craftsman*. London: Penguin.

Shannon, C.E. (1948) 'A mathematical theory of communication, Part I', *Bell Systems Technical Journal*, 27: 379–423.

Shapiro, C. and Varian, H. (1999) *Information Rules: A Strategic Guide to the Network Economy*. Boston, MA: Harvard Business School Press.

Sigfried, G. (1948) *Mechanization Takes Command: A Contribution to Anonymous History*. New York: Oxford University Press.

Sparrow, B., Liu, J. and Wegner, D.M. (2011) 'Google effects on memory: Cognitive consequences of having information at our fingertips', *Science*, 333(6043): 776–778.

Stomfay-Stitz, A. and Wheeler, E. (2007) 'Cyber bullying and our middle school girls', *Childhood Education*, 83(5): 308–314.

Surowiecki, J. (2004) *The Wisdom of Crowds: Why the Many are Smarter than the Few*. London: Abacus.

Sweetser, P. and Wyeth, P. (2005) 'GameFlow: A model for evaluating player enjoyment in games', *ACM Computers in Entertainment*, 3(3): 1–24.

Turkle, S. (1984) *The Second Self*. New York: Simon & Shuster.

Turkle, S. (1995) *Life on the Screen: Identity in the Age of the Internet*. New York: Simon & Schuster.

Varela, F.J., Thompson, E. and Rosch, E. (1991) *The Embodied Mind: Cognitive Science and Human Experience*. Cambridge, MA: MIT Press.

Vygotsky, L. (1978) *Mind in Society: The Psychology of Higher Mental Functions*. Cambridge, MA: Harvard University Press.

Watts, A. (1971) *Does It Matter*. Novato, CA: New World Library.

Weinberger, D. (2002) *Small Pieces Loosely Joined: A Unified Theory of the Web*. Cambridge, MA: Perseus Books.

Weiser, M. (1994) 'The world is not a desktop', *Interactions*, January: 7–8.

Zuboff, S. (1984) *In the Age of the Smart Machine*. New York: Basic Books.

Links

YouTube (2006) www.youtube.com/watch?v=xYqcMOKF4qM

PART 2

New Digital Technologies: Key Characteristics and Considerations

Armed with a broad understanding of where this book is positioned, our next section starts to cover some important areas of how we use digital technology. This section covers location, search, social connection, personalization and ethics. Each of these different 'takes' on digital media start to map out the space of research and how different scholars have come to think of and empirically examine digital technologies in use.

Chapter 3, by Matt Jones, engages with the role of location and context in digital technology – focusing on the attempts to define, measure and understand context. It starts by drawing on the history of attempts in artificial intelligence and machine learning to understand our context and location and make sense of that in systems behaviour. This has been a particular interest in mobile systems where location becomes much more pressing. Yet this is contrasted with the human meaning and sensemaking around location and context – in the ways

in which context comes to be interactively and communicatively created and shaped.

In Chapter 4, Gary Hsieh and Nicolas Friederici turn to questions of search and information access, engaging with enquiries into how we search for information online. As the rise of Google has shown, information search is no longer a specialized or esoteric activity and information retrieval is no longer limited to libraries or research institutes. This chapter also nicely covers the social nature of information search, underlining how search is not simply a solitary concern but is also something connected heavily with information sharing.

In Chapter 5 we move on to engage with questions of social media use and the ways in which our social lives are increasingly saturated with technology. Sonja Baumer's chapter discusses debates around 'addiction to social media', the interpersonal and emotional aspects of social media use and, lastly, the constraints and limits of social media. Rather nicely, through these points this chapter gives us a broad introduction to contemporary research on social media and its use, situating these debates in contemporary problems and arguments.

Heather Horst and Larissa Hjorth in Chapter 6 contrast these debates with arguments around the 'personalization' of technology. These authors take the notion of 'personalization' as the forming of an intimate relationship with an object or consumer item, such as a new technology. Rather than discussing how technical systems might adjust themselves to an individual, their target is how through platforms such as the iPhone meanings are reflected and made around particular technologies.

In the last chapter of Part 2, Anna Kouppanou and Paul Standish engage with the ethical issues around new digital technology. In particular, they lead the reader, via phenomenology, from ethical matters towards embedded ethical considerations concerning the nature of technology itself.

3

Context, Location and Mobility: A Human Story

Matt Jones

INTRODUCTION

Many digital researchers and practitioners, particularly those in the areas of mobile and ubiquitous computing, argue that context is a key design notion. This chapter is going to review what context is, why it has been so seductive a topic for researchers and the ways they have tried to exploit or accommodate it. We will see that many attempts have been made to provide systems with automatic ways of sensing and responding to context. The difficulties in such machine-orientated approaches lead us to consider the human-driven alternatives.

WHY CONTEXT?

First, though, let's consider why as a researcher you might be drawn to this topic in the first place. There are at least two main reasons for an emphasis on context: artefacts that do not show an empathy with context are likely to fail; and people seem to be context beings, constructing their everyday by adapting to and transforming the situations they encounter.

To illustrate these reasons to address context, let's consider some examples. First, then, how can context shape the form and features of an artefact? Good design clearly responds to context: the clock in Grand Central Station in New York is large and high up to enable rushing commuters to see the time at a glance, while the fashionable, breo watch is a minimalist, tiny timepiece designed for the 'active lifestyle'. Architects may also choose to fit in with a context – building, for example, a 'green' house that nestles into a hillside, the walls and roof dressed in grass and shrubs – but 'fitting in' does not always imply camouflage or polite compliance. Ghery's Guggennheim Museum in Bilbao, then, appears to tumble into the city side streets, rolling down from the mountains, disrupting the conventional city views.

Overlooking context at this basic level leads to failure. When smartphones with video calling capability first appeared, one manufacturer ran a print advert showing a father away on business; while many miles apart, he is able to read his child a bedtime story. Flipping open his videophone he can see them; they can see him and all is well. Or not. It is easy to find issue with this fiction – who will buy their four-year-old a smartphone? If one does, how would you avoid problems such as the child dropping it into their bedtime milk? How to synchronize the book the father is reading and the one the child holds (yes, two copies of the book are needed in the marketing story)? On top of all this, the vision of connecting father and child overlooks the point of the bedtime story ritual – it is as much as about the cuddle as the content. This book-reading design has been designed, then, out of context.

Like good designs, humans too are wonderfully adept at reacting to context. Think about impressive examples of humans acting in context-sensitive ways. Is it a surfer, feeling the turn of the wave under feet? Is it the comedian, riffing off an animated heckler, the performance growing and being negotiated as the banter rolls back and forth? Perhaps, it is supermarket checkout queues. Watch how people approach the paying area, assessing quickly the best line to join, taking in the type of customers, the number and sorts of purchase in their fellow shoppers' trolleys. Consider the rapid darts from one line to another as they reassess their chances or a new till opens up.

Now think about the best computer system you have used that is context-orientated. Was it a location-based service that found you a good restaurant based on your current position and knowledge of who you were with (Zeng et al. 2009)? Was it a music player that changed the audio as your running tempo increased (Marshall and Benford 2011)?

What about Facebook or Google+? People provide status updates that give others rich context information to respond to, play with or elaborate: a teasing update can elicit genuine follow-up comments asking for more, along with sarcastic ripostes in equal measure. In serious situations, the flow of information and comments can provide both solace and assistance. When people were stuck in the USA waiting for a flight home to the UK after the volcano eruption in April 2010, updates provided the latest news about atmospheric conditions along with morale-raising stories of successful escapes. Social media platforms are not context-aware systems in the conventional research use of the term but these simple frameworks thrive on users' ability to give and interpret diverse, subtle and appropriate cues about what they are doing, thinking and feeling to others.

From the surfer to the tweeter, humans demonstrate virtuoso skills in coping with and shaping context. The way people work with their circumstances to get jobs done and how these abilities should shape thinking about human–computer interaction and system design was first documented by Suchman in *Plans and Situated Actions* (Suchman 1987). At the time she did her PhD thesis work – work that led to the book – human behaviour was seen as something that could be formalized as a series of scripts. In this world, humans, like computers, follow programs, admittedly sophisticated ones. Suchman provided strong evidence of how actions and situations are woven together, with people adapting dynamically rather than operating robotically.

MACHINES THAT PROCESS CONTEXT

A desire to give machines intelligence comparable to humans spurred on so-called strong approaches to computer Artificial Intelligence. In a similar fashion, much has been done in an attempt to provide computing devices and services with a similar ability to cope or exploit their situatedness, beginning in 1994 with Schilit and Theimer's introduction of the notion of context-awareness (Schilit and Theimer 1994). Space, place – or just plain old location – time, temperature, heart rate and gait; all have been explored as cues to change a display, prompt an input or alter a process. We can call these attempts 'strong context adaptation' or machine orientated context approaches. The goal is to design in a way that does not simply respond to the broad setting – like a well-designed clock or parent–child reading device must – but can dynamically adapt.

In mobile devices and services, a focus for context research, such adaption is seen as a way to make up for the impoverished interface (smaller screens, fiddly input, the limited attention a user can pay to the device) and to tailor output to better suit the situation. Take the first case, then, and the increasingly important task of mobile search. Kamvar and Baluja (2007) have studied how to suggest query completions based on location (in San Francisco while typing 'Al'? Suggest 'Alcatraz'), time of day and carrier. As an example of the second case, consider Yamabe et al. (2008) and a framework to switch the output of a mobile system from being primarily screen-based, to audio and then finally haptic depending on the level of user activity sensed (are they walking, running or cycling?).

There are benefits in enabling a system to proactively adapt its state in some limited way such as seen in these examples. However, in even simple cases, and more so with more complex uses of context, there are potential pitfalls to seeing context as primarily an input to the device that then responds, taking control from the user.

PUTTING HUMANS BACK IN CONTROL

In this chapter we will look first at some of the difficulties that can occur with machine-oriented approaches. While these might trigger some researchers to address these technical challenges, perhaps the bigger problem is trying to do context in this way at all. How the problem is framed is important, and we go on to review Dourish's inspiring encouragement for re-seeing the issue of context in computing systems (Dourish 2004). Using his arguments as a springboard, we will consider the human-orientated framing and use of context in computer systems, illustrating our points with reference to place-based mobile systems.

Figure 3.1   What's inside makes it yours.

Several years ago, I bought a charming Iittala box in Finland (see Figure 3.1). With its fine-grained wood lower half and thick, clear blue top it is a stunning object; solidly simple in its appeal. The genius of the designer, though, is seen more in the small, silver piece of paper resting inside. On it, in gentle print is the phrase: 'What's inside makes it yours'. The message is clear – however wonderful the box is, its meaning and value come from what you put inside it; it is defined by its contents. Humans are context-sensing, shaping, sharing beings. Every situation we enter, we bring meaning to and try to make sense of. Just as with the box, a design approach to context without humans at the centre is a less meaningful one.

THE TROUBLE WITH MACHINE-ORIENTATED CONTEXT COMPUTING

Building successful context-aware or adaptive systems is undeniably hard. Take the simple sense–respond task of pausing music playback from a music player when headphones are removed. When I bought my first device with this feature, initial positive responses were soon tempered. Going for a jog, the device tucked into running shorts, the music paused after every other stride due to a too sensitive sensor.

The difficulties are threefold. First, sensing 'the context' accurately is difficult, even for relatively simple features such as location. Then, knowing what the sensed data means is non-trivial – does the fact that someone has stopped for two minutes indicate they are lost and need more navigational help or simply that they are enjoying an espresso at a stand-up pasticceria en route and would not welcome an interruption. Third, how to execute the adaption is not always straightforward – I once drove a car that, after its advice was ignored several times, promptly announced, 'OK you are now on your own', and switched itself off.

Researchers have worked on meeting each of these three challenges. Indoor positioning for mobile devices, for instance, is currently a very active area, with the goal of enabling these gadgets to know where one is while, for instance, navigating round a shopping mall. Some of these approaches are fully automatic (Belloni et al. 2009); others employ tags in the environment – such as QR codes (Taher and Cheverst 2011), RFID chips or other local standalone beacons (Ramirez et al. 2009) – that are scanned by a user to allow the mobile to orientate.

With the increasing availability of big data sets, machine learning algorithms are being deployed to provide better predictions of appropriate adaptations. Consider then the Google query completion–suggestion interface mentioned earlier; the system fuses knowledge about a location with models of query entry to give an n-best list of potential terms. Second guessing the user in even this constrained task, though, is difficult. As Kamvar and Baluja, note, outputs that are unpredictable for the user might confuse: 'Is it confusing if the letter "s" triggers the completion "sushi" when the user [is] in San Francisco, but triggers the word "soup" when that user visits North Dakota?' (2007).

Finally, in terms of effecting the adaption, negotiation or mediation between the system and user has been advocated; that is, interfaces should show the user what it thinks is a useful response but give them the option to override or choose between alternatives (Dey and Mankoff 2005).

Besides the difficulties of accommodating real-world messiness, context-aware systems can also fall down by overemphasizing one factor over others. I travel a fair deal and my laptop goes with me. This device holds giga-bytes of data about me – my e-mail, my talks and articles, tagged photos of my friends and family. Despite this potential intimate knowledge, when I connect to the Web overseas and use a major search engine, routinely, the interface and results are rendered in the local language. The system knows I am in a new country and assumes I can now speak the local language. This sort of interaction design tunnel vision was also seen dramatically when Facebook launched its Beacon system several years ago (Jamal and Cole 2009). This platform automatically updated a user's status to show products and services bought on partner websites. A cartoon captured the problems of such a 'useful' context-sharing system: a wife – having seen her husband's automatic updates – confronts him at Christmas, 'Wow!!' she says, 'Golden Earrings! Thank you Darling! That's so cute! And who has got the other pair you've bought?'[1]

THE BIGGER PROBLEM

The machine-orientated context-aware approaches – in the way characterized here – embody a design stance that is about assisting, pre-empting, nudging or shaping the user; providing answers rather than questions. The intentions are good: to save the users time; to provide them with better solutions, and so on. However, the viewpoint also undermines the users' agency and creativity. It can shut down a person's conversation with the world around them.

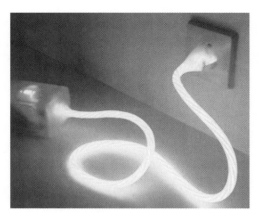

Figure 3.2   Energy monitors: the power-aware cord, developed by the Interactive Institute (© Interactive Institute).

To see how this framing can reduce rather than enhance human interaction, consider two types of home energy monitoring systems. The first, a conventional sort, senses energy consumption and presents this contextual information in terms of KwH and money (pounds and pence). Its precision and quantitative focus does not provide a rich soil for conversations about consumption and energy. In contrast, the power-aware cable proposed by Gustafsson and Gyllenswärd (2005) and shown in Figure 3.2 is designed to stimulate reflection on the nature of power as a finite and important resource. As energy is used, the lights in the cord pulse and flow, with a greater speed and intensity depending on the energy being drawn. Energy is not represented in abstract numerical terms, rather the observer is drawn to see it in the light of that most precious of resources, water.

REPRESENTATIONAL VERSUS INTERACTIONAL FRAMINGS

Dourish explores how a designer's framing of context can affect the sort of system they build, determining whether the priority is given to the machine or human. Two characterizations are considered – representational and interactional (Dourish 2004).

The representational framing is the one we've seen earlier – the problem is one of capturing as much as possible about 'the context' in precise ways, seeing interaction as then taking place within this meta-data wrapper. Many factors could be relevant, and all of these should be parameterized if possible; researchers with this perspective have gone to lengths to define the sorts of feature worthy of digitizing (Abowd et al. 1999). Meanwhile, from an interactional perspective, context is defined, or as Dourish puts it, *arises* from activity, changing dynamically as the action unfolds rather than being constant.

Think, then, of a lecture you have attended, you are caught up in the lecturer's narrative, enjoying the argument, the performance draws elements of both the external world – as she points out the twittering birds on the trees outside – and your internal one, as she leads you to reflect on an emotional memory. These – for that moment – make up the context that is being woven collaboratively by the speaker, the technology used in the presentation and yourself. Just at that moment, she stops and asks you to listen carefully. You hear the fan of the data projector. It was there all the time, now though it is part of the context, drawn in. The representational approach to that talk might, instead, attempt to monitor and encode the light and sound levels of the room, the location of each person relative to each other, the heartbeats of participants and the like.

From a representational perspective, in a photo tagging system, we'd want to tag a digital photo with data that allows the picture to be fully understood – the date, time, GPS coordinates, every face named, lighting conditions. From an interactional perspective, the photo could also be tagged so it can be understood; the difference being that a single descriptor (say, 'another glorious day', attached to a shot of a rider biking through a rainstorm gale) might be seen as sufficient, communicating more than all the other data ever could. The representational point of view helps a machine to get the picture; the interactional brings human meaning.

CONTEXT BEINGS

From a machine-orientated context perspective, then, the complexity of human life is a huge challenge. Re-seeing context as a phenomenon created and shaped by human activity turns people from a resource to be managed into one that affords great opportunities for rich digital–physical interactions.

In digital design, context is often characterized as an event-based feature. That is, a user is doing something and the system takes account of the context to help. People, as context beings, though, operate over the longer term, drawing context into their preparation for an event, during the activity itself, and, later, when the activity is over.

Consider, then, the planning I might do when preparing to travel to give a talk. Will I wear shorts or a business suit for the presentation? What are the background and expectations of the audience? Perhaps I should pack both sunglasses and a raincoat as the forecast is mixed. These are just some of the questions I might pose in imagining the contexts I will be within. The choices made for some of them will also certainly end up in altering the situations – a lecture in shorts versus in a suit, for instance. People obviously already use a range of Web content to assess places or situations they are planning to find themselves in – from weather forecasts to detailed map views. What additional future context-awareness services could be envisaged? (Brown and Chalmers 2003).

During an activity, people are able to interact with each other in sophisticated ways, picking up subtle cues and collaborating to perform effectively. Consider, then, two friends at a café table, laughing and chatting over lunch. Watch how their posture and body language mirrors each other during the meal and hear how their language – the choice of words and annunciation – accommodates to the other's style, their actions helping to smooth the flow of the conversation, making both more comfortable. Mobile technology at the moment is more likely to intrude and disrupt these sorts of flow rather than to support it (Tamminen et al. 2004).

Now, see them later, side by side in a car, still chatting, but as they approach a difficult junction and congested traffic, the passenger instinctively pauses, allowing the driver to focus their attention. Contrast this with current satnav devices that do not attend to distractions inside or outside of the car (Leshed et al. 2008).

Later that day, they are in a large noisy party, gathered in a group, so they are able to listen and engage in a conversation without being overwhelmed by all the competing voices, demonstrating the well-known 'cocktail party' effect (Moray 1959). They are able to select effectively what to focus on; everything else in this 'context' is filtered out. If we can access online real-time information about the people we are meeting, perhaps via in-ear or eye-mounted devices, can such devices be designed to allow us to filter this and the babble of the party effectively (Kurze and Roselius 2011)?

Late in the day, they join several others at a bar for an end-of-evening drink. One of the friends asks a mischievous question about the day's events, leading to a story being created, each person at the table adding their own elements and reflections. The emerging narrative is the context through which the group interacts playfully, reliving the day. Can (or indeed should) we design technology to enable people to toss in images, audio clips and the like into the mix during this sort of banter (Robinson et al. 2012)?

After an event, people often reflect on what they have experienced, integrating the different elements, filtering to find a way of expressing the past situation in ways meaningful to themselves. Online diaries, journals, blogs and social network posts are already widely used in this process (Durrant et al. 2012) and Facebook's recent timeline illustrates a more sophisticated way of bringing together the story of a person's experience over the longer term.

BEING HUMAN-CENTRED

If we turn, then, from the machine-orientated approaches, learning from what people can do to and with contexts before, during and after activities, how might we design more effectively for the phenomenon?

There are obviously lots of possibilities; here we will explore ones that are in the spirit of Dourish's design ethos that, 'ubiquitous computing [can] support the process by which context is continually manifest, defined, negotiated and shared' (Dourish, 2004).

In particular we will consider systems that:

- allow people to capture context in ways meaningful to them;
- present context engagingly, in a form capable of rich interpretations and use;
- enable people to transform their context; and,
- provide platforms for context sharing.

CONSTRUCTING CONTEXT FOR MEANING MAKING

A few years ago I spent some time working in Finland, living on a small island near Tampere. While I was there, my daily routine involved rowing to the mainland, running through a forest, and then cycling six kilometers to work. After a week or so, the journey became less of an exhausting routine and more of a ritual full of sensations. The splash of the water; the feel of the oars as they dipped and resisted the water; the bounce of the soft forest turf; the clack-clack of the raised walkway at the edge of the wood.

As I close my eyes now and think about that commute, I have a deep sense of peace; my breathing slows, and a smile crosses my face. How wonderful it would be to relive that journey in all its richness.

Technologies that help capture a patina of places, particularly places we pass through repeatedly, hold much potential. In contrast to the context-aware location systems that are becoming pervasive, putting the person back in the centre of the design in this case means not interrupting their activity by augmenting their reality. Rather, there is a need to provide lightweight tools to capture their experience for later reflection.

The Point-to-Geoblog system we developed illustrates the perspective; consider this scenario from Robinson et al. (2008):

Sam is in Singapore. Just across the road he notices some colourful, old houses, an interesting contrast to the shining newness of everything else around him. He takes his mobile out of his pocket and points at the area; he holds the phone almost vertically as the houses are so close by. Later he's downtown. Across the river he sees a statue – a cross between a lion and a mermaid. Bringing his mobile in front of him, he points, tilting it nearly horizontally as the statue is far away. When Sam returns to his hotel room, he enjoys re-tracing his journey and viewing the photos and web links associated with Arab Street and the Merlion on the automatically generated map.

Using either a very simple distance estimator allied to a GPS sensor (Figure 3.3(a)) or supplementing this with a visual display (Figure 3.3(b)), the user drops geo-tags at places they notice. Later, the system creates a map of the route along with content associated the points of interest (Figure 3.3(c)). These can be used as starting point for their own reflections.

Microsoft's SenseCam and the more recent Aided Eyes from Rekimoto Labs record much more than the simple points of the Geoblog system. SenseCam, then, captures a sequence of pictures from a gadget worn via a lanyard around the neck, conference badge-style (Hodges et al. 2006). Aided Eyes builds on the idea by combining the outward-facing camera with another that tracks the wearer's gaze, detecting what the eye fixes on (Ishiguro et al. 2010).

While showing potential to help us relive our rich experiences, such 'lifelogging' often seems a mechanistic, prosaic, capture-all process – systems to bottle the sum of our human existence. That people want to live – and relive – their lives more selectively and poetically can easily be seen in the imaginative,

(a)

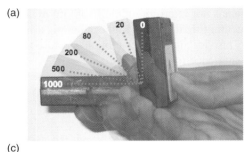

(b)

(c)

Figure 3.3   Clockwise from top left: (a) simple gestures to mark distance; (b) mobile map view for marking places of interest; (c) map generated after the journey.

inventive and witty use of status updates commonplace on social sites like Facebook. To meet such needs, researchers have recently argued for systems that are less memory substitutes and more sources of cues that trigger a person's own memories (Sellen and Whittaker 2010).

The power of a system that could capture the context of one's life and can deliver these evocative cues without becoming overwhelming is argued by Harper et al. in their study of how SenseCam images elicit responses (Harper et al. 2008). One of the examples they give shows a small girl standing up against a height measurer illustrated with a giraffe image. The authors explain that the mother remembers the day the image was taken as being one of frustration, trying to cope with the common, seemingly endless challenges of bringing up small children. The caption she writes for the image captures a sense of the stress and strain but she finishes by expressing succinctly the power of love and motherhood, something she could only feel later:

MUM, WHEN WILL I BE BIGGER?

But Mum when will I be bigger?

… When you eat all your meals …

But Mum, will I be as big as a giraffe next year?

… but Phoebe you don't want to be as big as a giraffe.

I do I want big long legs and a long neck.

… no you don't you are beautiful as you are.
(Harper et al. 2008)

As the authors put it, 'On retrieving she found that she was able to think about that moment differently, and to see it for "what it was". Gazing at the image led her into realising that the day in question was not as it seemed at the time.'

PRESENTING CONTEXT FOR APPROPRIATION

Think about two images of the night sky. One, taken by a high-definition camera shows a dark canvas pinpointed with discrete eyes of light, the stars. The other, is Van Gough's 'Starry Night': swirls of yellow and deep blue flow over the canvas, the sky alive with movement above a quiet, still town. The photograph aims at a definitive representation; the second is the artist's response that invites the viewer to bring the picture to life, taking their cue from the dynamics deftly styled on the canvas.

How can an Impressionist painting change our perspective on context-aware computing? Take the work of Håkansson et al. (2006) as an example. In this work the researchers illustrated how evocative representations can be used to inform mobile photography designs. In their prototype, images are adapted depending on sound and movement levels at the time the photo is taken. The aim is to conjure up more of the experience, the life of situation, than possible with a simple, exact image.

But, thinking more broadly, the contrast between the preciseness of the high-definition photo and an Impressionist painting can help us see the value of representations that demand a human response. If context data is seen simply as an input to a machine algorithm, precision is essential so that correct or good adaptions or information are furnished to the user. With a human-orientated perspective, though, the problems of perfectly sensing and representing contextual features can be reframed as an opportunity.

Researchers have explored the value of exposing the uncertainty in GPS and WiFi data, for instance. Benford et al. (2006), have shown how presenting good and poor GPS coverage affects the dynamics of a pervasive game, with players moving to areas of poor coverage to avoid detection. Visualizations of WiFi quality were seen to affect behaviour in a similar way in the Feeding Yoshi Game (Bell et al. 2006). The two gaming systems show how one can design for and reveal the workings of a context-sensing system, gaps and all, laying out the data the system has to work with.

This transparency in representation is in contrast to the ambiguity suggested by Gaver et al. (2003). In considering 'Ambiguity as a Design Resource', the authors present a series

of installations which use sensors to generate outputs which are intriguing, open to multiple interpretations. One, then, the Home Health Monitor is a design that senses activity in a home, measuring, amongst other factors, '… the stroke rate of a hairbrush, condensation on kitchen windows, or the state of the toilet seat', to generate a daily horoscope such as:

> Remind yourself that although you must fulfil routine obligations, you also need time to enjoy more romantic affairs. If you're too overwhelmed by duty, boredom sets in. The last thing you want is to make someone think it is their fault. No one is to blame. So make space for love; it won't work in a vacuum.

Imagine the household user of such a context-based system: what did I do today to generate this output? How can it know I am feeling so empty and bored? Did I really blame my friend? The awareness in this context-aware system is all about prompting the users to become aware of their life and situation.

Consider Web-based mapping services such as Google Maps and Bing Maps. These can show all of the digital content – photos, Wikipedia articles, videos, traffic flows, etc. – that is available about a place. It is about giving the users as much information as possible, providing them with the answers. If our context-sensing systems chart all our territories, do we lose the joy of exploration, the desire to seek for hidden gems?

Instead of presenting answers, location-aware systems can present questions and alternatives, allowing users to make sense of the presented context in their own ways. Let us turn to two of the systems my colleagues and I have built to illustrate this.

First, consider the Questions Not Answers mobile search system (Jones et al. 2007). Figure 3.4 illustrates the interface displayed on the user's handheld device. An aerial map of the location is shown overlaid with search terms that are being searched for at that location. Users can increase, decrease or pause the refresh rates. By showing real people's questions about a place, the aim is to give a sense of the changing nature of both the place

**Figure 3.4   The Questions Not Answers
mobile interface (Jones et al. 2007).**

and the people passing through it. Church and
Smyth (2008) have also explored these ideas
in their mobile social search systems.

Second, moving from the visual to the haptic domain, our team built a mobile navigation
system that presents the range of choices a
walker has when trying to get from A to B
(Robinson et al. 2010). Instead of providing
the users with a fixed, optimal route and turn-
by-turn instructions, the device simply vibrates
when they point it in a viable direction. If there
are many paths to a destination, the device
vibrates as the users sweep the device in an arc
in front of them; if there is a limited choice the
feedback occurs only in a narrow band, direct-
ing walkers more specifically.

TRANSFORMING THE CONTEXT

Think back to the little box in Figure 3.1 mentioned at the start of this chapter: 'What you
put inside it, makes it yours'. Remember, too,
Dourish's powerful argument that context is

an interactional property – it is dynamically
formed and changes through human activity
(Dourish 2004). Every situation we enter, we
change; in this section, let us explore some
examples of more dramatic interventions and
how they might inspire extravagant rather
than subtle digital context-changing systems.

Just outside of Waterloo railway station,
one of London's busiest, there's a concrete
tangle of underpasses that sprawl out under
some of the capital's most traffic-heavy roads.
The dismalness of this place is ameliorated by
a simple textual device. At the tunnel's
entrance, etched on a pillar, a poem begins:

> I dream of a green garden
> where the sun feathers my face
> Like your once eager kiss.

As the tunnels unfold, more of the poem is
revealed, and the place is changed. Holding
up an augmented-reality smartphone in that
location and seeing there's a Starbucks half a
mile away, isn't quite the same. How can we
enable people to use technology to change
the way they and those around them see the
locations they move through?

From text to music. Retailers have long
known the power of in-store music to inspire
or prompt purchases. For a powerful illustration of the technique in action you only have
to visit a branch of the youth-focused brand
Abercrombie & Fitch: the deep beat of the
music is designed to enhance the visceral
value of their wares. People also use private
music every day to complement or alter their
mood as they briskly walk their daily commute, amble as a tourist or power round a
route during a lunchtime run.

Oleksik et al. (2008) explore how sonic
interventions can be used at home to change
the experience of family living. One of these, the
sonic window, lets sound into a room when
the windowpane is raised. This tangible user
interface opens on to a virtual world, however,
bringing sounds from far away – the weather in
another geography or the babble from another
friend's or family member's home, for instance.

What about rich, visual transformations?
Many a street artist makes a good summer

living from sketching trompe l'oeil in chalk that change a pavement into, say, a breathtaking waterfall. Performance groups like NuFormer have been applying digital techniques to similar effect. Using high-definition laser mapping and projection systems, they have produced incredibly compelling animations that play with the architecture and surfaces of buildings as diverse as storefronts and medieval castles. As pico-projectors become more common, built into mobile phones, can we develop services that attempt similar adaptations, on a smaller scale (Dachselt et al. 2012)?

Bongers (2012) illustrates what might be possible in his Video Walks project. Over several years, he has experimented with a handheld projector – much brighter and larger than pico projectors but still portable – driven and powered by gadgetgery in a backpack. He has turned flat surfaces such as lawns and tarmac roads into tunnels into the earth; a bare wall into a tree replete with a parrot, its noise also intruding into the environment from the backpack speakers; and, a lake into a diving pool, the projection bringing the still surface to life with video footage of a diver entering into the water.

Utility-focused uses of context have a great deal of value; sometimes it is really helpful to know where the nearest coffee shop is given your current location and mode of transport. The purpose of these few examples, however, is to help us think about context in a more performative, extravagant way. Sometimes, perhaps more often than we imagine, people move through spaces open to affordances in the environment – be they a poem carved on a wall or inspiring music beating out from a shop door – that let them leap from mundane disengagement to switched on interaction.

SHARING CONTEXT

Some mobile research aims at hiding away a person's interaction with a system, preserving their privacy while enabling them to control function or information access. Take, for instance, the work of Costanza et al. (2007). In their intimate interfaces approach, sensors that can detect muscle contractions are placed

on a user's biceps. Shirtsleeves pulled down over these inputs, the wearers can simply flex their muscles to control, say, the music they are listening to. In this future, clothes are used to cover over interactions; in the past, hats, walking canes, kerchiefs and the like were used as props to play out meeting and greeting protocols, bonding people together (Candy, 2007).

If subtle approaches are used for location context-aware services, it is possible to give a user relevant place-based information without anyone around them being aware. Contrast this with the cues people give each other all the time. Has your head ever turned to see what another is looking at as you pass them? Can you resist allowing your gaze to follow that of the crowd whose fingers pointed towards something happening around you?

Pointing and gesturing in public as an interaction method is not yet widely evident and careful design is needed to promote acceptability (Rico and Brewster, 2010). However, just as talking loudly in the street is no longer simply a sign of eccentricity but the side-effect of hands-free mobile handsets, over time we can imagine people routinely pointing, gesturing and waving to perhaps push or grab digital content around them or to operate a device, capturing a picture simply by framing the picture with their hands (Mistry and Maes, 2009).

We developed Sweep-Shake to enable people to uncover and filter content (see Figure 3.5). Such point-to-discover (Fröhlich et al., 2007)

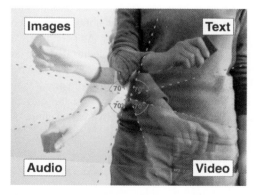

Figure 3.5 The Sweep-Shake system uses sweeping gestures to first locate content in a location and then – as shown – to filter the available resources.

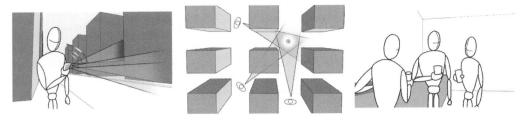

Figure 3.6   Using haptic feedback to share location information and to move towards a negotiated, dynamic meet-up (Williamson et al. 2010).

approaches may not only provide users with a heads-up style of interaction, allowing them to be aware of their context, but may also facilitate the sharing with those around them of what they see as important in the location.

In the Social Gravity prototype we turned from allowing co-located people to shape each other's understanding of context to providing a platform to do this for dispersed groups (Williamson et al. 2010). Figure 3.6 illustrates the approach.

A group of friends wants to meet up but they are in different parts of a city. It would be possible to show the location of all the people on a mobile map so that a mutual meeting point might be agreed by texting or calling each other. However, we wanted to both preserve the privacy of the individuals and provide a dynamic, lightweight way for the group to negotiate their eventual rendezvous.

So, the system automatically computes a central location and each friend can orientate themselves to it by pointing their mobile and receiving a buzzing, haptic notification when they are facing the right way. As people move towards the location, the central point is dynamically updated. If someone wishes to nudge the group to another meet-up point, they can deliberately walk away from the current centre, pulling the others with them, eventually.

that all successful design takes account of the ecology it is occurring within. Then, we moved on to explore the motivations for and ways of employing dynamic notions of context to enhance digital experiences.

A major consideration is the degree of agency given to the machine over the human. Machine-orientated context systems attempt to proactively adapt given their processing of the sensed situation, in order to better suit the environment in terms of what or how they communicate to or with the user.

Equipping mobile and ubicomp systems with these sorts of ability has been seen as a goal by researchers keen to support mobility. While this is important, we emphasized an alternative perspective, inspired by the work of Dourish, that places humans at the centre of designing for context. People are context beings, with sophisticated abilities to make sense, respond to and alter the situations they find themselves living through. Digital systems that are stages for these innate talents rather than being scripts to direct behaviour have, we argue, a more natural fit. In discussing the issues we have seen examples of systems that help people become aware of their context; empower them to make meaning of it later; allow them to shape it for themselves; and, to provide ways of sharing their interpretations with others.

CONCLUSION

In this chapter we have reviewed the role of context as a key resource in the design of digital artefacts. We began with a reminder

ACKNOWLEDGEMENTS

Parts of this article previously appeared or were adapted from: Matt Jones (2011) 'Journeying toward extravagant, expressive, place-based computing', interactions 18(1) (January): 26–31. DOI=10.1145/1897239. 1897247, available at: http://doi.acm.org/ 10.1145/1897239.1897247 © ACM Press. Some of the ideas were also developed during a series of invited talks between 2008 and 2011. I am indebted to my research colleagues who have designed, implemented and tested the systems discussed. In particular, thanks to Simon Robinson for his inspiring work. Some of the prototypes developed were funded by the EPSRC and Microsoft Research UK.

NOTES

1 http://geekandpoke.typepad.com/geekand-poke/2007/12/that-was-2007-3.html

REFERENCES

Abowd, Gregory D., Anind K. Dey, Peter J. Brown, Nigel Davies, Mark Smith and Pete Steggles (1999) 'Towards a better understanding of context and context-awareness', in Hans-Werner Gellersen (ed.), *Proceedings of the 1st International Symposium on Handheld and Ubiquitous Computing* (HUC '99). London: Springer-Verlag, pp. 304–307.

Bell, Marek, Matthew Chalmers, Louise Barkhuus, Malcolm Hall, Scott Sherwood, Paul Tennent, Barry Brown, Duncan Rowland, Steve Benford, Mauricio Capra and Alastair Hampshire (2006) 'Interweaving mobile games with everyday life', in Rebecca Grinter, Thomas Rodden, Paul Aoki, Ed Cutrell, Robin Jeffries and Gary Olson (eds), *Proceedings of the SIGCHI Conference on Human Factors in Computing Systems* (CHI '06). ACM, 417–426.

Belloni, F., V. Ranki, A. Kainulainen and A.Richter (2009) 'Angle-based indoor positioning system for open indoor environments', *Positioning, Navigation and Communication*. WPNC, pp. 261–265

Benford, Steve, Andy Crabtree, Martin Flintham, Adam Drozd, Rob Anastasi, Mark Paxton, Nick Tandavanitj, Matt Adams and Ju Row-Farr (2006) 'Can you see me now?', *ACM Transactions on Computer–Human Interaction*, 13(1) (March): 100–133.

Bongers, Bert (2012) 'The projector as instrument', *Personal Ubiquitous Computing*, 16(1) (January): 65–75. Available at: http://dx.doi.org/10.1007/s00779-011-0378-0

Brown, Barry and Matthew Chalmers (2003) 'Tourism and mobile technology', in Kari Kuutti, Eija Helena Karsten, Geraldine Fitzpatrick, Paul Dourish and Kjeld Schmidt (eds), *Proceedings of the Eighth European Conference on Computer Supported Cooperative Work* (ECSCW '03). Kluwer, 335–354.

Candy, F.J. (2007) 'Come on momma, let's see the drummer': Movement-based interaction and the performance of personal style', *Personal and Ubiquitous Computing*, 11(8) (December): 647–655.

Church, Karen and Barry Smyth (2008) 'Who, what, where and when: A new approach to mobile search', *Proceedings of the 13th International Conference on Intelligent User Interfaces* (IUI '08), ACM, 309–312.

Costanza, E., S.A. Inverso, R. Allen and P. Maes (2007) 'Intimate interfaces in action: Assessing the usability and subtlety of emg-based motionless gestures', *Proceedings of the SIGCHI Conference on Human Factors in Computing Systems. (CHI '07)*, ACM, 819–828.

Dachselt, Raimund, Jonna Häkkilä, Matt Jones, Markus Löchtefeld, Michael Rohs and Enrico Rukzio (2012) 'Pico projectors: Firefly or bright future?', *Interactions*, 19(2) (March): 24–29.

Dey, Anind K. and Jennifer Mankoff (2005) 'Designing mediation for context-aware applications', *Transactions on Computer-Human Interaction*, 12(1) (March): 53–80.

Dourish, Paul (2004) 'What we talk about when we talk about context', *Personal and Ubiquitous Computing* 8(1) (February): 19–30.

Durrant, Abigail, David S. Kirk, Steve Benford and Tom Rodden (2012) 'Pursuing leisure: Reflections on theme park visiting', *Computer Supported Cooperative Work*, 21(1) (February): 43–79.

Fröhlich, P., R. Simon, E. Muss, A. Stepan, A. and P. Reichl (2007) 'Envisioning future mobile spatial applications', *Proceedings of the 21st British HCI Group Annual Conference on people and computers: HCI … but not as we knowit.* (vol. 2), 35–38.

Gaver, W.W., J. Beaver and S. Benford (2003) 'Ambiguity as a resource for design', *Proceedings of the SIGCHI Conference on Human Factors in Computing Systems. (CHI '03)*, ACM, 233–240.

Gustafsson, Anton and Magnus Gyllenswärd (2005) 'The power-aware cord: Energy awareness through ambient information display', *Extended Abstracts on Human Factors in Computing Systems* (CHI EA '05), ACM, 1423–1426.

Håkansson, M., L. Gaye, S. Ljungblad and L.E. Holmquist (2006) 'More than meets the eye: An exploratory study of context photography', in A. Mørch,

K. Morgan, T. Bratteteig, G. Ghosh and D. Svanaes (eds), *Proceedings of the 4th Nordic Conference on Human-Computer Interaction: Changing Roles.* (NordiCHI '06, Vol. 189), ACM, 262–271.

Harper, R., D. Randall, N. Smyth, C. Evans, L. Heledd and R. Moore. (2008) 'The past is a different place: they do things differently there', *Proceedings of the 7th ACM Conference on Designing Interactive Systems* (DIS '08), ACM, 271–280.

Hodges, S., L. Williams, E. Berry, S. Izadi, J. Srinivasan, A. Butler, G. Smyth, N. Kapur and K. Wood (2006) 'SenseCam: A retrospective memory aid', in P. Dourish and A. Friday (eds), *Ubiquitous Computing: 8th International Conference, UBICOM 2006,* Springer-Verlag, 177–193

Ishiguro, Yoshio, Adiyan Mujibiya, Takashi Miyaki and Jun Rekimoto (2010) 'Aided eyes: Eye activity sensing for daily life', *Proceedings of the 1st Augmented Human International Conference* (AH '10), ACM, Article 25.

Jamal, Arshad and Melissa Cole (2009) 'A heuristic evaluation of the Facebook's Advertising Tool Beacon', *Proceedings of the 2009 First IEEE International Conference on Information Science and Engineering* (ICISE '09), 1527–1530.

Jones, Matt (2011) 'Journeying toward extravagant, expressive, place-based computing', *Interactions*, 18(1) (January): 26–31.

Jones, Matt, George Buchanan, Richard Harper and Pierre-Louis Xech (2007) '*Questions* not *answers*: A novel mobile search technique', *Proceedings of the SIGCHI Conference on Human Factors in Computing Systems* (CHI '07), ACM, 155–158.

Kamvar, Maryam and Shumeet Baluja (2007) 'The role of context in query input: Using contextual signals to complete queries on mobile devices', *Proceedings of the 9th International Conference on Human Computer Interaction with Mobile Devices and Services* (MobileHCI '07), ACM, 405–412.

Kurze, Martin and Axel Roselius (2011) 'Smart glasses linking real live and social network's contacts by face recognition', *Proceedings of the 2nd Augmented Human International Conference* (AH '11), ACM, Article 31.

Leshed, Gilly, Theresa Velden, Oya Rieger, Blazej Kot and Phoebe Sengers (2008) 'In-car GPS navigation: Engagement with and disengagement from the environment', *Proceedings of the Twenty-sixth Annual SIGCHI Conference on Human Factors in Computing Systems* (CHI '08), ACM, 1675–1684.

Marshall, Joe and Steve Benford (2011) 'Using fast interaction to create intense experiences', *Proceedings of the 2011 Annual Conference on Human Factors in Computing Systems* (CHI '11), ACM, 1255–1264.

Mistry, Pranav and Pattie Maes (2009) 'SixthSense: A wearable gestural interface', *ACM SIGGRAPH ASIA 2009 Sketches* (SIGGRAPH ASIA '09), ACM, Article 11.

Moray, N. (1959) 'Attention in dichotic listening: Affective cues and the influence of instructions', *Quarterly Journal of Experimental Psychology*, 11: 56–60.

Oleksik, Gerard, David Frohlich, Lorna M. Brown and Abigail Sellen (2008) 'Sonic interventions: Understanding and extending the domestic soundscape', *Proceedings of the Twenty-sixth Annual SIGCHI Conference on Human Factors in Computing Systems* (CHI '08), ACM, 1419–1428.

Rico, Julie and Stephen Brewster (2010) 'Usable gestures for mobile interfaces: Evaluating social acceptability', *Proceedings of the 28th International Conference on Human Factors in Computing Systems* (CHI '10), ACM, 887–896.

Robinson, Simon, Parisa Eslambolchilar and Matt Jones (2008) 'Point-to-GeoBlog: Gestures and sensors to support user generated content creation', *Proceedings of the 10th International Conference on Human-Computer Interaction with Mobile Devices and Services* (MobileHCI '08), ACM, 197–206.

Robinson, Simon, Parisa Eslambolchilar and Matt Jones (2009) 'Sweep-Shake: Finding digital resources in physical environments', *Proceedings of the 11th International Conference on Human-Computer Interaction with Mobile Devices and Services* (MobileHCI '09), ACM, Article 12.

Robinson, Simon, Matt Jones, Parisa Eslambolchilar, Roderick Murray-Smith and Mads Lindborg (2010) '"I did it my way": Moving away from the tyranny of turn-by-turn pedestrian navigation', *Proceedings of the 12th International Conference on Human-Computer Interaction with Mobile Devices and Services* (MobileHCI '10), ACM, 341–344.

Robinson, Simon, Matt Jones, Elina Vartiainen and Gary Marsden (2012) 'PicoTales: Collaborative authoring of animated stories using handheld projectors', *Proceedings of the ACM 2012 Conference on Computer Supported Cooperative Work* (CSCW '12), ACM, 671–680.

Schilit, B. and M. Theimer (1994) 'Disseminating active map information to mobile hosts', *IEEE Network*, 8(5): 22–32.

Sellen, Abigail J. and Steve Whittaker (2010) 'Beyond total capture: A constructive critique of lifelogging', *Communications of the ACM*, 53(5) (May): 70–77.

Suchman, Lucy A. (1987) *Plans and Situated Actions: The Problem of Human-Machine Communication.* New York: Cambridge University Press.

Taher, Faisal and Keith Cheverst (2011) 'Exploring user preferences for indoor navigation support through a combination of mobile and fixed displays', *Proceedings of the 13th International Conference on Human-Computer Interaction with Mobile Devices and Services* (MobileHCI '11), ACM, 201–210.

Tamminen, Sakari, Antti Oulasvirta, Kalle Toiskallio and Anu Kankainen (2004) 'Understanding mobile contexts', *Personal and Ubiquitous Computing*, 8(2) (May): 135–143.

Williamson, John, Simon Robinson, Craig Stewart, Roderick Murray-Smith, Matt Jones and Stephen Brewster (2010) 'Social gravity: A virtual elastic tether for casual, privacy-preserving pedestrian rendezvous', *Proceedings of the 28th International Conference on Human Factors in Computing Systems* (CHI '10), ACM, 1485–1494.

Yamabe, T., K. Takahashi and T. Nakajima (2008) 'Towards mobility oriented interaction design: Experiments in pedestrian navigation on mobile devices', *Proceedings of the 5th Annual International Conference on Mobile and Ubiquitous Systems: Computing, Networking, and Services*, ICST, 1–10.

Zheng, Yu, Lizhu Zhang, Xing Xie and Wei-Ying Ma (2009) 'Mining interesting locations and travel sequences from GPS trajectories', *Proceedings of the 18th International Conference on World Wide Web* (WWW '09), ACM, 791–800.

4

Online Information: Access, Search and Exchange

Gary Hsieh and Nicolas Friederici

INTRODUCTION

The amount of information in the Digital Universe was estimated to have arrived at 1.2 zettabytes, or 1,200,000,000,000,000 gigabytes, in 2010 (Gantz and Reinsel 2010). To put this into perspective, the complete works of Shakespeare take up only 0.5 gigabytes of storage. A pickup truck filled with books needs 1 gigabyte of storage. Even the whole print collection at the US Library of Congress, the largest library in the world, can be stored with just 10,000 gigabytes (Williams n.d.). Contrasting these numbers, it is clear that the amount of information available online is larger than in any information repository ever known to humankind.

In this chapter, we will explore how people utilize the Internet to meet their various information needs. First, we will examine how the access to information has changed with the rise of the Internet. Second, we will discuss how people search through this sea of information to find what they actually need. Finally, we will point out why and how, despite the vast amount of archived information accessible and improved search tools, people still turn to other people for information online.

ACCESS TO INFORMATION ONLINE

In the early 1990s, when the amount of information on the Internet was starting to skyrocket, many envisioned that the 'Information Superhighway' would revolutionize and change almost all aspects of society, advancing our education systems and access to information (Gore 1991). Some have likened the Internet to the invention of the printing press, dispersing information at a faster and cheaper rate than ever before (Goldsmith, as quoted in Katz and Rice, 2002: 2). Now, decades after the rise of the Internet, has our access to information improved?

Digital Divide

First of all, we must acknowledge that *not* everyone has access to online information.

As of March 2011, global Internet penetration has reached only around 30 per cent (www.internetworldstats.com/stats.htm). This has been a cause for concern as the so-called digital divide can lead to an opportunity divide – if certain social groups are unable to access the wealth of information and valuable services online, they may be at a great disadvantage. This is especially problematic if those who do have access to the Internet are already 'better off', and the digital divide may further widen the gap.

Research on the digital divide started by tracking and documenting demographic disparities in online access across age groups, gender, race, education, disabilities, global regions, etc. Initial findings were fairly consistent and showed that younger (18–24), male, employed, more highly educated (determined by the highest degree obtained), and affluent people in developed countries are more likely to be online (for more details, see Norris 2001). In addition, research has confirmed that, indeed, negative repercussions to the digital divide exist. For example, Internet use is associated with wages (DiMaggio and Bonikowski 2008), school test scores (Attewell and Battle 1999), social capital (Best and Krueger 2006), knowledge of current events (Tewksbury et al. 2001), and civic engagement (Haase et al. 2002).

More recent data suggests a somewhat positive trend and that the digital divide is narrowing. For example, the initial digital divide between women and men (Kennedy et al. 2003; Rice and Katz 2003) has become non-existent in many countries, including the US (Smith 2010a). Other examples are race and age: in recent years the gap in broadband adoption between Whites and Blacks has decreased (Smith 2010c) and the proportion of people born before 1982 compared to those who are part of the 'Net Generation' (born in or after 1982; Howe and Strauss, 2000) has increased as well, for example, in their usage of certain online applications such as travel reservations, (Jones and Fox 2009).

Technology advancement is one of the factors for the narrowing gap. Most notable is the advent of smartphones with Internet capabilities, which provide inexpensive and more flexible access. Mobile devices are cheaper than computers and require much less infrastructure (no physical lines) to connect to the Internet. Lower costs have enabled some lower-income populations to leapfrog traditional technology requirements for gaining online access. In the US, Blacks and English-speaking Latinos are *more* likely to own mobile phones than Whites (Smith 2010b) and to use them to access the Internet (Brown et al. 2011). In developing regions, the earliest adopters of mobile phones also tend to be part of the poorest segments of the population (Rice and Katz 2003).

Despite these positive trends, it is important to note that the digital divide is far from being closed. In the United States, major differences in Internet adoption persist between rural and urban/suburban regions. Across the world, developed countries are more highly represented online. For example, even though 14.1 per cent of the world's population lives in Africa, only 2.6 per cent of Internet users live there (Fuchs and Horak 2008).

While we can expect technological advancements to continue to lower the barrier of Internet access, the gap cannot be closed by technology alone. Much of the problem stems from underlying structural inequalities. In particular, there are still wide gaps with regard to income, both within and between countries. Studies have shown that income is the largest contributing factor to the global digital divide (Chinn and Fairlie 2006), and, unfortunately, the wealth gap between the rich and the poor is only increasing (Brown et al. 2011).

Second-level Digital Divide

Initially, the primary focus of digital divide research was on access to technology and network/broadband connections. However, it soon became clear that studying who has access to technology equipment is insufficient – the digital divide problem is also a problem of skills. Eszter Hargittai (2002) referred to this as the second-level digital divide and defined online skills as 'the ability to efficiently and effectively find information on the Web'.

In a study conducted in 2001, Hargittai (2002) found that there is a large variance in online skills among Internet users. Education and prior experience with the technology both improve participants' likelihood to find information and reduce the time they take. Hargittai and Hinnant (2008) later specified that those who report higher Web use skills are more likely to visit human, financial and social capital-enhancing sites. Related studies have also confirmed that education is a major predictor in the amount of capital-enhancing activities performed online (e.g. sending e-mail, searching for job-related, financial, political or government information, as well as online banking; Howard et al. 2001; Madden and Rainie 2003; Pew Research 2011).

While we have only briefly summarized key research on two levels of digital divide, we should point out that there might be more. Van Dijk and Hacker (2003) have outlined four levels of barriers to access: (1) lack of elementary digital experience ('mental access'); (2) no possession of computers and network connections ('material access'); (3) lack of digital skills ('skill access'); and (4) lack of usage opportunities ('usage access'). DiMaggio, Hargittai, Celeste and Shafer (2004) suggested five broad forms of inequality beyond equipment and skills: (1) *technical means* for Internet access; (2) *autonomy* in Web use; (3) *skill* for medium usage; (4) *social support* to motivate Internet use; and (5) *purposes* for Internet use. These taxonomies make evident that the digital divide is a complex issue and there are many levels and layers of information access which have yet to be examined more closely.

Changing Information Behaviors?

For those of us who have access, it is undeniable that the Internet has a big impact on our lives. Similarly to earlier information and communication technologies, the Internet has increased the volume of accessible information,

reduced the cost of access, and improved the speed at which we get information. The question is, has the Internet fundamentally changed how we approach information?

First, the Internet has changed where we get information. Many have observed that Internet usage comes at the expense of television viewing (e.g. Nie and Erbring 2000). For example, the Net Generation, compared to other generations, is more likely to log on to a news website for the latest information than to turn on the television (Windham 2005). The Internet has become the primary source for academic (Levin and Arafeh 2003; Weiler 2005), personal (Williams et al. 2008), and social information (Hoffman et al. 2004).

Early on, the immersion in technology was seen to make Internet users '*think and process information fundamentally differently from their predecessors*' (Prensky 2001a: 1, emphasis in the original). That is true. Studies have shown that the speed of online information has had a fundamental effect on information behaviors. Internet users, and the Net Generation in particular, prefer quick information and spend less time on evaluating information and comparing sites (Nicholas et al. 2011; Rowlands et al. 2008). Information seeking is often stopped when the articles are found, rather than when the articles are read.

Researchers also hypothesized that the Net Generation will be more accustomed to processing visual and dynamic information, learning at high speed, and making random connections (Prensky 2001a; 2001b). So far, researchers have not found evidence to support that the Net Generation is better at multitasking, however. In fact, a recent study actually showed that its members were less competent at multitasking and had poorer working memory (Nicholas et al. 2011). However, findings support the hypothesis that the Generation prefers visual information and interactive systems.

In summary, research suggests that where and how we access information has changed due to the Internet. For those who have access, the Internet has become the first and often the only stop for information. While it has enabled more and faster access to information, unfortunately, our information literacy has not improved. People have adapted to viewing more information more rapidly, but also more superficially. Some key challenges looking forward include how to further increase the rate of Internet adoption, improve library and information services to better match the needs of the Net Generation, and how to teach better online information skills to support capital-enhancing behaviors.

INFORMATION SEARCH – HOW PEOPLE GET TO INFORMATION

In the previous section, we discussed how our access to information has changed with the Internet. Once people have online access, a wealth of information is available. Yet, out of the 50 billion-plus sites online (www.worldwidewebsize.com), people often need to find just the few ones that are relevant to them. How do they do this? Here, we will discuss how people search for information online, focusing specifically on: (1) online search needs; (2) how searches are performed; and (3) the social aspect of searching.

In this section, we refer to online search as all active information-seeking behavior that uses online search systems. Users formulate a query and the system allows them to access indexed sites. Online search systems, for the most part, are keyword-based search engines, either querying a domain-specific catalog (e.g. Dialog, Medline, etc.) or the Web as a whole (e.g. Google, Bing, etc.).

Increasingly Diverse Online Search Needs

People do not search for the sake of searching. A search is only a means to an end; that is, satisfying a need a person has. The information search and seeking literature has traditionally only considered informational needs (Broder 2002; Case 2007). However, online, the types of search needs have become more diverse.

Broder (2002) identified three main types of online search needs – informational, navigational and transactional. An informational need is the intent to acquire *information assumed to be present* on some Web page. Many sites may contain the desired information, and the user's goal is to find the information (or links to the information), no matter on which site. In contrast, a navigational need represents the immediate intent to reach a website that the user *designates and already has in mind*. Typically, only the one desired page is the right result. Finally, a transactional need is the intent to perform a *Web-mediated activity*, where the goal is a further interaction, transmission of data or transfer of physical goods. This taxonomy has found a strong resonance in empirical work that has ensued (Chi 2009; Jansen et al. 2008; Rose and Levinson 2004). We should point out that Broder's taxonomy is quite broad. Informational needs in particular might differ along several sub-dimensions, such as complex v. simple, specific v. vague, fact- v. opinion-based, private v. public, determined v. exploratory, etc.

Performing an Online Search

Given the various search needs, how do people then proceed to actually perform their searches? Here, we synthesize key stages of search identified in several established process models, from both online and general search literature (Ellis 1989; Evans and Chi 2010; Kuhlthau 1991; Marchionini 1995; Pirolli and Card 2005), and highlight empirical findings for online searches at each step.

Once a need has been identified (as discussed in the prior section), the next step is the selection of search systems. When they can, people seem to rely on the systems they know and avoid exploration: when the task is specific enough (Kellar et al. 2007) or when the user is familiar with particular search systems already (Johnson et al. 2004), they will search using fewer systems. In contrast, Web-experienced people do not shy away from exploring, searching across more sites

(Johnson et al. 2004) and using more sophisticated queries (e.g. Aula and Nordhausen 2006; Hölscher and Strube 2000). Thatcher (2008) explains that these users have to spend less cognitive effort on one source, so that they can query several different sources in parallel. In addition, inexperienced searchers have a lower understanding of search systems, leading to less effective usage (e.g. Debowski 2001; Wang et al. 2000).

People then collect, read, examine and filter information to extract relevant material from their search results to satisfy their information needs – a process called foraging (Pirolli and Card 2005). Similarly to search system selection, less specific tasks will result in longer time and more effort spent on foraging (Kellar et al. 2007; Kim and Allen 2002). Web experience can also play a factor during foraging. More Web-experienced users in general need less time and use more complex, planned-out strategies (e.g. Aula and Nordhausen 2006; Hölscher and Strube 2000). Since the quality of online information varies greatly, careful evaluation of search results is critical. Unfortunately, research has found that people often rely on heuristic and superficial search strategies to judge the quality of information and to filter out irrelevant information (e.g. Fogg et al. 2001; Tamborello and Byrne 2005). Inexperienced Web users in particular are easily overwhelmed by the cognitive load of site evaluations and tend to use less efficient strategies (e.g. Aula and Nordhausen 2006; Hölscher and Strube 2000). Thus, one key challenge is to educate online searchers how to better assess the credibility and value of information online (see Metzger 2007 for a more detailed discussion).

Closely coupled to foraging and information evaluation, sensemaking (Russell et al. 1993) follows. This involves hypotheses management, evidentiary reasoning and decision-making. The searcher assembles the foraged information and contrasts it with prior beliefs. Here, users tend to once more apply flawed strategies: when evaluating which information is valuable and relevant, typically users are biased in favor of information that is compatible with what they knew already (Pirolli and Card 2005).

Finally, the online searcher has to decide when to stop their search process. For high-complexity tasks, users stop searching when they have developed a stable enough idea about the problem; whereas, for low-complexity tasks they stop only when they have ticked off a specified mental list or reached a concrete, self-set criterion (Browne et al. 2007). As the last step, online searchers often organize and schematize the search product (e.g. prepare a document, print the result, bookmark a site), and/or distribute it (Evans and Card 2008; Pirolli and Card 2005).

So far, we have taken a closer look at possible online search process steps, in the order that they would typically occur. However, admittedly, all of the steps interact with one another at and across different levels. Early steps might continue throughout the whole process (Marchionini 1995); depending on task specificity some steps might not be carried out at all (Choo et al. 1999); or reiteration might mainly happen during the foraging stage (Pirolli and Card 2005). Moreover, Web-experienced searchers more often go back to adapt the search process, based on the perceived success of the iterations done before (Brand-Gruwel et al. 2005).

The Social Nature of Online Search

We have, thus far, painted a picture of a Web search as a solitary activity, during which the user enters a search query to satisfy her or his individual needs. However, the most recent literature has highlighted the varied social nature of an online search. People do not always conduct it by themselves or for themselves.

An early area of research within the larger topic of the social search is the collaborative search (e.g. Hansen and Järvelin 2005; Twidale et al. 1997). The fundamental assertion is that a great deal of searching happens in the form of collaboration, either in the search process (e.g. watching over someone's shoulder as he or she searches the Web) or on the search product (e.g. e-mailing or displaying the results of a Web search to another person). In a survey of 204 knowledge workers by Morris (2008), 53 per cent of respondents reported having 'cooperated with other people to search the web'.

Building also on the collaborative search literature, Evans and Chi (2010: 657) more broadly define the 'social search' as 'an umbrella term used to describe search acts that make use of social interactions with others. These interactions may be explicit or implicit, co-located or remote, synchronous or asynchronous'. Incorporating prior research in library science (e.g. Kuhlthau 1991; Pirolli and Card 2005; Twidale et al. 1997; Wilson 1981), in addition to recent findings in the online context (Broder 2002; Evans and Card 2008), Evans and Chi present an *elaborated model of social search*. This model highlights how social interactions can occur before, during and after a search. Before a search, people may interact with others to better frame the context or refine their requirements. During a search, people may search with others collaboratively or rely on others to help with sensemaking. Finally, after the search, a user may distribute the findings to others.

This research on social search further improves our understanding of how people actually find the information they want from the sea of online information. In addition, it highlights the potential for better search technologies. Some researchers have begun to explore ways to make search technologies go beyond solo-use and be more social, including systems enabling co-located searches (Amershi and Morris 2008), social tagging (Kammerer et al. 2009), and social bookmarking (e.g. Chi et al. 2007).

For future research in this direction, we recommend broadening the research perspective even further and re-examining some of the established literature on information seeking. For example, Bates (2002), Kuhlthau (1991) and Wilson (1999) offer more comprehensive perspectives, for instance

incorporating notions of affect and stress, as well as passive seeking behaviors. These models account for psychological phenomena that might better explain why some online search strategies are avoided but others are pursued.

ONLINE INFORMATION EXCHANGE: RELYING ON OTHERS AS INFORMATION SOURCES

In the previous section, we described how the search has become an integral part of our information behavior. By typing keywords into a search engine, we are able to find much topically relevant information from the zettabyte of information archived online. However, despite the growing amount of information that can be retrieved through search, people are also heavily reliant on the Internet to seek information from other people. Here, we discuss this alternative strategy for locating information – exchanging information with others online.

Information Exchange as a Complement to Information Search

Information exchange between people has always been an integral part of our lives. We rely on others as valuable information sources for a wide range of purposes – to solve problems, to make decisions, to coordinate plans, and to gather news and gossip. This type of information behavior has also manifested online, asynchronously or synchronously, between just two people (dyads) or multiple people. For example, people can use natural language to seek information from friends who are knowledgeable about a certain topic through e-mail and instant messaging (IM) or broadcast their question to millions of strangers who have various expertise via online forums and question and answer (Q&A) sites.

Despite the convenience and the growing amount of information that is retrievable through information searches, information exchange remains an important part of our information behaviors for multiple reasons. First, using keywords in search boxes is not a natural way for people to find information – even the Net Generation, who grew up using the Internet, has a preference for using natural language rather than keywords for searches (Rowlands et al. 2008). Second, common search engines can only be used successfully when users know the correct set of keywords. However, as we all know from our own experiences, while we may know what information we want, we do not always know what keywords will get us the right results. Third, search systems can hardly factor in the searcher's context. Unless he or she explicitly specifies all the relevant parameters, the search results may not exactly answer the searcher's question. Finally, and perhaps most importantly, not all information is available online. Not only is there still lots of existing information that has yet to be digitized and made searchable (i.e. indexed by search engines), there is also much information that is constantly being created offline or on private networks. The Internet may have answers to a lot of questions, but it does not have answers to all questions.

Online Technologies for Information Exchange

Many communication technologies have arisen with the advent of the Internet, and, broadly speaking, they can all be considered technologies for information exchange. Two prime examples of more 'traditional' computer-mediated communication technologies are e-mail and instant messaging (IM). Despite their many other functions and uses, they are also heavily used for question and answer (Q&A) between people (e.g. Isaacs et al. 2002; Whittaker and Sidner 1996). Much research has closely studied these more 'traditional' computer-mediated communication technologies and has examined how features of these technologies, such as the enabling of real-time, remote, text-based communication between people, have affected

communication and social relationships (e.g. Kraut et al. 2002).

The Internet has also led to a number of online (or virtual) communities where much exchange of information takes place. They include newsgroups, forums, chat rooms, social networking sites, etc. Online communities represent an archetypical playground for the mix of interactive and passive, as well as interpersonal and informational exchanges that are combined in the online world (Burnett 2000).

More recently, a specific type of online community designed for question asking and answering has emerged – social Q&A services. Social Q&A services try to leverage the fact that 'everyone knows something' (Adamic et al. 2008: 665). Through these services, anyone can pose their questions to a community of users and can also answer questions asked by others. These services have become extremely popular. Yahoo! Answers, a leading social Q&A site, attracted visits from more than 27 million people in April 2011 alone (Quantcast 2011).

Challenges for Online Information Exchange

Despite the pervasiveness of information exchange in the online setting and the emergence of technologies designed specifically to facilitate it, there are limitations with current online information exchange that will need to be addressed. Most notable are the problems of interruptions and users' motivation to participate in information exchanges.

Limiting Interruptions in Information Exchanges

Being connected to more people remotely and exposed to real-time notifications of information exchange requests can lead to an increase in untimely interruptions (e.g. González and Mark 2004), which were shown to be detrimental to people's primary task performance and well-being (Bailey et al. 2001; Czerwinski et al. 2000). Therefore, many researchers have explored ways to minimize the cost of interruptions caused by existing technologies for information exchange.

One solution is to facilitate the timing of information exchange requests, either by deferring request notifications until a more opportune time (Adamczyk and Bailey 2004; Ho and Intille 2005; Horvitz et al. 2005) or until the receiver is in a more situationally appropriate context (Avrahami et al. 2007; Bellotti and Edwards 2001; Dabbish and Kraut 2004).

Another class of solutions is to utilize economic forces to combat unwanted interruptions. One idea is to increase the cost of sending information requests to reduce the number of frivolous requests and reduce the number of overall interruptions (van Zandt 2004). A similar approach is to use market mechanisms. Requesters of information can offer to pay potential helpers, and helpers can be financially compensated to offset their costs. This design, in theory, can benefit all involved exchange parties (Loder et al. 2006). Controlled experiments demonstrated the feasibility of this solution, but highlighted the need to consider additional cognitive costs that users may incur from using this type of market (Hsieh et al. 2008).

Much of the research in this area is only slowly incorporated into commercial products. It will be intriguing to see how online information exchange evolves in the near future as these features become more and more common.

Understanding and Increasing Motivation

Unlike an information search, a user must interact with another person in an information exchange. In other words, if no one is willing to answer and respond to the exchange request, then information exchange will not occur. Since responding to exchange requests incurs a cost (e.g. takes time away from other tasks), why would anyone be willing to participate in information exchanges? And how could we encourage more people to share their knowledge online?

Motivations of Answerers

At the outset, answering someone's question is a prosocial act, primarily benefiting someone else (see Batson et al. 2002, for a review). In line with this assertion, intrinsic motivation, or motivation that is driven by an enjoyment or interest in the task itself (Deci and Ryan 1985), seem to be the main drivers behind the sharing of information online. For example, in a study of Naver, a Korean Q&A site, the majority of motivations reported by its users are intrinsic (Nam et al. 2009). Many studies on user contributions in general online communities found a wide range of intrinsic factors for sharing online. These factors include: altruism, belonging, collaboration, egoism, egotism, emotional support, empathy, knowledge, power, reputation, self-esteem, self-expression (Moore and Serva 2007).

Beside intrinsic motivations, social motivators are also strong in the online context. In their study of Yahoo! Answers, Dearman and Truong (2010) found that recognition by the community and askers can be very motivating. For online communities in general, feedback from other users is a strong predictor of newcomers' likelihood to return (Lampe and Johnston 2005). In addition, social identity and social bonds are both factors that can draw people to participate in communities in the first place, and also can sustain their level of contribution (Ren et al. 2007).

Finally, extrinsic motivators exist. For services that offer points and money for answers, extrinsic rewards were shown to attract more answerers and encourage longer answers (Hsieh et al. 2010; Nam et al. 2009). However, aside from direct extrinsic rewards, users may also be motivated by indirect opportunities. For example, in a study by Nam et al. (2009), some users cited the ability to promote their business on Q&A sites or being recognized as an expert.

Increasing Participation

Understanding users' motivations for sharing information online also offers insights into how to increase participation. For example, services might target questions to individuals in order to ensure that they see all the questions they would find interesting to answer. Game mechanics can also make these services more fun (Antin and Churchill 2011). In terms of social factors, researchers have experimented with socializing with newcomers (Joyce and Kraut 2006) or introducing social norms to increase participation (Chen et al. 2010). Another strategy is to promote stronger social identity and bonds (Ren et al. 2007), but also points and payments can be useful (Farzan et al. 2008; Harper et al. 2008). However, current research shows that, while extrinsic rewards can attract more answers, the answers may not be of a higher quality (Hsieh et al. 2010). Furthermore, recent findings seem to suggest that extrinsic rewards, especially money, may undermine social interactions (Hsieh and Counts 2009).

Here, we have only presented a very brief overview of motivations and solutions to increase participation for online information exchange. For more about this relevant theme, a great follow-up is the forthcoming book *Building Successful Online Communities* by Kraut and Resnick (in press).

CONCLUSION

The Internet has become the center of our information behaviors. The amount of information available online eclipses anything that humankind has hitherto seen, and many of us have this wealth of information at our fingertips. In this chapter, we presented findings from research on three key aspects of online information – access, search and exchange – to explain how people obtain the information they need using the Internet.

We showed that socio-economic factors continue to represent major hurdles to Internet access, but for those who are fortunate enough to have it, the Web has become the primary source of information for a wide range of purposes. We then explained how people actually find the information they need from the vast amount of information available online. We

highlighted how users are able to use search engines to satisfy their various needs, but also that they behave in ways that could be detrimental, such as relying on simple heuristics when evaluating search results. Finally, due to various limitations of information searches, people still rely greatly on other people for online information exchange.

Many challenges lie ahead in improving the Internet and educating its users to better support information needs. We have highlighted some of these challenges in the chapter. For example, how can we increase Internet access for everyone? How do we educate people to use the Internet more effectively? How can we design search engines that better fit search needs and practices? And how do we better facilitate information exchange between people to account for their limited attention and diverse set of motivations?

However, perhaps the biggest challenge for researching in this domain is the speed at which online behaviors are changing. Not only are there novel technologies emerging on a daily basis, but people are gaining more experience and their skills are also gradually improving. Scholars interested in studying online information behaviors must be proactive in staying up-to-date on current research in this exciting and fast-changing field.

REFERENCES

Adamczyk, P.D. & Bailey, B.P., 2004. If not now, when? *Proceedings of the 2004 Conference on Human Factors in Computing Systems* (CHI '04), ACM, 271–278.

Adamic, L.A. et al., 2008. Knowledge sharing and Yahoo! Answers: Everyone knows something. *Proceedings of the 17th International Conference on World Wide Web*, ACM, 665–674.

Amershi, S. & Morris, M.R., 2008. CoSearch: a system for co-located collaborative web search. *Proceedings of the Twenty-sixth Annual SIGCHI Conference on Human Factors in Computing Systems*, ACM, 1647–1656.

Antin, J. & Churchill, E.F., 2011. Badges in social media: A social psychological perspective. UX Scientist, May 2011. Available at http://uxscientist.com/public/docs/uxsci_2.pdf

Attewell, P. & Battle, J., 1999. Home computers and school performance. *The Information Society*, 15(1), pp. 1–10.

Aula, A. & Nordhausen, K., 2006. Modeling successful performance in web searching. *Journal of the American Society for Information Science and Technology*, 57(12), pp. 1678–1693.

Avrahami, D. et al., 2007. Improving the match between callers and receivers: A study on the effect of contextual information on cell phone interruptions. *Behaviour & Information Technology*, 26(3), pp. 247–259.

Bailey, B.P., Konstan, J.A. & Carlis, J.V., 2001. DEMAIS. *Proceedings of the Ninth ACM International Conference on Multimedia (MULTIMEDIA '01),* 241.

Bates, M.J., 2002. Toward an Integrated Model of Information Seeking and Searching. (Keynote Address, Fourth International Conference on Information Needs, Seeking and Use in Different Contexts.) *New Review of Information Behaviour Research*, 3, pp. 1–15.

Batson, C.D., Ahmad, N. & Tsang, J., 2002. Four motives for community involvement. *Journal of Social Issues*, 58(3), pp. 429–445.

Bellotti, V. & Edwards, K., 2001. Intelligibility and accountability: Human considerations in context-aware systems. *Human-Computer Interaction*, 16, pp.193–212.

Best, S.J. & Krueger, B.S., 2006. Online interactions and social capital: Distinguishing between new and existing ties. *Social Science Computer Review*, 24(4), pp. 395–410.

Brand-Gruwel, S., Wopereis, I. & Vermetten, Y., 2005. Information problem solving by experts and novices: Analysis of a complex cognitive skill. *Computers in Human Behavior*, 21(3), pp. 487–508.

Broder, A., 2002. A taxonomy of web search. *ACM SIGIR Forum*, 36(2), p. 3.

Brown, K., Campbell, S.W. & Ling, R., 2011. Mobile phones bridging the digital divide for teens in the US? *Future Internet*, 3(2), pp. 144–158.

Browne, G.J., Pitts, M.G., & Wetherbe, J.C., 2007. Cognitive stopping rules for terminating information search in online tasks. *MIS Quarterly*, 31(1), pp. 89–104.

Burnett, G., 2000. Information exchange in virtual communities: A typology. *Information Research*, 5(4).

Case, D.O., 2007. *Looking for Information: A survey of research on information seeking, needs, and behaviour*, 2nd edn, Oxford: Academic Press.

Chen, Y. et al., 2010. Social comparisons and contributions to online communities: A field experiment on MovieLens. *American Economic Review*, 100(4), pp. 1358–1398.

Cheshire, C. & Antin, J., 2008. The social psychological effects of feedback on the production of internet information pools. *Journal of Computer-Mediated Communication*, 13(3), pp. 705–727.

Chi, E.H., 2009. Information seeking can be social. *Computer*, 42(3), pp. 42-46.

Chi, E.H., Pirolli, P. & Lam, S.K., 2007. Aspects of augmented social cognition: Social information foraging and social search. *Proceedings of the 2nd International Conference on Online Communities and Social Computing,* Springer-Verlag, 60–69.

Chinn, M.D. & Fairlie, R.W., 2006. The determinants of the global digital divide: A cross-country analysis of computer and internet penetration. *Oxford Economic Papers*, 59(1), pp. 16–44.

Choo, C.W., Detlor, B. & Turnbull, D., 1999. Information seeking on the web: an integrated model of browsing and searching. *Proceedings of the 62nd ASIS Annual Meeting*, ACM, 3–16.

Czerwinski, M., Cutrell, E. & Horvitz, E., 2000. Instant messaging and interruption: Influence of task type on performance. Available at: ftp://ftp.research. microsoft.com/pub/ejh/ozchi2000.pdf (accessed May 25, 2011).

Dabbish, L. & Kraut, R.E., 2004. Controlling interruptions: Awareness displays and social motivation for coordination. *ACM Conference on Computer Supported Cooperative Work (CSCW 2004)*, ACM, 182–191.

Dearman, D. & Truong, K.N., 2010. Why users of Yahoo! Answers do not answer questions. *Proceedings of the 28th International Conference on Human Factors in Computing Systems*, ACM, 329–332.

Debowski, S., 2001. Wrong way: Go back! An exploration of novice search behaviours while conducting an information search. *The Electronic Library*, 19(6), pp. 371–382.

Deci, E.L. & Ryan, R.M., 1985. *Intrinsic Motivation and Self-Determination in Human Behavior*, New York: Plenum.

DiMaggio, P. & Bonikowski, B., 2008. Make money surfing the web?: The impact of internet use on the earnings of U.S. workers. *American Sociological Review*, 73(2), pp. 227–250.

DiMaggio, P. Hargittai, E., Celeste, C. and Shafer, S., 2004. From unequal access to differentiated use: A literature review and agenda for research on digital inequality. In K. Neckerman (ed.), *Social Inequality*. New York: Russell Sage Foundation, pp. 355–400.

Ellis, D., 1989. A behavioural approach to information retrieval system design. *Journal of Documentation*, 45(3), pp.171–212.

Evans, B.M. & Card, S., 2008. Augmented information assimilation. *Proceedings of the Twenty-sixth Annual CHI Conference on Human Factors in Computing Systems (CHI 2008)*, ACM, 989.

Evans, B.M. & Chi, E.H., 2010. An elaborated model of social search. *Information Processing & Management*, 46(6), pp. 656–678.

Farzan, R. et al., 2008. Results from deploying a participation incentive mechanism within the enterprise. *Proceedings of the Twenty-sixth Annual CHI Conference on Human Factors in Computing Systems (CHI '08)*, ACM, 563.

Fogg, B.J. et al., 2001. What makes web sites credible? *Proceedings of the SIGCHI Conference on Human Factors in Computing Systems (CHI '01)*, ACM, 61–68.

Fuchs, C. & Horak, E., 2008. Africa and the digital divide. *Telematics and Informatics*, 25(2), pp. 99–116.

Gantz, J. & Reinsel, D., 2010. The digital universe decade – are you ready? Available at www.emc.com/collateral/analyst-reports/idc-digital-universe-are-you-ready.pdf

González, V.M. & Mark, G., 2004. Constant, constant, multi-tasking craziness. *Proceedings of the 2004 Conference on Human Factors in Computing Systems (CHI '04)*, ACM, 113–120.

Gore, A., 1991. Information superhighways: The next information revolution. *The Futurist*, 25, pp. 21–23.

Haase, A.Q., et al., 2002. Capitalizing on the internet: Social contact, civic engagement, and sense of community. In B. Wellman & C. Haythornethwaite (eds), *The Internet and Everyday Life*. Oxford: Blackwell.

Hansen, P. & Järvelin, K., 2005. Collaborative information retrieval in an information-intensive domain. *Information Processing & Management*, 41(5), pp. 1101–1119.

Hargittai, Eszter, 2002. Second-level digital divide: Differences in people's online skills. *First Monday*, 7(4).

Hargittai, E & Hinnant, A., 2008. Digital inequality: Differences in young adults' use of the internet. *Communication Research*, 35(5), pp. 602–621.

Harper, F.M. et al., 2008. Predictors of answer quality in online Q&A sites. *Proceedings of the Twenty-sixth Annual SIGCHI Conference on Human Factors in Computing Systems*, ACM, 865–874.

Ho, J. & Intille, S.S., 2005. Using context-aware computing to reduce the perceived burden of interruptions from mobile devices. *Proceedings of the SIGCHI Conference on Human Factors in Computing Systems (CHI '05)*, ACM, 909.

Hoffman, D.L., Novak, T.P. & Venkatesh, A., 2004. Has the internet become indispensable? *Communications of the ACM*, 47(7), pp. 37–42.

Horvitz, E. et al., 2005. Bayesphone: Precomputation of context-sensitive policies for inquiry and action in mobile devices. In L. Ardissono, P. Brna & A. Mitrovic (eds), *User Modeling 2005*. Berlin, Heidelberg: Springer, pp. 251–260.

Howard, P.E.N., Rainie, L. & Jones, S., 2001. Days and nights on the internet: The impact of a diffusing technology. *American Behavioral Scientist*, 45(3), pp. 383–404.

Hölscher, C. & Strube, G., 2000. Web search behavior of internet experts and newbies. *Computer Networks*, 33(1–6), pp. 337–346.

Howe, N. & Strauss, W., 2000. *Milennials Rising: The Next Great Generation*. New York: Random House.

Hsieh, G. & Counts, S., 2009. mimir: a market-based real-time question and answer service. *Proceedings of the 27th International Conference on Human Factors in Computing Systems*, ACM, 769–778.

Hsieh, G. et al., 2008. Can markets help?: Applying market mechanisms to improve synchronous communication. *ACM Conference on Computer-Supported Cooperative Work (CSCW 2008)*, ACM, 535–544.

Hsieh, Gary, Kraut, Robert E. & Hudson, Scott E., 2010. Why pay?: Exploring how financial incentives are used for question & answer. *Proceedings of the 28th International Conference on Human Factors in Computing Systems (CHI '10)*, ACM, 305.

Isaacs, E. et al., 2002. The character, functions, and styles of instant messaging in the workplace. *Proceedings of the 2002 ACM Conference on Computer Supported Cooperative Work (CSCW '02)*, ACM, 11.

Jansen, B.J., Booth, D.L., & Spink, A., 2008. Determining the informational, navigational, and transactional intent of Web queries. *Information Processing & Management*, 44(3), pp. 1251–1266.

Johnson, E.J. et al., 2004. On the depth and dynamics of online search behavior. *Management Science*, 50(3), pp. 299–308.

Jones, Sydney & Fox, S., 2009. Generations online in 2009. Available at: www.pewinternet.org/~/media//Files/Reports/2009/PIP_Generations_2009.pdf [accessed June 2, 2011].

Joyce, E. & Kraut, Robert E., 2006. Predicting continued participation in newsgroups. *Journal of Computer-Mediated Communication*, 11(3), pp. 723–747.

Kammerer, Y. et al., 2009. Signpost from the masses. *Proceedings of the 27th International Conference on Human Factors in Computing Systems (CHI '09)*, ACM, 625.

Katz, J.E. & Rice, R.E., 2002. *Social Consequences of Internet Use: Access, involvement, and interaction*. Cambridge, MA: MIT Press.

Kellar, M., Watters, C. & Shepherd, M., 2007. A field study characterizing web-based information-seeking tasks. *Journal of the American Society for Information Science & Technology*, 58(7), pp. 999–1018.

Kennedy, T., Wellman, B. & Klement, K., 2003. Gendering the digital divide. *IT & Society*, 1(5), pp. 149–172.

Kim, K-S. & Allen, B., 2002. Cognitive and task influences on web searching behavior. *Journal of the American Society for Information Science and Technology*, 53(2), pp. 109–119.

Kraut, R.E. & Resnick, P., in press. *Building Successful Online Communities: Evidence-based social design*. Cambridge, MA: MIT Press.

Kraut, R.E. et al., 2002. Understanding effects of proximity on collaboration: Implications for technologies to support remote collaborative work. In P.J. Hinds & S. Kiesler (eds), *Distributed Work*. Cambridge, MA: MIT Press, pp. 137–162.

Kuhlthau, C.C., 1991. Inside the search process: Information seeking from the user's perspective. *Journal of the American Society for Information Science*, 42(5), pp. 361–371.

Lampe, C. & Johnston, E., 2005. Follow the (slash) dot. *Proceedings of the 2005 International ACM SIGGROUP Conference on Supporting Group Work (Group '05)*, ACM, 11.

Levin, D. & Arafeh, S., 2003. The digital disconnect: The widening gap between internet-savvy students and their schools. In C. Crawford et al. (eds), *Society for Information Technology & Teacher Education International Conference 2003*. Albuquerque, NM: AACE, pp. 1002–1007.

Loder, T., Van Alstyne, M. & Wash, R., 2006. An economic response to unsolicited communication. *Advances in Economic Analysis & Policy*, 6(1), Article 2.

Madden, M. & Rainie, L., 2003. America's online pursuits. Available at www.pewinternet.org/Reports/2003/Americas-Online-Pursuits.aspx

Marchionini, G., 1995. *Information Seeking in Electronic Environments*. Cambridge: Cambridge University Press.

Metzger, M.J., 2007. Making sense of credibility on the Web: Models for evaluating online information and recommendations for future research. *Journal of the American Society for Information Science and Technology*, 58(13), pp. 2078–2091.

Moore, T.D. & Serva, M.A., 2007. Understanding member motivation for contributing to different types of virtual communities. *Proceedings of the 2007 Conference on the Global Information Technology Workforce (SIGMIS-CPR '07)*, ACM, 153.

Morris, M.R., 2008. A survey of collaborative web search practices. *Proceedings of the Twenty-sixth Annual SIGCHI Conference on Human Factors in Computing Systems*, ACM, 1657–1660.

Nam, K.K., Ackerman, M.S. & Adamic, L.A., 2009. Questions in, knowledge in? *Proceedings of the 27th International Conference on Human Factors in Computing Systems (CHI '09)*, ACM, 779.

Nicholas, D. et al., 2011. Google generation II: Web behaviour experiments with the BBC. *Aslib Proceedings*, 63(1), pp. 28–45.

Nie, N.H. & Erbring, L., 2002. Internet and Society: A Preliminary Report. *Stanford Institute for the Quantitative Study of Society*. Available at www.bsos.umd.edu/socy/alan/webuse/handouts/Nie%20and%20Erbring-Internet%20and%20Society%20a%20Preliminary%20Report.pdf

Norris, P., 2001. *Digital Divide: Civic engagement, information poverty, and the Internet worldwide*. New York: Cambridge University Press.

PEW Research, 2011. Trend Data (Adults). Available at: www.pewinternet.org/Static-Pages/Trend-Data-(Adults) (accessed June 2, 2011).

Pirolli, P. & Card, S., 2005. The sensemaking process and leverage points for analyst technology as identified through cognitive task analysis. *The Analyst*, 2005, pp. 2–4.

Prensky, M., 2001a. Digital natives, digital immigrants, Part I. *On the Horizon*, 9(5), pp. 1–6.

Prenksy, M., 2001b. Digital natives, digital immigrants, Part II. Do they really think differently? *On the Horizon*, 9(6), pp. 1–6.

Quantcast, 2011. answers.yahoo.com. Available at: www.quantcast.com/answers.yahoo.com (accessed June 2, 2011).

Ren, Y., Kraut, R. & Kiesler, S., 2007. Applying common identity and bond theory to design of online communities. *Organization Studies*, 28(3), pp. 377–408.

Rice, Ronald E. & Katz, James E., 2003. Comparing internet and mobile phone usage: Digital divides of usage, adoption, and dropouts. *Telecommunications Policy*, 27(8–9), pp. 597–623.

Rose, D.E. & Levinson, D., 2004. Understanding user goals in web search. *Proceedings of the 13th Conference on the World Wide Web (WWW '04)*, ACM, 13.

Rowlands, Ian et al., 2008. The Google generation: The information behaviour of the researcher of the future. *Aslib Proceedings*, 60(4), pp. 290–310.

Russell, D.M. et al., 1993. The cost structure of sensemaking. *Proceedings of the SIGCHI Conference on Human Factors in Computing Systems (CHI '93)*, ACM, 269–276.

Smith, A., 2010a. Home broadband 2010. Available at: www.pewinternet.org/Reports/2010/Home-Broadband-2010.aspx

Smith, A., 2010b. Mobile access 2010. Available at: www.pewinternet.org/Reports/2010/Mobile-Access-2010.aspx

Smith, A., 2010c. Technology trends among people of color. *PEW Research Center's Internet & American Life Project*. Available at: www.pewinternet.org/Commentary/2010/September/Technology-Trends-Among-People-of-Color.aspx (accessed May 5, 2010).

Tamborello, F.P. & Byrne, M.D., 2005. Information search. *CHI '05 Extended Abstracts on Human Factors in Computing Systems (CHI '05)*, ACM, 1821.

Tewksbury, D., Weaver, A.J. & Maddex, B.D., 2001. Accidentally informed: Incidental news exposure on the world wide web. *Journalism and Mass Communication Quarterly*, 78(3), pp. 533–554.

Thatcher, A., 2008. Web search strategies: The influence of web experience and task type. *Information Processing & Management*, 44(3), pp. 1308–1329.

Twidale, M.B., Nichols, D.M. & Paice, C.D., 1997. Browsing is a collaborative process. *Information Processing & Management*, 33(6), p. 761.

van Dijk, J. & Hacker, K., 2003. The digital divide as a complex and dynamic phenomenon. *The Information Society*, 19(4), pp. 315–326.

van Zandt, T., 2004. Information overload in a network of targeted communication. *The RAND Journal of Economics*, 35(3), pp. 542–560.

Wang, P., Hawk, W.B. & Tenopir, C., 2000. Users' interaction with world wide web resources: An exploratory study using a holistic approach. *Information Processing & Management*, 36(2), pp. 229–251.

Weiler, A., 2005. Information-seeking behavior in Generation Y students: Motivation, critical thinking, and learning theory. *The Journal of Academic Librarianship*, 31(1), pp. 46–53.

Whittaker, S. & Sidner, C., 1996. Email overload. *Proceedings of the SIGCHI Conference on Human Factors in Computing Systems Common Ground (CHI '96)*, ACM, 276–283.

Williams, P. et al., 2008. Digital lives: Report of interviews with the creators of personal digital collections. *Ariadne*, April (55).

Williams, R. (n.d.) Data powers of ten (archived from the original). Available at: http://web.archive.org/web/19990508062723/http://www.ccsf.caltech.edu/~roy/dataquan (accessed May 5, 2011).

Wilson, T.D., 1981. On user studies and information needs. *Journal of Documentation*, 37(1), pp. 3–15.

Wilson, T.D., 1999. Models in information behaviour research. *Journal of Documentation*, 55(3), pp. 249–270.

Windham, C., 2005. Father Google and Mother IM: confessions of a Net Gen learner. *EDUCAUSE Review*, 40(5), pp. 42–59.

5

Social Media, Human Connectivity and Psychological Well-being

Sonja Baumer

With the emergence of social media, it became increasingly easy for people to connect and interact, create and share knowledge, and broadcast their personal messages to large and small audiences. According to Kaplan and Haenlein (2010), social media are Web- and mobile-based platforms that facilitate interaction between people and the sharing of media contents, including collaborative projects (e.g. Wikipedia), blogs and microblogs (e.g., Twitter), content-sharing communities (e.g. Flickr, YouTube), social networking sites (e.g. Facebook, LinkedIn), virtual game worlds (e.g. World of Warcraft), and virtual social worlds (e.g. Second Life). These media typically include a combination of technologies that enable instant and asynchronous messaging, creation of online profiles, befriending, blogging, microblogging, sharing of multimedia, tagging, commenting, rating, etc. Typically, social media are characterized by 'user-friendliness' in making content creation and sharing easy and quick for users of all levels of digital literacy. Furthermore, social media also incorporate some major innovations of the human–computer interaction field, including user-centred design, personalization and user-tailorability.

As a result of their ubiquitous accessibility and their appeal to users, social media have permeated all spheres of society, including mainstream media (journalism and television), business, politics and religion. Nowadays it is hard to find a business or media outlet that does not have social media feeds, or a public persona who does not possess some kind of an online profile. Moreover, social media are no longer associated with youth and Internet-savvy early adopters. In June 2011, several news agencies broadcast an image of Pope Benedict XVI tweeting prayers and praises of Jesus on his iPad. Another compelling image that comes to mind is that of an Egyptian demonstrator broadcasting images of devastation during the 2011 anti-regime protests.

Alongside these powerful images that demonstrate the utility and the pervasiveness of social media, there has been a number of concerns among scholars and members of the broader public about the negative psychological effects of social media. The chief concern is that there is an over-consumption of social media, especially among teenagers and young adults, and that many users have become addicted to them, to the point of detriment to their life and emotional well-being. A related concern is that social media also increase loneliness and isolation. It is said that social media act as a social displacer by providing users with a false sense of connection and intimacy, and by reducing the time they spend in offline interactions (e.g. Nie 2001). The third concern is that social media foster a false sense of anonymity, safety and risk-free engagement. Presumably, these distorted perceptions are then likely to lead to a loosening of behavioral inhibitions and impulsive acts (e.g. cyber-bullying, inappropriate self-disclosure, promiscuity, etc.).

This chapter will bring these and similar voices into a conversation with those who hold opposing views. Inevitably, as social media continue to evolve, our exploration remains 'under construction', but will continue to revolve around the idea that social media have been both endorsed and criticized with an 'unbearable lightness', often without careful analysis and scrutiny. Our intention is to go beyond 'media panic' v. 'media utopianism' ideological divisions, and to inform this discussion by relevant empirical evidence.

SOCIAL MEDIA ADDICTION

In the last few years, with the rapid proliferation and diffusion of social media, there seems to be a growing concern about social media addiction. Phrases such as 'the twenty-first century epidemic' and 'social media addiction' have quickly made headlines in major news outlets.

A good example is the 2010 study by Susan Moeller and her colleagues from the International Center for Media and the Public Agenda (ICMPA), University of Maryland (Moeller 2010). It is important to note that the study was in fact an extension of a class assignment, and was not originally designed as a research project aiming to investigate social media addiction. For 200 students who took a journalism course with professor Moeller, the assignment was to go without social media for 24 hours and then write about their experience. The results of the analysis of student submissions included quotes that were illustrative of their experience. Although some students even used the word 'addiction' in their submissions, most comments reflect the students' over-reliance on social media in order to stay connected to friends, family and world events. In discussion of their findings, the authors do not draw conclusions about social media addiction, but make points about the way social media fit into the students' social matrix, their expectations about frequency of contact, and how that impacts on the way they relate to the world. Nevertheless, the press release issued by the University of Maryland's news desk reframed the study under the headline of social media addiction, and diffused it as such to major news outlets, including the *Washington Post* and *New York Times*. As a result, the study received a large amount of media attention. The press coverage omitted the information that the study was an offshoot of a class assignment, giving the impression that the study was a major research project on social media addiction out of the University of Maryland.

The concern about social media addiction follows a similar public flurry about 'Internet addiction' that started during late 1990s, and was presumably fueled by the ongoing evolution of Internet use and rapid growth in the amount of time people spend online. Unlike 'social media addiction', the concept of Internet addiction (IAD) has received some empirical verification, starting with Young (1996, 1998) who pioneered this concept and developed an instrument to measure IAD.

Young (1998) explored the proposal that excessive Internet use needs to be considered as a mental disorder. Her findings indicate that addictive Internet users experience tolerance, withdrawal, and negative academic and occupational consequences, in a way consistent with the symptoms of a substance abuse disorder. Young's instrument for measuring Internet addiction – 'Internet Addiction Test' (IAT) subsequently underwent multiple psychometric evaluations in several different languages (Ferraro et al. 2007; Khazaal et al. 2008; Widyanto and McMurran 2004). Although these studies indicate relatively good psychometric characteristics of the IAT instrument, they disagree as to whether IAT reveals a single or multiple factors of Internet addiction. In addition, the studies also suffer from an inadequate sampling procedure (self-selection and bias) and a relatively small sample size (Byun et al. 2009).

Concerns about Internet addiction have continued to persist up to the present (e.g. Block 2008; Kim and Haridakis 2009), yet there still does not exist an agreement among researchers as to the definition of Internet addiction. The problem seems to be that many researchers have used different terms to describe very similar types of behavior. For example, some researchers talk about 'problematic Internet use' (Caplan 2002; Davis et al. 2002), while others prefer 'pathological Internet use' (Morahan-Martin and Schumacher 2000), or 'Internet dependency' (Wang 2001).

Further controversy that surrounds the concept of Internet addiction is that terms such as dependency and addiction have been primarily used in the context of substance abuse. In contrast, as Kim and Haridakis pointed out, the term 'dependence' in media studies is typically associated with a reliance on a particular medium or channel and has been viewed as a normal consequence of using a medium to satisfy one's communication needs, regardless of the excessiveness of use (Kim and Haridakis 2009).

In addition, the concept of Internet addiction confounds with other forms of behavioral addictions, such as compulsive gambling and hypersexual disorder, since the symptoms of excessive viewing of online pornography or online gambling have already been included in the diagnosis of these disorders (DSM-IV). Moreover, mental health specialists also warn that for many individuals, overuse of the Internet serves as a maladaptive coping strategy for alleviating symptoms of depression, anxiety and boredom, and thus Internet addiction seems to be a manifestation of another mental disorder, rather than a diagnostic category of its own.

With regard to potential abuse of social media, the controversy is even bigger, since the main activities on social media (chatting, interacting, befriending, content sharing) do not seem to have much in common with pathological gambling, which the IAD notion heavily parallels. Social media purportedly promote interaction, connectivity and pro-social causes, while gambling is widely seen as a self-destructive and anti-social behavior, and is outlawed in many places.

Byun and his colleagues carried out a meta-analysis of all empirical studies for the period 1996–2006 (Byun et al. 2009). They concluded that Internet addiction is in these studies defined extremely broadly with little commonality and guidance. They identified three main methodological and theoretical flaws in studies of Internet and social media addictions: (1) inconsistent criteria to define Internet addiction and Internet addicts; (2) problematic recruiting methods that result in a serious sampling bias; and (3) the use of exploratory rather than confirmatory data analysis techniques that establish merely associative rather than causal relationships among variables.

In the light of the methodological flaws discussed above, the American Psychiatric Association (APA) did not add 'Internet Addiction Disorder' to the pool of officially recognized mental disorders in the 5th edition of the Diagnostic and Statistical Manual (DSM-5) that was issued in 2013. Instead, APA appended 'Internet Gaming Disorder' in Section III, together with other mental health conditions that require further research. The diagnostic criteria for 'Internet Gaming Disorder' do not seem to be directly relevant for excessive use of social media.

This decision should encourage looking closely into the psychological consequences of increased use of social media. In that respect, the term 'addiction' does not seem suitable to describe our over-reliance on the Internet in general, and social media in particular.

Sherry Turkle, an MIT professor and a prominent researcher of digital media uses the term 'tethering' to describe our increased dependency and over-reliance on digital media. She warns that children and adolescents are growing up 'tethered' to social media in 'the text driven world of rapid response' and 'always on' communication, leaving little time for self-reflection, privacy and intimacy (Turkle 2011: 174). The next section examines this and related concerns about social media and psychological well-being.

INTERPERSONAL AND EMOTIONAL CONSEQUENCES ASSOCIATED WITH SOCIAL MEDIA USE

Turkle's main thesis is that digital media dramatically alter our social lives. While acknowledging that social media allow people to connect and communicate, Turkle critically examines the quality of online connections. She argues that we have become so wrapped up in the seemingly limitless possibilities of our digital devices that we completely overlook the decreasing standards of human interaction. In other words, we accept 'friends', disclose personal information or break up relationships with an apparent lightness, and without having to bear the predicaments and consequences of offline relationships (intimacy, commitment, loyalty, etc.). According to Turkle, digital connections provide the illusion of companionship without the demands of friendship. Our networked life allows us to hide from each other, even as we are tethered to each other. Turkle argues that we would rather text than talk and would

rather break up a relationship by sending a text message than having to face our partner's gaze of pain and anguish. As a result, according to Turkle, we tend to expect more from technology than from each other, and our human relationships are thus becoming increasingly disposable and dispensable.

Turkle's research is based on the ethnographic method. She collects data by interviewing people about how they feel about those technologies. This method is typically used to shed light on people's experiences. In the present case the method provides a picture of what people think about the effects of technology. It uncovers the angst that some young people have about the impact of technology on their lives, but it does not provide a definite answer as to what those effects actually are.

Yet, some empirical studies that have employed different research methods do yield support to Turkle's ideas. Morahan-Martin and Schumaker (2000, 2003) investigated the relationship between the frequency of use of digital media and loneliness among 277 undergraduate students. They found that loneliness was associated with increased use and that lonely users were more likely than the non-lonely users to seek emotional support online, find more satisfaction with online versus offline friends, and to be experiencing more disturbances in their daily lives. In a similar vein, Niemz, Griffins and Banyard (2005) studied 371 British college students and found that excessive users reported more perceived academic, social and interpersonal problems as well as lower self-esteem.

The main theoretical question that this kind of research revolves around is whether social media use extends the social networks of individuals who already have very active social lives (extroverts) or if it is used by individuals with low levels of social contact offline to compensate for a lack of 'real life' social interaction. These two perspectives have been described as 'The Rich Get Richer Approach' and the 'Social Compensation Approach' (see Stamoulis and Farley 2010 for discussion).

In their review of empirical studies on digital media use and users' social and emotional well-being, Stamoulis and Farley (2010) recognize that past studies have yielded inconsistent findings (Livingstone and Helsper 2007; Peter et al. 2005; Sheldon 2008), yet they found more evidence for the Social Compensation Approach. This includes above-mentioned studies by Morahan-Martin and Schumacher (2000, 2003), Niemz et al. (2005), Chak and Leung (2004) and Fortson et al. (2007), which are also consistent with the Social Compensation model. However, studies by Ellison et al. (2007), Sheldon (2008), Park (2010) and Peter et al. (2005) support the 'Rich Get Richer' model. For example Park (2010) found that wireless Internet use among college students was positively associated with face-to-face interactions with friends and acquaintances. Similarly, Sheldon (2008) found that college students who were more willing to communicate offline had more online friendships. Peter et al. (2005) also found that extraverted adolescents self-disclosed more and spent more time communicating online.

In a similar vein, a study from the Michigan State University conducted on a random sample of 800 undergraduate students, indicates that Facebook users have more social capital than the students who are not on Facebook, and that membership of the site increased measures of psychological well-being, especially in those suffering from low self-esteem (Ellison et al. 2007). This study found that the membership in certain online communities mirrors people's social networks in their everyday lives; thus online actions and interactions cannot be seen as independent of existing offline identities.

A recent study by Shields and Kane (2011) examined the extent of Internet use (and types of use) and the relationship between Internet use and a variety of social and psychological variables such as depression, face-to-face interaction, and stress related to interpersonal relationship. The participants in the study were 215 college students. In lieu of the mixed findings in the literature, the researchers anticipated both positive and negative associations with Internet use. The main rationale for the positive associations was Internet use for the purpose of connecting with others, for example e-mail (LaRose et al. 2001; Morgan and Cotton 2004), leading to fewer symptoms of depression and interpersonal problems. In congruence with the 'Rich Get Richer' view, the expectation was that individuals who are socially active and well adjusted offline, would show similar adaptive patterns online. The rationale for the negative associations was a possible connection between excessive use of the Internet and frequency of use of alcohol, marijuana, and other illicit drugs. The reasoning was that if excessive use of the Internet serves a similar purpose as the excessive use of alcohol and drugs (i.e. maladaptive coping behaviors to relieve depressive or anxiety symptoms), it should also be associated with drug or alcohol use. Finally, the Shields and Kane study (2011) also investigated the relationship between Internet use (for work, school and personal use) and academic performance. Based on the literature, the researchers also expected higher levels of Internet use for non-academic purposes to be associated with lower academic performance.

Overall, the results of the study provide support for the 'Rich Get Richer' perspective.

The findings indicate that Internet use for personal and school-related purposes was higher than for work-related purposes. This is probably related to the integration of an Internet course delivery system at the particular college, and the fact that the students were young and probably did not hold professional-level work positions that might require the use of the Internet. Similar to findings from other research (PEW 2002), the use of e-mail was the most common way to communicate with others, but visiting networking sites was almost as frequent. The use of search engines was also common. Other uses of the Internet were much less common, especially visiting a sexually explicit website, and this in contrast to the growing concern over consumption of online pornography emphasized by the media and some analyses of popular culture. Consistent with other research on depression among college students (e.g. Voelker 2003), Shields and Kane also found symptoms of depression to be relatively high in their sample. However, the frequency of Internet use was not related to symptoms of depression, and most specific types of Internet use were also unrelated, suggesting that Internet use is not a major cause or consequence of depression. The significant relationships that did emerge were primarily positive relationships. The findings lend some support to the authors' original hypothesis that use of the Internet for social connection would be associated with fewer symptoms of depression and social problems. Specifically, 'starting the day on the Internet' was related to fewer symptoms of depression, possibly because starting the day on the Internet is an indication of a desire for social connectivity. It is interesting that the two specific types of Internet use that were related to fewer symptoms of depression were visiting news sites and viewing videos. Although the authors did not ask about the content of the videos their participants were viewing, it is possible that some of them were news related. In addition, many videos on sites such as YouTube are designed to be entertaining and humorous. The only digital media behavior that was associated with more symptoms of depression was ending the day on the Internet. It was also associated with 'stress from arguments with parents' and 'dissatisfaction with a significant other'. Ending the day on the Internet might indicate a desire to escape from interpersonal problems and thus serve as a maladaptive coping behavior.

With regard to the relationship between the use of substances (alcohol, illicit drugs and marijuana) and Internet use, the findings indicate that substance use was not associated with Internet use, for the most part. However there were a few exceptions, which were all related to specific types of Internet use. Binge drinking was associated with visiting a networking site, suggesting that certain types of Internet use may be used to promote social activities or those individuals with wider social networks are more likely to engage in binge drinking. The latter interpretation might be more likely, since in colleges binge drinking often takes place at parties and other social activities. Illicit drug use was associated with using e-mail, but since the researchers did not ask for specific purposes of using e-mail, it is not clear what the nature of this relationship is, except that e-mail might be used to promote social activities with others that involve drug use (similar to binge drinking). Of the other specific types of Internet use, only visiting a sexually explicit website was associated with marijuana and illicit drug use. The authors argue that only this specific type of Internet use (i.e. porn viewing) might be similar to actual use of substances. A similar argument has been made by Meerkerk, Van Den Eijnden and Garretsen (2006), who found in a longitudinal project that viewing erotica had the highest potential for the development of compulsive Internet use.

Finally, the study also found that Internet use did not have a strong impact on academic performance. However 'starting the day on the Internet' was positively associated with GPA (i.e. grade point average), perhaps because the Internet is being used to obtain academically related information, or it is being used to complete course assignments. In contrast, 'listening to music' was negatively correlated with the number of hours the student spent studying and with GPA.

In general, the Shields and Kane study found that Internet use is associated with both positive and negative social and psychological variables. Their findings suggest the connection between Internet use and the social and emotional well-being of users should be understood in terms of specific types of Internet use rather than simply time spent on the Internet. In other words, the reasons why individuals use the Internet must be taken into account in order to understand associations.

The studies that examined the relationship between increased online connectivity and social and emotional well-being did not yield conclusive evidence. However the majority of studies seem to provide support for the 'Rich Get Richer' perspective, indicating that social media use was generally related to more face-to-face interaction, and suggesting that social media are being used to augment rather than replace social interaction. In the following section we will review a body of research that examines communicative affordances and the ways in which social media mediate human interaction and connection. This may shed more light on the findings discussed above that social media facilitate relationship building and augment our social lives.

COMMUNICATIVE AFFORDANCES ON SOCIAL MEDIA

Nancy Baym (2010) examines human connectivity on social media from the perspective of communication and media studies. She explores the core communication processes on digital media, and in particular, how we use mediated language to develop and maintain relationships. Baym argues against the technological determinism that juxtaposes the online with the offline, looking for clear lines of influences. She sees digital connections as part and parcel of our everyday lives in offline contexts, rather than as an agent of radical transformation. In other words, she sees social media as tools that enrich the ways in which we self-express and communicate with others. In her view, social media provide some unique affordances. For example, by eliminating shared location as a prerequisite for first meeting, social media widen the range of potential relationships. Furthermore, social media encourage people to form new relationships, even when they are not explicitly looking for them. Following Walther et al. (1994) Baym identifies three major characteristics that are responsible for this increased connectivity: the communication imperative (i.e. a drive to maximize communication satisfaction and interaction), assumed similarity (the idea that users who use the same media have shared affinities), and reduced social cues

(lack of nonverbal and paraverbal cues of communication). Baym further argues that in online communication contexts people adopt communication strategies that solicit the exchange of personally revealing information, that is designed to reduce the uncertainty in the online context. In other words, digital media not only facilitate connection making, but they also seem to encourage people to reveal personal information faster than in offline contexts. Just like people on airplanes who readily spill their secrets to strangers, the anonymity of some online interactions promotes a greater self-disclosure. As a result, Baym argues, social media enable us to quickly find out important facts about each other, and thus either get closer and more connected, or move on to other relationships when the compatibility is lacking (Baym 2010).

In the case of romantic relationships, social media are in an apparent discrepancy with conventional understandings of relationship building, which emphasize the importance of physical attraction and nonverbal signals. For example, in face-to-face interactions, people use nonverbal cues (eye contact, smile, touch, physical proximity) to indicate attraction and deepen connection. Online media apparently do not typically offer such communication affordances, although some recent developments (e.g. voice and video chat, emoticons on textual media) attempt to remedy this limitation.

On the other hand, the very lack of nonverbal cues on digital media and the control offered in message construction (especially when it is asynchronous and editable), afford selective self-presentation, partner idealization, editing and attention advantages, and mutually enhancing reciprocal feedback (Walther et al. 1994). Typically, nonverbal cues are 'given off', that is, unintentionally leaked through communication and are often undesirable (Goffman 1959). By eliminating nonverbal cues, digital media increase our ability to manage our self-presentation and eliminate sources of potential embarrassment and social inhibition. Furthermore, researchers have also demonstrated that digital media can be useful in handling potentially stressful interpersonal tasks in romantic relationships such as initiating a relationship or rejecting a pursuer (Tong and Walther 2011a, 2011b). For example, digital 'wink' on Match.com allows users to make contact with potential partners without having to extend any specific personal message. Digital wink generates a scripted message to the recipient ('X has winked at you'), containing a link to the sender's profile. Since the message content is pre-written and generated by the system, it is experienced as an impersonal first attempt at initiating communication. Thus, it is also likely to be perceived as less stressful and risky than when a first attempt has to be made in face-to-face interaction.

A recent study by Tong and Walther (2011a) examined how online daters use tools provided by dating website messaging services to reject their romantic pursuers (Tong and Walther, 2011a). These tools include options for users to reject pursuers, either by one-click automated rejection or passively by remaining unresponsive, and this in addition to the option of composing a personal message and tailoring it to the specific pursuer. The study shows that online daters fully take advantage of these new communicative tools and are choosing to use them to simplify what can often be a difficult, painful and stressful process of romantic refusal. The option of automated rejection makes it easier for the sender, by reducing emotional risk and cognitive effort while still fulfilling the main functional goal of romantic refusal. The authors also suggest that in addition to the mechanism for automated rejection, the selective editing and asynchronous nature of online communication enables rejectors to devise additional interpersonal tactics to deliver rejection.

From research reviewed above it appears that communicative tools provided by social media are likely to facilitate communication. Social media also seem likely to alleviate the social inhibitions that face-to-face communication typically entails, as they enable us to better control self-presentation. The next section will examine some implications associated with anonymity and the increased impression management options that social media provide for.

BEHAVIORAL DISINHIBITION AND SOCIAL MEDIA: A REMEDY OR SOCIAL PATHOLOGY?

Social anxiety and shyness are often taken to be incapacitating conditions that prevent individuals from leading fully productive lives and satisfying their needs for companionship and affiliation (Crozier 2001). Shyness is a tendency to feel anxious during social interactions and to be overconcerned about managing one's self-presentation. Shy people tend to be intimidated in face-to-face interactions. Shyness is a form of anxiety and it differs from introversion, since introverts prefer to be alone, whereas shy people desire social interaction but are held back by their insecurities (Larsen and Buss 2008).

Empirical work on shyness and social media is relatively scarce, compared to the amount of research on shyness in face-to-face interaction, and also having in mind that social media may be a context where shy individuals could potentially experience better-quality relationships (Baker and Oswald 2010; Roberts et al. 2000; Yen et al. 2012).

Bardi and Brady (2010) examined the relationship between shyness and the use of instant messaging (IM) in an explorative study that included 55 college students. The instruments used in the study included a shyness scale, an instant messaging use index, and a motivational scale developed specifically for that study. The study found that the strongest motives for IM use were to increase personal contact as opposed to gain social ease or to decrease loneliness. An interesting finding is that there was no direct association between shyness and the intensity of IM use. In other words, shy people were as attracted to IM as non-shy people. However, the difference emerged in the motives for the use of IM. The study found that the primary reason to use instant messaging for shy people was the motivation to decrease loneliness. Unfortunately, though the study has suggested that instant messaging may be used by shy individuals to reduce feelings of isolation, it did not provide information on whether IM was indeed an effective way to reduce loneliness.

Bonetti and his colleagues (2010) have also investigated the link between the usage of digital communication devices and self-reported loneliness and social anxiety. Their sample included 626 students, ages 10–16 years. The findings of the study indicated that children and adolescents who self-reported being lonely communicated online significantly more frequently about personal and intimate topics than those who did not report being lonely. Another important difference between the two groups is that students who reported being lonely were motivated to use online communication significantly more frequently to compensate for their poor social skills to meet new people. Similarly to the study by Baker and Oswald (2010), this study clearly lends support to the Social Compensation approach, but it does not provide information as to how effective social media are in reducing the sense of loneliness.

Yen and his colleagues (2012) compared the amount of social anxiety that is produced in offline and online social interactions. Their study included over 2,000 Taiwanese college students. Their main finding was that online social interactions generate lower levels of social anxiety in comparison with offline social interactions. They also found that social anxiety and behavioral inhibitions decreased even more in online interaction among subjects with high social anxiety, depression and behavior inhibitions. This result suggests that social media may have a potential as a tool for the treatment of social anxiety.

The Baker and Oswald (2010) study provides specific insight about how effective social media are in supporting the social life of shy individuals and helping them find friends and decrease their sense of loneliness. The participants (241 undergraduate students)

completed a questionnaire that assessed their use of Facebook, degree of shyness, perceived social support, loneliness, and the quality of their friendships.

The study yielded a myriad of interesting findings evidencing the utility of Facebook as a tool that supports communication and friendship-building among shy individuals. First, the researchers found that shyness was positively correlated with reporting that information obtained from Facebook helped get to know other students better. The authors suggest that having the ability to gain additional information might be especially important to shy people, as they are likely to know less about their peers due to avoiding or withdrawing from social situations. Further, shyness was positively associated with reporting that Facebook allowed shy users to feel closer to their peers. Because shy individuals find it difficult to achieve intimacy through face-to-face interaction, it is likely that they find Facebook particularly useful for achieving intimacy because they feel more comfortable using media than engaging in face-to-face interactions (Roberts et al. 2000). Next, Facebook usage was also associated with greater social support for shy individuals and better-quality peer relationships among shy individuals.

However, despite the finding that Facebook usage was associated with better friendship satisfaction, perceived closeness, and importance for relatively shy individuals, as well as greater perceived social support, the study did not find that Facebook usage actually helped shy people decrease loneliness. The authors suggest that this unexpected finding could be due to the cross-sectional design of the study. They explain that previous research (Kraut et al. 2002) has shown that although Internet use is positively associated with initial loneliness, over time it leads to decreases in loneliness. Thus, as suggested by the authors, it is quite possible that similar decreases in loneliness could be observed in a longitudinal study of Facebook usage.

Another important finding is that shyness was not correlated with reporting that

Facebook was useful for feeling comfortable with others offline. In other words, Facebook did not seem to help shy individuals transfer the feelings of comfort they feel talking with others online to talking offline.

This conclusion is congruent with the findings of Pollet's study (Pollet et al. 2011). Pollet and his colleagues examined the association between using social media (IM and social networking sites – SNS), the number of relationships at different layers of the offline social network (i.e. support group, sympathy group, outer layer) and the quality of these relationships. The findings indicate that there was no relationship between time spent using IM or SNS and the size of any of the three layers of the offline network. Furthermore, spending more time on IM or SNS did not increase the emotional closeness of relationships in any of these layers. However, the authors recognize that their study faced some limitations with regard to the size of sample, the sampling procedure and the overall design of the study. They indicate that a study with a larger randomized sample could provide more generalizable findings, and a longitudinally designed study could allow the researchers to examine the relationship between online and offline networks over time.

While a large body of research suggests the benefits of social media for people who suffer from shyness and social anxiety, future research is needed to establish how and under what conditions social media can remedy vulnerabilities associated with shyness and social anxiety in face-to-face interactions. Alongside the potential benefits of social media, there is a growing concern that the social disinhibition facilitated by digital media can be destructive for certain types of individuals, such as youth and those with impulse control problems. The next question that we will examine is whether social media encourage certain users to engage in negative social behaviors, such as online risk taking and cyber-bullying.

A study by Stamoulis and Farley (2010) attempted to identify user characteristics that are associated with online risk taking. The study was a secondary data analysis of information obtained from the 2006 'Parents and Teens Survey' of the PEW Internet & American Life Foundation Project. The survey included items about computer and technology usage, computer and technology ownerships, general perception of the Internet, social/after school activities, privacy issues, reasons for engaging in social networking sites, effects of Internet usage, stranger contact and video game use. Multiple regression analysis was conducted on the survey data to determine the predictors of online risk-taking behaviors, such as posting cell phone numbers on SNS, befriending a stranger, responding to a stranger, etc. The predictor variables used in each regression include: Age, Participation in an Extracurricular Activity, Time Spent with Friends In-Person, and Time Spent Talking to Friends on a Landline Phone, Time Spent Talking on a Cell Phone, and Holding a Part-Time Job.

The Stamoulis and Farley study provides evidence that supports the Social Compensation approach discussed above. This approach postulates that the Internet primarily benefits introverted people. For example, the study finds that common predictors for risk-taking factors were 'infrequent time spent socializing with friends in person' and 'lack of involvement in extracurricular activities'. More specifically, 'lack of involvement in extracurricular activities' was a significant predictor of 'Four or More Online Risks' for boys, while 'less time spent socializing with friends' offline was a significant predictor for girls. The authors propose that adolescents may be motivated to interact with strangers online in the hope of making friends and social connections. Therefore, they may be more willing to assume the risks associated with communicating with strangers and posting personal and contact information online.

Yet, one finding seems to support the Rich Get Richer approach. This finding is that girls who post photographs online were also more likely to spend time socializing with friends offline, participate in extracurricular activities and spend less time online. In other words, girls who spend a lot of time with friends and in activities are likely to be extrovert and thus likely to showcase that side of themselves online. It is possible that for girls, posting photographs online is not an inherently risky behavior.

It should also be noted that time spent talking with friends on either a cell phone or a landline telephone and time away from a computer at a part-time job did not serve as a protective measure against online risk taking and in some cases predicted certain risk-taking behaviors. The authors speculate that behaviors which are most preventative against adolescent online risk taking are forms of face-to-face socializing or involvement in activities. In other words, socially isolated youths lacking participation in activities are more likely to be lonely, isolated or depressed than those that participate in extracurricular activities and are not socially isolated. Consequently, adolescents who are lonely, isolated or depressed may be going online in search of connection with others.

A related concern about youth interactions on social media is that social media act as 'enablers' of bullying. Bullying that takes place via digital media (e-mail, IM, SNS, cell phone) is termed cyber-bullying. Social media provide a few affordances that potentially could make cyber-bullying even more serious than traditional bullying. Willard (2007) argued that cyber-bullying is less predictable than traditional bullying and can occur any time, which leaves victims feeling helpless, knowing that they may receive a harassing message every time they turn on their cell phone or go online. Conversely, victims of traditional bullying can potentially feel safe at home or elsewhere outside of school. A further affordance is that cyber-bullying messages and images can be distributed widely and quickly. Finally, being anonymity allows for reduced social accountability for bullies, which may encourage individuals to engage in inappropriate behavior online.

Cyber-bullying typically takes the form of relational and verbal aggression. Relational aggression is a type of aggression intended to harm others through deliberate manipulation of their social standing and relationships. Forms of relational aggression that are associated with online interactions on social media range from excluding others from social activities, damaging victims' reputations by spreading rumors and gossiping about the victim, or humiliating them publicly. Withdrawing attention and rejection of friendship are also considered to be forms of relational aggression. It is said that social media not only have facilities that support relationship building, but they also make it easier to spread rumors, gossip or de-friend a person, and that adolescents are particularly vulnerable to cyber-bullying.

Cyber-bullying seems to be quite prevalent, especially among children and adolescents. Studies examining cyber-bullying estimate that 4 to 15 per cent of US youths engage in cyber-bullying, with more cases reported in late middle school and high school (Twyman et al. 2010). The percentage of children and adolescents that have been targets of cyber-bullying ranges from 19 to 42 per cent, depending on the study reported.

Schneider and her colleagues (2012) used data from a regional census of US high school students to evidence the prevalence of cyberbullying and school bullying victimization and their associations with psychological distress. Their pool of data included responses from over 20,000 ninth- through twelfth-grade students. The survey assessed bullying victimization and psychological distress, including depressive symptoms, self-injury and suicidality. The findings indicate that 15.8 per cent of students experienced some form of cyber-bullying and that victimization was even higher among non-heterosexually identified youths and among girls. Victims of cyber-bullying were also more likely to report lower school performance than non-victims and were also more likely to have depressive symptoms and suicide attempts requiring medical treatment. The authors also found a significant overlap between cyber-bullying and offline school bullying (i.e. 59.7 per cent of cyber-bullying victims were also school bullying victims, and 36.3 per cent of school bullying victims were also cyber-bullying victims), but the study did not shed light on the nature of that relationship.

Schneider and her colleagues (2012) conclude that cyber-bullying victimization is strongly associated with multiple forms of psychological distress along the continuum from depression to suicide attempts. Most importantly, the study found that victims of cyber-bullying reported more distress than did victims of traditional school bullying. Specifically, reports of depressive symptoms were highest among victims of both cyber- and school bullying (47.0 per cent), followed by cyber-bullying-only victims (33.9 per cent), and school-bullied-only victims (26.6 per cent), compared with 13.6 per cent of non-victims. Similarly, attempted suicide was highest among adolescents who were victims of both cyber- and school bullying (15.2 per cent). However, it was also elevated among cyber-only victims (9.4 per cent) and school-only victims (4.2 per cent), compared with students reporting neither form of victimization (2.0 per cent). These findings imply that cyber-bullying is potentially more damaging than traditional school bullying and is consistent with the above-mentioned notions that electronic media take bullying to another level, by creating a sense of helplessness in victims who are constantly under stress, expecting to be victimized whenever they turn on their electronic devices.

INSTEAD OF A CONCLUSION

The research reviewed above paints a complex picture about the impact of social media on our psychological well-being and on society in general. The question that social media researchers nowadays ask is not whether social media are beneficial or harmful. Instead, contemporary researchers attempt to identify who can benefit from social media, and in what ways, as well as, who is most vulnerable to potential harm from them.

The concern about the overconsumption of social media continues to be on the rise among members of the broader public, as well as among researchers (e.g. Young and Nabuco de Abreu 2011). However, the conceptualization of excessive use of social media as behavioral addiction might not be particularly heuristic, since it confounds with other types of behavioral addictions (gambling, shopping, sexual/romantic addiction, etc.). Nonetheless, research has demonstrated that in some individuals, an excessive use of social media is associated with detrimental conditions, similar to those linked to substance abuse, such as tolerance, withdrawal and negative academic and occupational consequences (Young 1998).

Studies have also identified social media affordances that allow users to have a better control over their self-presentation (e.g. Baym 2010; Walther et al. 1994). They also demonstrated the utility of social media in eliminating stress associated with some interpersonal tasks such as rejecting friendship or initiating a romantic relationship (Tong and Walther 2011a, 2011b).

Overall, the above-reviewed research does not yield conclusive evidence about psychological consequences associated with the use of social media. The main drawback of existing research seems to be the lack of studies that provide information about the causal relationships between the use of social media and different psychosocial variables. The majority of studies reviewed above are correlational studies that are good at identifying associations between variables, but are unable to determine what causes a particular phenomenon. Thus, even though there is convincing evidence that online risk taking is strongly associated with a lack of offline social experiences, we cannot determine that the former causes the latter or vice versa. Nor we can determine that having many Facebook friends and Twitter followers will make us less prone to depression and anxiety, and this despite the observed strong correlation between these variables.

Furthermore, the majority of studies reviewed above have utilized the cross-sectional design.

As a result, they do not provide insight into how patterns of social media use change over time. One possibility is that the excessive use of social media might have been spurred by their novelty and popular cultural trends. Over time, patterns of social media use may evolve to become less excessive and better suited to users' needs, reflecting their accumulated experiences and saturation with this kind of media.

Finally, the studies reviewed above were conducted on nonrandomized samples that predominantly included college students from middle-class backgrounds in industrialized countries. As such, those samples are extremely likely to be biased. Consequently, the above-discussed research findings cannot be generalized to a broader public, nor be considered culturally universal. Future research might benefit from the utilization of longitudinal design and from the selection of samples that better represent the general public.

However, even with the methodological improvements suggested above, research on the psychological effects of social media is likely to encounter a controversy that other media effects studies have been facing in the last two decades. For example, a large body of research exists on the psychological effects of violent video games, including research on long- and short-term effects, cross-sectional and longitudinal studies, as well as correlational and experimental studies. Yet, despite ample evidence having been collected, there does not seem an unequivocal conclusion about the causal link between playing violent video games and aggressive behaviors. Some researchers argue that, contrary to common belief, the current literature does not provide strong support for the conclusion that media violence causes aggressive behaviors (e.g. Ferguson 2009; Ferguson and Kilburn 2009). However, other prominent researchers of media effects (e.g. Anderson et al. 2010; Bushman and Gibson 2011) have arrived at a quite opposite conclusion. For instance, Craig Anderson and his colleagues performed a meta-analysis on a similar set of data and suggested that exposure to violent video games is

a causal risk factor for increased aggressive behavior, aggressive cognition and aggressive affect, and for decreased empathy and prosocial behavior (Anderson et al. 2010).

Returning to the issue of the psychological effects of social media, an additional methodological complication is that users are not simply media consumers as in the case of video games. In the case of social media, users are not just an audience, but are also content creators and media producers. Consequently, social media cannot be conceptualized as a true independent variable, and experimental methods developed for measuring media effects cannot be applied for the purpose of identifying causal relationships between variables.

Some contemporary researchers (Ahn 2011; Kling 2007) are opposed to the technological determinism, seen in previous media effects research, that attempts to identify the causal link between the use of social media platforms and the psychological well-being of users. Instead, they argue for an emergent theoretical perspective that combines technology affordances with social adoption, known as 'social informatics' (Kling 2007). The proponents of the social informatics perspective argue that one cannot study effects of the technology alone without taking into account the communication behaviors within the system, including both the technical affordances of a particular media and human factors, such as users' characteristics, as well as cultural and institutional contexts of use.

Clearly, social media research shows that the technical features and infrastructure of a particular social media impacts users' behavior. However, it is the human–media interaction that ultimately serves as a causal factor and leads to various psychological and social consequences. Users bring already existing psychological characteristics into the online community. For example, users with rich and supportive offline social networks will be likely to have positive wall posts on Facebook and will continue to feel good about themselves. However, for users who suffer from peer isolation or bullying, 'wall posts' may be yet another source of misery.

Studies on social media effects must integrate perspectives from social and clinical psychology with those from human computer interaction and social informatics. Future experimental research should focus on establishing causal relationships between different patterns of user–media interactions and measures of psychological well-being. The field could also benefit from studies that include participants from socially and culturally diverse backgrounds.

REFERENCES

Ahn, J. (2011). The effect of social network sites on adolescents' social and academic development: Current theories and controversies. *Journal of the American Society for Information Science and Technology*, 62(8), 1435–1445.

American Psychiatric Association. (2000). *Diagnostic and statistical manual of mental disorders* (4th ed., text rev.). Washington, DC: Author.

American Psychiatric Association. (2013). *Diagnostic and statistical manual of mental disorders* (5th ed.). Washington, DC: Author.

Anderson, C.A., Shibuya, A., Ihori, N., Swing, E.L., Bushman, B.J., Sakamoto, A., Rothstein, H.R. and Saleem, M. (2010). Violent video game effects on aggression, empathy, and prosocial behavior in Eastern and Western countries. *Psychological Bulletin*, 136, 151–173.

Baker, L.R. and Oswald, D.L. (2010). Shyness and online social networking services. *Journal of Social and Personal Relationships*, 27, 873–889.

Bardi, C.A. and Brady, M.F. (2010). Why shy people use instant messaging: loneliness and other motives. *Computers in Human Behaviour*, 26(4), 1722–1726.

Baym, N. (2010). *Personal Connections in the Digital Age*. Cambridge: Polity Press.

Block, J.J. (2008). Issues for DSM-V: Internet addiction. *American Journal of Psychiatry*, 165, 306–307.

Bonetti, L., Campbell, M. and Gilmore, L. (2010). The relationship of loneliness and social anxiety with children's and adolescent's online communication. *CyberPsychology, Behavior & Social Networking*, 13, 279–285.

Bushman, B.J. and Gibson, B. (2011). Violent video games cause an increase in aggression long after the game has been turned off. *Social Psychological and Personality Science*, 2, 29–32.

Byun, S., Ruffini, C., Mills, J.E., Douglas, A.C., Niang, M., Stepchenkova, S., Lee, S.K., Loutfi, J., Lee, J.K., Atallah, M. and Blanton, M. (2009) Internet addiction: Metasynthesis of 1996–2006 quantitative research. *Cyberpsychology & Behavior*, 12(2), 203–207.

Caplan, S.E. (2002). Problematic internet use and psychosocial well-being: Development of a theory-based cognitive-behavioral measurement instrument. *Computers in Human Behaviors*, 18, 553–575.

Chak, K. and Leung, L. (2004). Shyness and locus of control as predictors of internet addiction and internet use. *CyberPsychology & Behavior*, 7(5), 559–570.

Crozier, W.R. (2001). *Understanding Shyness: Psychological perspectives*. Houndmills, Basingstoke: Palgrave Macmillan.

Davis, R.A., Besser, A. and Flett, G.L. (2002). Validation of a new scale for measuring problematic internet use: Implications for pre-employment screening. *Cyberpsychology & Behavior*, 5(4), 331–345.

Ellison, N.B., Steinfield, C. and Lampe, C. (2007). The benefits of Facebook 'friends': Social capital and college students' use of online social network sites. *Journal of Computer-Mediated Communication*, 12(4), article 1. Available at: http://jcmc.indiana.edu/vol12/issue4/ellison.html

Ferguson, C.J. (2009). Media violence effects: Confirmed truth or just another X-File? *Journal of Forensic Psychology Practice*, 9(2), 103–126.

Ferguson, C.J. and Kilburn, J. (2009). The public health risks of media violence: A meta-analytic review. *Journal of Pediatrics*, 154(5), 759–763.

Ferraro, G., Caci, B., D'Amico, A. and Di Blasi, M. (2007) Internet addiction disorder: An Italian study. *CyberPsychology & Behaviour*, 10(2), 170–175.

Fortson, B.L., Scotti, J.R., Chen, Y., Malone, J. and Del Ben, K.S. (2007). Internet use, abuse, and dependence among students at a southeastern regional university. *Journal of American College Health*, 56(2), 137–144.

Goffman, E. (1959). *The Presentation of the Self in Everyday Life*. Garden City, NY: Doubleday.

Kaplan, A.M. and Haenlein, M. (2010). Users of the world, unite! The challenges and opportunities of social media. *Business Horizons*, 53(1), 59–68.

Khazaal, Y., Billieux, J., Thorens, G., Kahan, R., Louati, Y., Scarlatti, E., Theintz F., Lederrey, J., Van Der Linden, M. and Zullino, D. (2008). French validation of the Internet addiction test. *CyberPsychology & Behaviour*, 11(6), 703–706.

Kim, J. and Haridakis, P.M. (2009). The role of Internet user characteristics and motives in explaining three dimensions of Internet addiction. *Journal of Computer-Mediated Communication*, 14, 988–1015.

Kling, R. (2007). What is social informatics and why does it matter? *The Information Society*, 23, 205–220.

Kraut, R., Kiesler, S., Boneva, B., Cummings, J., Helgeson, V. and Crawford, A. (2002). Internet paradox revisited. *Journal of Social Issues*, 58(1), 49–74.

LaRose, R., Eastin, M.S. and Gregg, J. (2001). Reformulating the internet paradox: Social cognitive explanations of internet use and depression. *Journal of Online Behavior*, 1(2).

Larsen, R.J. and Buss, D.M. (2008). *Personality Psychology: Domains of knowledge about human nature* (3rd edn). New York: McGraw-Hill.

Livingstone, S. and Helsper, E. (2007). Taking risks when communicating on the internet: The role of offline social-psychological factors in young people's vulnerability to online risks. *Information, Communication & Society*, 10(5), 619–644.

Meerkerk, G., Van Den Eijnden, R. and Garretsen, H. (2006). Predicting compulsive internet use: It's all about sex! *CyberPsychology & Behavior*, 9(1), 95–103.

Moeller, S. (2010). A day without media. Available at: http://withoutmedia.wordpress.com

Morahan-Martin, J. and Schumacher, P. (2000). Incidence and correlates of pathological internet use among college students. *Computers in Human Behavior*, 16, 13–29.

Morahan-Martin, J. and Schumacher, P. (2003). Loneliness and social uses of the internet. *Computers in Human Behavior*, 19, 659–671.

Morgan, C. and Cotton, S.R. (2004). The relationship between internet activities and depressive symptoms in a sample of college freshmen. *CyberPsychology & Behavior*, 6(2), 133–142.

Nie, N.H. (2001). Sociability, interpersonal relations, and the internet: Reconciling conflicting findings. *American Behavioral Scientist*, 45(3), 420–435.

Niemz, K., Griffins, M. and Banyard, P. (2005). Prevalence of pathological internet use among university students and correlations with self-esteem, the General Health Questionnaire (GHQ) and disinhibition. *CyberPsychology & Behavior*, 8(6), 562–570.

Park, N. (2010). Integration of internet use with public spaces: College students' use of the wireless internet and offline socializing. *Cyberpsychology: Journal of Psychosocial Research on Cyberspace*, 4(2). Available at: www.cyberpsychology.eu/view.php?cisloclanku=2010112501&article=4

Peter, J., Valkenburg, P. and Schouten, A. (2005). Developing a model of adolescent friendship formation on the internet. *Cyberpsychology & Behavior*, 8(5), 423–430.

PEW Internet and American Life. (2002, September 15). The internet goes to college: How students are living in the future with today's technology. Available at: http://eric.ed.gov/PDFS/ED472669.pdf

Pollet, T.V., Roberts, S.G.B. and Dunbar, R.I.M. (2011). Use of social network sites and instant messaging does not lead to increased offline social network size, or to emotionally closer relationships with offline network members. *CyberPsychology, Behavior and Social Networking*, 14, 253–25.

Roberts, L.D., Smith, L.M. and Pollock, C.M. (2000). 'U r a lot bolder on the net': Shyness and internet use. In W.R. Crozier (ed.), *Shyness: Development, consolidation and change* (pp. 121–138). New York: Routledge.

Schneider, S.K., O'Donnell, L., Stueve, A. and Coulter, R.W.S. (2012). Cyberbullying, school bullying, and psychological distress: A regional census of high school students. *American Journal of Public Health*, 102(1), 171–177.

Sheldon, P. (2008). The relationship between unwillingness-to-communicate and students' Facebook use. *Journal of Media Psychology: Theories, Methods, and Applications*, 20(2), 67–75.

Shields, N. and Kane, J. (2011). Social and psychological correlates of internet use among college students. *Cyberpsychology: Journal of Psychosocial Research on Cyberspace*, 5(1). Available at: http://cyberpsychology.eu/view.php?cisloclanku=2011060901

Stamoulis, K. and Farley, F. (2010) Conceptual approaches to adolescent risk-taking. *Cyberpsychology: Journal of Psychosocial Research on Cyberspace*, 4(1). Available at: www.cyberpsychology.eu/view.php?cisloclanku=2010050501

Tong, S.T. and Walther, J.B. (2011a). Just say 'no thanks': Romantic rejection in computer-mediated communication. *Journal of Personal and Social Relationships*, 28(4), 488–506.

Tong, S.T. and Walther, J.B. (2011b). Relational maintenance and CMC. In K.B. Wright and L.M. Webb (eds), *Computer-Mediated Communication in Personal Relationships* New York: Peter Lang Publishing. pp. 98–118.

Turkle, S. (2011) *Alone Together: Why we expect more from technology and less from each other*. New York: Basic Books.

Twyman, K., Saylor, C., Taylor, L.A. and Comeaux, C. (2010). Comparing children and adolescents engaged in cyberbullying to matched peers. *Cyberpsychology, Behavior, and Social Networking*, 13(2), 195–199.

Voelker, R. (2003). Mounting student depression taxing campus mental health services. *Journal of the American Medical Association*, 289(16), 2055–2056.

Walther, J.B., Anderson, J.F. and Park, D.W. (1994). Interpersonal effects in computer-mediated interaction: A meta-analysis of social and antisocial communication. *Communication Research*, 21, 460–487.

Wang, W. (2001). Internet dependency and psychosocial maturity among college students. *International Journal of Human-Computer Studies*, 55, 919–938.

Widyanto L. and McMurran, M. (2004) The psychometric properties of the Internet Addiction Test. *Cyberpsychology and Behavior*, 7(4), 443–450.

Willard, N. (2007). *Cyberbullying and Cyberthreats: Responding to the challenge of online social aggression, threats, and distress*. Eugene, OR: Research Press.

Yen, J.Y., Yen, C.F., Chen, C.S., Wang, P.W., Chang, Y.H. and Ko, C.H. (2012). Social anxiety in online and real-life interaction and their associated factors. *Cyberpsychology Behavior and Social Networking*, 15(1), 7–12.

Young, K.S. (1996) Internet addiction: The emergence of a new clinical disorder. *Cyberpsychology and Behavior*, 3, 237–244.

Young, K.S. (1998) *Caught in the Net*. New York: John Wiley.

Young, K.S. and Nabuco de Abreu, C. (eds) (2011) *Internet Addiction: A handbook and guide to evaluation and treatment*. Hoboken, NJ: John Wiley.

6

Engaging Practices: Doing Personalized Media

Heather A. Horst and Larissa Hjorth

INTRODUCTION

A schoolgirl updates her Facebook timeline. Two friends take a hipstamatic photo on their iPhone and upload it to their Kakao *I love coffee* social media game. A teenage boy chooses skin color and clothing for his avatar on the latest *Halo* game. An *anime* (animation) music video fan selects a screensaver and changes the ringtone on her phone. A mum uses locative media settings to surreptitiously check in to see where her daughter is. A young entrepreneur types in a username from which to upload restaurant reviews on Yelp. These practices represent just a few ways in which the personalization of digital artifacts and environments has become an everyday part of life for people around the world.

Within the context of digital and social media, scholars and others approach personalization as an intimate relationship with an object or consumer item as well as a marketing strategy wherein individuals identify with a particular brand or lifestyle. The creation of an intimate relationship with an object can take place through user customization (Ito et al., 2005), the use of cultural rituals and practices (Hjorth, 2009), forms of use and repair, and other forms of engagement. Such practices transform objects into signifiers of identity and/or even transform the objects into part of the self. While forms of personalization have always taken place in relation to other forms of material and consumer culture and the companies or individuals who create them, digital media has enabled personalization to occur with greater speed and frequency. The ability to easily copy and paste code has given people without technical skills greater access to the tools of production, enabling users to become active users or, in Toffler's (1980) terms, prosumers (Perkel, 2008). For marketers and others operating in the era of metadata or 'Big Data' (boyd and Crawford, in press), the personalization of technologies, profile pages, avatars and other activities has become a way through which companies can create and sustain niche markets and create brand loyalty. Whereas customization often involves users choosing or controlling the content, layouts, colors and other aspects of pre-set options to reflect the users' notions of the self and identity, personalization increasingly involves the use of (predictive) algorithms that anticipate preferences and patterns of use based upon previous behaviors.

Only a decade ago, ownership of digital media and technology like mobile phones primarily conveyed status; today the personalization of digital artifacts and environments has become mundane. Personalization often takes place in 'networked public culture' – that is, in spaces where 'those cultural artifacts associated with "personal" culture (like home movies, snapshots, diaries, and scrapbooks) have now entered the arena of "public" culture (like newspapers, cinema, and television)' (Russell et al., 2008, 88). For some individuals, the decisions around personalization represent annoyances wherein opportunities to personalize artifacts and environments are at odds with the struggle to remember and manage the range of digital identities we are now forced to inhabit. Andrejevic (2011) and others contend that the world of digital personalization is an example of Terranova's 'social factories' whereby commercial interests are 'colonizing' narratives of personal self-presentation and sociality (Gopinath, 2005). Here, the social is not only being commodified (Andrejevic, 2011), but also simplified through a conflation with media (Lovink, 2012). For others, personalization that occurs in and through online media spaces helps to define and redefine a sense of self and community (boyd, 2011; Ellison et al., 2011; Hjorth and Kim, 2005). Such individuals view decisions around personalization as an opportunity to express themselves and the different dimensions of their personality and play with the capabilities or affordances of each device and environment.

In this chapter we explore the emergence of the 'personalization' of digital media. Drawing from disciplines ranging from anthropology, cultural studies, sociology and others, we begin this chapter by examining the ways in which personalization has been approached through different disciplinary and interdisciplinary debates. After discussing four key analytic approaches – Social Shaping of Technology (STS)/Social Construction of Technology (SCOT); Cultural Studies; Material Culture Studies; and Domestication Theory – that explicitly examine the role of media, technology and consumption in social and cultural life, we then introduce two case studies of digital environments and artifacts. The first case study explores digital environments, exploring how [personal] history and meaning is made through the possibilities and constraints of different social network site platforms. The second case study considers the implications of mobile media convergence and branding through platforms such as the iPhone by examining how these forms of mobile media personalization 'settle' (or are domesticated) into everyday life. Through these case studies, we reflect upon how the focus upon personalization in digital artifacts and environments returns us to classic tensions in society between agency and structure and choice and constraint, as well as new questions around the role of openness and closure (Lessig, 2001), generated v. tethered (Zittrain, 2008) freedom and control (Chun, 2006), and customization v. standardization (Castells, 1999) in our digital worlds.

APPROACHES TO PERSONALIZATION

Social Shaping of Technology (STS) and Social Construction of Technology (SCOT)

While it is said that STS started in the 1960s, Science and Technology Studies (STS) as we know it today – what Steve Woolgar calls the 'turn to technology' – is attributed to the groundbreaking books, *Social Shaping of Technology* (MacKenzie and Wajcman, 1985) and *The Social Construction of Technological Systems* (Bijker et al., 1987). Within STS, there have been three main approaches – substantive, social constructivism and affordances. The substantive approach has often been associated with technological determinism or 'media effects' and has contributed to moral panics around new media like TV and video games; media scholar Marshall McLuhan's work is often described as 'media effects'. The substantive approach influenced the early development of cyberculture studies (Bell and Kennedy, 2000). Sci-fi writers like William Gibson and feminist scholars such as Donna Haraway, Sherry Turkle and N. Katherine Hayles were important in much of the imaginaries about the role of the Internet and in debating the impact the Internet would have on our offline lives. However, the substantive approach was criticized for its simplistic conception of technologies and users; critics argued that substantive approaches often negated the multi-dimensional agency of the user by ignoring the context in which the technologies/media were deployed and the myriad ways in which the user participated in, and even ignored, the media. The model was inadequate for exploring *how* and *why* users engaged in media in complex and diverse ways.

A second approach informed by STS frameworks is the concept of 'affordances', developed by Donald Norman (1988). Affordances draws upon the principles of human-centered design and attempts to consider not only the actors'/users' physical capabilities but also their motivations, plans, values and history. Focusing upon a relational rather than subjective or essentialist position, the affordances approach seeks a more 'ecological' understanding of the ways in which media and technologies become – or can be designed to become – part of users' everyday lives. The affordances approach has been widely influential in the fields of interaction design, cognitive design and human–computer interaction (HCI) that are interested in the development of practical design.

A third approach informed by STS is social constructivism, or what is often described as the social construction of technology (SCOT). Early pioneers in the area of technologies, such as Raymond Williams and Judy Wajcman,

influenced and informed what would become the SCOT approach that sought to move away from technological determinism. Overlaps between the work around audience and user agency in cultural studies and STS could be found at this time and informed what would become the 'participatory' model of contemporary media (Jenkins, 2006).

As technologies have become increasingly more integral in social relations and a defining feature of the user's identity and lifestyle, so too the models in which to conceptualize the dynamic relationship between the users and their technologies have needed to become more complex. Technologies operate across various levels, encompassing both the symbolic and material, and extend existing rituals of social interaction while providing new forms of expression and media literary. As Sherry Turkle notes, 'technology serves as a Rorschach over a lifetime, a projective screen for our changing and emotionally charged commitments' (2008: 11). While scholars in STS studies were moving into a more complex understanding of the different factors shaping our understanding and use of media and technology, SCOT theorists were increasingly in conversation with cultural studies scholars in trying to comprehend the role power and identity played in media practice.

Cultural Studies

The establishment of cultural studies is largely attributed to Richard Hoggart and the Birmingham School at the Centre for Contemporary Cultural Studies (CCCS). Operating between 1964 and 1988, the Birmingham School analyzed popular culture (especially the relationship between mass media and youth subcultures), the politics of difference (gender, race and class) and hegemony (ideologies of power). While Williams and Hoggart are viewed as the key figures in this early stage of cultural studies, it was under Stuart Hall's directorship (1968–79) that cultural studies flourished. Focusing upon analyzing tacit power relationships – especially hegemony – cultural studies increasingly explored the various tensions (resistance and control) around popular and material cultures. Key texts of this period were Hebdige's *Subculture: The Meaning of Style* (1979) and *Resistance through Rituals* (Hall and Jefferson, 1976). Feminist scholars such as Angela McRobbie were key in highlighting the ways in which subcultures often replicated the hegemonic (and sexist) structures.

Concurrent to the work conducted by anthropologists such as Clifford Geertz and others interested in cultural texts, Stuart Hall's (1996) emphasis on encoding/decoding began to contest the 'fixed' meaning of texts in favor of interpretive models that gave agency to the reader. Extending the 'death of the author' paradigm posed by Roland Barthes, David Morley (Morley and Brundson, 1980) extended Hall's approach through his empirical case study of *Nationwide*; the focus upon readers'/audiences' agency represented a shift from textual analysis into what was dubbed an 'ethnographic' turn (Ang, 1985). It also contributed to the growth of fan studies and a range of other studies of the relationship between media and public/popular culture (Jenkins, 2006).

Building upon Morley's work, Paul du Gay et al.'s (1997) 'circuits of culture' approach invited researchers to map the dynamics of culture – described by the core categories of consumption, production, regulation, representation and identity – as they co-influenced one another to produce the meanings of a particular cultural object (in that case, the Sony Walkman). While the Sony Walkman case study itself is now decades old, the ongoing significance of this approach continues (Goggin, 2006). *Doing Cultural Studies* (du Gay et al., 1997) used an analysis of a significant technology of the time (the Sony Walkman) for dual purposes: to analyze new forms of media practice (consumption and production); and at the same time to consolidate and communicate the theoretical and methodological approaches of what was still an emerging discipline (cultural studies) for the benefit of students and researchers. The ongoing relevance of this work was that it captured the *zeitgeist* of cultural analysis in the 1990s and provided a benchmark for cultural studies research beyond that moment; Hjorth, Burgess and Richardson (2012) recently employed a 'circuit of culture' approach to the analysis of the iPhone. While the circuit of culture approach was an important moment in cultural studies, it also identified the ways in which the relationship to, and understandings of how, approaches to material culture and consumption were changing.

Material Culture Studies

Whereas cultural studies draws our attention to texts and the tensions between structure and agency, material culture studies focuses upon the processes and organizing principles behind people's engagement with things, in all of their material form. Historically situated within the museums and 'artifact' collections where objects were ordered and examined in terms of degrees of technological and thus social progress (Buchli, 2002), material culture studies developed out of British social anthropology and is characterized by interdisciplinary discussions and methods. While material culture studies projects and approaches vary, a fundamental assertion of material culture studies is its distinction from linguistic and other related forms of communication. As Tilley observes, 'material culture does not communicate meaning content in the same way as speech or phonetic script' (Tilley, 1991: 17).

As a theoretical, conceptual and methodological framework, material culture studies utilize objects as a vantage point for understanding social change and social interactions. Central to this perspective is an engagement with the specific materiality of the objects that acknowledges the relationality between persons and objects. In one of the seminal works of material culture studies, Miller (1987) explores the manner in which individuals, couples, families, communities or societies create or are influenced by material forms, a process called objectification. As Miller summarizes, objectification is:

> A dual process by means of which a subject externalizes itself in a creative act of differentiation, and in turn reappropriates this externalization through an act which Hegel terms sublation (*aufhebung*). This act eliminates the separation of the subject from its creation but does not eliminate this creation itself; instead, the creation is used to enrich and develop the subject, which then transcends its earlier state. (Miller, 1987: 28)

Like Miller, Pierre Bourdieu's (1977) work on consumption highlights the processes through which culture is created and, eventually, becomes hidden from consciousness through the socialization of habits and routines, as well as the order and structures of the material and social worlds, or, the habitus.

How people construct things, how things construct individuals and, in turn, how the various creations continue to change and interact over time, represent fundamental questions for contemporary material culture studies. As Victor Buchli observes, 'the processes of materialization are more significant than materiality itself and in fact variable constitutive of it – material culture itself is just a peculiar moment in these processes' (2002: 16). Igor Kopytoff's (1986) biography of the commodity demonstrates how commodities change over time and place and become alienated from the process of production (Appadurai, 1986). Understanding the processes that govern this tendency to normalize and/or re-impose normativity as rapidly as media and technology changes (Horst and Miller, 2012), continues to represent one of the perennial questions of contemporary work on digital artifacts and environments.

Domestication Theory

The domestication approach focuses on the use and symbolic role of technologies in everyday life after their acquisition. The word 'domestication' comes from 'taming the wild animal' and this was then applied to describing the processes involved in 'domesticating

ICTs' when bringing them into the home. Founded by Roger Silverstone, the British tradition of domestication theory grew out of media studies and interest in consumption studies, as well as anthropological work by Daniel Miller, Eric Hirsch and others that explored how objects become part of identity and social life. The Norwegian version integrated British approaches to domestication with conceptual frameworks from SCOT (Ling, 2004) in order to understand the evolution of mobile media and game studies, locating media within broader cultural and technological practices. The ongoing influence of the domestication approach is tied to its ability to elucidate the socio-cultural and individualistic symbolic power of commodities, especially communication technologies as an extension of older rituals and cultural practices. For example, rituals around mobile communication often involve older cultural practices like gift giving. As Silverstone and Hirsch observe in their now classic contribution to domestication theory, contemporary technological artifacts must be viewed as essentially material objects, capable of great symbolic significance, investment and meaning, while domestic technologies are 'embedded in the structures and dynamics of contemporary consumer culture' (1992: 20).

The approach identifies new technologies as having become embedded in everyday life and household social relations (i.e. family power relations); this results in new technologies and media not only being a site (a space or context) for making meaning, but also a place in which meaning can be gleaned. In other words, objects have their own meanings that are then put into the dynamics of cultural practice, which, in turn, redefines meaning. This is why objects take on different meanings dependent upon context. While a number of critiques have emerged around the application of domestication theory outside of the home and Western contexts (Lim, 2006), domestication theory brings together the focus upon the relationship between objects or ecologies represented by SCOT, the focus upon text and meaning in cultural studies, and the focus upon logics of change and processes characteristic of material culture studies.

THE PLACES OF PERSONALIZATION: CASE STUDIES

The four perspectives introduced in the previous section shape the content, theory and analysis of the following case studies of personalization. The first case study of the domestication of social network sites draws most explicitly on material culture and domestication theory through its examination of the processes of engagement with the sites and the changes that emerge over time. Focused upon broader power relations and society shifts that shape use, the second case study integrates STS and cultural studies approaches in its exploration of iPhone users and the broader structural shifts of design personalization in mobile media platforms. In both cases we see the interplay between the affordances and capabilities of the artifacts and environment and its implications for use. Both case studies illustrate some of the tensions around conceptualizing personalization as a process that encompasses notions of control/freedom, empowerment/exploitation and customization/standardization.

Case Study: Aesthetics of the Self in and through Social Network Sites

In most Western contexts, youth is a phase of life where establishing independence and a sense of autonomy from families and other institutions gives way to the influence of friends and peer groups. The creation of social network site profiles with pictures, personal details and other information is designed to connect and enhance communication between classmates, friends, family, co-workers and other acquaintances. Although the connections and interactions between participants qualitatively varies with particular websites, interests and activities, it is clear that a broad spectrum

of youth and adults now participate in creating, maintaining and negotiating an expanded range of connections using these sites. Excited by the possibility of 'hanging out' with their friends in social spaces which are largely (though not exclusively) outside the purview of adults and parents (Horst et al., 2010), youth view social network sites as important tools and spaces to develop relationships with their friends and peers.

During her junior and senior years of high school, 18-year-old Ann was an active MySpace user who uploaded pictures and commented on friends' comments on a daily basis. Ann also participated in what she and her friends called 'MySpace parties', or sleepovers that involved dressing up and taking photographs to post on their respective MySpace pages. In addition to trying on clothes and posing for pictures, Ann and her friends also began to make videos of themselves doing 'funny stuff', such as dancing or imitating celebrities. While the pictures, songs, personality quizzes and other content on her MySpace page changed on an intermittent basis, Ann's favorite part of her page, and the most consistent feature of her MySpace page and profile, involved the incorporation of Ann's signature colors, brown and pink. As Perkel (2008) observes, copying and pasting has become a prevalent practice among those who want to personalize, customize or update and alter their social network site profiles, acts that simultaneously signal the domestication of digital media and technology.

Many of the tips or guides for changing one's profile and MySpace page (such as embedding images and videos and uploading pictures) are online – on other people's profiles, in online guides and a range of social network sites. Describing her MySpace page, Ann notes, 'It's actually the colors of my room so it's like brown and pink. And then I don't know. I had … a default pink so it's like what everyone sees when they see a comment'. As work in cultural studies on the relationship between consumer and popular culture and identity foregrounds, Ann's personal page mirrors the private space of her bedroom at home. The walls of her room are painted a matte chocolate brown, and the main features of her room – such as her twin-sized quilt, a large desk and a large French bulletin board – are pink. Other pink and brown accents – such as pillows on the bed, the ribbon on her bulletin board, the cushion on her desk chair and picture frames – have been carefully selected and arranged throughout her bedroom. For Ann, brown and pink constitute the backdrop to her daily life in both online and offline spaces.

After accepting an offer to attend a small liberal arts college, however, Ann received an invitation from her future dorm's resident assistant (RA) to participate in Facebook. Ann's RA sent her an invitation to be a member of the 'Crystal Mountain' wing, part of a wider network of 90 dorm residents attending her new college. Ann began spending hours at a time perusing different people's sites, looking for familiar names and faces and checking out friends of friends. As the summer progressed, Ann increasingly felt that she was becoming 'addicted' to Facebook, checking it whenever she had a free moment for status updates about four or five times per day. Through this brief, repetitive engagement, Ann started to meet the other students slated to live in her dorm, the most important and exciting of these new connections being her future roommate, Sarah.

Over the summer, Ann and Sarah sent each other short messages and comments. Some of these messages were pragmatic, such as solidifying plans to move into their dorm room, what furnishings they would be bringing or which classes they planned to take. In addition to using Facebook to communicate, Ann delved into the details of Sarah's Facebook page for insight into what she imagined would be shared interests. Decisions around what to bring to college were aligned with a desire to construct an aesthetic balance. With the aim of decorating their new dorm room, Ann and Sarah decided to upload a few pictures of their bedrooms at home on to their Facebook pages to get a sense of each other's

style and tastes. Ann was excited when she looked at the photographs and saw Sarah's signature colors. 'I'm brown and pink stuff and she's brown and blue stuff!'. As scholars of material culture studies have highlighted (Young, 2005), Ann surmised that this aesthetic harmony would also signify a harmonious relationship.

Individuality is highly valued in the United States, particularly in a place like Silicon Valley where culture and competition are closely intertwined (Horst, 2009). In American society, adolescence is segmented as a particularly important time for discovering and expressing a sense of self that seems 'uniquely' one's own, an identity that is separate and autonomous from given social relationships, such as families, neighborhoods and communities. The locations of self-making and, in the language of Erving Goffman (1959), the 'presentation of the self' have roughly corresponded with the interplay between the front stage (public) and the back stage (private). In the age of networked public culture, the boundaries between the public and private presentations of the self are increasingly blurred.

The focus on 'face' and 'presentation' have remained central to the study of the constitution of the self and individual identity on the Internet, especially the formal (and often static and textual) presentations on webpages and other online sites (Miller 1995). However, the material properties of new media and social network sites like MySpace and Facebook shape the way that these are expressed and, increasingly, the very terms and definitions of self. In addition to maintaining a collection of 'friends' (boyd, 2008; Ito et al., 2010), MySpace enabled Ann to customize the background color and font of her profile page in the same color palate as her bedroom. It was also possible to add favorite songs, videos and a range of other features. Indeed, MySpace makes it easy to 'copy and paste' html code from others' profile pages and websites in order to customize and copy the style on one's own webpage, this ability to customize ultimately undermined Ann's friends' attempts to recreate a profile after Ann's parents forced her to delete her profile after a very public case of a girl being abducted (Perkel, 2008). The ability to delete and recreate profiles thus structures a very different engagement with digital spaces.

Ann views MySpace and Facebook as places where the physical and material – relationships, tastes and connections – are reaffirmed. For Ann, MySpace and Facebook are tangible spaces where she establishes and asserts her sense of self. In a consumer culture like the United States, Ann essentially constructs herself as different configurations of predetermined selections (customization) within the generally acceptable genres of her peers, a person who likes pink and brown, someone who likes a particular kind of music and someone who maintains a balance and order in all the 'environments' she inhabits. Also like many American teenagers, her sense of self in the world hinges upon asserting a material presence in these different environments.

Ann's aesthetic is based on the balance and continuity between a variety of key relationships. These may be objects, persons or places. In many ways Ann uses her roommate as a critical background relationship for her Facebook profile, in much the same fashion she used her bedroom as the critical background relationship for her MySpace profile. The social network sites where Ann chooses to participate extend the mirror in which she comes to see herself and gain a sense of who she might be (Miller, 1995; Robinson, 2007; Strathern, 2004). Youth like Ann create order in and through the construction, alteration and appropriation of their interconnected media worlds. For Ann, and many individuals like her, social network sites such as MySpace and Facebook play or have played an important role in structuring and sustaining social worlds, including Ann's ability to imagine her future college life in the dorm and establish relationships with new individuals and communities. Social network sites also provide Ann with opportunities to understand and assert her own sense of who she is and who

she might become in the transition from high school to college through spaces of networked public culture.

Case Study: Mobile Media as Personalization Par Excellence

The mobile phone is one of the most intimate devices today (Fortunati, 2002; Lasén, 2004), a repository for the personal (Bull, 2007). Personalization as a process is harnessed by industry as well as the practices of the user – epitomized by user-created content (UCC). In each location, the practices of personalization differ, highlighting the growing importance of place in shaping networked and mobile media. Far from eroding the importance of place, mobile media amplify a sense of locality and belonging. As mobile media moves into the undulating terrain of smartphones we see tensions around open and closed systems coming to the fore (Goggin, 2012). On the one hand, smartphones like iPhones afford users new applications and convergent platforms to create content and develop 'vernacular creativity' (Burgess, 2009). On the other hand, with the rise of big data – that is, users' profiles taken by companies like Google and sold to marketers – media personalization may become little more than a node in the social factory of capitalism (Andrejevic, 2011).

The iPhone, with its motto of 'it's not a phone, it's a platform' (Grossman, 2007), epitomizes contemporary debates and tensions over contemporary forms of personalization. The escalation of the iPhone's popularity has been very much due to the ability to customize applications. Paralleling the customization of Japan's i-mode, the iPhone's success has been contingent upon applications, and in particular the way in which Apple has managed to rebrand personalization to such an extent it seems to be an Apple invention. Yet, the personalization of technology occurred long before mobile media and the iPhone phenomenon. Indeed, countries such as Japan have excelled globally in their ability to spearhead the 'personal technologies' revolution from the Sony Walkman onwards. Clearly signposting her SCOT lineage, Ito (2005), for example, argues that it is the notion of the 'personal', along with the pedestrian and portable, that has characterized Japanese technologies for decades. Indeed, part of Apple's success results from their deployment of high-level personalization, particularly apparent in what anthropologist Brian McVeigh (2000) has called 'techno-cute' which makes the coldness of new technologies 'warm' and 'friendly'. Personalization continues to be about domesticating, locating and humanizing new technologies; it is this ongoing and localized practice that is indicative of the tension between users and industry as personal technologies increasingly become public.

The closed platform world offered by the iPhone notably borrows from its precursor, the i-mode in Japan. Sawhney describes this paradox of personalization in the iPhone's precursor – a tension between user creativity (i.e. UCC) and openness, and industry's commercializing of personalization in the form of applications – as a gated community version of the Internet. Moreover, whereas Zittrain (2008) contends that the iPhone is part of a constellation of devices he calls 'tethered applications', Goggin (2011) argues that 'tethered devices' fail to be generative platforms because they are configured to be actively inimical to user experimentation and co-creation. For Goggin, part of this shift lies in the way in which Apple managed to brand the smartphone evolution into an Apple revolution. As he notes:

> Perhaps what most distinguished the iPhone from many other adaptations of cell phones was its rapturous reception, and, hand-in-hand with this, Apple's phenomenally successful marketing campaign. Herein lies the paradox of adaptation that the iPhone represents. The iPhone is clearly an adaptation of the cell phone. As *Wired* magazine's Geekipedia points out, the iPhone is an obvious descendant of the smartphone – the multimedia cell pone that combines various computer programs with entertainment options. Yet the 'biggest launch since the Apollo'

> rebadges this evolution as a revolution … (Goggin, 2011: 142)

The paradoxes surrounding this repurposing and relocation of the mobile media evolution into the iPhone revolution is perhaps best understood by moving beyond media images and rhetoric to the personalization practices of everyday users. For example, can we say there is such a thing as an iPhone affect? What are the realities of personalization practice beyond the initial excitement of 'newness' and novelty?

Given that our first case study explored the world of a young women's rites of passage vis-à-vis social media, in this second case study we reflect upon women's experiences of mobile media personalization and its relationship to work and family life. While our first case study considered personalization in relationship to a young person's identity formation, this case study reveals the multigenerational character of personalization practices through an analysis of surveys and interviews among users in Melbourne, Australia between 2009 and 2010. Focused upon the relationship between personalization and affect, the study revealed that mobile media practice is a highly gendered activity (Fortunati, 2009; Hjorth, 2009). Indeed, where personalization was concerned, women seemed to be most vocal in discussing the struggles between work and leisure, home and public – what Gregg (2011) has called 'presence bleed'. In the study, iPhone users spoke in detail about the apps that seemed to create tensions around control and freedom, customization and standardization. However, despite the initial downloading of a plethora of apps, many of the respondents noted a decline in use after the 'honeymoon' was over.

In order to understand iPhone perceptions and practices *after* the honeymoon phase, we surveyed users after they had initially bought the iPhone and the types of media practice they participated in, then again one year later, to see how these practices had been domesticated (or not) into the users' lives. This surveying occurred when the iPhone 3 was first introduced in 2009 and then again, around the time of the iPhone 4 launch in 2010. For this study we surveyed 40 women of varying ages and across different socio-economic and cultural backgrounds in Australia. One of the first things that became apparent was the fact that many respondents either had IT experience or viewed new media as necessary for work or as a hobby. This media literacy phenomenon is not the impact of the iPhone *per se*, but highlights how the timing of the iPhone can be viewed as part of a broader new media landscape in which women's labor and creativity feature. As Fortunati (2009) has observed, there needs to be more studies of gender gradations, especially in terms of age and class, to understand the role of gendered mobile media practice. In addition, we see how the phone has become the multimedia device akin to a miniature caravan, which houses all one's personal details, much like the function of one's home as a symbol for domesticity, privacy and family. While this 'caravan' affords much mobility, its symbolic weight, and wait (i.e. the temporality), operates as a perpetual reminder of the various tasks and work in need of doing. This phenomenon has been called many things, including the 'wireless leash' (Qiu, 2008) whereby the mobile sets us free at the same time as it creates more limits. For example, one is 'free' to roam but also available. This miniature caravan also reveals people's increasing proclivity towards working at home. Just as the intimate goes public, the public – and especially work – goes private.

The tensions between the pros and cons surrounding mobile intimacy – that is, the overlay of the physical, geographic and electronic with the social, emotional and cultural – are amplified in the case of working mothers. For working mothers, the tethering of the phone to the domestic is clear – symbolized by the always-on mothering feature that Matsuda has aptly called 'mum in the pocket' (2009). As Lim (2006) identifies in the case of mainland Chinese parents, many see technologies like mobile media as not only important for their children's education but also a

way in which to keep a perpetual eye on them. Indeed, the levels and layers of tethering (that often pull) afforded through the so-called mobility of mobile media come at a price. Given that many women negotiate mothering along with paid work, there is a tension between mobile media decreasing yet also adding to the daily workload. Not only is the miniature caravan an embodiment of domestic labor and tethering to home, it is also about the home office. Gregg's (2011) notions of 'work's intimacy' and the 'presence bleed' highlights how workers struggle to differentiate between work and leisure time and spaces in light of the presence of online and social media. Despite initially downloading many 'cool' apps about creativity (i.e. photo apps like Hipstamatic), play (standard puzzle or simple games like *Angry Birds* and haptic games like Balloonimals), socializing (Facebook, Twitter, Foursquare) and lifestyle, it was often the basic work tasks like answering and writing e-mails and surfing the Internet that featured in respondents' usage. In short, their usage of the iPhone demonstrates no clear 'iPhone affect'.

One of the key features that emerged was the difference in personalization practices between those who had and didn't have children. This quality played out across a variety of personalization practices, from respondents' creativity and UCC to the types of affordances and applications used. For this discussion we focus upon a few of the female respondents who were working, some of whom had children and some of whom did not, and either worked in or had immense interest in the area of new media. The need to create clearly defined boundaries between work and life was something that the respondents grappled with as work and intimacy became increasingly mobile and performed both at home and away. On the one hand, the iPhone functioned as a well-designed mobile with Internet, allowing them to be connected to work whilst at home. On the other hand, this flexibility also meant that many felt the pressure to always be 'on' and in work mode whilst also doing domestic work. For Zoe, like other respondents, the main use of the iPhone was e-mails and Internet on the run. Whilst she had initially started using the calendar she stopped after a few weeks and reverted back to the old mode of writing on paper, noting that the multimedia capacity of the iPhone was 'probably wasted on me'. Viewing the iPhone as fun, useful and about connectivity, she described the iPhone as just a 'glorified iPod'.

For women who had children, the role of media such as iPhones clearly helped to forge work/home fusions in what was an already highly multitasking environment. Many of these women did a lot of their work from home (so that they could effectively do at least two jobs: looking after the children and working), demonstrating the role of personal technologies in not only outsourcing domestication outside the physical home but also bringing work back into the home. The iPhone could be seen to be part of the casualization of labor whereby work becomes and is squeezed into micro-moments between other activities in the home. Being a repository for, rather than a cause of, these work/home fusions, the iPhone highlights how being 'always on' comes at a cost (Gregg, 2011; Wajcman et al., 2009). Just as mobile technologies set us free to roam wherever we wish, they also create new types of restrictions, whereby one is always on call, and impact upon the types of co-presence (being simultaneously here and there, online and offline) we participate in as the personal and intimate fuses with the public, and work bleeds into the private. Indeed, mobile media personalization highlights many of the paradoxes within contemporary life.

CONCLUSION: PERSONALIZATION, 'PRODUSERS', PRIVATIZATION AND BEYOND

In this chapter we have discussed the genealogy of digital media personalization through various disciplinary trajectories. In the first

case study, we explored the interplay between the capabilities and affordance of social network sites (the ability to personalize user profiles, the capacity to upload pictures and the use of the site as a platform for other activities), the motivations for youth to use social network sites (e.g. to create and maintain friendships, share information and present oneself) and the broader identity work that teenagers in the US and other Western contexts often engage in as part of the process of coming of age. Through the case study of mobile media we highlighted some of the ways in which personalization is an unavoidable and deeply paradoxical experience that reflects broader socio-cultural inflections about control and freedom (Chun, 2006). Mobile media as digital artifacts are repositories for, and of, intimacy by reflecting the user's relationships and communication patterns with others. In this section we draw from a case study of the iPhone and the role of applications (apps). Apps represent industry-generated rather than user-created personalization.

According to Campbell and Park, the shift towards a new 'personal communication society' can be 'evidenced by several key areas of social change, including symbolic meaning of the technology, new forms of coordination and social networking, personalization of public spaces, and the mobile youth culture' (2008: 371). Digital media environments and mobile media spur us to reconsider approaches such as domestication, given the increasing collapsing boundaries between a definitive space called 'home' and the artifacts and environments that accompany us throughout the day. As Morley (2003) suggests, various artifacts and environments provide the capacity to tether the user to notions of the home at various levels: emotional, social, and psychological, to name a few.

Just as intimacy has become public in and through networked public culture, so too has the notion of the personal taken on new cartographies. In the first case study, we focused upon how youth construct a sense of order in and through digital environments and the interplay between these new media and their relationships with places, persons and objects. We further revealed the ways in which these media spaces suggest an act of self-construction that is highly social, but also constrained, whereby individuality emerges through the ordering and configuring of space in relation to peers and parents. What is significant is not the degree of individualism Ann exhibits, but the ways in which Ann and other individuals exist in alignment with highly socialized media of expression. This is perhaps even more evident with social network sites, which enable youth to make public the bedroom, a space often viewed as a highly privatized and personal domain. This case study also highlights how the idea of the domestic – through the conflation between domestic and personal technologies – has become increasingly mobile across technological, geographic and socio-economic terrains. In turn, this requires a reworking of such approaches as the domestication theory, given that the dynamism of the home is now increasingly pervasive outside the physical space of the home, leading many to suggest that 'home' may be as much a social network site or mobile phone as it is a physical location with a street address.

The personalization of media also reflects broader debates about user agency and the increasing commodification of data. As the second case study revealed, underlying the so-called User created content (UCC) revolution has been the fact that the data is then mined and sold to advertisers (Lovink, 2012; Vaidhyanathan, 2011). This has led some, like Andrejevic to argue that, 'social networking sites don't publicize community, they privatize it' (Andrejevic, 2011: 97). While Andrejevic sees the personalization of media as part of users' creative, social and emotional labor that are exploited by corporations for financial gain, Humphreys and Banks (2008) view online game users/players as co-creators. Extending Toffler's concept of prosumers, wherein consumers start to take on elements of the production for free (i.e. filling their own petrol tanks), Bruns (2008) utilizes the term 'produsers' to describe the phenomenon of producing users. This debate around the exploitation and empowerment elements of online media today is only going to amplify with the increasing integration of 'Big Data' into our everyday lives. Increasingly, questions about agency, power and labor continue to haunt the field as it moves unevenly into convergent social, locative and mobile media terrains. As personalization migrates across different platforms, contexts and media, its cartographies become more political in their expression of inner capitalist paradoxes. Whether it is empowerment v. exploitation, control v. freedom or customization v. standardization, personalization in an age of data mining has become a vexed issue that is as ubiquitous as it is ambiguous.

SUGGESTIONS FOR FURTHER READING

Hinton, S. and L. Hjorth (2013) *Understanding Social Media*. London: Sage.

Hjorth, L. (2009) *Mobile Media in the Asia-Pacific: Gender and the art of being mobile*. Abingdon, Oxon: Routledge.

Horst, H. and Miller, D. (eds.) (2012) *Digital Anthropology*. Oxford: Berg.

Gerard, G. (2011) *Global Mobile Media*. Abingdon, Oxon: Routledge.

Ito, M., et al. (2010) *Hanging Out, Messing Around and Geeking Out*. Cambridge, MA: MIT Press. Available at: http://mitpress.mit.edu/books/hanging-out-messing-around-and-geeking-out

Perkel, D. (2008) 'Copy and paste literacy?: Literacy practices in the production of a myspace profile', in K. Drotner, H.S. Jensen and K. Schroeder (eds.), *Informal Learning and Digital Media: Constructions, contexts, consequences*. Newcastle: Cambridge Scholars Press.

Papacharissi, Z. (ed.) (2011) *A Networked Self: Identity, community, and culture on social network sites*. Abingdon, Oxon: Routledge.

REFERENCES

Andrejevic, M. (2011) 'Social network exploitation', in Z. Papacharissi (ed.), *A Networked Self: Identity,* community, and culture on social network Sites. Abingdon, Oxon: Routledge, 82–101.

Ang, I. (1985) *Watching Dallas: Soap opera and the melodramatic imagination*. London: Routledge.

Appadurai, A. (ed.) (1988). *The Social Life of Things: Commodities in Cultural Perspective*. Cambridge: Cambridge University Press.

Bell, D. and Kennedy, B. (eds.) (2000) *The Cyberculture Reader*. Abingdon, Oxon: Routledge.

Berlant, L. (1998) 'Intimacy: A special issue', *Critical Inquiry*, 24/2: 281–88.

Bijker, W., Hughes, T. and Pinch, T. (eds.) (1987) *The Social Construction of Technological Systems: New directions in the sociology and history of technology*. Cambridge: MIT Press.

Bloustein, G. (2003) *Girl Making: A cross-cultural ethnography on the processes of growing up female*. New York: Berghahn Books.

Bourdieu, P. (1977) *Outline of a Theory of Practice*. Translated Richard Nice. Cambridge: Cambridge University Press.

Bovill, M and Livingstone, S. (2001) 'Bedroom culture and the privatization of media use', in *Children and Their Changing Media Environment: A European comparative study*. Mahwah, NJ: Lawrence Erlbaum, pp. 179–200.

boyd, d. (2008) 'Why youth (heart) social network sites: The role of networked publics in teenage social life', in D. Buckingham (ed.), *Youth, Identity, and Digital Media*. John D. and Catherine T. MacArthur Foundation Series on Digital Media and Learning. Cambridge, MA: MIT Press, pp. 119–42

boyd, d. (2009) 'Friendship', in Ito, et al. (eds.) *Hanging Out, Messing Around and Geeking Out: Living and learning with new media*. Cambridge: MIT Press.

boyd, d. and Crawford, K. (in press) 'Critical questions for big data: Provocations for a cultural, technological, and scholarly phenomenon', *Information, Communication, & Society*.

Boyd, d. (2010). 'Social Network Sites as Networked Publics: Affordances, Dynamics, and Implications,' in Zizi Papacharissi (ed.) *Networked Self: Identity, Community, and Culture on Social Network Sites*, pp. 39–58.

boyd, d. and Ellison, N. (2007) 'Social network sites: Definition, history, and scholarship', *Journal of Computer-Mediated Communication*, 13(1), Available at: http://jcmc.indiana.edu/vol13/issue1/boyd.ellison.html

Bruns, A. (2005) 'Some Exploratory Notes on Produsers and Produsage', snurblog, 3 November 2005, Available at: http://snurb.info/index.php?q=node/329

Bruns, A. (2008) *Blogs, Wikipedia, Second Life, and Beyond: From production to produsage* (*Digital Formations; v. 45*). New York: Peter Lang.

Buchli, V. (2002) *The Material Culture Reader*. London: Berg.

Bull, M. (2007) *Sound Moves. iPod Culture and Urban Experience*. Abingdon, Oxon: Routledge.

Burgess, J. (2008) '"All Your Chocolate Rain Are Belong to Us"?: Viral video, YouTube and the dynamics of participatory culture', in G. Lovink et al. (eds.), *The Video Vortex Reader*. Amsterdam: Institute of Network Cultures.

Burgess, Jean E. (2009) Remediating vernacular creativity: Photography and cultural citizenship in the Flickr photosharing network, in Edensor, T., Leslie, D., Millington, S., and Rantisi, N., (eds.), *Spaces of Vernacular Creativity: Rethinking the cultural economy*. Abingdon, Oxon and New York: Routledge, pp. 116–126.

Campbell, S. and Park, Y. (2008) 'Social implications of mobile telephony: The rise of personal communication society', *Sociology Compass*, 2(2): 371–87.

Cassell, J. and Cramer, M. (2008) 'High tech or high risk: Moral panics about girls online', in T. McPherson's (ed.), *Digital Youth, Innovation, and the Unexpected*. The John D. and Catherine T. MacArthur Foundation Series on Digital Media and Learning. Cambridge, MA: MIT Press, pp. 53–76.

Castells, M. (1999) *The Rise of The Networked Society*. New York, MA: Oxford University Press.

Chun, W. (2006) *Control and Freedom*. Cambridge, MA: MIT Press.

Du Gay, P., Hall, S., Janes, L., Mackay, H. and Negus, K. (eds.) (1997) *Doing Cultural Studies: The story of the Walkman*. London: Sage.

Ellison, N., Lampe, C., Steinfield, C., and Vitak, J. (2011). 'With a little help from my Friends: Social network sites and social capital,' in Z. Papacharissi (ed.), *A networked self: Identity, community and culture on social network sites*. New York: Routledge. Pp. 124–145

Fortunati, L. (2002) 'Italy: Stereotypes, true and false', in J.E. Katz and M. Aakhus (eds.), *Perpetual Contact: Mobile communications, private talk, public performance*, Cambridge: Cambridge University Press. pp. 42–62.

Fortunati, L. (2008) 'Gender and the mobile phone', in G. Goggin and H. Hjorth (eds), *Mobile Technologies*. Abingdon, Oxon: Routledge, pp. 23–35.

Fortunati, L. (2009) 'Gender and the cell phone', in Goggin, G. and Hjorth, L. (eds), *Mobile Technologies*. New York: Routledge, pp. 23–36.

Gerard, G. (2011) *Global Mobile Media*. Abingdon, Oxon: Routledge.

Goffman, E. (1959) *The Presentation of Self in Everyday Life*. New York: Anchor Books.

Goggin, G. (2006) *Cell Phone Culture: Mobile technology in everyday life*. Abingdon, Oxon: Routledge.

Goggin, G. (2011) *Global Mobile Media*. Abingdon, Oxon and New York: Routledge.

Gopinath, S. (2005) 'Ringtones, or the auditory logic of globalization', *First Monday* 10(12) (December), Available at: http://firstmonday.org/htbin/cgiwrap/bin/ojs/index.php/fm/article/view/1295/1215

Gregg, M. (2011) *Work's Intimacy*. Cambridge: Polity Press.

Grossman, L. (2007) 'Invention of the year: The iPhone', *Time*. Available at: www.time.com/time/specials/2007/article/0,28804,1677329_1678542,00.html (accessed 1 December 2008).

Hall, S. (1973) [1996] 'Encoding/decoding', in P. Marris and S. Thornham (eds.), *Media Studies: A reader*. Edinburgh: Edinburgh University Press.

Hall, S. and Jefferson, T. (eds.) (1976). *Resistance through Rituals*. London: Routledge.

Hebdige, D. (1979) *Subculture: The meaning of style*. London: Metheun.

Hjorth, L. (2009) *Mobile Media in the Asia-Pacific: Gender and the art of being mobile*. Abingdon, Oxon: Routledge.

Hjorth, L., Burgess, J. and Richardson, I. (eds) (2012) *Studying Mobile Media: Cultural technologies, mobile communication, and the iPhone*. Abingdon, Oxon: Routledge.

Hjorth, L. and Kim, H. (2005). 'Being There and Being Here: Gendered customising of mobile 3G practices through a case study in Seoul', Convergence: The International Journal of Research into New Media Technologies, 11 (2): 49–55.

Horst, H. (2009) 'Aesthetics of the self: Digital mediations', in D. Miller (ed.), *Anthropology and the Individual*. Oxford: Berg, pp. 99–113.

Horst, H. (2010). 'Families,' in Ito, M., et al. *Hanging Out, Messing Around and Geeking Out*. Cambridge, MA: MIT Press, 149–194. Available at http://mitpress.mit.edu/books/hanging-out-messing-around-and-geeking-out

Horst, H. and Miller, D. (eds) (2012) *Digital Anthropology*. London: Berg.

Humphreys, S. and Banks, J. (2008) 'The labour of user co-creators', *Convergence* 14(4): 401–18.

Ito, M. (2005) 'Introduction: Personal, portable, pedestrian', in M. Ito, D. Okabe and M. Matsuda (eds), *Personal, Portable, Pedestrian: Mobile phones in Japanese life*. Cambridge, MA: MIT Press, pp. 1–16.

Ito, M., Matsuda, M. and Okabe, D. (eds.) (2005) *Personal, Portable, Pedestrian: Mobile phones in Japanese life*. Cambridge, MA: MIT Press.

Ito, M., Baumer, S., Bittanti, M., boyd, d., Cody, R., Herr-Stephenson, R., Horst, H., Lange, P.G., Mahendran, D., Martinez, K.Z., Pascoe, C.J., Perkel, D., Robinson, L., Sims, C. and Tripp, L. (2010) *Hanging Out, Messing Around, and Geeking Out: Kids living and learning with new media*. Cambridge, MA: MIT Press.

Jenkins, H. (2006) *Fans, Bloggers, and Gamers: Essays on participatory culture*. New York: New York University Press.

Kopytoff, I. (1986) 'The cultural biography of things: Commoditization as process', in A. Appadurai (ed.), *The Social Life of Things: Commodities in cultural perspective*, Cambridge: Cambridge University Press, pp. 64–91.

Lasén, A. (2004) 'Affective technologies: Emotions and mobile phones', *Receiver*, 11. Available at: www.academia.edu/472410/Affective_Technologies._Emotions_and_Mobile_Phones

Lessig, L. (2001) *The Future of Ideas*. New York: Random House.

Lim, S. (2006) 'From cultural to information revolution: ICT domestication by middle-class chinese families', in T. Berker, M. Hartmann, Y. Punie and K. Ward (eds.), *Domestication of Media and Technology*. Maidenhead: McGraw-Hill, pp 185–201.

Ling, Richard S. (2004) *The Mobile Connection: The cell phone's impact on society* (interactive technologies). San Francisco, CA: Morgan Kaufmann.

Lovink, G. (2012) *Networks Without a Cause: A Critique of Social Media*. Cambridge: Polity Press.

MacKenzie, D. and Wajcman, J. (eds) (1985). The social shaping of technology. London: Open University Press.

MacKenzie, D. and Wajcman J. (eds.) (1999) *The Social Shaping of Technology: How the refrigerator got its hum*. Milton Keynes: Open University Press.

McRobbie, A. and Garber, J. ([1978] 2000) 'Girls and subcultures', in A. McRobbie (ed.), *Feminism and Youth Subcultures* (2nd edn). London: Routledge, pp. 12–25.

McVeigh, B. (2000) 'How Hello Kitty Commodifies the Cute, Cool and Camp: "Consumutopia" versus "Control" in Japan', *Journal of Material Culture*, 5(2): 291–312.

Matsuda, M. (2009) 'Mobile Media and the Transformation of Family', in G. Goggin and L. Hjorth (eds.), *Mobile Technologies*. New York: Routledge, pp. 62–72.

Mazzarella, S. (2005) 'Claiming a Space', in. S. Mazarella (ed.), *Girl Wide Web*. New York: Peter Lang Publishing, pp. 141–60.

Miller, D. (1987) *Material Culture and Mass Consumption*. Oxford: Blackwell.

Miller, H.W. (1995) 'Goffman on the Internet: The Presentation of Self in Electronic Life'. Paper presented at the *Embodied Knowledge and Virtual Space Conference*. Available at: www.dourish.com/classes/ics234cw04/miller2.pdf

Morley, D. (2003) 'What's "Home" Got To Do With It?', *European Journal of Cultural Studies*, 6(4): 435–58.

Morley, D. and Brundson, C. (1980) *The Nationwide Television Studies*. London: Routledge.

Norman, D. (1988) *The Design of Everyday Things*. London: Basic Books.

Pascoe, C.J. (2009) 'Intimacy', in M. Ito et al. (eds.), *Hanging Out, Messing Around and Geeking Out: Living and Learning with New Media*. Cambridge, MA: MIT Press.

Perkel, D. (2008) 'Copy and Paste Literacy?: Literacy Practices in the Production of a MySpace Profile', in K. Drotner, H.S. Jensen and K. Schroeder (eds.), *Informal Learning and Digital Media: Constructions, Contexts, Consequences*. Newcastle: Cambridge Scholars Press.

Qiu, J. (2007) 'The Wireless Leash: Mobile Messaging Service as a Means of Control', *International Journal of Communication*, 1: 74–91.

Qiu, J. (2008) 'Wireless Working-Class ICTs and the Chinese Informational City', *Journal of Urban Technology* 15(3): 57–77.

Robinson, L. (2007) 'The Cyberself: The Self-ing Project Goes Online: Symbolic Interaction in the Digital Age', *New Media and Society*, 9(1): 93–110.

Russell, A., Ito, M., Richmond, T. and Tuters, M. (2008) 'Culture: Networked Public Culture', in K. Varnelis (ed.), *Networked Publics*. Cambridge, MA: MIT Press.

Sawhney, H. (2004) 'Mobile Communication: New Technologies and Old Archetypes', in A. Lin (ed.), *Proceedings of the Mobile Communication and Asian Modernities I*, City University of Hong Kong, June.

Silverstone, R. and Haddon, L. (1996) 'Design and Domestication of Information and Communication Technologies: Technical Change and Everyday Life', in R. Silverstone and R. Mansell (eds.), *Communication by Design: The Politics of Information and Communication Technologies*. Oxford: Oxford University Press, pp. 44–74.

Silverstone, R. and Hirsch, E. (eds.) (1992) *Consuming Technologies: Media and Information in Domestic Spaces*. London: Routledge.

Strathern, M. (2004) 'The Whole Person and Its Artifacts', *Annual Review of Anthropology*, 33: 1–19.

Tilley, C. (1991) *Material Culture and Text*. London: Routledge.

Toffler, A. (1980) *The Third Wave*. New York: Bantam Books.

Turkle, S. (ed.) (2008) *The Inner History of Devices*. Cambridge, MA: MIT Press.

Vaidhyanathan, S. (2011) *The Googlization of Everything*. Berkeley, CA: University of California Press.

Varnelis, K. (ed.) (2008) *Networked Publics*. Cambridge, MA: MIT Press.

Wajcman, J., Bittman, M. and Brown, J. (2009) 'Intimate Connections: The Impact of the Mobile Phone on Work/Life Boundaries', in G. Goggin and L. Hjorth (eds.), *Mobile Technologies*. New York: Routledge, pp 9–22.

Woodward, S. (2005) 'Looking Good, Feeling Right: Aesthetics of the Self', in S. Kuechler and D. Miller (eds.), *Clothing as Material Culture*. Oxford: Berg.

Young, D. (2005) 'The Colours of Things', in P. Spyer, C. Tilley, S. Kuechler and W. Keane (eds.), *Handbook of Material Culture*. Thousand Oaks, CA: Sage.

Zittrain, J. (2008) *The Future of the Internet – and How to Stop It*. New Haven, CT: Yale University Press.

7

Ethics, Phenomenology and Ontology

Anna Kouppanou and Paul Standish

We begin by considering how ethics and technology and their interrelation are commonly understood, specifically in terms of neutral tools that are means to ends. Social determinism regarding these matters is shown to collude with simplistic understandings of ethics and tidy but unwarranted compartmentalization. Conversely, modern technology cannot be understood without recognition of the way that the technological extends back to the use of simple tools, to the beginnings of civilization. Writing tools are of particular significance in this. The approach we advocate, therefore, is in line with 'disclosive ethics', but we give this a clear phenomenological twist. This shows how, in new technologies especially, the human being gets into the picture. New ethical aspects produced by digital ecologies, especially with the predominance of representation effected by virtual presence, are then brought to light, as is the ethical import of the construction of selves and environments by such seemingly benign technological innovations as the search engine. These are ontological issues to which our exploration

will lead, but we know that we must begin elsewhere.

WHY ETHICS AND TECHNOLOGY?

From flaming to hacking, from identity theft to cyber-rape, new technology brings ethical issues to the fore in unprecedented ways. There is no denying the range of importance here, from the manner of our daily interaction with one another to major breaches of security, potentially affecting our whole way of life. It would seem appropriate to foreground such issues, and we do not doubt their importance, but to see the ethics of new technology this way would also be to miss the complex nature of what is at stake. Technology is necessarily implicated in the ethical, so pervasively so that we lose sight of it. The headline importance of the issues cited above may hide from view more far-reaching ways in which technology continually reconstructs our world and being, the ethical inherent at every point. *New* technology's innovations are truly spectacular, but to overplay the revolution it has brought about is to obscure the reworking of the fabric of our lives that extends back through civilization to the advent of tools.

Ethics is deeply embedded in technology and pertains to every stage of its design, production and use. It is implicated, further, in the very discussion of that design, production and use. And, finally, it is constitutive for the transmission processes of such discussion. This can be illustrated by numerous examples of political debate – about, say, nuclear power, genetic engineering and neurotechnology, or the ever-changing project of artificial intelligence. Debates over such matters inevitably involve beliefs regarding the nature of the human being, the nature of technology, and the nature of nature itself. They prompt us to assert what something is (i.e. its ontology), what our mode of access to the thing in question is (its epistemology), and also what this thing, and our relation to it, ought to be (its ethics). Technology raises specific kinds of philosophical question, demanding attention from different philosophical fields, precisely because technology is so deeply implicated in human experience. Now it may seem that, when it comes to those aspects of technoscience which affect what it is (or should be) to be human, such questions are plainly in evidence, whereas in the realm of 'pure' technology they arise no more than is the case with the making of simple writing implements. These, it will be claimed, are mere instruments, neither good nor bad, but is this really the case? Is there such a thing as a non-value-laden instrument? Can technology be a mere means to an end? Is technology ever 'pure'?

SOCIAL DETERMINISM AND ITS DISCONTENTS

There are, for sure, those who argue that technological objects are socially determined. 'What matters is not technology itself', as Langdon Winner characterizes this, 'but the social or economic system in which it is embedded'

(1980: 122). By contrast, we want to investigate the view, associated especially with Martin Heidegger (1977a), that social determinism cannot be sustained because *technology is inherently value-laden and for this reason implicated in ethical matters*. Marshall McLuhan refers to social determinism as 'the voice of the current somnambulism' (2009: 11). He writes:

> Suppose we were to say, 'Apple pie is in itself neither good nor bad; it is the way it is used that determines its value.' Or, 'The smallpox virus is in itself neither good nor bad; it is the way it is used that determines its value.' Again, 'Firearms are in themselves neither good nor bad; it is the way they are used that determines their value.' That is, if the slugs reach the right people firearms are good. If the TV tube fires the right ammunition at the right people it is good. (2009: 11)

Similarly, Winner's now classic 'Do Artifacts Have Politics?' claims that 'technical things' do 'embody specific forms of power and authority' (1980: 121). He notes the way that many bridges on Long Island are surprisingly low, in comparison with the standard American highway bridge, but this is not mere coincidence since the bridges were designed and built by architect Robert Moses in such a way as to 'discourage the presence of buses on his parkways'. This reflects:

> Moses's social-class bias and racial prejudice. Automobile-owning whites of 'upper' and 'comfortable middle' classes, as he called them, would be free to use the parkways for recreation and commuting. Poor people and blacks, who normally used public transit, were kept off the roads because the twelve-foot tall buses could not get through the overpasses. One consequence was to limit access of racial minorities and low-income groups to Jones Beach, Moses's widely acclaimed public park. Moses made doubly sure of this result by vetoing a proposed extension of the Long Island Railroad to Jones Beach. (Winner, 1980: 123–124)

But surely, it might be said, it is obvious that urban planning has an impact on the lives of citizens, and so is plainly value-laden, because it is not exactly technology but *building* which inevitably affects relations between people and their environment. A bridge is never a mere instrument like, let's say, a pencil. A pencil is just a tool. Yet it is exactly around such seemingly innocuous examples that a constrained understanding of technology will solidify, firmly separating technology and ethics. This is evident from the very beginnings of philosophy when Aristotle differentiated between *phronēsis* (practical wisdom) and *technē* (art or 'technics'). The 'prudent man', the one possessing practical wisdom, is 'able to deliberate rightly about … what is conductive to the good life generally', and this involves action and not production, which falls of course in the domain of art (Aristotle, 2004: 150). As Webster F. Hood puts it:

> The goal of *technē*, its work or product – the article of clothing, the house, or whatever – which the activity of making posits as its object, is strictly instrumental to something else from which it receives its complete justification. And this 'something' else is the use to which it is put – wearing the article of clothing, living in the house – for the sake of some activity that ultimately is its own end, namely moral or intellectual activity. Accordingly, technology is subordinate to practical wisdom, to moral and intellectual activities which are their own justification. (1983: 349)

According to this conceptualization, technology and its products do not bear any inherent value. They are mere means to an end. It is the end that can potentially be ethically wrong or right, and this is unaffected by the means. The theoretical disconnection between technology and ethics flourishes in philosophy, and in many respects this is reinforced by the manner in which ethics is understood. How does this come about?

ETHICS MADE SIMPLE

Introductory courses in ethics typically exacerbate the situation with their tendency to present matters as a choice between relatively clear-cut, alternative 'positions' – most notably, deontology and utilitarianism. Let us see where these lead.

In the eighteenth century Immanuel Kant (1724–1804) formed a theory of deontological ethics in which the moral life is understood primarily in terms of *duty*. Two key principles must obtain. First, there is the principle of universalizability: any judgement that x is (morally) right or that y is (morally) wrong must carry with it the implication that this will be the case for anyone under comparable circumstances. Second, there is the principle that human beings be treated as ends-in-themselves, never solely as means. Of course someone can be employed to do a job, but they cannot be used as a slave. These two principles are at the heart of Kant's understanding of ethics in terms of duty – of what one ought to do.

Utilitarianism, by contrast, which is associated with the nineteenth-century English philosophers Jeremy Bentham (1748–1832) and John Stuart Mill (1806–1873), holds that the ethical or moral worth of an action is to be judged in terms of its consequences – that is, in terms of a calculus of happiness and harm. It yields the principle: the greatest happiness of the greatest number. Such a principle needs to be carefully qualified, however, as the proponents of the theory acknowledged; otherwise, this would seem to legitimate atrocities to the few for the sake of gains for the many. Utilitarianism arose partly in response to social change, in the context of the Industrial Revolution, where the need to manage rapidly growing urban populations led to the formation of major social institutions, relating, for example, to health and sanitation, and to education. While the problems with excesses of social engineering have become more apparent over the past century, it is clear that utilitarian ways of thinking have become deeply embedded in policy and practice, as well as in our private lives.

Understanding ethics in this way, as positions we self-consciously opt into and 'apply', reinforces both the tendency to compartmentalize our lives and an insidious instrumentalism. It fixes the focus on 'headline issues', complacently hiding from view the permeation of values through all that we say and do. The consequences of this in the face of technology are particularly problematic. Thus, it is easy to apply ethical theories along these lines to questions concerning, say, the availability of education to women or the use of corporal punishment – worthy questions in themselves but questions that typically leave the values embedded in technology quietly out of the picture. So let us turn more directly to examine the *means* of education, its various technologies in their most everyday forms, in order to see where this might lead.

THE EFFECTS OF WRITING TECHNOLOGIES ON THINKING

In fact, as we shall see, questions concerning books, writing implements and writing in general present themselves very early in philosophy, raising ethical considerations that cannot be reduced to the theories and positions we have just entertained.

Here is a *locus classicus* for such thoughts. At a late stage in the dialogue of Plato known as the *Phaedrus*, Socrates expresses his suspicions about the 'dangerous invention' of writing. (Socrates, who is identified with the very founding of philosophy, wrote nothing, it is primarily through Plato's writings that we have inherited his thoughts.) He recounts an ancient myth concerning its origins. The Egyptian god Theuth invents writing and presents it to the god-king Thamus, arguing that his invention 'will make the Egyptians wiser and give them better memories' (Plato, n.d.). Much to the consternation of Theuth, however, King Thamus replies that writing 'will create forgetfulness in the learners' souls', because they will not use their memories: they will trust to the external written characters and 'not remember of themselves'. 'The specific which you have discovered,' he explains:

> is an aid not to memory, but to reminiscence, and you give your disciples not truth, but only the semblance of truth; they will be hearers of many things and will have learned nothing; they will appear to be omniscient and will generally know nothing; they will be tiresome company, having the show of wisdom without the reality.

Hence, in this text inaugurating philosophy's discussion of writing, we find, ironically enough, doubts about the very medium it uses. Writing, argues Thamus, is not a mere tool and is certainly a supplement of memory: it is external to the human faculty, shifting human attention from its primary focus, moving it into the realm of recollection, and, worst of all, diverting it from the truth. The idea of writing as technology – because of its nature (its ontology, mediating and indirect) and because of its effects on the mode of access we have to the world (in other words, its epistemology) – raises serious ethical concerns regarding its desirability.

Writing, then, with its various implements, has similar functions to those of the bridge: both construct our context by shaping our thinking, and both become limits of our world. This is the line taken repeatedly by Heidegger. The tool opens up our world, allows us to live in our space-time. Even something as apparently non value-laden as the common hammer allows time and space to take shape in a certain way. Such a tool is ready-to-hand. Unobtrusive and inconspicuous, it allows the discovery of space by making room for our projects. It exists in a network with other tools, people and purposes. The nodes of such a network, Heidegger will say, have the characteristics of reference and assignment: the hammer refers to the wood and the wood to the nail, and thus is the workshop-space discovered by the craftsperson. Nails refer to the making of a table, a table that will be used or sold. Hence, the tool is always orientated '*towards* something', our time lived towards something in the future.

But if this true, if tools construct our world and our thinking, then ethical matters concerning their characteristics and use arise at every point. Any account of ethics that purports to address questions of technology must address ontological and epistemological concerns. Where should we turn to pursue this?

DISCLOSIVE ETHICS WITH A PHENOMENOLOGICAL TWIST

The more ubiquitous computer and digital technologies become, the more intensified this need will be. Computer ethics, as Philip Brey argues, must be concerned with 'disclosing and evaluating embedded normativity in computer systems, applications and practices' (2000: 127). A disclosive ethics can reveal how, during the design, production and use of any technological artefact, some 'particular interests, and not others' are enclosed (Introna, 2005: 78). To recognize this enclosure of interests and values is to reveal something of the politics at work in technologies. Hence, a disclosive ethics aspires:

> (a) To disclose the nondisclosure of politics by claiming a place for ethics as being always and immediately present in every actual operation of power
>
> (b) To trace and disclose the intentional or unintentional enclosure of values and interests from every minute technical detail through to social practices and complex social-technical networks. (Introna, 2005: 79)

In the case of informational and digital technology, the disclosure of techniques favourable to particular social groups or ways of thinking becomes progressively harder because the design, production and use of these technologies is increasingly impenetrable for the consumer: paradoxically, a technology aspiring to maximize access becomes in itself inaccessible. This only increases the political and ethical import of what is happening.

Our own approach is aligned with disclosive ethics, but it incorporates a *phenomenological* perspective: we wish to stress the inseparability of the person from their environment, a relation that is always mediated technologically. This inseparability is given the greatest emphasis by Heidegger, who describes human existence as 'Being-there' (*Dasein*): the self is part of its own milieu and that milieu part of the self.[1] This thoroughly situated existence is 'constitutive for knowing the world' (Heidegger, 2008a: 90/61).

Digital technologies in effect illustrate this: they offer the tele-presence of online worlds and transform offline presence, our general being-in-the-world; and their information-access techniques – as the examination of search engines reveals – construct limits and modalities of knowing our world.

INSTRUMENTALITY, REPRESENTATION AND THE CHALLENGES OF THE DIGITAL TOOL

As we have seen, instrumentality is not a neutral, pure means employed towards a particular end: on the contrary, it partly constitutes the very activity it supposedly serves. Belief in the instrument's neutral, exterior and secondary nature is unsettled in particularly telling ways by such seminal texts as Jacques Derrida's (1930–2004) *Of Grammatology* (1976). Writing, commonly understood as a supplement, is defined by mediation in contrast to the supposed immediacy of its origin, speech and thought, but in fact that same mediation proves essential to their operation too, for signs are necessarily open to interpretation. Writing displays, writ large, the characteristics of language *per se* and this mediation, contrary to traditional conceptions, extends through our thought and action.

Deborah Johnson and Thomas Powers satirize the widespread tendency to understand the ethics of an action as independent of the means employed. Theories of action addressing normative questions are, they claim, typically constructed in the following manner:

> First, there is a potential agent with an internal state. The internal state consists of intentional mental states, one of which is, necessarily, an intending to act. Together, the intentional states (e.g., belief that X is possible, desire to X, plus an intending to X) constitute a reason for X-ing. Second, there is an outward, embodied event – the agent does something, moves his or her body in some way. Third, the internal state is the cause of the outward event; that is, the movement of the body is rationally directed and is an action insofar as it is caused by an internal state. Fourth,

the outward action has an outward effect. Finally, the effect has to be on a patient – the recipient of an action that can be harmed or helped. Moral patients are typically human beings, but the class may include other beings or things as well. (Johnson and Powers, 2005: xxix)

The tool (the technology), on this understanding, does not even come into the equation. Johnson and Powers overcome this by extending the moral responsibility usually attributed to the human agent to technological artefacts. We agree, but want to emphasize, in accord with phenomenology, that this is so because of the way that the environment, context or milieu within which the intentional action takes place emerges as it does because of the tool. The hammer, for example, as we saw, is a shaping factor of the workshop-world and the craftsperson's lived time-space. Its referential character allows a network of tools to emerge. The hammer, the bridge and the pencil, then, share the characteristic of opening-up-a-world. They influence our perceiving of the world, our experiencing of time and space, and the very processes of our thinking.

With digital technologies the opening-up-of-a-world is intensified: relying on visual markers – such as, minimally, the cursor – they allow users to represent themselves as present in other representations. Usually referred to as 'tele-presence', this is to be understood in terms of immersion or simply being-there. These extended multiform representations, the user's presence being one of them, become the user's environment. What is more, the representations of user and environment often coexist such that it becomes increasingly difficult to separate user, environment and tool. This is most obviously evident in the avatar's virtual presence but surely apparent also in various forms of simulation. Attributing this insight to Heidegger may seem far-fetched, given that he died in 1976, before the advent of the technological changes we are describing, but he showed remarkable prescience regarding the logic of representation that inheres in these developments, a logic that, Heidegger

argues, can result in the objectification of both world and human being. It is to the question of how this can happen that we now turn.

GETTING INTO THE PICTURE

We are now, Heidegger claims in the 'The Age of the World Picture', now thrown into an age in which 'whatever is comes into being in and through representedness' (1977b: 130). The modern world – in contradistinction, say, to that of Ancient Greece – is one in which representation comes to the fore in unprecedented ways. Here he is referring not to political representation but to the manner of our perception of and engagement with the world – with things and with people. These are then said to 'stand over against us'. It is in relation to them that we 'get into the picture', in precedence over what is. Things are understood in terms of the way they are pictured: in order to *be* they must present themselves, and this becomes a requirement for human beings no less. So, instead of a more fluid or dynamic involvement with things, we have objectification governed by a logic of representation.

The word processor attests to a very primitive stage of this process. A very basic trace, the cursor, stands for *you*. This minimal representation of your self is already there spatially among your words, precisely there where your thought is emerging. Through spatial representation, the computer screen already retains, in the form of traces, the just-passed moment of your thinking – in this case the beginning of your sentence – and allows you to anticipate and technically to extend your thought with the completion of the sentence. In this respect, the word processor, like any means of inscription, mediates your thoughts in the very action of writing. The cursor's blinking apprises you of the potentiality of carrying on.

With virtual worlds, social networking sites or mobile context-sensitive computing, this experience is intensified. In the face of such circumstances, the theory of action as commonly understood – where, as Johnson and Powers show, certain distinctions are maintained, distinctions between internal and external states (mind and body) and between the agent, the action and the object – is indeed to be found wanting: digital technologies are situated ontologically in a middle space that traverses the distinctions on which it relies. Existence in 'digital ecologies' usually takes the form of 'presence'; that is, it involves a feeling of 'being-there'. When, in Second Life, we make use of an avatar, which is to say we interact and form relations and histories with other users and objects, this avatar that constitutes our online presence does not appear as a mere tool or extension of the self, but rather as our self itself. In fact, in cases where such identities have been hacked, victims do not so much feel that something has been taken away from them as that their existence, their self or even their home, has somehow been violated: what has been affected is the very tool they use to construct themselves. Such a tool is what defines selves and environments; it is not merely a means of action originating in some realm of the self external to the technology. Let us consider how this construction takes place and how it connects with ethics.

DIGITAL ECOLOGIES AND ETHICS

In order to see how this construction takes place, let us turn to the French philosopher Bernard Stiegler (1952–). In *Technics and Time, 1*, Stiegler (1998) tries to show that technology and language belong to the same apparatus that constructs our selves and our milieu such that we cannot separate them. This, he claims, is a deeply ethical and political matter since control of what means we have at our disposal is control of the kind of people we can become, of the kind of thinking processes of which we are capable, and of the way we can perceive the world.

The argument goes something like this. Perception of a temporal object, let's say a melody, takes place note by note. With every passing of a note a retention is attached to the current note, which gives meaning to what is received at any moment but also allows anticipation of what is to come.[2] What has passed contributes to the anticipation of what is to be received, experienced and imagined. It is not merely, however, what has *just* passed that contributes to this process but *all* past experience: our 'already-there' is part of the process of perception (Stiegler, 1998: 248). Understanding this becomes possible only with the advent of identical repetition. Before the gramophone a human being was not able to hear an identical temporal object twice. With such technology, however, infinite repetitions are possible. This does not mean that we receive the same temporal object, however since when we listen to the same melody for the second time our past experiences shape our selection criteria of what is to be heard: the already-there constructs our anticipation of the future. For this reason, Stiegler views the tool as a 'mnemotechnics': it forms memory and attention as a 'pros-thesis' that is already set in front of us not only spatially but also temporally; it guides our future by supplementing our selves and constructing possibilities of becoming something different (1998: 152, 217). Today, however, Stiegler (2011) argues, it is the culture industries, the media or even technologies in general that form our experience, imposing on us their own selection criteria and thus constituting the milieu and the terms by which we live.

Yet, at the heart of our human ontology lies a possibility for thinking that interacts with technology and plays its part in constructing our milieu or our habitat. We have the capacity to think through and perhaps to contest what forms our thinking and structures our world. To inhabit a world is, thus, also to bear an original responsibility for thinking normatively about that world. Ethics then comes into view. In the 'Letter on Humanism', Heidegger writes:

> If the name 'ethics,' in keeping with the basic meaning of the word ēthos, should now say that

'ethics' ponders the abode of man, then that thinking which thinks the truth of Being [ontology] as the primordial element of man, as one who ek-sists, is in itself the original ethics. However, this thinking is not ethics in the first instance, because it is ontology. (2008b: 176)

To break down the word 'exist' in this way is to show the way that, paradoxically, we are always outside ourselves, there in the relations-between, and there beyond ourselves in our projects and purposes. This is how, as human beings, we dwell. Ethics then requires us to think the *oikos* (Gk, 'home'); it requires a kind of economy (*oikos* + *nomos*, meaning ordering or law); it structures our being in the world and produces normative discourse, an ecology in its original sense (*oikos* + *logos*).[3]

Perhaps, then, in the light of this, we are more ready to ponder what digital technology is, what it means that it is ubiquitous, and through what processes it is able to open worlds, to open worlds where the user feels at home. Such technology does incorporate elements of the older tools we discussed above: it creates networks of reference, conserves memories and traces, constructs the very criteria of selection, and constitutes our milieu. But sociality or setting the selection criteria, otherwise considered by-products of instrumentality, now emerge as capital in the fullest sense of the word. Let us venture a comparison. Not so long ago, a television programme could regulate the schedule of a family or nation, in a manner that Stiegler calls 'synchrony', a synchronization of consciousnesses absorbed in the reception of the same temporal object (2011: 163). Family or national schedules might be synchronized to a televised football match, for example. When TV becomes digital, on the other hand, this gatheredness decreases: viewers appear to develop and exercise their own selection criteria, choosing between programmes and times, and regulating their own milieu and involvement in the world. Digitization, then, decreases conformity in sociality, allowing personal choice to be exercised in a new production of difference. Simultaneously, what once seemed a by-product, sociality, now becomes the very thing that is

capitalized – say, by social networking sites, search engines and online communities. A very dubious game is then played out between personalization and filtering. Some digital technologies have filters in every aspect of their design, production and use, and these become selection criteria of what is to enter human consciousness – even where the user is unfamiliar with, or unaware of, the criteria themselves, where, that is, they are not the product of the user's choices and past experience. A brief discussion of search engines illustrates the point.

SEARCH ENGINES AND THE LIMITS OF KNOWLEDGE

Search engines work on the premise that a disparate and abundant field of online information needs to be filtered in order to provide relevant information to be provided for a single user. This information access is not, however, merely a matter of representation but a movement determined by a structure of specification that selects the criteria determining relevance. Google's search engine works through the following three steps: first, the Googlebot, a Web crawler, accesses pages, either by crawling from link to link or via their URL, and downloads them in Google's browsers. Then the indexer 'sorts every word on every page and stores the resulting index of words in a huge database' so that the query processor 'compares your search query to the index and recommends the documents that it considers most relevant'.[4] Relevance of texts is determined by a 'judgement' made by PageRank, which is an agglomeration of more than a hundred selection criteria 'including the popularity of the page, the position and size of the search terms within the page, and the proximity of the search terms to one another on the page'.[5] As a patented algorithm, Google's PageRank cannot be examined in its entirety, but we do know that it 'gives more priority to pages that have search terms near each other and in the same order as the query'.[6] Emphasis on popularity, proximity and measured presence suggests that PageRank relies on an identity principle of representation. Information resembling what is searched for is judged more relevant and, hence, graced with more presence, whereas information more obliquely related though potentially more significant is marginalized. That the popularity of a page is one of the selection criteria suggests that search engines maintain a hierarchy of information interpreted in the modality of the representation of results. This means, however, not only that the top ten are considered the most relevant but also that their being visited on this basis empowers them with yet more visibility and, thus, presence. Even though individual searches vary, results tend to homogenize and, therefore, to define what is to be searched for. It is for this reason that, as Lucas D. Introna and Helen Nissenbaum argue, 'not only are most users unaware of these particular biases [selection criteria], they seem also to be unaware that they are unaware' (2000: 176).

Earlier we spoke of the limits of our world, referring not only to limits of knowledge but also to modalities of its access. Online worlds are constructed through an incessant movement of indexing and, despite the common belief that online information is accessed without constraints, the truth is that the modality of its access takes on the characteristics of 'presence' offline: things that are hard to find offline probably will remain so online. Otherwise put, the presentation of information by search engines is a re-presentation that hides the fact it is a representation: it excludes information in order to give the selected item greater presence. Representation thus becomes a matter of political importance since, in concealing its own nature, it masquerades as presencing. 'Memory is', as Stiegler (2009: 9) puts this, 'always the object of a politics, of a criteriology by which it selects the events to be retained'.

Online retentions are inscribed by each and every search. This is at the heart of the criteriology they employ. Where search engines offer additional services, such as e-mail

accounts and social networking (including such applications as profiles, photos, posts, etc.), the user's online activity is turned into index material that enables the customization of the search. This is supposedly personalization, as are those 'autocomplete' Google functions that not only provide results but also generate queries by completing search words even before they are typed. Those retentions that are supposedly inscribed by you, constituting your 'already-there', become criteria of your search: they synthesize your protentions and construct your anticipation, displacing the exercise of imagination. Google tells us, 'If you're signed in to your Google Account and have Web History enabled, the algorithm may show some predicted queries based on searches that you've done in the past', adding the weak reassurance that 'Data that you send to Google is protected by Google's privacy policy'.[7] Despite possible gains in time, this process can lock you into a certain temporality in which the has-been does not simply contribute in the formation of anticipation but fully sets it in order: it suspends possibilities of difference, deferral and deviation in the search, leaving the searcher doomed to return to the same, to an identity routinely reinforced by recurrent retentions. This is a technique common to Web portals. Robert Luke invokes the *panopticon* in order to explain this more fully:

> As we are enticed to increase browsing and/or spending habits according to what is on the network, we are also influenced by (while we influence) the market. We become the market even as we track the market via W3 and W4 portals: 'the technical structure of the *archiving* archive also determines the structure of the *archivable* content in its very coming into existence and in its relationship to the future. The archivization produces as much as it records the event' (Derrida, 1995, p. 17). The archive has a direct relationship to the future in so far as the past habits are re-presented for future consumption. That is, the habit is an archive wherein, while we watch stock quotes filter through our portals, we are tracked in this watching in a panoptic, pan *info* con becomes the archive even as we create it, add to it, participate within it. (Luke, 2003: 335)

Selection is indeed manipulated by economic and political criteria since influential organizations, companies or even historical events, can easily gain presence. This is illustrated by the fact, noted by Introna and Nissenbaum, that 'of the top 100 sites – based on traffic – just 6 are not .com commercial sites' (2000: 177). This puts into question the very idea of the Internet as a public good:

> [S]earch engines … raise political concerns not simply because of the way they function, but also because the way they function seems to be at odds with the compelling ideology of the Web as a public good. This ideology portrays the fundamental nature and ethos of the Web as a public good of a particular kind, a rich array of commercial activity, political activity, artistic activity, associations of all kinds, communications of all kinds, and a virtually endless supply of information. In this regard the Web was, and is still seen by many as, a democratic medium that can circumvent the hegemony of the traditional media market, even of government control. (Introna and Nissenbaum, 2000: 178)

This modality of access is at the heart of the Web's ongoing naturalization. This is a point that came to public attention recently when the US Stop Online Piracy Act (SOPA) and the Protect Intellectual Property Act (PIPA) were debated in the US Congress. These bills were introduced in order to battle online piracy and protect intellectual property, but they were opposed on the grounds that they constituted a restriction of free speech and a violation of the core principle of the Internet: its enabling of the free flow of information. 'The legislation', it was claimed, 'would allow the Justice Department and content owners to seek court orders requiring search engines to block results associated with piracy'.[8] This, however, would mean that many websites that depend precisely on the contributions of users, such as Wikipedia or YouTube, would not be able to function as open communities since each and every user, and the content they provided, would need to be checked, with infringement possibly resulting in the blocking of the relevant page. Although this might be understood as filtering

to ensure the protection of information, it runs the risk of turning the Internet into a completely controlled space: the generation of criteria for what can be uploaded would be in the hands of the few, including no doubt media interests. Debate of this kind shows how the Internet's liquid architecture can be construed as either sacred knowledge accessed and manipulated by the few or public discourse representative of human rights.

The traces users leave behind and filters of what is to be retained raise ethical and political questions. They shape the nature of where we live and the limits of our world. When we enter the picture, we become part of this world and subject to its limits, and this intensifies in the case of other digital technologies. It determines what comes near.

NEARNESS AS REPRESENTATION

The movement of nearness through representation comes to its apogee with social networking sites. The user gets into the picture, (re)constructing themselves in the digital image. Dichotomies between self and tool, tool and environment, and representation and presentation intertwine in a single system of self-construction. As we have already seen, this is felt most acutely in moments of violation, when, say, a user's account is hacked and comments are posted on their behalf, words uttered as if by them, or when their avatar performs as if eerily activating their own body. Thus, cyber-rape is a violation of a certain representation of the self that the person considers precious. If the violation were restricted to a representation separate from the self, then this effect on the actual self would scarcely be intelligible. Plainly this is not the case. The rape is of the site realized in the avatar, which is a site in which the tool allows the human being to be a Dasein (being-there) of a certain kind. Tool and site are incorporated into the self whose emergence they have allowed in the first place. What seems supplementary – the avatar, the site, the tool – is not separable from that self, very much in the way that, in a physical assault, the body is not understood as simply separated from the mind. Tools' capacity to be extensions of the self in real life is experienced in its more radical aspect in these moments of violation, moments in which clear-cut separations of self and tool would be most beneficial. This why bullying can be effectively carried out on online.

Thus, the representation, the tool itself, becomes the image of a self that has no image in the first place but only traces of a presence in the process of becoming. The image, the avatar, the online movement, becomes a representation not of an originary auto-affecting self but of another image that participates in this schema. Like the cane of the blind-man, the digital tool becomes prosthetic; it does this through its imaging. With digital tools this connection is inscribed, registered and often recorded and archived. In Second Life the user becomes literally part of the picture, exposing herself to whatever vulnerabilities it harbours. My traces are inscribed retentions in the very environment I inhabit, they cannot be separated from me. As a result, my words and actions constitute the archive that is my ecology, an archive making me available as a target for commerce or an object of research.

Social networking sites similarly offer infinite possibilities for self-inscription, and these can become objectifications of the self. Their manner of bringing close derives from the inseparability of the self from its representation, the inseparability of Dasein (being-there) from its Da (there). Facebook, for example, demands the constant renewal of presence:

- Create your profile (by posting of personal details about your place of residence, work placement, education, date of birth, relationship status, sexual interest, languages spoken). Include an 'about me' section showing what is unique about you. Automatic notifications will be sent to your friends each time you update your profile.
- Update your status by posting a comment about your life.
- Post photos and create albums.

- Follow threads by posting comments on a friend's wall. Log in to read the comments other users have left.

To a significant extent, this constant representation of the self and personalization of the page happens involuntarily. Facebook sets its own demands and criteria for representation. All these techniques, constantly updated, and many others, constantly added, allow you to customize your page and thus create your own personalized habitat. Given longer duration or granted presence on the Web, things appear to come closer. You post a photo and a comment about what is going on in your life, and a thread of reactions has begun. For this to happen, however, that first post must be easily, quickly intelligible; it must be user-friendly. Pensive remarks or complicated political views, for example, rarely provoke the responses that other posts do.

If there is a stream of response, this first post will recur in your News Feed, and, if you choose 'top stories' instead of 'most recent' as your criterion of presentation, this will remain visible on Facebook. Top stories 'are stories published since you last checked News Feed that we think you'll find interesting. They're marked with a blue corner and may be different depending on how long it's been since you last visited your News Feed'.[9] As with search engines, specific selection criteria for a story to be considered 'top' are not made apparent. Yet what they result in is a synthesized construction of temporality (what is more recent) and significance (what is most important), and this is not a judgement made by you. Techniques such as these have real effects: Facebook makes decisions concerning the types of representations that build your supposedly personalized environment. This is effected through a constant archiving and representing of your traces. It not only allows the inscription of traces as a construction of your self but also constantly rearranges these traces in order to strengthen your presence – always on the pretext of bringing you closer to your friends and interests. This is the there where you are.

With applications that allow you to view your page as it would be viewed by a friend or by the public – offering spatial representations of your interaction with other users (your friendship history), collecting your traces and turning them into the retentions that create your Facebook history, and presenting this as a 'timeline' – this constitutes a complicated new profiling where posts, photos and preferences are presented as your life. These applications have one common imperative: connect and be connected; activate potential connections to make your online presence stronger and more interesting. This has been there from the start with 'People you may know' and 'Find friends', but there is increasing urgency to make visible the greater network of which you are evidently a part by representation online. The notion of bringing close is increasingly turned into the potentiality for representations and the insertion of the constructed self into this greater picture. You then become, so it seems, creator of, and focal point within, the network or picture to which you belong.

Yet nothing could be further from the truth. Everybody and no one is at the centre, and our personalized ecology is the product of what are in effect commercial decisions determining selection criteria for our representations, criteria in turn re-inscribed in the selection criteria.

The 'straightforward' representation of geographical space turns into the sci-fi fantasy of the avatar of Second Life (SL). The user is offered the means to build a self through the construction of an ecology. The building blocks of that ecology, however, already contain the possibilities of the structure to be made. Bringing close is redefined according to a logic of representation that indexes you and puts you into a picture already painted for you. This new ecology is, in effect, a new kind of *polis*, the marketing of which is predicated on the belief that every citizen has participated in this design, that every user has individuated themselves.

By the same token, global positioning systems synthesize in map-like form representations of the user's position and the affordances that position provides (say, possible routes, speed limits, restaurants). Even though this kind of technology employs an extraterrestrial vantage point, which turns the earth itself into a representable object, the users experience themselves as the centre of the movement represented on the screen. The users are *in* the representation, experiencing themselves as interacting with the representation, instead of with the world. Conversely, inferences regarding a town's layout are no longer needed since navigating this space is turned into an index search. Enter the name of the place you need, and a route appears in the representation, in which, as a trace, you are already installed: 'man gets into the picture'; Dasein gets into the index. In Stiegler's words:

This grid, and the consequent digital representation of territory, is happening now, and the general availability of the infrastructure for localized information emission has witnessed the instauration of a 'second generation' of digital navigation techniques: geo-information. The digitization of territory means simultaneously having systems for navigating the geo-referenced data on digital maps that have also integrated photographs, videos, reproductions of the country or territory in all genres and directions, telephone relay beacons, usage guides, and more generally the management of portable appliances, roaming devices, and all other such instruments. This also means that the device user becomes a datum circulating in a 'data stream': electronic data physically localized and situated in interfaces simulating actual territorial spaces. (2011: 137–138)

Digital representation, the grid and second-generation navigation techniques combine in this process of 'geo-information'. Information, as the word begins to suggest, not only affects us as something external to ourselves, selves that remain intact, but forms us within: thus, they (in-)form our dwelling-place, the device-user becoming a datum in a constant datastream.

So where does this leave us? We have not reached this point without changes in registers of expression, culminating in Stiegler's highly charged, allusive words, and this we think is necessary in order to disturb preconceptions embedded in our habitual phrasing of these matters. This, however, is not a counsel for despair but an opening towards a better, more sensitized, more realistic understanding of the human condition. Understanding is empowering. In our concluding remarks, we shall endeavour, more prosaically, to draw together the strands of our argument, not so much through reiteration as by placing them against an influential contemporary account, the broad theoretical basis of which resonates with our own, but from which we part company in significant respects.

BETWEEN INFORMATION ETHICS AND PHENOMENOLOGY

To move beyond traditional understandings of ethics and action is, we have seen, to expand the possibilities of understanding the intentionality inherent in technology. Viewing technology in this way enables its moral import to become more readily apparent. This is not an optional extra. Not to do this is to fail to respond to the ethical demands of our lives; it is to fail to recognize the ecology of our lives – that is, the embeddedness of users and objects in digital environments.

The idea of 'information ethics' or of 'information ecology' has been developed by a number of contemporary theorists, amongst whom Luciano Floridi has been particularly influential. Information ethics suggests, he writes:

> that there is something even more elemental than life, namely being – that is, the existence and flourishing of all entities and their global environment – and something more fundamental than suffering, namely entropy ... Entropy here refers to any kind of destruction, corruption, pollution, and depletion of informational objects (mind, not of information), that is, any form of impoverishment of being. (Floridi, 2008: 12)

One implication of such a line of thought is that the place we inhabit is best construed not simply as the earth or land itself, but in terms of the entirety of informational objects (the 'infosphere'), of which cyberspace is a part (2008: 14). Informational objects are all objects informationally understood, and ethics is implicated across this entire range. The infosphere denotes 'the whole informational environment constituted by all informational entities (thus including informational agents as well), their properties, interactions, processes, and mutual relations' (2008: 3).

Much of this is in accord with our emphasis on the merging of the user and the environment, but Floridi's approach is partly at odds with our own 'disclosive ethics with a phenomenological twist'. We emphasize that the environment is already us – that is, already Dasein. As we saw with cyber-rape, it can now be seen that the self that is the victim is not a mere representation but a newly created self. Phenomenology, thus, opens the way to new types of investigation in information ethics.

Like Floridi, we would emphasize that, if information ethics is to develop on a sound basis, work at a relatively abstract, metaphysical level is essential. It is not sufficient to tackle particular problems (including computer ethics problems) head-on: problem-solving procedures and case-orientated analyses must be guided by this conceptual and metaphysical groundwork (2008: 19). Without this our attempts to grapple with the practical problems new technology presents will be hopelessly blinkered and we will fail to see the extent to which technology changes our world and ourselves. The familiar headline 'issues' can blind us to the multiple ways in which our world is enriched. Cyber-feminism did much to enhance appreciation of the communicative possibilities in new technology, which the imagery of programming and databases is apt to obscure.[10] Do we not see every day new possibilities of interaction? Blogs and fora allow thoughtful conversation and reflection on the most intimate matters. The twittering of mobile devices enables a political activism sufficient to trouble even the most repressive regimes.

NOTES

1. Heidegger adopts the term *Dasein* on the ground that the terms 'human being', 'man' and 'human subject' have become burdened by ways of thinking that prevent us from seeing precisely the structures he seeks to describe.

2. Stiegler's argument owes much to Edmund Husserl's (1859–1938) theory of perception, but Stiegler diverges from the latter's approach by emphasizing the imagination's modificatory role in perception. Stiegler does not maintain the sharp distinctions between imagination and perception, retention and recollection, or even presentation and representation, that exist in Husserl and in other philosophers that influenced him such as Kant and Heidegger. This paves the way for his account of the technological realm of the already-there within our perception and imagination.

3. Recognizing connections between ecology and information ethics, Rafael Capurro has coined the term 'information ecology' (1990: 189, cited in Takenouchi, 2006: 189), whilst Luciano Floridi argues that 'IE translates environmental ethics into terms of infosphere and informational objects, for the land we inhabit is not just the earth' (2008: 14). Our position is somewhat different: we do not merely draw analogies between physical and digital ecologies but elaborate an ecology in terms of the phenomenology of being-in-the-world, which is always mediated technologically.

4. Online at www.googleguide.com/google_works.html (Retrieved 25/4/2012).

5. Ibid.

6. Ibid.

7. Online at support.google.com/websearch/bin/answer. py?hl=en&answer=106230 (Retrieved 25/4/2012).

8. Online at www.bbc.com/news/technology-16590585 (Retrieved 25/4/2012).

9. http://en-gb.facebook.com/help/glossary (Retrieved 18/4/12).

10. See Standish (1999). Donna Haraway develops the idea of cyber-feminism especially in her 'A manifesto for cyborgs: Science, technology and socialist feminism in the 1980's (1991).

REFERENCES

Aristotle (2004) *Nicomachean Ethics*. Trans. J.A.K. Thomson, Rev. H.Tredennick, Intro. J. Barnes. London: Penguin.

BBC (2012) 'Wikipedia Joins Blackout Protest at US Anti-piracy Moves', www.bbc.com/news/technology-16590585

Blachman, Nancy and Peek, Jerry (n.d.) 'Google Guide: How Google Works', www.googleguide.com/google_works.html

Brey, P. (2000) 'Method in Computer Ethics: Towards a Multi-level Interdisciplinary Approach', *Ethics and Information Technology*, 2: 125–129.

Capurro, R. (1990) *Towards an Information Ecology*, Contribution to the NORDINFO International seminar "Information and Quality", Royal School of Librarianship, Copenhagen, 23-25 August 1989. Proceedings: I. Wormell (ed.): Information Quality. Definitions and Dimensions. London, Taylor Graham 1990, p. 122-139. www.capurro. de/nordinf.htm

Derrida, J. (1976) *Of Grammatology*. (Trans. G. C. Spivak). Baltimore, and London: MD, Johns Hopkins University Press.

Derrida, J. (1995) *Archive Fever: A Freudian Impression*. Trans. (E. Prenowitz). Chicago, IL: University of Chicago Press.

Facebook (2012) 'Help Centre: Top Story', www.facebook. com/help/?faq=164850830273714

Floridi, Luciano (2008) 'Foundations of Information Ethics', in K.E. Himma and H.T. Tavani (eds), *The Handbook of Information and Computer Ethics*. Hoboken, NJ: John Wiley. pp. 3–24.

Google (2012) 'Inside Search: Autocomplete', http:// support.google.com/websearch/bin/answer.py?hl=en& answer=106230

Haraway, D. (1991), 'A Manifesto for Cyborgs: Science, Technology and Socialist Feminism in the 1980s', in S. Seidman (ed.), *The Postmodern Turn*. Cambridge: Cambridge University Press.

Heidegger, Martin (1977a) 'The Question Concerning Technology', in *The Question Concerning Technology and Other Essays*. Trans. & Intro. W. Lovitt. New York: Harper. pp. 3–35.

Heidegger, Martin (1977b) 'The Age of the World Picture', in *The Question Concerning Technology and Other Essays*. Trans. & Intro. W. Lovitt. New York: Harper. pp. 115–154.

Heidegger, Martin (2008a) *Being and Time*. Trans. J. Macquarrie and E. Robinson. Oxford: Blackwell.

Heidegger, Martin (2008b) 'Letter on Humanism', in D.F. Krell (ed.), *Basic Writings*. Abingdon, Oxon and New York: Routledge Classics. pp. 147–181.

Hood, Webster. F. (1983) 'The Aristotelian Versus the Heideggerian Approach to the Problem of Technology', in C. Mitcham and R. Mackey (eds), *Philosophy and Technology: Readings in the Philosophical Problems of Technology*. New York: Free Press. pp. 347–363.

Introna, L.D. (2005) 'Disclosive Ethics and Information Technology: Disclosing Facial Recognition

Systems', *Ethics and Information Technology*, 7: 75–86.

Introna, L.D. and Nissenbaum, H. (2000) 'Shaping the Web: Why the Politics of Search Engines Matter', *Information Society*, 16(3): 169–186.

Johnson, Deborah G. and Powers, Thomas M. (2005) 'Ethics and Technology: A Program for Future Research', in C. Mitcham (ed.), *Encyclopaedia of Science, Technology and Ethics, Vol. 1*. New York: Macmillan. pp. xxvii–xxxv.

Luke, Robert (2003) 'Signal Event Context: Trace Technologies of the habit@online', *Educational Philosophy and Theory*, 35(3): 333–348.

McLuhan, Marshall (2009) *Understanding Media: The Extensions of Man*. Abingdon, Oxon and New York: Routledge.

Plato (n.d.) *Phaedrus*. Trans. B. Jowett, in D.C. Stevenson (ed.), *The Internet Classic Archives*, http://classics.mit.edu/Plato/phaedrus.html

Standish, P. (1999) 'Only Connect: Computer Literacy from Heidegger to Cyberfeminism', *Educational Theory*, 49(4): 417–435.

Stiegler, B. (1998) *Technics and Time, 1: The Fault of Epimetheus*. Trans. R. Beardsworth and G. Collin. Stanford, CA: Stanford University Press.

Stiegler, B. (2009) *Acting Out*. (Trans. David Barison, Daniel Ross and Patrick Crogan. Ed. Werner Hemacher). Stanford, CA: Stanford University Press.

Stiegler, Bernard (2011) *Technics and Time, 3: Cinematic Time and the Question of Malaise*. (Trans. S. Barker). Stanford, CA: Stanford University Press.

Takenouchi, T. (2006) 'Information Ethics as Information Ecology: Connecting Frankl's Thought and Fundamental Informatics', *Ethics and Information Technology*, 8(4): 187–193.

Winner, L. (1980) 'Do Artifacts Have Politics?' *Daedalus*, 109(1): 121–136.

PART 3

Research Perspectives for Digital Technologies: Theory and Analysis

The previous two parts have provided a clear positioning of this book in the field of digital research and outlined some key characteristics and considerations for digital research, including how different scholars have come to think of and empirically examine digital technologies in use. This third part builds on this work to present a comprehensive and diverse range of research perspectives for digital technologies. Each chapter provides an introduction to the theory, its rationale, key terms and debates, and illustrates its application within different research contexts.

Chapter 8 (Sara Grimes and Andrew Feenberg) introduces readers to the critical theory of technology. It explains the background and rationale of this approach, its framework – used to analyse the socio-political dimensions of technologies and technological systems, and key terms. The chapter briefly describes previous applications of the critical theory of technology approach within three different studies of information technologies.

Chapter 9 (by Jeffrey Bardzell) outlines the use of critical and cultural intellectual approaches to support HCI. It clarifies some key differences and similarities between cultural scholarship and scientific scholarship and points to the potential for both to coexist and complement each other in HCI. The author explores how humanistic HCI research skills can be developed by HCI researchers and practitioners and argues for a more central role for *criticism* in future human–computer interaction (HCI).

Chapter 10 (by Paul Marshall and Eva Hornecker) focuses on theories of embodiment in HCI and highlights different traditions respecting how our bodies and active experiences shape our perceptions and how we feel and think. It provides an overview of the core theoretical underpinnings of work on embodied interaction in HCI and how this has been developed in phenomenology and embodied cognition, including discussion of the role of the environment. The chapter discusses selective examples from the literature to illustrate the diversity of work that has drawn on embodiment theories in technology development, analysis and evaluation.

Chapter 11 (by Luigina Ciolfi) provides an overview of theories of place and space that are relevant to digital technology research, which is increasingly concerned with technology that is pervasive of physical spaces and thus requires an understanding of theoretical aspects related to the role of the physical, experientially, socially and culturally constructed environment. The chapter also provides examples of how spatial concepts from different theoretical traditions have been used by researchers in the interactive systems design field.

Chapter 12 (by Kristina Höök) extends this theme of 'physical' experience to address issues arising when designing for affective and experiential interactions with digital technologies. In particular, how to understand and articulate different kinds of 'experience' to inform design of digital artefacts, and an exploration of theoretical foundations for such designs.

Chapter 13 (by Barry Brown) outlines ethnographic approaches, which are, he argues becoming pervasive in HCI, where the ethos of understanding users has moved, as the other chapters in this book demonstrate, beyond seeing users as measurable units of information processing to focus on getting as close to people as possible, understanding how they see the world, working through their problems and with their ideas. The chapter reviews the history of ethnography in HCI, documents some recent ethnographic work, and provides an introduction to the methods of ethnographic research.

Chapter 14 (by Victor Kaptelinin) adopts a particular perspective on digital technology, namely the *mediational* perspective, according to which technology is considered a mediating means that affects, and even shapes, the structure, functioning and development of human minds and actions. Notably, the Vygotskian cultural-historical tradition is examined, although pragmatism, phenomenology and actor-network theory are also discussed. The

chapter provides an overview of existing digital technology research from the point of view of the mediational perspective on digital technology and explores key challenges and implications for the analysis and design of interactive technologies.

Ethnomethodology and conversation analysis are the focus of Chapter 15 (by Robert Moore), both of which have played a prominent role in the area of HCI and computer supported cooperative work (CSCW). This chapter elaborates on the aim of these approaches – to understand how humans *achieve* social order in and through their concrete, observable behaviours, and maps their guiding theoretical principles in order to provide a practical framework for these empirical approaches and their application within studies of digital technology. Aspects of these approaches are illustrated through a concrete example on the topic of *virtual embodiment* in avatar-mediated human interaction.

Chapter 16 (by Cliff Lampe) concerns behavioural trace data for analysing online communities. While online communities have been studied using a variety of research techniques for many years, this chapter shows that one of the new opportunities afforded by the medium is the ability to track the behavioural traces of users of these systems as they interact with one another. These can create rich, complete records of people's activities in these online communities, allowing us to understand individual and social processes on a scale previously unreachable in the social sciences. The challenges of accessing such data, its limitations, and the ethical and legal issues involved are discussed, as is the potential for mixed method approaches to mitigate many of these issues and create holistic approaches to answering research questions.

Chapter 17 (by Carey Jewitt) provides an introduction to the field of multimodality and discusses its potential application for researching digital data and environments. It outlines what multimodality is, its theoretical origins, underlying assumptions and key concepts. This is followed by a discussion of the scope and potential of multimodality for researching digital technologies. The remainder of the chapter sets out an illustrative example of one of the many ways in which multimodality can be applied to the study of digital technology use. It concludes with a discussion of the limitations and challenges of a multimodal approach for digital technologies.

Chapter 18 (by Steven Dow, Wendy Ju and Wendy Mackay) concerns the paradigm of research through design (RtD). This combines the forward-thinking, artefact-generating practices of design with the knowledge-generating goals of research. In RtD, artefacts are not the end goal, but a means for framing an alternative future and uncovering human needs, desires, emotions and aspirations. The chapter defines RtD, explores the origins and rationale for RtD and its methodological implications, and discusses some of the standards the research community has adopted to evaluate RtD contributions.

Chapter 19 (by Laurel Swan and Kirsten Boehner) is focused on critical design as a type of design research that uses design to research both the status quo and alternate realities. The chapter provides an historical overview of critical design and takes an in-depth look at three iconic projects in order to highlight several salient features of critical design.

8

Critical Theory of Technology

Sara M. Grimes and Andrew Feenberg

INTRODUCTION

Critical theory of technology argues that technologies are not separate from society but are adapted to their social and political environment. Since technologies are implicated in the socio-political order they serve and contribute to shaping, they cannot be characterized as either neutral or as embodying a singular 'essence'. The following chapter will introduce readers to an innovative and practical framework for applying this approach as the basis for qualitative and mixed methods research on digital technologies (Feenberg, 1991, 1995, 1999, 2002). Critical theory of technology is derived from Marcuse's version of Frankfurt School critical theory, concretized through a constructivist approach to the analysis of specific technologies, artifacts and systems.

The chapter begins with a summary of the philosophical positions that critical theory of technology seeks to reconcile. It then proposes a dual-level framework for analyzing the socio-political dimensions of technologies and technological systems: a primary level at which objects and people are decontextualized to identify affordances, and a secondary level of recontextualization within natural, technical, social and cultural environments. Several key terms are then introduced. These include the notions of the *technical code* and *formal bias*, which are used to uncover features of technologies that reflect the values and beliefs that prevail in the design process, and which form a background of unexamined cultural assumptions literally designed into technology itself. Additionally, the notion of *democratic rationalization* will be presented as an analytic tool for understanding user innovation and public interventions through which technically subordinate groups challenge technical codes. The remainder of the chapter briefly describes previous applications of the critical theory of technology approach to three studies of information technologies: the Minitel, online education and massively multiplayer online games (MMOGs).

BACKGROUND AND RATIONALE: WHY CRITICAL THEORY OF TECHNOLOGY?

There are currently two dominant approaches to technology. These approaches are each useful in their own way, but in the usual formulations they are diametrically opposed. This has the unfortunate consequence of limiting our ability to adopt a comprehensive, critical approach to the study of technology.

The first, *substantivism*, emerges out of philosophy of technology and argues that technology is autonomous and inherently biased towards domination. This argument is most notably found in the works of Heidegger, Ellul and the Frankfurt School where it is formulated in various critiques of instrumental rationality. These substantivist philosophers observe that the consequences of modern technology are not the harmonious and prosperous society promised by the Enlightenment tradition but ever larger disasters and threats. Technology seems to develop according to its own intrinsic logic without regard for democratic and humanitarian ideals. A solution is difficult to imagine since none of these philosophers advocates returning to a society without modern technology and yet they see little hope of change in the established institutions. They end up questioning the viability of modernity as a form of life.

The second main trend in the study of technology, *constructivism*, comes out of contemporary social science. Social studies of technology pursue empirically grounded investigations of technological design and development. Case histories show the important role of social 'actors', groups with influence, in the 'construction' of technology. The famous case of the VHS v.s Beta video recording technology illustrates the constructivist argument. Neither technology was absolutely superior, although Beta did have some important advantages. Yet in the end, the design of video recording equipment was determined by the market power of the backers of VHS and not by technical criteria.

Other cases reveal other sources of influence and other reasons for the triumph of one alternative over another. There are no intrinsic forces driving technological progress, but, rather, shifting criteria of various social groups who decide on the next stage of development. Constructivism demonstrates the highly contingent nature of technological devices, innovations and choices (Pinch & Bijker, 1987). It sticks close to its cases and largely avoids generalizations about modernity.

Both views can be understood as responding to the largely outmoded instrumentalist argument that technology is neutral. According to that view technologies are mere tools and have no independent power to influence social life. This view is now widely rejected. It seems clear that modern technology has all sorts of social implications, that it is a transformative force in its own right, regardless of the ends it serves. The argument between constructivism and substantivism is primarily over how to interpret the design process and the value-laden character of technology.

Substantivism is concerned with the general consequences of technology. It holds that technology as such has definite social implications, which show up in the character of contemporary life. Constructivism, however, focuses on the specific social factors involved in individual case histories. Instead of generalizing about technology, it explains the development of particular technologies. The two views can be contrasted in other ways as well. For instance, when compared to the rich complexity uncovered in social studies of technology, philosophy of technology can appear overly abstract and lacking adequate socio-historical grounding. Substantivist philosophy of technology tends to be heavily critical of modernity, whereas most constructivist technology research can appear uncritical in its disregard for the larger issue of modernity (Feenberg 2003). It is in contrasting these two views that their underlying and complementary strengths and limitations are made visible.

Critical theory of technology seeks to reconcile the schism between substantivist and constructivist theories of technology. It argues that while technologies materialize in socially determined ways depending on the contexts, people and practices involved, they also have larger normative implications that demand critical attention. Foremost among these implications are the enduring issues of power and inequality raised by substantivist philosophy of technology and by political theories of modernity. A key concern is how, despite advances in human rights, individual freedoms and markets, the ongoing centralization of ever more powerful public and private institutions shapes a technology compatible with the capitalist rationalization of production (Marx), of the prison (Foucault), and of the public sphere (Habermas).

Marcuse summarizes these concerns, arguing, in 'advanced industrial society', the dominant powers impose designs that narrow the range of interests and concerns which can be served by the normal functioning of technology. This narrowing distorts the structure of experience and causes human suffering and damage to the natural environment. A critical questioning of technology's role in the grand narrative of progress enables deeper consideration of the ways in which technological design and development come to serve as the material base of a distinctive social order.

In critical theory of technology, technologies are viewed in terms of a larger system, a lifestyle, a collection of social values, meanings and cultural cues, some of which become inscribed in technological design. Many of these contain political implications that also demand attention. In this sense, technologies are envisioned as functioning in a similar way to laws and customs, 'shaping' their inhabitants through the representation and privileging of certain interests while excluding others. Like laws and customs, technologies are deeply political, even though the full meaning of their politics may remain hidden until questioned and analyzed. Critical

theory of technology brings together social study and philosophical enquiry in order to include this political dimension.

The theoretical framework of critical theory of technology is based in part on Marxism, not only through recognition of the primacy of capitalism within modern society and its associated processes of social rationalization but also in terms of method. Marx's method for studying rational systems such as the market can be applied more generally in a critique of social rationality, including technological rationality. Marx focuses his critique on the discriminatory effects of a rational order, thereby demonstrating that these effects are due not to some external, non-rational cause such as Western ideology, patriarchy, bourgeois greed, etc., but are, rather, the outcome of the rational institution of the market, based on equal exchange. In this way, Marx helps devise a critical approach to studying rational systems that addresses the broader social and political implications raised by the substantivists, while acknowledging the constructivist emphasis on particular socio-historical contexts.

Drawing inspiration from Marx's approach, critical theory of technology argues that technology is both rational and biased. The design, development and eventual implementation of a technology are seen as normative processes through which interests and values are delegated to technological devices and systems. Critical theory of technology calls this the 'formal bias' of technology in contrast with the more familiar type of bias based on personal feelings and preferences. Formally biased technologies usually embody and reproduce the social, economic and political conditions within which they are constructed. Sometimes, however, dominated groups gain new levers of influence through their technical involvements. The concluding section of this chapter describes examples of such subversive interventions.

To explain the formal bias of technology, critical theory of technology employs the constructivist idea of 'underdetermination', according to which the technical properties of a device are insufficient to determine its design. Social factors must intervene to select from among the various alternative designs available in the course of the development of a technology. The option that 'fits' best within the particular social context, and that best satisfies the (sometimes conflicting) interests of 'relevant social groups' undergoes a 'process of stabilization' through which it eventually comes to define the standard form of the technology in question (Pinch & Bijker, 1987). The ultimate path a technology takes is thus determined by various intersecting social factors, which have as much to do with politics, cultural norms, user agency and everyday practices as they do with the technical limitations and economic imperatives cited in deterministic accounts.

Another key concept in understanding this process is the 'technical code', which is used to uncover features of technologies that reflect the values and beliefs prevailing in the design process, and which form a background of unexamined cultural assumptions literally designed into technology itself (Feenberg, 1995). Examples of technical codes are all around us, for example, in the adaptation of everyday life to the automobile and the automobile itself to social demands for status, pollution control and safety.

The concepts of formal bias, underdetermination and technical code enable critical theory of technology to consider the role of technology as an important – as well as particularly pervasive and powerful – dimension of society, without giving up the constructivist commitment to empirical research.

INSTRUMENTALIZATION THEORY: A DUAL-LEVEL FRAMEWORK

Although constructivist and substantivist approaches are usually presented as totally opposed, critical theory of technology finds something right in both. They can be reconciled through the development of a common framework, the *instrumentalization theory.*

Instrumentalization theory analyzes technologies at two levels – at the level of our original functional relation to reality, as well as at the level of technological design and implementation. The dual approach reflects the fact that one cannot be in a technical relation to the affordances of a technology without also being in a specific social relation to its context. In this way, instrumentalization theory enables critical analysis that addresses both the substantivist emphasis on hermeneutic questions about what technology *means* for modern life (as opposed to focusing too narrowly on practical questions about what technologies *do*), as well as constructivist questions concerning who makes technology, how and why. The answers to these questions are understood to contain significant social and political implications, which are addressed at both levels of the analysis.

The first level, called 'primary instrumentalization', explains the functional constitution of technical objects and subjects. Primary instrumentalization describes the processes through which functions are separated from the continuum of everyday life and subjects positioned to relate to them. This involves an imaginative leap by which a natural object is perceived outside its normal context as useful in a new context established by a user. Even higher animals are capable of this: a monkey can use a fallen branch to knock a piece of fruit off a tree. The branch is subject to a primary instrumentalization through which it acquires the potential to reach a high place. Human beings can build far more complex systems of instrumentalized objects in the construction of technological devices.

This process of primary instrumentalization reacts back on the user. Using a technology requires establishing a certain relationship to it. This relationship is to some extent determined by the structure of the technology, which dictates the practices that make it useful. The user is thus not entirely external to the technological system but shares in it through the practices it 'scripts'. These practices may also have decontextualizing implications, for example, substituting a modern technical operation for a traditional social ritual that accomplished similar goals. Fast food and the family dinner both deliver calories, but fast food does it through a technologically sophisticated process cut off from the home and its rituals, while the family dinner is an essentially social practice.

This level of analysis is inspired by categories introduced by Heidegger and other substantivist critics of technology. They focus on the effects of technical mediation of human activity regardless of its goals. Critical theory of technology does not ontologize technology and thus avoids the dystopian master narrative assumed by thinkers such as Heidegger.

The second level, called 'secondary instrumentalization', examines the ways in which actors experience and make sense of these functional affordances, emphasizing how they perceive and construct the meanings of the devices and systems they design and use. No sooner is a technical affordance identified in some material than it is recontextualized socially. Every technology belongs to a social world in which it has a meaning. This meaning is constituted by such things as the nature of the problem to which it is addressed, its different effects on the social groups that coexist with it, its appearance and the connotations that attach to it, and so on. This second level of the analysis is inspired by the empirical studies favored by constructivist scholars of technology, modified, however, to consider the macro-social influences they often dismiss.

For examples of secondary instrumentalization, let's return to the contrast between fast food and the family dinner once again. It is true that fast food decontextualizes eating and reduces it to a kind of refuelling, yet that is not the whole story. Fast food technology has a context, too. It belongs to an enterprise that generates profits by distributing cooked food. It conforms to legal requirements for cleanliness and other amenities. In some cases, it is also ornamented by branding (advertising and public relations activities) through which associations with certain cultural signifiers are suggested. By contrast, the family dinner

appears rich in meaning, yet it too depends on the primary instrumentalization of the ingredients and cooking implements. Fast food is decontextualized with respect to the family dinner, to be sure, but both involve primary and secondary instrumentalizations. There is no pure technology.

The distinction between the two 'levels' of instrumentalization is analytic rather than real. They form a research framework within which various facets of a technology can be explained, rather than describing separate existing things. The levels *presuppose one another*, and each of the various 'steps' entailed in the development of a given technology involves both. A key aspect of the dual-level framework is the iterative interaction between primary and secondary instrumentalization. The meaning technologies take on in the lifeworld feeds back into their design from one stage in their development to the next.

In sum, all technologies are characterized by a particular configuration of primary and secondary instrumentalizations. The subject of technical action is always also a social subject engaged in two kinds of operations – imaginatively grasping a device's affordances in terms of socially constructed criteria of usability, that is, function and meaning. In terms of methodological approach, instrumentalization theory enables a hermeneutic study of technology that makes explicit the implicit cultural meanings embedded in the devices we use and in the rituals they script. Thus, while the social histories produced by sociology of technology (including Pinch and Bijker's bicycle or Schwartz Cowan's stove) provide a useful entry point, applications of critical theory of technology must also surpass the limitation of case-specific studies by making the linkages to larger normative issues.

DEMOCRATIC RATIONALIZATION

The concept of democratic rationalization refers to the effects of public interventions in the formation of technical codes and their realization in the design and redesign process. These codes translate between social attitudes and demands and the technical specifications of devices. Democratic rationalizations arise out of a specifically technological form of social participation.

Bruno Latour (2005) has argued that technologies enrol individuals in networks. These networks associate the individuals in various roles, for example as users of the technology or workers building it, or as victims of its unanticipated side effects. Designs represent some of these groups better than others. Interests thus both preside over design choices and emerge from the consequences of these choices. For example, it may happen that users are well served by a technology that causes pollution affecting third parties. The victims of the pollution are also enrolled unwittingly in the network created by the technology. They discover what critical theory of technology calls 'participant interests' in cleaner air that would never have occurred to them had they not suffered from the unintended consequences of the technology.

Once enrolled in a network, individuals are motivated to address its failings and in some cases they also acquire potential power over its development. That power may have no formal outlet and it may even be suppressed. Nonetheless, it is a basis from which struggles can emerge. Also, the power of individuals within a network is quite different from that of individuals who have no connection to it. Because they are on the inside they learn about the technologies that support the network and can identify vulnerabilities and bring pressure to bear. This gives them a platform for changing the technical codes that shape the network.

Public interventions into design can be conceived as an ongoing process that is gathering strength as technology is implicated in more and more domains of social life and causes a wide variety of problems to which solutions must be found. This process is unfolding around issues in environmental, medical and information technology today.

A new picture of technical politics can be drawn through the study of these events, philosophy of technology and contemporary constructivist research. The dominant technological rationality is based on a fatally simplified understanding of its objects. This is a specific problem of the capitalist heritage of technological development. Most technology was shaped for success on markets where many externalities could be ignored for generations. Reduced to raw materials and disconnected from their natural background, the materials incorporated into technical systems have unanticipated side effects that become significant as the technologies become more powerful. Despite various obstacles to recognizing and compensating these effects in appropriate secondary instrumentalizations, eventually they cause such destruction and disease that ordinary people are affected and protest. The protests feed back into technological design through markets and regulation, and can result in modifications that reflect a more realistic understanding of the objects.

This process can be conceived as a technological rationalization, but because it depends on public debate and intervention it is appropriate to call it a *democratic* rationalization. The overall dynamic of this process leads to increased awareness of the social character of technology and a weakening of the technocratic and determinist ideologies that have been associated with the standard concept of rationalization since Weber. Predictably, technical politics will become part of mainstream politics as the process unfolds.

Meanwhile, in other domains such as information technology a related dynamic is at work. Technologies introduced in the context of military and business enterprise are colonized by ordinary people in pursuit of personal fulfilment. The communicative opportunities opened by the technologies have a role parallel to that of environmental side effects, revealing complex potentials of the systems unsuspected by their original designers. In this case the potentials are not threatening but, on the contrary, benign and deserving of development. They enable new forms of sociability and multiply creative possibilities for ordinary people. The democratic implications of these technologies are also emerging clearly, for example, in the case of the Arab uprisings, which are nonetheless unfolding against a background of commercial and political exploitation and surveillance.

The ultimate reality check for technology is public acceptance since the public must deal not only with each particular technology in its ideal setting but all of them together in the chaotic world of daily life. Feedback from 'reality' as it is experienced by ordinary people is thus not extraneous to technology but indeed essential to its successful development. In a differentiated society this feedback takes the form of a circulation of information and products between the technical disciplines and society at large. This is the process of democratic rationalization. It potentially leads to a new form of modern society in which values embodied in technical codes are visible and publicly debated, and technical power is increasingly distributed across the populations concerned rather than concentrated at the top of administrative hierarchies.

PREVIOUS APPLICATIONS

The critical theory of technology was developed through a combination of case studies and theoretical reflection. These studies have concerned a broad cross-section of subjects, from clinical research and environmentalism, to Japanese modernization and several types of information technology. In conclusion, we will briefly describe three of these applications in the field of information technology. These cases are intended to serve not only as illustrative examples but also as entry points for future applications of the method of critical theory of technology.

The Minitel and Democratic Rationalization

The Minitel system was introduced in 1981 on the then standard videotex model. The French State hoped that this initiative would prepare France for the coming information age. The system was based on the free distribution of dumb terminals, called Minitels, which gave access to a packet-switching network operated by France Telecom. France Telecom conceived the system as a virtual marketplace and database. Human communication was an afterthought. The success of the network was startling. It eventually reached 6 million households and 15 million users. At its height, it generated a billion dollars a year in income, divided between France Telecom and the service providers.

This success was due in large part to user intervention. In 1982 users hacked the system and introduced instant messaging. Network operators introduced programs to support this new demand. The new service was primarily employed to seek dates and sex and so was dubbed 'pink messaging' by journalists. There were other applications such as medical advice and homework services, political discussion, and even the organization of a student strike in 1986. Demand for messaging was so great that it soon absorbed half the online usage.

This case illustrates the role of user innovation as a form of democratic rationalization in transforming a computer network from a cold information source into a hot communication service. Users did not simply promote a particular program, but changed the very meaning of the computer through their intervention by transforming its technical code. This had repercussions at the level of primary instrumentalization, affecting the configuration of the software running on the system. The demand for human communication has flowered on an even larger scale on the Internet, definitively changing the meaning of computer technology (Feenberg, 2010: Chapter 5).

Online Education and Technical Code

This second case introduces a very different type of democratic intervention in a very different social environment. In 1982 the Western Behavioral Sciences Institute (WBSI) launched the first distance learning system on a computer network. Its unique feature was a discussion-based pedagogy in online forums led by university faculty. This program provoked considerable interest in the business press and in universities in the English-speaking world and Scandinavia. However, large-scale interest in online education only appeared at the end of the 1990s, during a crisis in university funding. Paradoxically, what computer companies and college administrators understood by this term by that time was something quite different from the early programs such as the one at WBSI. Where WBSI added communication to a traditional distance learning system that lacked it, the new advocates of online education hoped to automate education on the Internet, eliminating the existing interaction in the classroom.

At issue in this case was the struggle for control of the meaning of online education between actors with different agendas; automation in one case, electronic extension of traditional education to new constituencies in the other. The same basic equipment configured in different ways could support completely different social relations and a completely different concept of education.

The automation of education responds to the industrial technical code, going back to the early nineteenth century. The effect of this code is to centralize control of the workforce and to lessen labor costs by substituting machines tended by unskilled labor for skilled labor. The last great advance along these lines occurred in the machine tool industry. At the end of the 1990s it seemed that education would follow suit.

An alternative design of online education, based on the original WBSI model, emphasizes human interaction online. For this alternative, the computer is a medium of communication rather than a simulacrum of the classroom. This alternative employs writing rather than video as the backbone of the system.

The WBSI model has its own limitations and problems, related to the pragmatics of online communication. Online communication differs from its face-to-face equivalent through asynchronicity and the absence of paralinguistic (or nonverbal) signs. Focused group interaction under these unusual conditions requires skilled leadership and appropriate software to accomplish complex goals such as learning. In the educational field, teachers working closely with programmers have devised original solutions to the problem of achieving traditional pedagogical goals in a new environment. This 'participatory design' represents another type of democratic intervention. Participatory design contrasts with technocratic design by isolated experts charged with centralizing power and enhancing control over a dependent and deskilled network of users or workers.

So far neither alternative has prevailed. Examples of communicative usages of computers in education coexist with attempts at automation. Online education is still a technology in the process of formation (Feenberg, 2002: Chapter 5; Hamilton & Feenberg, 2012).

Massively Multiplayer Online Games and Ludification

The video game industry offers another example of the power relations that emerge in technical networks. The industry is now larger than Hollywood and engages millions of subscribers in online multiplayer games. A study of massively multiplayer online games (MMOGs) builds on the dual-level framework of critical theory of technology through a game-specific adaptation of instrumentalization theory (Grimes & Feenberg, 2009). The players' gaming activities are structured by the game code and signified through game narratives of one sort or another, but, once online together, the players participate in online communities that organize them in informal relationships the industry does not control. These communities form within and often in reaction to the rationalized structures of game technology.

Once activated, game communities struggle to reconfigure aspects of the game, mobilizing code and items from the game in new ways and contexts. Markets appear in goods won during play as players auction them off for money. The games themselves are modified by players skilled at hacking. Vast databases of information, cheats and lore about the games are compiled by the players as they engage in knowledge sharing and meta-participation on wikis and other forums. Companies may protest these unauthorized activities but in the end they usually give in and attempt to co-opt what they cannot control. Interaction between game designers and players, and among the players themselves, creates feedback cycles that foreshadow a very different world in which technological citizenship prevails. This world is unlike the mass audiences created by television broadcasting and requires analytic approaches such as those employed by critical theory of technology.

CONCLUSION

Critical theory of technology brings together a general theory of modernity and its democratic potential with constructivist methods of empirical study of technologies. In its application to information technology, it makes possible normatively and politically informed critique. The meaning of these technologies is not determined by their technical properties but is an object of contestation and struggle. Critical theory of technology avoids both the exaggerated hype surrounding such concepts as Web 2.0, and also the equally exaggerated denigration of the Internet as an electronic mall. When technologies are understood as terrains of struggle rather than as fixed and finished things, they are dereified and exposed to criticism, and opened to transformation.

REFERENCES AND FURTHER READING

Bakardjieva, M. & Feenberg, A. (2002) 'Community technology and democratic rationalization', *The Information Society* 18(3): 181–192.

Feenberg, A. (1991) *Critical Theory of Technology.* Oxford: Oxford University Press.

Feenberg, A. (1995) *Alternative Modernity: The Technical Turn in Philosophy and Social Theory.* Berkley, CA: University of California Press.

Feenberg, A. (1999) *Questioning Technology.* Abingdon, Oxon and New York: Routledge.

Feenberg, A. (2002) *Transforming Technology: A Critical Theory Revisited* (2nd edition of *Critical Theory of Technology*). Oxford: Oxford University Press.

Feenberg, A. (2003) 'Modernity theory and technology studies: reflections on bridging the gap', in T. Misa, P. Brey and A. Feenberg (eds), *Modernity and Technology.* Cambridge, MA: MIT Press.

Feenberg, A. (2010) *Between Reason and Experience: Essays in Technology and Modernity.* Cambridge, MA.: MIT Press.

Feenberg, A. & Barney, D. (eds) (2004) *Community in the Digital Age.* Lanham, MD: Rowman & Littlefield.

Feenberg, A. and Friesen, N. (eds) (2012) *(Re)Inventing the Internet.* Rotterdam: Sense Publishers.

Flanagin, A.J., Farinola, W.J. and Metzger, M.J. (2000) 'The technical code of the internet/world wide web', *Critical Studies in Mass Communication* 17: 1–19.

Gratham C. (2009) 'Incorporating critical theory of technology into educational leadership: examining the technical codes of the open journal system', *Access to Knowledge: A Course Journal* 1(2): 1–11.

Grimes, S.M. & Feenberg, A. (2009) 'Rationalizing play: a critical theory of digital gaming', *The Information Society* 25(2): 105–118.

Hamilton, E. & Feenberg, A. (2012) 'The technical codes of online education', in Feenberg, A. and Friesen, N. (eds), *(Re)Inventing the Internet.* Rotterdam: Sense Publishers.

Latour, B. (2005) *Reassembling the Social: An Introduction to Actor Network Theory.* Oxford: Oxford University Press.

Pinch, T. & Bijker, W. (1987) 'The social construction of facts and artefacts', in W. Bijker, T. Hughes and T. Pinch (eds), *The Social Construction of Technological Systems.* Cambridge, MA.: MIT Press.

Veak, T. (ed.) (2006) *Democratizing Technology: Andrew Feenberg's Critical Theory of Technology.* New York: SUNY Press.

9

Critical and Cultural Approaches to HCI

Jeffrey Bardzell

INTRODUCTION

The field of human–computer interaction (HCI) has had to be as fast-moving as the technological changes that have marked the past few decades. Since the early 2000s, HCI entered its so-called 'third wave' (Bødker, 2006; Harrison et al., 2011; Kuutti, 2009), emphasizing non-workplace-based forms of computing, including ubiquitous, mobile, domestic, cross-cultural and entertainment computing and, more generally, experience design, among others.

Designing interactive software for these new contexts – the home, a village halfway around the world, the automobile – revolves around different problems from those of designing for professionals in the workplace. Consider the traditional goals of usability: low task completion time, low error rates, high learnability, high accessibility, and high subjective satisfaction (e.g. Gould & Lewis, 1985; Schneiderman & Plaisant, 2009). To what extent are these goals sufficient to guide the evaluation of massively multiplayer online role-playing games (MMORPGs), ebooks, and mobile apps? Presumably we want our games and apps to be usable, but we also want them to be fun, cool and exciting. It's fairly straightforward to measure error rates and task completion times, but how does one evaluate whether an interaction is 'cool' (Holtzblatt et al., 2010; Read et al., 2011)?

Some of the questions we now use to evaluate interactions are similar to the sorts of questions we use to evaluate art, literature, film, fashion or industrial products. HCI researchers also engage in ethical and political questions, for example, of designing for the developing world, confronting digital divides, and designing technologies for use in ecologically responsible ways. Not surprisingly, many researchers and practitioners in HCI have turned to design and cultural disciplines that have generations (and in some cases, millennia) of experience responding to such questions.

Within the broad field of HCI, a number of subdomains in particular have turned to and innovated on cultural approaches to HCI.

These include the following: experience design (Blythe et al., 2003; McCarthy & Wright, 2004), ubiquitous computing (Dourish & Bell, 2011), design-based approaches to HCI (Gaver et al., 1996; Löwgren & Stolterman, 2004; Zimmerman et al., 2007), reflexive HCI (Sengers et al., 2005), craft (Goodman & Rosner, 2011; Rosner & Ryokai, 2008), aesthetics (J. Bardzell, 2009, 2011; Fishwick, 2006; Udsen & Jørgensen, 2004), feminist and queer HCI (S. Bardzell, 2010; Light, 2011), HCI4D (Ho et al., 2009), affective computing (Boehner et al., 2007), intimate and sexual interaction (Bardzell, J. & Bardzell, S. 2011; Kannabiran et al., 2011; Kaye, 2006), and domestic computing (Bell et al., 2005), among others.

The goal of this chapter is to offer an overview account of what cultural theory is; to explore cultural theories' intellectual practices, domains and concerns; and above all to explore both existing and as yet unexplored linkages between cultural scholarship and HCI, with an eye to enriching the development of HCI's third wave.

KNOWLEDGE AND/OF CULTURE

The idea of 'culture' is famously difficult to define. Rather than offering my own attempt at a definition, I will let those in cultural studies speak for themselves. Two introductory books on cultural theory (Milner & Browitt, 2002; Smith, 2001) begin with definitions of culture based on the seminal work of cultural theorist Raymond Williams. Williams, as paraphrased by Smith, offers three common understandings of culture.

- To refer to the intellectual, spiritual, and aesthetic development of an individual, group, or society.
- To capture a range of intellectual and artistic activities and their products (film, art, theatre). In this usage culture is more or less synonymous with 'the Arts,' hence we speak of a 'Minister of Culture.'
- To designate the entire way of life, activities, beliefs, and customs of a people, group, or society. (Smith, 2001: 2)

These definitions collectively suggest that culture refers to the world of ideas and their expressions in a given social group. In Milner and Browitt's formulation, culture refers to 'that entire range of institutions, artefacts and practices that make up our symbolic universe' (Milner & Browitt, 2002: 5). Culture can be understood in terms of an intersubjectively participatory inner life that is simultaneously constituted by and given expression through symbolic systems (or 'texts' as they are often called, though they may be novels, films, paintings, or even advertisements, comic books, and other forms of popular culture).

This formulation of culture has implications for how researchers study it, a point that is clearest when set in contrast to other scholarly approaches. For example, much knowledge in the natural sciences is derived from the discovery of facts in the physical world, for example the chemical composition of different types of volcanic material, the mating behaviors of a certain kind of animal or patterns of celestial movements. Much work in the social sciences seeks similarly to discover facts of the human social world, such as the relationships between organizational structures and the creative output of professional teams, the effects of different pedagogical strategies on fourth grade reading scores or relationships between economic incentives and social behavior.

In contrast to science's emphasis on the *discovery* of facts, humanistic questions are often not so much fact-based as they are based in rational accounts of the *judgments* we form as to whether a work is good, what it means, how its formal qualities bear certain meanings, how innovations in production techniques evolve an expressive symbolic vocabulary, how a work reflects and comments upon the circumstances under which it was made, and so forth.

Several philosophers and cultural scholars (e.g. Barnard, 2001; Taylor, 1971) argue that knowledge of cultural phenomena is of a different kind from the knowledge of the natural world and, accordingly, the relationship of the knowing subject (i.e. the scientist, the cultural critic) to the objects of enquiry also differs. Another way to express this idea is that scientific and humanistic disciplines tend to prioritize different epistemologies. An *epistemology* is a theory of knowledge, that is, an account of how knowledge is acquired, how it is evaluated, what doesn't count as legitimate knowledge (e.g. the rejection of astrology as pseudo-science), who is able to acquire or create knowledge and under what circumstances, etc.

One challenge for HCI has been that the predominantly science-trained HCI community has not always been in a strong position to perceive or evaluate the rigor or quality of humanist contributions to the field. In other words, humanistic research runs the risk of coming across as bad science because its knowledge practices and forms of rigor differ. I hope this chapter can contribute to bridge building within HCI between scientific and humanistic knowledge traditions. In the following sections, I will first elaborate on ways that humanist and scientific epistemologies differ, then I will elaborate on ways that they are similar.

SCIENTIFIC VERSUS HUMANISTIC EPISTEMOLOGIES

Malcolm Turvey, a philosopher of film, explores the differences between scientific and humanistic epistemologies in an essay in which he challenges the fitness of scientific models for the study of film. Turvey offers a summary of how empirical research and the development of theory have been so successful in science, then he points out that film theorists do not generally collect any data or attempt to verify empirically their own claims. In light of all this, Turvey wonders, 'how is it that theories about [films] ever convince anyone … ?' (Turvey, 2005: 25).

His answer to this question is, whereas science seeks to discover laws and patterns that are hidden from everyday observation, 'film theories concern *what human beings already know and do*' (2005: 25, italics in

original). For example, we already know what a romantic comedy is, and we know how to respond to it, and how to behave in the theatre during it; all of this is present to everyday observation and awareness and need not be discovered. The problem is that our awareness of such norms is fuzzy: we often sense, for example, that a film, novel, or song is beautiful or resonates with us in some very personal way, but we struggle to explain why or how. We have all been in a museum or a class and experienced that moment of revelation when the docent, or, teacher points to a detail, a phrase or a stroke and tells us what it means (e.g. symbolically, its historical significance, how it was made, etc.). That detail had been right in front of our eyes all along, and we needed no microscope or DNA test to see it, yet we just overlooked the detail and its significance; now, we can't see the work without focusing on that detail. The ability to perceive such details and relate them to an account of a work's overall meaning is one of the paradigmatic skills of the humanities scholar. As aesthetic philosopher Marcia Eaton puts it, critics 'are in the business of *pointing*. They get us to *perceive* some features of an object or event' (Eaton, 1988: 116).

The ability to perceive and point to the right details is not something one is born with, since it depends on one's awareness of norms and conventions, that is, things we learn throughout life. Alexander Baumgarten, the eighteenth-century philosopher who first coined the term 'aesthetics', defines it as 'the perfection of sensory cognition', a human faculty that includes keenness of sensation, imaginative capacity, penetrating insight, good memory, poetic disposition, good taste, foresight and expressive talent (Shusterman, 2000: 264–65). This faculty is cultivated or improved through years and years of practise by engaging both canonical texts (again, 'texts' is used in an inclusive sense) and generations of scholarship on and criticism of those texts.

The humanist epistemological foregrounding of perception, insight, imagination, etc., has strong implications for what constitutes good humanist knowledge. A successful critical interpretation, in the words of philosopher Richard Eldridge, captures 'genuine aspects of a work's meaning, of how [the work] distinctly presents a subject matter as a focus for thought and emotion, fused to the [critic's] imaginative exploration of material' (Eldridge, 2003: 147). Eldridge's statement here nicely summarizes humanistic attitudes towards the object of their enquiry (i.e. a text that 'distinctly presents a subject matter as a focus for thought and emotion') and the subject of enquiry (i.e. the scholar-critic engaged in an 'imaginative exploration of material'). 'Text' and 'data set' are two different kinds of objects of enquiry, and the critic and the objective scientist are two different kinds of knowing subject.

THE COMMONALTIES OF SCIENTIFIC AND HUMANISTIC WORK

In the preceding section, I drew contrasts between scientific and humanistic ways of knowing. Drawing such contrasts is constructive because it helps readers understand the different intellectual problems, goals and priorities of scientific versus humanistic disciplines. The disadvantage of drawing such contrasts is that it seems to suggest there is an unbridgeable gulf between science and the arts, as if they are completely incommensurate. However, I do not want to reinforce a 'two cultures' narrative, especially when I am advocating that scientific and humanistic approaches be used together in HCI! (It is also worth noting that some academic fields – anthropology in particular – have productively blended these epistemologies for decades.)

The prioritization of perceptiveness and keenness of insight in humanist works should not imply that scientific works lack any of these. The distinctions drawn here between science and humanistic approaches have to do more with their *primary* epistemologies, as opposed to their whole processes or products.

In other words, the primary means by which knowledge is achieved may differ for scientific versus humanistic disciplines, but all disciplines nonetheless display qualities of the other. For example, a rigorous empirical study has critical or humanistic aspects inasmuch as its researchers make decisions or judgments about what data and in which domain they should conduct their research as well as in making decisions about how to interpret their data and present it to their peers. Submitting a successful scientific paper also requires a mastery of the norms of a given journal's or conference's readership. Likewise, a rigorous humanistic project involves discoveries of fact: what is in a given text, the conditions of its production, and the history of its reception. Understanding a cultural text also entails the development of generalizations based on evidentiary particulars, and such understandings are evaluated by the community in part on their ability to explain those particulars.

Philosopher of science Dagfinn Føllesdal points out a more fundamental similarity between humanistic and scientific scholarship. Both, he claims, use the *hypothetico-deductive method*. This method entails 'an application of two operations: the formation of hypotheses and the deduction of consequences from them in order to arrive at beliefs which – although they are hypothetical – are well supported' (Føllesdal, 1994: 234). Individual hypotheses have credibility inasmuch as they are both logically consistent and fit with our experience. Also, a collection of related hypotheses is compelling inasmuch as it achieves theoretical simplicity. From such a perspective, we might say that the following two claims are equivalent.

- Hypothetical inferences derived from biological data can be explained by the theory of evolution.
- Hypothetical inferences derived from Federico Fellini's films can be explained with the theory that Fellini's Italian Catholic upbringing deeply affected his views of the world, his sexual relationships, and himself.

From the perspective of how we reason about our objects of enquiry, scientific and humanistic approaches aren't so different after all. Neither is merely opinion-based, and both produce claims that have an evidentiary basis that is persuasive to an intellectual community.

AN EXAMPLE: THE SCIENCE AND CULTURE OF USER EXPERIENCE

The broad-brush philosophical summary I have just offered between scientific and humanistic epistemologies might come across as irrelevant to HCI. Yet recent developments in the field are recapitulating the types of issues described here.

One of the most visible examples is (what is in my view) HCI's schizophrenic take on user experience, which can be summarized as follows.

- One body of UX research treats user experience as a discoverable phenomenon, out there in the world. This broadly scientific approach uses theories, such as Csikszentmihalyi's 'flow', and diverse methods to discover it, for example self-report techniques using the Positive and Negative Affect Schedule (PANAS), questionnaires with Likert scales, and the use of physiological sensors to detect user engagement, frustration, and other experiential qualities. Examples of this kind of user experience research include Hassenzahl et al. (2010), Law et al. (2008), Law et al. (2009), Liao et al. (2006), Picard (1997), Picard and Klein (2002) and Sutcliffe (2010).
- Another large body of research treats experience as a form of felt life to be interpreted. This broadly cultural approach leverages philosophy, literary theory and design methods to help the UX researcher critically interrogate a user experience situation. Examples of this kind of user experience research include McCarthy and Wright (2004), Buchenau and Fulton Suri, (2000), Forlizzi and Battarbee (2004), Nardi (2010) and Boehner et al. (2005).

The HCI community accepts user experience (UX) research contributions from both scientific and cultural approaches, which I regard as a very positive development. Less encouraging is the fact that these two literatures on UX seem to be developing almost in isolation of one another: references from one side to the other are few, and when they do occur, they are often unsympathetic (Bardzell, 2012).

If we want to build links between these two approaches, we might begin by observing that UX research seems to involve two different underlying questions.

1. What do actual users like about, and how do they feel about, given experiences involving interaction design?
2. How can user experience researchers and designers perceive what is most worthy of perceiving in an experience involving an interaction design?

The first question is fundamentally empirical in nature: the desired information is 'out there' (i.e. users do have preferences, feelings and perspectives about their experiences), and the job of the UX researcher or designer is to discover that information in the best way possible. The second question is about distinguishing relevant elements of an experience to perceive and to clarify their significances and meanings. For example, a film critic can identify links among editing or *mise en scène*, broader cinematic or cultural movements, available production technologies and techniques, and experiential qualities. By the same reasoning, an interaction designer should be able to 'read' relationships among the visual languages of an interaction, its functionality, and the most important use qualities that contribute to its experience.

Is there any doubt that competence in answering both questions would support the design of better interactive experiences? Also, answers to each of the two questions should be able to inform the other. A rigorous empirical data set could effectively expose issues of interest – aspects of an experience which were, for example, surprisingly good or bad – that a more cultural perspective could then be leveraged to interpret in a more culturally sensitive and perceptively insightful way.

Conversely, more critical understandings of how experiential qualities arise from interacting with certain types of systems and interfaces, given relevant socio-technical norms surrounding such interactions, could lead to innovation in the development of empirical measures and research instruments.

It seems that the HCI community is ready for both cultural and scientific approaches to coexist, not just in experience design but in any number of areas (e.g. affective computing, embodied interaction, HCI4D, etc.). Yet if each of these topics replicates a 'two cultures' pattern, we will have missed out on the potential benefits of a more synergistic approach. My hope is that a better understanding of humanist scholarly norms in the HCI community will help prevent excellent critical work from looking like it is just bad science and hopefully decrease some of the confrontational rhetoric that sometimes occurs, while, ideally, stimulating mutually beneficial collaborations among researchers and practitioners from both intellectual traditions.

DOING CULTURAL HCI

For the balance of this chapter I will focus on 'doing cultural HCI' as a knowledge practice. One aspect of this is the core competencies or foundational abilities of cultural approaches to HCI, including design thinking and criticism. Another aspect is how critical and cultural theory has moved into HCI practice. I explore each of these in turn.

Developing an Ability to Do Cultural HCI

In both the design and humanities literatures, the designer or critic is understood to be a special kind of person, what I characterize as an expert subject. By *expert subject* I refer to people who, like master crafters or important filmmakers, are better than the average person at what they do, and yet importantly they are better at it in a very individual or personal way; for example, a

Raymond Loewy design or an Alexander McQueen dress. This means that the products of their efforts – be they designs, crafts or essays – are recognized as valuable to a community but also recognized as unique to them, that is, not replicable.

The value that design brings is not a fact to be discovered. As Nelson and Stolterman describe it:

> The success of the design process can best be determined when those being served experience the *surprise of self-recognition*. This comes when that which emerges from a design process meets and exceeds the client's original expression of that which they (usually only dimly) perceived as desirable in the beginning. This original expression of what is desired is known as the client's *desiderata*. The designer's role is to midwife that desiderata, which could not have been imagined fully from the beginning, by either client or designer … (2003: 48)

The key vocabulary in this quote – perception, imagination, and the metaphor of midwifery to describe the collaborative labor of bringing something new of value into the world – unmistakably echoes the notions of aesthetics as 'the perfection of sensory cognition' described earlier in this chapter. The role of the designer is to serve as the expert subject who creatively and insightfully perceives and pursues solution opportunities within the desiderata. This subjective expertise is sometimes referred to as *design thinking*, that is, the idea that professional designers share a knowledge practice, a certain way of reasoning about design problem domains, solution generation, and solution evaluation that enables them to be skilled designers (Cross, 2007; Rowe, 1987).

This thinking is not a rare cognitive faculty that only some people get to have: design thinking is an outcome of training. In design school, students are immersed in a design culture and gradually learn to tacitly appropriate its knowledge and skills (Snodgrass & Coyne, cited in Greenberg & Buxton, 2008). Löwgren and Stolterman describe the cultivation of design ability through, among other activities, developing a 'sense of quality', a position that clearly resonates with the notion of criticism as pointing and as a form of skilled perception, as described earlier. This sense of quality includes an insightful appreciation of and ability to create quality, and it 'has to be developed, continuously challenged, and improved' (Löwgren & Stolterman, 2004: 58).

A primary way that designers and scholars of culture develop, continuously challenge and improve this ability is through criticism. *Criticism* refers to expert acts of judgment, that is, the practice that leads to the acts of perceptive pointing and the clarifying of norms and meanings that I described earlier in this chapter. It is often motivated by vague yet intriguing intuitions that we have when we respond to human works. Philosopher of film Stanley Cavell (Cavell & Klevan, 2005: 181) describes how certain moments in a film speak to him in this way: 'I didn't only think the shots were unusual, or striking, I thought they were gently mysterious, and that they were significant. They asked questions of me'. Once certain moments in a film capture his attention, he then rationally interrogates them to try to understand why they affected him in that way; he tries to solve their gentle mystery. Schön and other design theorists also have described this idea of designs and materials speaking back to them (Schön calls it 'backtalk'). In this way, the framing of problems and the proposing of design solutions happen iteratively in a 'dialogue' between designers and their materials and designs.

Summarizing, criticism can be characterized as the rational investigation of wisdom or intuition, initially enabled by a receptivity to or perceptiveness with regard to how human works speak to us and then developed through learned interpretative strategies made possible through exposure to comparable examples, theory and knowledge about the contexts of creation and experience. This rational investigation clarifies the relationships between objects and their subjective experiences; intensifies and beautifies those experiences (Dewey, 1934; Shusterman, 2000);

and even reveals the false pleasures of those experiences, that is, the ways in which works are 'subjecting you to false views of yourself and of the world' (Cavell & Klevan, 2005: 178). Critical interpretation and backtalk in this sense is also inexpensive (compared, at least, to scientific enquiry) and practised daily; my experience of working with designers is that they critically interpret designs more or less constantly and reflexively, both at work and in everyday life.

In my own work, I have sought to explore criticism as it applies to interaction design (Bardzell, 2009, 2011). In Bardzell (2011), I define *interaction criticism* as

> rigorous interpretive interrogations of the complex relationships between (a) the interface, including its material and perceptual qualities as well as its broader situatedness in visual languages and culture; and (b) the user experience, including the meanings, behaviors, perceptions, affects, insights, and social sensibilities that arise in the context of interaction and its outcomes.

I have argued that in critical fields – literature, art, design, philosophy, media studies – these interrogations tend to emphasize one of the following four fundamental perspectives.

- *The creator and the act and situation of creation.* Critical reasoning from this perspective foregrounds creative agency, including creative intentions, skills, desires and influences. When we talk about 'a typical Hitchcock film', we are reasoning from this perspective. Studies of designers, design processes, crafters, and the notion of a designer's unique signature or voice also relate to this perspective.
- *The artifact itself.* This perspective focuses on qualities of the artifact, including its content and arrangement in form, its material qualities, and its visual language. When we think about how a work references other styles, genres or movements, we are reasoning from this perspective. Much traditional design criticism emphasizes this perspective, and HCI would do well to develop this kind of critical vocabulary more explicitly for interaction.
- *The individual consumer, reader, or user and the experience.* A work is only meaningful when it enters into human experience, including

perception, imagination, emotional reaction, and so forth. When we talk about the emotional qualities or meanings of works, we are generally speaking from this perspective. User experience research is often situated in this perspective.

- *The socio-cultural context.* Works are created, distributed and become meaningful not just in individual minds but also more broadly in socio-cultural contexts. National, political, economic, gendered contexts influence meanings and perceived value, often in ideological ways. Gender and HCI, HCI4D, and sustainable interaction design are a few areas that are increasingly exploring HCI from this perspective.

I want to stress from the outset that we all use these four perspectives all the time. Everyone at one point or another has chosen to watch a movie based on the director, or on its a genre (e.g. horror) in anticipation of having a certain kind of experience, or has chosen a movie because it reflects or challenges their politics or values. Everyone has tried to interpret what an author or painter might have meant in a novel or painting. We can all easily recognize gender stereotypes in old advertisements. Again, cultural approaches are about clarifying what we already know and do; I view them as different in degree, not in kind, from the everyday critical judgments we all make.

What cultural scholarship adds to this are thousands of theories and concepts applied to works from nearly every culture. Such theories offer models that support our attempts to rationally interrogate our intuitive reactions from each of these perspectives more rigorously and insightfully. The challenge for interaction design professionals thus lies in choosing and leveraging such theories to cultivate their intuitive reactions into intellectually useful design understandings.

From Cultural Theory to HCI Practice

In the previous section, I sketched some of the background ability common to arts and humanities-influenced knowledge workers, including designers. I stressed that the lifelong practice of criticism helps with the cultivation of a perceptive and imaginative eye, and I offered some broad perspectives commonly used to frame such activities. Yet such descriptions fall far short of a methodology. How does cultural theory translate into HCI practice?

If we understand a *methodology* as the link between an epistemology (i.e. a theory of knowledge-making) and individual research methods, then a methodology would seem to be just what we need to connect the rich theoretical tradition of the humanities with the practice-oriented demands of interaction design. However, the notion of a 'methodology' is seldom used in the humanities, which emphasizes instead the creativity and insightful perception of the individual critic over any replicable approach. I do not mean to imply that linking cultural theory design practice is impossible; it just suggests it may be a little bit difficult and it will take some work. In the following three subsections, I summarize instances where this has occurred, beginning with one or more critical theories from the humanities, moving into HCI by way of theory, and from there influencing HCI research and practice methodologically.

Aesthetic Experience: From Dewey to HCI

The first example involves the work of American pragmatist philosopher John Dewey, who wrote extensively on art and experience in the middle of the twentieth century (Dewey, 1925, 1934). Concerned with the ways that industrialization and factory labor created deadening environments, Dewey sought to improve the quality of life for people through better education and a more just society. These philosophical explorations led to one of his most enduring contributions: a theory of aesthetic experience. In contrast to traditional notions of aesthetic experience, which centered on the appreciation of fine arts by the leisured classes or limited them to museum experiences for everyone else, Dewey sought to articulate a notion of everyday aesthetic experience.

A very basic distinction in Dewey's analysis is the difference between the constant flow of experience and 'an experience', that is, a grouping of experience-flow together in the form of a coherent entity, for example a memory of an experience or the story one tells of an experience. Dewey wants to understand what makes a good 'an experience', and he proposes a model framework. As a coherent entity, an aesthetic experience has a beginning, a middle and an end. Each 'an experience' is directed towards some end or purpose; that is, it has directedness. Within this directed experience, there is 'doing' and 'undergoing', that is, actions by and reactions to the experiencing subject. These doings and undergoings are characterized by variety and movement, which are pleasurable. All of this leads to a consummation, which satisfactorily concludes the aesthetic experience.

Dewey's work has been picked up in HCI and design fields several times. Donald Schön's dissertation was on Dewey, and design theorist Erik Stolterman has suggested that Dewey's thought pervades Schön's *The Reflective Practitioner* (private communication). In HCI, McCarthy and Wright (2004) wrote a highly influential book on experience design, *Technology as Experience* (alluding to Dewey's book, *Art as Experience*). McCarthy and Wright apply Dewey's theory of experience within the experience design domain. In doing so, McCarthy and Wright propose norms by which researchers, practitioners and users alike might evaluate the experiential qualities of interaction. Recalling how humanistic scholarship tends to stress the clarification of norms surrounding what we already know, it is easy to see McCarthy and Wright's work as fitting squarely in this tradition. Their application of Dewey offers a theory that makes UX practitioners and researchers more perceptive and imaginative in their engagement with users, interaction designs and experiences. Thus, one HCI outcome of an engagement with cultural theory is the emergence of new HCI theory.

A more recent example in HCI is Bonnie Nardi's (2010) use of Dewey's theory of aesthetic experience in her ethnographic work on World of Warcraft (WoW). Nardi applies three theoretical frames – Skinnerian behavioralism, Deweyan aesthetic experience, and activity theory – to construct an understanding of the ways in which WoW play is pleasurable. Nardi takes on issues that resonate with the WoW player community, such as the buying and selling of player accounts, and constructs an insightful understanding of them using these theoretical constructs. In effect, her work is strikingly similar to Cavell's process of moving from initial intuition to theoretically informed rational interrogation. Nardi's use of Dewey suggests another HCI outcome of engaging with cultural theory: a critical interpretation, or 'reading', of a complexly layered, yet important interaction in its particular context.

From Critical Theory to Critical Design

A common goal for most strands of critical theory is an effort to use theory to reveal hidden, yet important, aspects of culture, and through that revealing, to subject them to intentional change. In this regard, critical theory and design seem like natural partners. This linkage has been explored in critical design (Dunne, 2006; Dunne & Raby, 2001). Dunne offers the following definition of critical design: It is:

> a design approach for producing conceptual electronic products that encourage complex and meaningful reflection on inhabitation of a ubiquitous, dematerializing, and intelligent environment: a form of social research to integrate critical aesthetic experience with everyday life. ... I hope in my approach I have retained the popular appeal of industrial design while using it to seduce the viewer into the world of ideas rather than objects. Industrial design locates its object in a mental space concerned with identity, desire, and fantasy and shaped by media. ... Again, I hope this remains intact but is subverted to challenge the aesthetic values of both consumers and designers. (2006: 147)

Critical design promises an approach that, rather than serving needs as they are presently understood, instead seeks to disrupt, to transgress such constructions of need and encourage people to 'enrich and expand our experience of everyday life', to stage dilemmas 'that force a decision onto the user, revealing how limited choices are usually hard-wired into products for us' (Dunne & Raby, 2001: 45–6). Examples of works that feature critical designs are Gaver et al. (2004, 2010) and Boehner et al. (2005).

Critical design is 'critical' in the sense that it foregrounds perception and imagination, rather than the discovery of facts and their translation into needs and requirements. By challenging users' expectations, critical designs invite users and researchers to perceive and reason about norms in new ways. As Cavell – and generations of Marxist and feminist cultural researchers – have warned, art and design can also provide false pleasures, by entertaining us or serving our functional needs in a way that gives us a false or unjust understanding of the world: critical designs can subversively challenge these false pleasures. By provoking those involved into discussion, critical designs also stimulate an imaginative discourse that is oriented towards the articulation of new desiderata, simultaneously leading to new designs, but also, more radically, leading to demand for new kinds of design. At the same time, critical design strives to remain grounded in everyday life; Dunne (2006: 84) writes, 'But if [critical design] is to avoid accusations of escapism this design thinking must also develop strategies for linking itself to everyday life'.

Cultural HCI Methodological Innovations

Interaction criticism and critical design are not the only methodological innovations that have origins in cultural HCI. I will sketch four more in this section and then explore their underlying commonalties.

The first is 'pastiche scenarios', as presented in Blythe and Wright (2006). The basic idea behind pastiche scenarios is to blend together the conventional HCI notion of a scenario with a character or scene from a literary narrative, such as a novel. Whereas the goal of a traditional scenario is to put a human face and everyday situation on empirical data sets, the goal of a pastiche scenario is to interrogate the felt life of interaction in much richer ways – albeit at the expense of strict fidelity to the data itself. For example, the authors used Bridget Jones to construct a pastiche scenario about iPod use, not only because she meets the core demographic criteria of iPod users, but also because 'she brings more to the scenario than her age, occupation, and gender', including a 'distinctive narrative voice' and very specific habits, such as her anxieties about how she is perceived, her habits of linking specific songs to memories of prior boyfriends, and a general sense of disorganization in her everyday habits (Blythe & Wright, 2006: 1146). Pastiche scenarios push the underlying logic of personas and scenarios – to put a human face on data to help us understand it – to another level, one of creative empathy with the felt life of others.

Another methodological innovation is the concept of *defamiliarization*, introduced in Bell, Blythe and Sengers (2005). The concept is originally derived from literary theory, specifically the work of the Russian formalists in the early twentieth century, who argued that the *telos* (or end) of art 'is to change the mode of human perception, to render [hitherto] imperceptible formulas [so that they become] unusual and palpable' (Steiner, 2005: 18). In their paper, Bell, Blythe and Sengers (2005) describe one of the chief difficulties in designing new technologies for the home as the fact that the home is so overly familiar to us it is hard for us to see it critically. They offer defamiliarization as a methodological approach to dealing with this problem. One strategy is to use an historical and cultural analysis of the home, to identify historical trends – such as the rise of industrialized efficiency in the kitchen, which is not combined with a rise in the quality of food – and situate them alongside emerging designs in HCI as a means to exposing their

> Comment
> [RC6]:
> Not in Refs?

assumptions. Another strategy is to study homes in different countries; studies of homes in Asia helped the researchers perceive numerous differences in the ways families use them and how family members relate to each other within their homes.

The third methodological innovation I wish to introduce is *feminist HCI*, which seeks to bring together feminist theory and HCI research and practice. In a recent paper, Shaowen Bardzell (2010) introduced the term and then motivated and defined it with an interesting rhetorical strategy. She began by introducing feminism itself – its core values, positions and epistemologies. Next, she explored its influence in design fields that are comparable to HCI, including industrial design, urban planning, game design. Next, she synthesized research in HCI that revolved around key themes of feminism, even though it did not use that vocabulary; she then turned her synthesis into an analysis of opportunities for feminism to contribute to HCI, before concluding with a handful of qualities of feminist HCI. Her technique of tracing feminism from its origins in critical theory through other design fields and into HCI is perhaps the most explicitly genealogical importation of cultural theory into HCI. Building on this work, we subsequently proposed an outline of a feminist HCI methodology that offers concepts to support researchers' navigations between the scientific imperative to 'know truth' and the moral imperative to 'do good' (Bardzell, S. & Bardzell, J. 2011).

The final methodological innovation I will describe here is slightly different in character but equally important. It is a particular form of analysis developed by Phoebe Sengers and her colleagues, in particular in Boehner et al. (2007) and DiSalvo, Sengers and Brynjarsdóttir (2010), which Sengers calls the *epistemological survey*. An epistemological survey collects a corpus of the works in HCI on a given topic; it then *critiques* that corpus to expose the unsaid, the assumptions, and even the ideologies implicit in the corpus. In DiSalvo, Sengers and Brynjarsdóttir (2010, 1976), for example, they asked the same set of questions of each of the roughly 70 papers in the corpus.

(1) how does the paper define and justify attention to sustainable HCI? (2) what disciplinary orientation is used? (3) how is the problem of sustainability and its solution framed? ... (4) How is the role of the researcher framed? (5) Who takes action, or is supposed to take action? (6) Who is considered the 'expert,' and whose point of view is questionable? (7) How do the authors deal with political disagreements about the environment? (8) Does the paper aim to establish a definitive truth, or does it leave open the possibility of serious differences of opinion about its subject matter? (9) What constitutes success?

One of the primary goals of this critical analysis is to explicate the subject positions that are constructed through the corpus. Who is the researcher? Who gets to act? Who is framed as part of the problem? If design is about effecting change, and change is inherently political, then it would follow that HCI is a political field. However, HCI does not always acknowledge its own political dimensions, the ways that it asserts power, introduces divisions, or establishes who is helping and who is fixing whom. As the authors write, 'Our goal is to provide a reflective lens for practitioners of sustainable HCI which will allow for principled, reflective discussion of how we have, until now, defined sustainable HCI, and how we might best choose to do so in going forward' (DiSalvo et al., 2010, 1975)

It is worth taking a moment to summarize some features that are common across two or more of these works. The most basic feature is the use of critical theory and its approaches in HCI; that much is obvious. Another feature that they share is a notion of the researcher as an expert subject, someone who is responsible for creative perception and insightful interpretation of some aspect of HCI. The pastiche scenarios paper relies heavily on the researcher's ability to identify an appropriate literary source and to apply it thoughtfully. The notion of reflexivity (Sengers et al., 2005) is also crucial to this work: critical HCI researchers do not only see themselves as incrementally adding knowledge to the literature that has come before but also, significantly, as re-evaluating that literature and redefining ourselves as a result. Bardzell's feminist HCI paper suggests that much HCI research was already feminist, even though that term ('the F-word' as one attendee at her talk put it) was not used: in this sense, feminist HCI brought about a 'surprise of self-recognition'. The methodologies described here are also historically and/or spatially situated; for example, the defamiliarization paper uses history and different countries as broad strategies to defamiliarize the here and now. Another feature common to these methodologies is a political awareness, which is made explicit in both the feminist HCI and the epistemological survey research.

CONCLUSION

I have introduced culture, cultural scholarship and the application of cultural scholarship in design and HCI. Though it is clear that HCI wants to (and must!) understand culture better, I want to conclude this chapter by stressing that cultural HCI should have less to do with *telling us about* culture and more to do with *helping us improve* culture. It would be wrong, I argue, to see cultural approaches primarily as another research lens to tell us what is out there in the world; the social sciences are a better fit for this direction of enquiry. Cultural approaches should be used to help HCI improve our lived environment and improve ourselves. I see two broad strategies for pursuing these goals.

First, HCI should orientate itself towards the production of technologies and interactive experiences that are of genuine and lasting value. That means our work should meet 'human needs' in the broadest, deepest and most robust sense possible: social, emotional, physical and spiritual needs at both individual and societal scales. We should contribute to the making of artifacts, spaces, technologies, practices and social arrangements that challenge and ennoble us to be more empathic and more vigilantly participatory and democratic. We

should be careful that we are not contributing to false pleasures. This all sounds like a tall order, but I do not propose it as a literal agenda – rather, as an orientation to our research agendas; let this, not efficiency and productivity, be our ultimate goal.

Second, *HCI – like the best art, craft, and traditional design – should cultivate us as appreciators of what is worthy*. This includes the education of our senses and imagination, as proposed in the eighteenth century by Baumgarten and developed more recently by Shusterman. It includes increasing our capacity for empathy, which philosopher Richard Rorty suggests can be achieved through our encounters with literature and art – why not interaction design? It includes contributing to community aesthetics at the local, national, and even global levels, because what people appreciate, they want. As we saw with critical design, helping users perceive the world more freshly, provoking and surprising them, can spur both imagination and demand, improving quality of life today and generating the desiderata for a better tomorrow. The philosopher Michel Foucault suggests that each of us ought to treat our self as a work of art. Let us HCI researchers and practitioners challenge ourselves to be a part of that project.

REFERENCES

Bardzell, J. (2009) Interaction criticism and aesthetics. *Proceedings of CHI '09*, ACM, 2357–66.

Bardzell, J. (2011) Interaction criticism: An introduction to the practice. *Interacting with Computers* 23(6), 604–21.

Bardzell, J. (2012) Commentary on: Tractinsky, Noam (2012): Visual aesthetics: in human–computer interaction and interaction design. In: Soegaard, Mads and Dam, Rikke Friis (eds), *Encyclopedia of Human-Computer Interaction*. Aarhus, Denmark: The Interaction-Design.org Foundation. Available www.interaction-design.org/encyclopedia/visual_aesthetics.html

Bardzell, J. & Bardzell, S. (2011) 'Pleasure is your birthright': Digitally enabled designer sex toys as a case of third-wave HCI. *Proceedings of CHI '10*, ACM, 257–66.

Bardzell, S. (2010) Feminist HCI: Taking stock and outlining an agenda for design. *Proceedings of CHI '10*, ACM, 1301–10.

Bardzell, S. & Bardzell, J. (2011) Towards a feminist HCI methodology: Social science, feminism, and HCI. *Proceedings of CHI '11*, ACM, 675–84.

Barnard, M. (2001) *Approaches to Understanding Visual Culture*. New York: Palgrave Macmillan.

Bell, G., Blythe, M. & Sengers, P. (2005) Making by making strange: Defamiliarization and the design of domestic technologies. *ACM Transactions on Computer–Human Interaction* 12(2), 149–173.

Blythe, M., Overbeeke, K., Monk, A. & Wright, P. (2003) *Funology: From Usability to Enjoyment*. Dorderecht: Kluwer Academic Publishers.

Blythe, M. & Wright, P. (2006) Pastiche scenarios: Fiction as a resource for user centred design. *Interacting with Computers* 18, 1139–64.

Bødker, S. (2006) When second wave HCI meets third wave challenges. *Proceedings of NordiCHI '06*, ACM Press.

Boehner, K., DePaula, R., Dourish, P. & Sengers, P. (2005) Affect: From information to interaction. *AARHUS '05*, ACM, 59–67.

Boehner, K., Vertesi, J., Sengers, P. & Dourish, P. (2007) How HCI interprets the probes. *Proceedings of CHI '07*, ACM, 1077–86.

Buchenau, M. & Fulton Suri, J.F. (2000) Experience prototyping. *Proceedings of DIS*, ACM, 424–33.

Cavell, S. & Klevan, A. (2005) 'What becomes of thinking on film?': Stanley Cavell in conversation with Andrew Klevan. In: Read, R. & Goodenough, J. (eds), *Film as Philosophy: Essays on Cinema after Wittgenstein and Cavell*. New York: Palgrave Macmillan, 167–209.

Cross, N. (2007) *Designerly Ways of Knowing*. Basel: Birkhauser.

Dewey, J. (1925) *Experience and Nature*. Chicago, IL: Open Court Publishing Company.

Dewey, J. (1934) *Art as Experience*. New York: Perigee Books.

DiSalvo, C., Sengers, P. & Brynjarsdóttir, H. (2010) Mapping the landscape of sustainable HCI. *Proceedings of CHI '10*, ACM, 1975–84.

Dourish, P. & Bell, G. (2011) *Divining a Digital Future: Mess and Mythology in Ubiquitous Computing*. Boston, MA: MIT Press.

Dunne, A. (2006) *Hertzian Tales: Electronic Products, Aesthetic Experience, and Critical Design*. Cambridge, MA: MIT Press.

Dunne, A. & Raby, F. (2001) *Design Noir: The Secret Life of Electronic Objects*. Basel: Birkhäuser.

Eaton, M. (1988; 1999) *Basic Issues in Aesthetics*. Long Grove, IL: Waveland Press.

Eldridge, R. (2003) *An Introduction to the Philosophy of Art*. Cambridge: Cambridge University Press.

Fishwick, P. (ed.) (2006) *Aesthetic Computing*. Cambridge, MA: MIT Press.

Føllesdal, D. (1994) Hermeneutics and the hypothetico-deductive method. In: Martin, M. & McIntyre, L. (eds), *Readings in The Philosophy of Social Science*. Cambridge, MA: MIT Press.

Forlizzi, J. & Battarbee, K. (2004) Understanding experience in interactive systems. *Proceedings of DIS '04*, ACM, 261–8.

Gaver, W., Blythe, M., Boucher, A., Jarvis, N., Bowers, J. & Wright, P. (2010) The prayer companion: Openness and specificity, materiality and spirituality. *Proceedings of CHI '10*, ACM, 2055–64.

Gaver, W., Bowers, J., Boucher, A., Pennington, S., Schmidt, A., Steed, A., Villars, N. & Walker, B. (2004) The drift table: Designing for ludic engagement. *Extended Abstracts of CHI*, ACM.

Gaver, W., Dunne, T. & Pacenti, E. (1996) Cultural probes. *Interactions*, ACM, 21–9.

Gould, J. & Lewis, C. (1985) Designing for usability: Key principles and what designers think. *Communications of the ACM* 28(3), 300–11.

Goodman, E. & Rosner, D.K. (2011) From garments to gardens: Negotiating material relationships online and 'by hand'. *Proceedings of CHI '11*, 2257–2266.

Greenberg, S. & Buxton, B. (2008) Usability evaluation considered harmful (some of the time). *Proceedings of CHI '08*, ACM, 111–20.

Harrison, S., Sengers, P. & Tatar, D. (2011) Making epistemological trouble: Third-paradigm HCI as successor science. *Interacting with Computers* 23(5), 385–92.

Hassenzahl, M., Diefenbach, S. & Göritz, A. (2010) Needs, affect, and interactive products: Facets of user experience. *Interacting with Computers* 22(5), 353–62.

Ho, Melissa R., Smyth, Thomas N., Kam, Matthew & Dearden, Andy (2009) Human computer interaction for international development: Past present and future. *Information Technology and International Development* 5(4), 1–18.

Holtzblatt, K., Rondeau, D. & Holtzblatt, L. (2010) Understanding 'cool'. *Proceedings of the 28th International Conference: Extended Abstracts on Human Factors in Computing Systems (CHI EA '10)*. ACM, 3159–62.

Kannabiran, G., Bardzell, J. & Bardzell, S. (2011) How HCI talks about sexuality: Discursive strategies, blind spots, and opportunities for future research. *Proceedings of CHI '11*, ACM, 695–704.

Kaye, J. (2006) I just clicked to say I love you: Rich evaluations of minimal communication. *Proceedings of*, ACM *CHI '06 EA.*

Kuutti, K. (2009) HCI and design: Uncomfortable bedfellows? In: Binder, T., Löwgren, J. & Malmborg L. (eds), *(Re)searching the Digital Bauhaus*. London: Springer, 43–59.

Law, E.L., Roto, V., Hassenzahl, M., Vermeeren, A.P.O.S. & Kort, J. (2009) Understanding, scoping and defining user experience: A survey approach. *Proceedings of CHI '09*, ACM, 719–28.

Law, E., Roto, V., Vermeeren, A., Kort, J. & Hassenzahl, M. (2008) Towards a shared definition of user experience. *CHI '08 Extended Abstract*, ACM, 2395–8.

Light, A. (2011) HCI as heterodoxy: Technologies of identity and the queering of interaction with computers. *Interacting with Computers* 23(5), 430–8.

Liao, W., Zhang, W., Zhu, Z., Ji, Q. & Gray, W. (2006) Toward a decision-theoretic framework for affect recognition and user assistance. *International Journal of Human-Computer Studies* 65, 847–73.

Löwgren, J. & Stolterman, E. (2004) *Thoughtful Interaction Design*. Cambridge, MA: MIT Press.

McCarthy, J. and Wright, P. (2004) *Technology as Experience*. Cambridge, MA: MIT Press.

Milner, A. & Browitt, J. (2002) *Contemporary Cultural Theory: An Introduction*, (3rd edition). Abingdon, Oxon: Routledge.

Nardi, B. (2010) *My Life as a Night Elf Priest: An Anthropological Account of World of Warcraft*. Ann Arbor, MI: University of Michigan Press.

Nelson, H. & Stolterman, E. (2003) *The Design Way: Intentional Change in an Unpredictable World*. Englewood Cliffs, NJ: Educational Technology Publications.

Picard, R. (1997) *Affective Computing*. Cambridge, MA: MIT Press.

Picard, R. and Klein, J. (2002) Computers that recognize and respond to user emotion: Theoretical and practical implications. *Interacting with Computers* 14(2), 141–69.

Read, J., Fitton, D., Cowan, B., Beale, R., Guo, Y. & Horton, M. (2011) Understanding and designing cool technologies for teenagers. *Proceedings of the 2011 Annual Conference: Extended Abstracts on Human Factors in Computing Systems (CHI EA '11)*, New York: ACM, 1567–72.

Rosner, D.K. & Ryokai, K. (2008) Spyn: Augmenting knitting to support storytelling and reflection. *Proceedings of Ubicomp '08*, ACM.

Rowe, P. (1987) *Design Thinking*. Boston, MA: MIT Press.

Sengers, P., Boehner, K., David, S. & Kaye, J. (2005) Reflective design. *Proceedings of CC '05*, ACM, 49–58.

Schön, D.A. (1983) *The Reflective Practitioner: How Professionals Think in Action*. New York: Basic Books.

Shneiderman, B. & Plaisant, P. (2009) *Designing the User Interface*. Boston, MA: Addison-Wesley.

Shusterman, R. (2000) *Pragmatist Aesthetics: Living Beauty, Rethinking Art*. Lanham, MD: Rowman & Littlefield.

Smith, P. (2001) *Cultural Theory: An Introduction*. Malden, MA: Blackwell.

Steiner, P. (2005). Russian Formalism. In: Selden, R. (ed.) *The Cambridge History of Literary Criticism: Volume VIII: From Formalism to Poststructuralism*. In: Brooks, P., Nisbet, H.B. & Rawson, C. (vol. Eds.) *The Cambridge History of Literary Criticism*. Cambridge: Cambridge University Press, 11–29.

Sutcliffe, A. (2010) Designing for user engagement: Aesthetic and attractive user interfaces, *Synthesis Lectures on Human-Centered Informatics*. San Rafael, CA: Morgan & Claypool Publishers.

Taylor, C. (1971) Interpretation and the sciences of man. *The Review of Metaphysics* 25(1), 3–51.

Turvey, M. (2005) Can scientific models of theorizing help film theory? In: Wartenberg, T. & Curran, A. (eds.), *The Philosophy of Film: Introductory Text and Readings*. Malden, MA: Blackwell, 21–32.

Udsen, L. & Jørgensen, A. (2004) The aesthetic turn. *Digital Creativity* 16(4), 205–16.

Zimmerman, J., Forlizzi, J. and Evenson, S. (2007) Research through design as a method for interaction design research in HCI. *Proceedings of CHI '07*, ACM, 493–502.

10

Theories of Embodiment in HCI

Paul Marshall and Eva Hornecker

INTRODUCTION

The concept of embodiment is increasing in prominence in thinking about the design of digital technologies, particularly in the decade since the publication of Paul Dourish's *Where the Action Is* (Dourish, 2001). In this chapter, we focus on how it has been applied in the area of human–computer interaction (HCI).

Embodiment typically refers to our being living, feeling, bodily entities situated in a physical world. This contrasts with a view of human cognition as grounded in abstract information processing. Theories of embodiment focus on how our bodies and active experiences shape how we perceive, feel and think. However, rather than being a single coherent theoretical perspective, there are a number of different traditions and emphases.

Here, we provide an overview of the core theoretical underpinnings of recent work on embodied interaction in HCI and show how this work has been developed in primarily two related directions through the description of work drawn from the literature. We introduce these two main branches of theory: phenomenology and embodied cognition. In the first main section, we provide a short historical overview of the cognitivist theories that were once ascendant in HCI, and follow the development of alternative views of thinking and acting, drawing from phenomenology through the work of Terry Winograd and Carlos Flores, Lucy Suchman and Paul Dourish. In the next section we provide an overview of work in cognitive science described by the umbrella term *embodied cognition*, which has been influenced by discussions in phenomenology, but has followed a quite different trajectory. This comprises a number of theoretical perspectives, including that cognition is offloaded on to the environment (e.g. Scaife and Rogers, 1996); the environment is part of the cognitive system (e.g. Hutchins, 1995); and abstract thinking is grounded in bodily experience (e.g. Lakoff and Johnson, 1999). In the final section, we attempt to show, by discussing (highly selective) examples from the literature, some of the diversity of work that has drawn on theories of embodiment in technology development, analysis and evaluation. In particular, we describe four perspectives. First, we discuss Daniel Fällman's use of Merleau-Ponty's work in developing a perspective on mobility and bodily interaction that is shaped by a focus on engagement with the immediate context. Second, we provide an overview of Toni Robertson's analysis of the public availability of socially situated action. Third, we discuss Eva Hornecker's framework on tangibility and social interaction. Finally we describe Hurtienne's application of Lakoff and Johnson's image schema theory to the design and evaluation of both new and traditional interfaces.

BACKGROUND

Cognitivism

Many theoretical models in early HCI adopted a perspective, drawn from the then dominant approach in the cognitive sciences (especially in artificial intelligence, philosophy of mind and cognitive psychology), that is now described as *cognitivism* (e.g. Fodor, 1975). The central claim of the cognitivist approach is that thinking is information processing – the manipulation of physical symbols, representing facts and things in the world, according to syntactical rules in order to make inferences and to guide action. A central claim of this approach was that an information-processing system is both necessary and sufficient for general intelligent action (Newell and Simon, 1972). Furthermore, it is the *function* of the representations and rules used to process them that are of prime importance rather than the details of their implementation. The power of this *functionalist* approach (cf. Putnam, 1975) is to treat cognition as a formal system that can be studied and modelled separately from the details of neuronal organization or manifest behaviour; the implementation doesn't matter as long as it can perform the same function of processing physical symbols. Cognition can

therefore potentially occur in a human brain, a digital computer or another medium. Mental processes and physical interaction are treated as separate domains, a philosophical position known as *dualism*. The theorist most associated with dualism is René Descartes, and thus the cognitivist perspective is often described as *Cartesian*.

The cognitivist programme of research has been successful in modelling cognitive processes such as planning and abstract problem solving, associated with higher-level and specifically human cognition. In HCI, it is perhaps best represented in Card, Moran and Newell's (1983) *The Psychology of Human-Computer Interaction*. A well-known model described by Card et al. is GOMS (Goals, Operators, Methods and Selection rules). This model represents interaction with a computer as abstracting and processing information from the environment through the perceptual system, processing it in a separate cognitive system to select an appropriate action according to a pre-specified goal and then sending a message to the appropriate body parts (usually the fingers) to carry out the interface action via a motor control system. Another well-known approach in this tradition is requirements specification through a top-down hierarchical task analysis.

Critiques of cognitivism

The cognitivist model of thinking and interaction has come in for sustained criticism from a variety of positions. The philosopher John Searle (1980) has argued that a model based only on the manipulation of abstract symbols doesn't have an account of how meaning is attached to the symbols in the first place. Therefore, it is questionable whether this is a good account of human reasoning.

Dreyfus (1979) critiqued cognitivism from the perspective of Heideggerian philosophy, arguing that systems which represent knowledge about the world as just a collection of symbolic facts are never going to be able to respond flexibly to changing real-world contexts. The system will never be able to work out which of the changes are pertinent to the ongoing task and which to ignore. Instead, Dreyfus argues that intelligent action is grounded in a complex history of skilful bodily experiences in the world – knowing how-to do things rather than just knowing-that:

> To say a hammer has the function of being for hammering leaves out the defining relation of hammers to nails and other equipment, to the point of building things, and to our skills ... and so attributing functions to brute facts couldn't capture the meaningful organization of the everyday world. (Dreyfus, 2007: 248)

Dreyfus's critique draws support from the perceived failure of classical AI to move beyond reasoning in very constrained artificial environments, with simple semantics that are specified in advance to flexible responses in complex changing environments (e.g. Brooks, 1991).

INTRODUCTION OF EMBODIED THEORIES INTO HCI

Although there are many precursors to the current focus on embodied interaction in the design and evaluation of human interface technology, two in particular have had a profound impact in HCI and the broader cognitive sciences that they served to critique: Winograd and Flores's *Understanding Computers and Cognition* and Suchman's *Plans and Situated Actions*

Winograd and Flores's Understanding Computers and Cognition

Winograd and Flores (1986) presented an influential alternative to the view of cognition as a formal system, which has had significant influence in HCI, and introduced a new vocabulary to talk about thinking about and acting with technology. Drawing in particular from

phenomenology (Heidegger, 1927; Dreyfus, 1991), they argued against the cognitivist position that separates thinking from the context in which it occurs. Instead, they propose an understanding of technology use that is inherently historical, material and social. On this view, completely detached reasoning is impossible, as it always depends on a tradition or pre-understanding that derives from a history of interactions with others who share the tradition. In this characterization, all understanding derives from the state of being-in-the-world, which Heidegger terms *Dasein*.

Winograd and Flores also adopted Heidegger's concept of *thrownness*, which emphasizes the experience of being fully engaged in skilfully coping within a particular context, where there is no way to predict exactly what the outcome of your actions will be and no stable objective representation of the situation. It should be clear that this kind of intrinsically *online* reasoning (Wheeler, 2005) is very different from the abstract thinking characteristic of cognitivism. Thrownness is closely linked to the idea of *readiness-to-hand* or *transparency*, which refers to the way that things 'disappear' in the course of everyday action. This doesn't mean that they literally vanish of course – rather, they cease to be the focus of attention. The canonical example, taken from Heidegger, is of using a hammer. Here the focus is on the activity of hammering. The hammer ceases to be viewed as an object in its own right, in the same way that the tendons of the arm are used transparently during the activity, disappearing into the web of background understandings of relationships between bodies, hammers, nails, activities of making things, and so on.

Readiness-to-hand is seen to be the primary mode of being-in-the-world in Heidegger's phenomenology. Again, this is in contrast to the rationalist, cognitivist approach outlined above, where knowledge is represented as a collection of objective facts, separate from the context in which they are used. Winograd and Flores adopt the perspective that nothing can be viewed separately from interpretation.

Presence-at-hand refers to the case where a situation is attended to theoretically and objects or properties are viewed as things in their own right. Heidegger argues that this has been the typical way of viewing the world in scientific analysis, disregarding readiness-to-hand, which is the fundamental way of experiencing the world. Presence-at-hand occurs in the event of a *breakdown:* an 'interrupted moment of our habitual, standard, comfortable "being-in-the-world"' (Winograd and Flores, 1986: 77). Returning to the example of hammering, the hammer might become present-at-hand in the event of the handle becoming loose, appearing as an object of attention in its own right and available for theoretical reflection. However, again this is not the same as the rationalist cognitivist view of objective knowledge, as present-at-hand reflection will always be related to the background of ready-to-hand experience. Learning is an engaged practice of present-at-hand reflection in a context of use and application.

> We do at times engage in conscious reflection and systematic thought, but these are secondary to the pre-reflective experience of being thrown in a situation in which we are already acting. We are always engaged in acting within a situation, without the opportunity to fully disengage ourselves and function as detached observers. Even what we call 'disengagement' occurs within throwness: we do not escape our throwness, but shift our domain of concern. (Winograd and Flores, 1986: 71)

While their work has been criticized for retaining some aspects of rationalism when applying their perspective in system development (Suchman, 1994) Winograd and Flores have had a very significant influence in introducing ideas from phenomenology into the cognitive sciences and HCI. A second significant perspective that also drew upon aspects of phenomenology was developed at around the same time by Lucy Suchman.

Suchman's Plans and Situated Actions

Lucy Suchman (1987) presented another influential critique of the cognitivist conception of mind as applied to the design of interactive systems. She studied interactions with a photocopier that had been designed to model human cognition, using a planning-based model of cognition in interactions with users: one that treats cognition as the generation of a blueprint of steps to be taken based on goals, which are then used to guide behaviour. Suchman adopted an ethnomethodological orientation (Garfinkel, 1967), drawing in particular from studies of conversation (Sacks et al., 1974). Ethnomethodology is concerned with the everyday practices by which mutual intelligibility and social order are achieved. It treats the objectivity of social facts as an ongoing achievement of the members of a social group. A central concept here is the accountability of members' methods – that is, the ways they are made observable and reportable to others. Ethnomethodology also draws on ideas from phenomenology that emphasize the everyday practical engagement of social interactions (Schutz, 1970).

Suchman argued that plans are a representation, but not a specification of behaviour. Thus, they can act as a projection of what will happen or a retrospective account of what is happening or what did happen. However, behaviour itself is generated as *situated action*: in interaction with the contingencies of the physical and social environment. In this model, a plan is only one resource used to help guide action. Language use is also situated, relying on its indexicality – a relationship to the changing circumstances in which it is used rather than any abstract collection of shared meaning. Ethnomethodologically informed conversation analysis (e.g. Sacks et al., 1974) has shown how the same utterance can have a multitude of context-dependent meanings. For example, 'that's brilliant!' could be either an enthusiastic response or a sarcastic put-down depending upon the context in which it is used.

Suchman presents a case study of people struggling to use a photocopier designed with an 'intelligent' interface help system. Using detailed video analysis, she shows how problems emerge relating to disparities between the fixed plan implemented in the system of how to complete an action and the user's actual situated behaviour, which is far more ad hoc and flexible. A particular problem relates to the lack of accountability of the system's behaviour and its lack of responsiveness to the behaviour of the users.

Suchman's contribution to HCI has been very significant, introducing ideas from ethnomethodology to many in the community, providing a powerful critique of cognitivism and emphasizing the importance of studying practice as it occurs in real situations, thus providing part of the intellectual foundation of computer-supported co-operative work.

Paul Dourish (2001) has drawn significantly from both Winograd and Flores and Suchman in developing a view of embodied interaction as a foundational concept for HCI. We describe this approach in the next section.

EMBODIED INTERACTION AS A FOUNDATIONAL FRAMEWORK

Building on the research outlined in the previous section, Dourish (2001) has suggested that embodied interaction should be seen as a foundational concept for HCI. He draws upon and expands the phenomenological perspective that underpins both Winograd and Flores's and Suchman's work, including an overview of Husserl's introduction of ideas in phenomenology and Heidegger's analysis of *Dasein* as the inseparability of being and the world. He also draws significantly on Schutz's (1970) development of social phenomenology. Dourish highlights three elements that are common to this work. First, embodiment – meaning 'grounded in everyday, mundane experience' (Dourish, 2001: 125) – is central to all of them; second they focus on practice: 'everyday engagement with the world directed towards the accomplishment of practical tasks' (2001: 125); and finally, this embodied practice is the source of meaning: 'we find the world meaningful primarily with respect to the ways in which we act within it' (2001: 125).

Dourish structures his argument by showing how recent work in the seemingly disparate fields of tangible and social computing can both be viewed as having a common concern with embodiment, in the sense of a focus on engaged activity in the world rather than abstract theorizing: 'Embodied interaction is the creation, manipulation and sharing of meaning through engaged interaction with artefacts' (Dourish, 2001: 126). For example, Underkoffler and Ishii's (1999) urban planning workbench (Urp) enables users to flexibly explore interactions between wind, reflection and shadow effects for different configurations of buildings by manipulating tangible models. Dourish cites Bowers, Button and Sharrock (1995), who describe how the skilled practices through which workers in a print shop manage their activities often involve stepping outside formalized procedures. Introducing a computerized system that formalized idealized procedures still further, ignoring the situated processes by which things were actually carried out, had a negative impact on the work done.

A key issue in the characterization of embodied interaction is how meaning is understood and mapped on to things in the world. Dourish focuses on three different senses of meaning: ontology, intersubjectivity and intentionality. *Ontology* deals with the nature of being: how it is structured into different kinds of things and the relationships between them. The phenomenological underpinning of the embodied interaction approach emphasizes how ontology is derived through purposeful interactions in the world rather than being objectively defined. Therefore, it can differ significantly between individuals. Thus, if a designer embeds a particular set of ontological commitments into the design of a piece of software or other technology, this can cause problems for users who may not share them, and may have a set of quite different purposes to put the system to than the designer had in mind: 'a design may *reflect* a particular set of ontological commitments on the part of a designer, but it cannot *provide* an ontology for the user' (Dourish, 2001: 130).

The second sense of meaning that Dourish discusses is *intersubjectivity*, which refers to the ways that two or more people can come to a shared understanding without having direct access to each other's mental states. Dourish highlights two ways that intersubjectivity is relevant to the design of technology: first in the ways that the designer is able to communicate to the user how they envisage that the technology will be used (Suchman's photocopier example is a case of this going wrong); and second, the extent to which systems enable different users to communicate through them to develop shared ways of using software systems and appropriate them for shared patterns of practice.

The third sense, *intentionality*, refers to the directedness of meaning; the property (of a thought, action, etc.) of being about something. Intentionality is a core characteristic of embodied interaction, as people act on and through computational representations to enact effects on the world – we are always already directed 'to' or 'towards the world' since it is our habitat (into which we are thrown) and primary source of meaning, as described by Merleau-Ponty. The way that intentionality is expressed in embodied interaction is through a process of *coupling*: 'By coupling, I mean the way that we can build up and break down relationships between entities, putting them together or taking them apart for the purpose of incorporating them into our action' (Dourish, 2001: 138). Technologies that support embodied interaction well are ones that make clear how they are coupled to the world, allowing users to orientate to them in a variety of ways. Thus, a tangible interface, for example, might be attended to as an iconic representation of digital information, a controller to manipulate digital information or as a physical object in its own right that can be picked up and shared with others, left on a desk as a reminder to complete a task the next day or moved out of the reach of others, depending upon the way it is coupled to the ongoing concerns of those who are using it.

Dourish's theoretical account of embodied interaction has been very influential, paving the way for work that has explored new forms of tangible (Hornecker and Buur, 2006), mobile (Oulasvirta et al., 2005), ubiquitous (Chalmers and Galani 2004) and movement-based (Hummels et al., 2007) interaction. The primary contribution of Dourish's work has been to analyse how meaning is fluidly negotiated in interaction with technology, the world and other people. His perspective is primarily a phenomenological one, building on the insights of Winograd and Flores and Suchman to present embodiment as a foundational concept for HCI. It has inspired a range of research that often goes back to its own reading of the original phenomenological authors. In particular, Merleau-Ponty's work is increasingly being referred to, which has a stronger focus on the body and its felt experience.

However, there has been another strand of work that has rejected aspects of the cognitivist view of thinking and interaction, but has not drawn so explicitly from phenomenology. It is to this embodied cognition approach that we turn in the next section.

EMBODIED COGNITION

Embodied cognition refers to a diverse group of theories and approaches that challenge different aspects of the Cartesian view of cognition (Clark, 1997; Wheeler, 2005). Much of this work has drawn on the phenomenological critique of cognitivism, but it has also come about as a response to particular technical challenges or experimental findings. Thus, the influence of phenomenology should be seen as more implicit than in the work described in the previous section (cf. Wheeler, 2005). There have also been several precursors to the recent growth of embodied cognition. In particular, Gibson's (1979) ecological approach to visual perception, which presents perception and action as inseparable has been very influential. Gibson introduced the concept of affordances as action possibilities picked up from the environment, in a relationship to the physical capacities and ongoing concerns of an organism. The concept of affordance was adapted by Donald Norman in HCI (e.g. 1999) to account for the ways in which physical and graphical artefacts suggest how they should be used (in the latter case the affordances are 'perceived' rather than real).

There is currently little agreement on the core concepts of the embodied cognition approach, with some approaches rejecting Cartesian cognitive science completely, and others retaining some parts of it, such as symbolic representations or some aspects of functionalism. Rohrer (2007) offers a broad survey of the literature on embodiment, describing (at least) 12 different uses of the term. These include: the usage in socio-cultural studies where it refers to the environment in which the body is situated (e.g. Hutchins, 1995); the usage relating to morphology, which looks at how the physical characteristics of a cognitive agent can influence the types of cognitive processing it can carry out; and the use of the term to refer to grounding, or, how abstract concepts are linked to a history of concrete, physical experience. Lakoff and Johnson (1999) suggest that abstract concepts are related by metaphorical mappings to basic-level image schemas – mental structures that are formed through sensorimotor interaction with the world to guide our action. These metaphorical mappings preserve the inferential structure of the original domain. For example, Lakoff and Núñez (2000) suggest that Boolean logic is an extension of a sensorimotor container schema, with the same inferential structure of IN, OUT and transitivity (e.g. a ring contained within a box that is held in a hand is also inside the hand), originally developed through experiences with real containers.

Wilson (2002) focuses more narrowly on the literature identified as embodied cognition, identifying six claims. (i) *Cognition is situated*: it takes place in a real-world environment and inherently involves perception and action. (ii) *Cognition is time pressured*: it functions under the pressures of having to interact in real time with a dynamic environment. This and the previous claim were presented by Brooks (1991) as criticisms of robotic systems designed to first build up a model of the environment, devise a plan and then act. He argued that in reality, an organism would not have this luxury, as the world would have changed by the time it decided what it was going to do. Brooks recommends instead building systems that 'use the world as its own best model' and which use simpler more responsive architectures. (iii) *We offload cognitive work on to the environment*: cognitive workload is alleviated by holding or manipulating information in external structures (cf. Kirsh, 2010; Scaife and Rogers, 1996). (iv) *The environment is part of the cognitive system*: the links between internal and external representations and processing are so fundamental that they should be considered the same unit for analysis. This claim has been described as distributed cognition (Hollan et al., 2000; Hutchins, 1995) and the extended mind hypothesis (Clark and Chalmers, 1998). It is also related to a number of approaches within cognitive science that resist traditional explanations in terms of internal representations (Thelen and Smith, 1994; van Gelder, 1995). (v) *Cognition is for action*: the function of the mind is to guide action, so cognition should be understood in terms of its contribution to behaviour. Glenberg (1997) claims that cognition evolved to coordinate interaction with a three-dimensional world, enhancing survival and hence reproductive success. In this characterization, the meaning of a situation to an organism is a coordinated set of possible actions, which are determined by physical form, learning history and goals. (vi) *Off-line cognition is body-based*: even when decoupled from the environment, mechanisms evolved for interaction with it play a role in cognition; sensorimotor systems are involved in processing even in the absence of task-relevant perceptual input (cf. Lakoff and Johnson, 1999).

Theoretical work that makes up the new approaches of embodied cognition has also been very influential in HCI. Hutchins (1995) introduced the theory of *distributed cognition*, arguing that what classical cognitive science took to be internal, individual acts of information processing, were in fact outcomes of a socio-cultural system.

> Having failed to notice that the central metaphor of the physical-symbol-system hypothesis captured the properties of a socio-cultural system rather than those of an individual mind, AI and information-processing psychology proposed some radical conceptual surgery for the modeled human. The brain was removed and replaced with a computer. The surgery was a success. However, there was an apparently unintended side effect: the hands, the eyes, the ears, the nose, the mouth, and the emotions all fell away when the brain was replaced by a computer. (Hutchins, 1995: 363)

The idea of cognition as information processing is retained in distributed cognition. However, the process is analysed as propagating through a variety of representational media, including other minds, physical artefacts and technologies, and parts of the body. Distributed cognition has been used to analyse the flow of information through a variety of socio-technical systems, typically using an ethnographic approach. For example, in critiquing the idea of organizational memory through showing the details of information flow involved in a telephone hotline call (Ackerman and Halverson, 1998).

A second perspective, related to Hutchin's description of distributed cognition, is the analysis of how interaction with external representations can support cognition, change the nature of a cognitive task or form an intrinsic part of thinking (Kirsh, 2010; Scaife and Rogers, 1996). Kirsh and Maglio (1994) for example, described how expert users of the video game Tetris solved the problem of fitting irregularly shaped blocks together by physically rotating them on a screen and using the computationally cheap mental processes of pattern matching and recognition. They call physical activity to reduce the burden of problem solving *epistemic action*.

A third approach attempts to use Lakoff and Johnson's (1999) work on image schemas and embodied conceptual metaphor in the design and evaluation of technology. Image schemas are representations, abstracted from recurrent patterns of sensorimotor experience. For example, the *front-back* schema is derived from movement in the world and the structure of the human body where the eyes are positioned on the front side of the body. Embodied conceptual metaphors are extensions of image schemas to think about other entities, which can be more abstract. For example, good and bad can be thought of in terms of up and down (e.g. 'things are looking up'), whereas time can be thought of in terms of front and back (e.g. 'your future is ahead of you'). A recent trend in HCI research has been to try to use insights from this work to design novel interface applications and to improve interface usability (e.g. Antle et al., 2009a; Hurtienne et al., 2008)

A final example of embodied cognition theory that has found application in the design and evaluation of technology is work which has shown physical changes to the body, such as adopting a different posture, can in some circumstances induce changes in emotional or attitudinal states and social perceptions (e.g. Niedenthal et al., 2005). Bianchi-Berthouze et al. (2007) have experimented with trying to increase user engagement in gameplay by increasing the level of bodily involvement in interaction.

CASE STUDIES OF WORK IN EMBODIED INTERACTION

In the previous sections we described some of the theoretical underpinnings of recent work in embodied interaction. Strong and consistent themes have yet to emerge, but we have shown how recent work can often be described as deriving from work on phenomenology or on embodied cognition. In this section, we describe in more detail four examples of recent projects. No attempt is made here to give a comprehensive overview of work that has attempted to apply concepts drawn from embodied interaction – indeed it is unclear if an exhaustive framework could be proposed at this stage. Rather, four quite different projects were selected to give a flavour of the diversity of work in this area.

Fallmän: Supporting Skilful Engagement with the World in Mobile Interaction

Fällman used embodiment as a guiding perspective for a series of projects that formed part of his PhD research (Fällman, 2003b). His work is significantly influenced by Merleau-Ponty, but also by other work in phenomenology. Merleau-Ponty's influence is evident in the emphasis that Fällman gives to the subjective experience of technology through first-person accounts. Fällman takes a phenomenological perspective to the analysis of mobile technologies, focusing on how they are experienced. His approach is design-orientated research, where research is driven by design and designing is the means for producing new knowledge (Fällman, 2003a). Theories of embodiment were utilized in the conceptual design of a series of practical projects, in the design of a support tool for mobile service technicians, a slide scroller on a small screen, and a wearable 'alternate' reality helmet.

Fällman (2003b) argues that the traditional cognitivist HCI-perspective of a disembodied mind is a poor model for understanding or designing mobile interaction. This is because mobility is 'strongly situated and rooted in a world … ' (Fällman, 2003b: 7). Analysing his own subjective experience of using mobile devices, he finds that these all have the common characteristic of being embedded in a relatively small physical form which relates to the human body – they are held close, felt, worn, etc. Moreover, when using mobile devices our focus is often on the world (e.g. taking photos, checking the network connection, taking a call). As mobile device use is strongly related to context, Fällman argues

that it and mobility should be seen as a mode of being-in-the-world. People 'become mobile in different ways – not only corporeal – to be able to get involved in different physical and social contexts' (Fällman, 2003b: 157). The design of mobile technologies should therefore not interfere with this involvement or, better still, it should support engagement with the world. This also means that desktop interface metaphors are inappropriate for mobile devices, since they tend to make it difficult to engage with the real world while interacting with the interface.

Motivated by this theoretical background, one of Fällman's projects aims to support 'skillful coping' with a mobile support system for service technicians. The resulting device is worn on the arm, allowing for hands-free interaction and always being there, but receding into the background of attention when not in use. It is interacted with primarily by pointing at, for example, a broken component in order to access its data sheet, thereby connecting the physical work of industrial components directly with the virtual world of data on the device. Users then interact by tilting to slide the interface and tapping the screen. Tilting makes use of the notion of embodied interfaces (Fishkin et al., 2000), exploiting the fact that humans have an embodied understanding of gravity (Lakoff and Johnson, 1999).

Robertson: The Mutual Availability of (Embodied) Action

Robertson (1997, 2002) also applies a phenomenological perspective to computer-supported cooperative work (CSCW) research, investigating the role of embodied action in supporting awareness and coordination. Specifically, she aims to show that Merleau-Ponty's phenomenology of perception provides a new perspective on cooperative work, by emphasizing how the public availability of actions and artefacts provides environmental support for participants in cooperation. For designers of novel technologies such as those used in distance communication, one of the most important lessons is that perception is learned – over time we gain skill, adapt our perception and body image, and the research question thus becomes what kinds of cues and feedback mechanisms might support us in this process? A focus on lived cognition and perception furthermore highlights the agency of users, who should be supported with resources for action.

Robertson (1997) analysed how embodied action supports cooperative work, based on a field study of collaborative work within an educational game design company where staff often work from home. The study highlights how much cooperative design and development of software relied on communicative interactions between staff. Sharing physical space within company premises 'enabled communication by supporting the mutual perception of their embodied actions', of talking and making or using artefacts within this space. When staff members were not on-site, workers relied on complex work practices that had evolved to support communication. Video analysis of embodied activities revealed several categories of actions. Individual actions could relate to physical objects, as in moving them (for oneself or to make them available for somebody else), producing new representations (drawing, writing), highlighting aspects of the object or personal use. Individual actions could also relate to other human bodies, either through communicative gesture, facial expression, talk etc., or through enacting user behaviour or the design object itself (a frequent activity in design). Further actions related to the physical workplace, such as moving around, pointing at things or changing direction of gaze. Group activities such as conversation, shared attention on an object, creation of shared resources for conversation and shared representations (clearing the table and then sketching on it), shared use of objects and so on, in turn are comprised of individual actions.

These kinds of embodied actions were not specific to any particular phase of the software development process. Thus, distance communication technology in support of distributed design should enable and mediate the mutual perception of embodied actions and negotiation of meaning rather than supporting specific design processes or phases. This would be greatly facilitated by the reversibility of perception: the ability to anticipate how actions will be seen by others (which in turn enables the 'actor' to shape their actions for the observer).

Hornecker: Group Facilitation Mechanisms to Influence Social Interaction

Hornecker's (2005) notion of 'embodied facilitation' highlights how physical, spatial and software-determined configuration of a system affects group interaction patterns and influences the social formations and interaction patterns that emerge with and around the system. Hornecker's argument is that both physical and software design define a structure and this structure may facilitate, prohibit or hinder some actions. Specific behaviours are easily feasible, and may even be invited, while other behaviours are made difficult or even prevented. Systems can thus embody structure and thereby styles, methods and means of facilitation, very similar to how meeting facilitators, educators and professional group work facilitators steer group processes by imposing structure (with an agenda and rules of discussion), staging the setting (placing tables, chairs, flipcharts or projection surfaces) and providing work materials (e.g. a number of markers for writing). This perspective encourages the analysis of technological systems in terms of the resources they provide for participants to engage in, or join, an activity and to collaborate and the ways in which they influence how people will coordinate and collaborate.

This concept is useful for understanding the effects of UbiComp technologies, which are often embedded in furniture and everyday objects, on social interaction patterns and collaboration. For example, in the design of interactive multitouch tables, the shape, size and height of the table influence how well groups tend to engage with each other, share an activity and how many people may take part. Hornecker refers to tangible interaction systems in particular as 'embodying facilitation'. These are comprised of physical structures embedded in space that users interact with through some form of bodily interaction (Hornecker and Buur, 2006). The notion of 'embodied facilitation' draws in particular from Merleau-Ponty in that it is through bodily interaction that we use these systems, seeing and pre-reflectively interpreting the resources and constraints they provide for action.

Hurtienne: Using Image Schemas to Design Intuitive Interfaces

Hurtienne, Israel and Weber (2008) conducted an empirical analysis of the utility of image schemas (Johnson, 1987; Lakoff and Johnson, 1999) in interface design. This work has shown how image schemas and embodied conceptual metaphors, can be productively utilized in interface design and provide a basic vocabulary for consistent and intuitive mapping.

Hurtienne et al. showed that violating embodied metaphorical extensions of image schema results in increased reaction times and error rates when interacting with simple GUI interfaces. User interfaces that were congruent with image schemas (e.g. moving a lever upwards to evaluate a hotel as being 'better quality' or to indicate that its staff is friendly) were not only judged by participants as being better, but also resulted in faster decisions and greater accuracy (Hurtienne, 2011; Hurtienne and Blessing, 2007). This phenomenon could also be demonstrated with two buttons, where the positive (or 'more') rating was on top.

An example of how a complex interface metaphor builds on a range of image schemas is 'putting files into the trashcan', which employs the schemas of path (dragging),

compulsion, containment and full/empty. Yet, unfortunately, sometimes there can be competing image schemas, which then interfere with unconscious information processing. Image schemas thus cannot be implemented in a mechanical way. Furthermore, as some image schemas relate and depend on each other (a container can also block content from leaking out), there may be competing image schemas, which equally could describe the users' mental model and thus there may be several ways of implementing this in an interface.

In related work Antle and colleagues have explored the potential of embodied metaphor to support the design of less traditional interfaces, such as a whole body interaction system to encourage reasoning about social justice (Antle et al., 2009b, in press) and tangible and whole body interactive systems to control different properties of sound (Antle et al., 2009a)

DISCUSSION

Embodied interaction has been taken up and developed enthusiastically in the design, analysis and evaluation of interactions with technology in recent years. However, while theoretical work on embodied interaction has been united in challenging the cognitivist model of thinking and its application in system design, this recent explosion of interest has resulted in a sometimes bewildering range of approaches, techniques and claims. In this chapter, we have summarized two broad strands of theory that are being used in HCI. The first rejects the cognitivist model completely, building instead on insights from phenomenology to emphasize the engaged, direct ways that people normally participate in activities and the flexible ways that meaning can be ascribed and negotiated in ongoing practice. Analysis of the details of situated practice is well known in HCI, particularly through the influence of ethnomethodologically informed ethnography, which traces its lineage, in part, to Schutz's (1970) analysis of the phenomenology of the social world. More recently, authors have drawn from other phenomenologists, such as Merleau-Ponty (1962), and have begun to use their insights to design novel technology interactions as well as new methods of analysis.

The second strand of theory derives from cognitive science and has attempted to respond to problems with the cognitivist model by adapting, extending or replacing different aspects of it. This work is very diverse. Some, such as those arguing for a greater role for external representations in cognitive processing, have been influential in HCI and interaction design for at least 15 years (e.g. Scaife and Rogers, 1996). Work that characterizes cognition as extending beyond the boundary of the skull to include physical movements of the body, other brains, physical artefacts and technologies, has also been used for some time in analysing existing socio-technical systems, as part of the distributed cognition approach. We are also beginning to see more work looking at how these insights can be used in designing novel technologies, (cf. Bird et al., 2009; Nagel et al., 2005). More radical approaches, such as those which model cognition as a *dynamical system* spanning brain, body and environment (e.g. Beer, 2000), dispensing with the idea of internal representations altogether, and the *enactive* approach (e.g. Thompson, 2005), which focuses on the autonomous agency and lived subjectivity of an agent (and has many links with the phenomenological approach), have so far had little influence in HCI.

A focus on different aspects of embodied interaction is increasingly popular in HCI and interaction design. However, as in other areas, such as cognitive science, there is still no consensus on the core concepts of embodied interaction or the best ways to apply these concepts. In this chapter we have highlighted two strands of work – one that focuses primarily on insights from phenomenology, and one that focuses primarily on ideas from embodied cognition, which tends to be influenced only implicitly by phenomenology. There is, however, much work still to be done to develop this theory into the kinds of frameworks, methods and perspectives needed in the design, analysis and evaluation of interaction with technology.

GLOSSARY OF TERMS

Cartesian: relating to the work of René Descartes. In particular, this often refers to the philosophical position of dualism – treating the mind as non-physical and therefore separate from the body.

Embodied conceptual metaphor: suggests that concepts are typically understood through metaphorical mappings to concrete sensorimotor experiences.

Cognitivism: an approach in cognitive science that treats cognition as information processing on discrete internal symbols.

Epistemic action: a physical action that changes the nature of a cognitive operation necessary to carry out a particular task.

Functionalism: a philosophical approach that considers mental states in terms of the role they play in a cognitive system rather than in terms of their constitution.

Image schema: a cognitive structure that abstracts across a number of sensorimotor experiences: for example, the containment schema abstracts across many concrete experiences of something being contained within something else.

Phenomenology: the study of the structure of conscious experience as experienced (although not necessarily studied) from a first-person perspective

Presence-at-hand: the state of attending to an object, tool or representation itself as the focus of an activity.

Readiness-to-hand: this concept refers to how, when working with a tool or representation we treat it almost as if it were invisible, focusing instead upon the task it is used for.

FURTHER READING

Dourish, P. (2001). *Where the Action Is: The Foundations of Embodied Interaction*. Cambridge, MA: MIT Press.

Suchman, L. (1987) *Plans and Situated Actions. The Problem of Human–Machine Communication*. Cambridge: Cambridge University Press.

Winograd, T. and Flores, F. (1986): *Understanding Computers and Cognition: A New Foundation for Design*. Norwood, NJ: Ablex Publishing.

REFERENCES

Ackerman, M.S. and Halverson, C. (1998) Considering an organization's memory. *Proceedings of the 1998 ACM Conference on Computer Supported Cooperative Work* (CSCW '98). ACM, 39–48.

Antle, A.N., Corness, G., Bakker, S., Droumeva, E., van den Hoven, E. and Bevans, A. (2009a) Designing to support reasoned imagination through embodied metaphor. *Proceedings of Creativity and Cognition* (C&C '09). ACM, 275–284.

Antle, A.N., Corness, G. and Droumeva, M. (2009b) What the body knows: exploring the benefits of embodied metaphors in hybrid physical digital environments. *Interacting with Computers*, 21(1–1), 66–75.

Antle, A.N., Corness, G. and Bevans, A. (in press) Balancing justice: comparing whole body and controller-based interaction for an abstract domain. *International Journal of Arts and Technology*.

Beer, R.D. (2000). Dynamical approaches to cognitive science. *Trends in Cognitive Sciences*, 4, 91–99.

Bird, J., Marshall, P. and Rogers, Y. (2009) Low-Fi Skin Vision: A Case Study in Rapid Prototyping a Sensory Substitution System. *Proceedings of BCS-HCI '09*, 55–64.

Bianchi-Berthouze, N., Kim, W.W. and Darshak, P. (2007) Does body movement engage you more in digital game play? And Why? *Proceedings of ACII '07*, Springer, 102–113.

Bowers, J., Button, G. and Sharrock, W. (1995) Workflow from within and without: technology and cooperative work on the print industry shopfloor. *Proceedings of the Fourth Conference on Computer-Supported Cooperative Work* (ECSCW '95), 51–66.

Brooks, R.A. (1991) Intelligence without representation. *Artificial Intelligence*, 47, 139–159.

Card, S., Moran, T. and Newell, A. (1983) *The Psychology of Human-Computer Interaction*. Hillsdale, NJ: Lawrence Erlbaum.

Chalmers, M. and Galani, A. (2004) Seamful interweaving: heterogeneity in the theory and design of interactive systems. *Proceedings of the 5th Conference on Designing Interactive Systems: Processes, Practices, Methods, and Techniques* (DIS '04), ACM, 243–252.

Clark, A. (1997). *Being There: Putting Brain, Body and World Together Again*. Cambridge, MA: MIT Press.

Clark, A. and Chalmers, D. (1998) The extended mind. *Analysis*, 58: 7–19.

Dourish, P. (2001) *Where the Action Is: The Foundations of Embodied Interaction*. Cambridge, MA: MIT Press.

Dreyfus, H.L. (1979) *What Computers Can't Do: The Limits of Artificial Intelligence*. New York: Harper & Row.

Dreyfus, H.L. (1991) *Being-in-the-world: A Commentary on Heidegger's Being and Time, Division I*. Cambridge, MA: MIT Press.

Dreyfus, H.L. (2007) Why Heideggerian AI failed and how fixing it would require making it more Heideggerian. *Philosophical Psychology*, 20(2), 247–268.

Fällman, D. (2003a) Design-oriented human–computer interaction. *Proceedings of the Conference on Human Factors in Computing Systems* (CHI '03), ACM, 225–232.

Fällman, D. (2003b) In romance with the materials of mobile interaction: a phenomenological approach to the design of mobile information technology. PhD thesis, Umea University, Sweden.

Fishkin, K.P., Gujar, A., Harrison, B.L., Moran, T.P. and Want, R. (2000) Embodied user interfaces for really direct manipulation. *Communications of the ACM*, 43(9), 74–80.

Fodor, J.A. (1975) *The Language of Thought*. Cambridge, MA: Harvard University Press.

Garfinkel, H. (1967) *Studies in Ethnomethodology*. Englewood-Cliffs, NJ: Prentice Hall.

Gibson, J.J. (1979) *The Ecological Approach to Visual Perception*. Boston, MA: Houghton-Mifflin.

Glenberg, A.M. (1997) What memory is for. *Behavioral and Brain Sciences*, 20(1), 1–55.

Heath, C. and Luff P. (2000) *Technology in Action*. Cambridge: Cambridge University Press.

Heidegger, M. ([1927]1990) *Being and Time* (J. Macquarrie and E. Robinson, Trans.) Oxford: Blackwell.

Hollan, J., Hutchins, E. and Kirsh, D. (2000) Distributed cognition: toward a new foundation for human–computer interaction research. *Transactions on Computer–Human Interaction*, ACM, 7(2), 174–196.

Hornecker, E. (2005) A design theme for tangible inter-action: embodied facilitation. *Proceedings of the 9th European Conference on Computer-Supported Cooperative Work* (E-CSCW '05), Kluwer/Springer, 23–43.

Hornecker, E. and Buur, J. (2006) Getting a grip on tangible interaction: a framework on physical space and social interaction. *Proceedings of the SIGCHI Conference on Human Factors in Computing Systems* (CHI '06) ACM, 437–446.

Hummels, C., Overbeeke, K. and Klooster, S. (2007) Move to get moved: a search for methods, tools and knowledge to design for expressive and rich movement-based interaction. *Personal Ubiquitous Computing*, 11(8), 677–690.

Hurtienne (2011) Image schemas and design for intui-tive use: exploring new guidance for user interface design. PhD thesis, TU Berlin. Available at: http://opus.kobv.de/tuberlin/volltexte/2011/2970/pdf/hurtienne_joern.pdf

Hurtienne, J. and Blessing, L. (2007) Design for intuitive use: testing image schema theory for user interface design. *Proceedings of the International Conference on Engineering Design*, Paris, Design Society. P_386, 1–12 [CD-ROM].

Hurtienne, J., Israel, J.H. and Weber, K. (2008) Cooking up real world business applications combining physi-cality, digitality, and image schemas. *Proceedings of TEI '08*, ACM, 239–246.

Hutchins, E. (1995) *Cognition in the Wild*. Cambridge, MA: MIT Press.

Johnson, M. (1987) *The Body in the Mind: The Bodily Basis of Meaning, Imagination, and Reason*. Chicago, IL : University of Chicago Press.

Kirsh, D. (2010) Thinking with external representations. *AI and Society*, 25: 441–454.

Kirsh, D. and Maglio, P. (1994) On distinguishing epis-temic from pragmatic action. *Cognitive Science*, 18(4), 513–549.

Lakoff, G. and Johnson, M. (1999) *Philosophy in the Flesh: The Embodied Mind and its Challenge to Western Thought*. New York: Basic Books.

Lakoff, G. and Núñez, R. (2000) *Where Mathematics Comes From: How the Embodied Mind Brings Mathematics into Being*. New York: Basic Books.

Merleau-Ponty, M. (1962) *Phenomenology of Perception*, C. Smith, Trans. London: Routledge & Kegan Paul.

Moran, T.P. and R.J. Anderson (1990) The workaday world as a paradigm for CSCW design. *Proceedings of the Conference on Computer-Supported Cooperative Work* (CSCW '90), ACM, 381–393.

Nagel, S.K., Carl, C., Kringe, T., Martin, R. and Konig, P. (2005) Beyond sensory substitution: learning the sixth sense. *Journal of Neural Engineering*, 2, 13–26.

Newell, A. and Simon, H. (1972) *Human Problem Solving*. Englewood Cliffs, NJ: Prentice Hall.

Niedenthal, P.M., Barsalou, L.W., Winkielman, P., Krauth-Gruber, S. and Ric, F. (2005) Embodiment in attitudes, social perception, and emotion. *Personality and Social Psychology Review*, 9(3), 184–211.

Norman, D.A. (1999) Affordances, conventions and design. *Interactions*, 6(3), 38–43.

Oulasvirta, A., Tamminen, S., Roto, R. and Kuorelahti, J. (2005) Interaction in 4-second bursts: the frag-mented nature of attentional resources in mobile HCI. *Proceedings of CHI 2005*, ACM, 919–928.

Putnam, H. (1975) Philosophy and our mental life. In H. Putnam (ed.), *Mind, Language and Reality* (Vol. 2). Cambridge: Cambridge University Press, 291–303.

Robertson, T. (1997) Cooperative work and lived cogni-tion: a taxonomy of embodied actions. In *Proceedings of the Fifth European Conference on Computer-Supported Cooperative Work* (E-CSCW '97). Kluwer, 205–220.

Robertson, T. (2002) The public availability of actions and artefacts. *Computer-Supported Cooperative Work*, 11, 299–316.

Robertson, T. and Loke, L. (2009) Designing situations. *Proceedings of OZCHI '09*, ACM, 1–8.

Rohrer, T. (2007) The body in space: dimensions of embodiment. In T. Ziemke, J. Zlatev and R. Frank (eds), *Body, Language and Mind* (Vol. 1: Embodiment). Berlin: Mouton de Gruyter, 339–378.

Sacks, H., Schegeloff, E.A. and Jefferson, G. (1974) A simplest systematics for the organization of turn-taking conversation. *Language*, 50, 696–735.

Scaife, M. and Rogers, Y. (1996) External cognition: how do graphical representations work? *International Journal of Human-Computer Studies*, 45, 185–213.

Schutz, A. (1970) On phenomenology and social rela-tions: selected writings. In H. Wagner (ed.), *The Heritage of Sociology Collection*. Chicago, IL: University of Chicago Press.

Searle, J.R. (1980) Minds, brains, and programs. *Behavioral and Brain Sciences*, 3(3): 417–457.

Suchman, L. (1987) *Plans and Situated Actions: The Problem of Human–Machine Communication*. Cambridge: Cambridge University Press.

Suchman, L. (1994): Do categories have politics?: The language/action perspective reconsidered. *Computer-Supported Cooperative Work* 2 (3): 177–190.

Thelen, E. and Smith, L.B. (1994) *A Dynamic Systems Approach to the Development of Cognition and Action*. Cambridge, MA: MIT Press.

Thompson, E. (2005). Sensorimotor subjectivity and the enactive approach to experience. *Phenomenology and the Cognitive Sciences*, 4, 407–427.

Underkoffler, J. and Ishii, H. (1999) Urp: a luminous-tangible workbench for urban planning and design. *Proceedings of Conference on Human Factors in Computing Systems* (CHI '99). ACM, 386–393.

van Gelder, T. (1998) The dynamical hypothesis in cognitive science, *Behavioral and Brain Sciences*, 21(5), 615–665.

Wheeler, M. (2005) *Reconstructing the Cognitive World*. Cambridge, MA: MIT Press.

Wilson, M. (2002) Six views of embodied cognition. *Psychonomic Bulletin and Review*, 9(4), 625–636.

Winograd, T. and Flores, F. (1986) *Understanding Computers and Cognition: A New Foundation for Design*. Norwood, NJ: Ablex Publishing.

11

Space and Place in Digital Technology Research: A Theoretical Overview

Luigina Ciolfi

INTRODUCTION: THE CASE FOR SPACE AND PLACE

Digital technology research is increasingly concerned with theoretical and methodological aspects related to the physical environment, its study and its role in shaping interactions between people and computer systems. A regard for spatiality is seen as necessary both to aid the study of systems in use in particular contexts and to support their successful design.

Why do notions of space and place matter to digital technology research? First, the current trend in technology development is to be pervasive of physical spaces, if not actually embedded in them, overcoming the traditional 'desktop model' of computing and moving closer to Weiser's vision of ubiquitous computing, where computational elements become diffused into everyday objects, portable devices and the built environment (Weiser, 1994). This notion of technology as diffused within physical environments engenders a shift in how technology developers

think of spaces, from something to be modelled and replicated (e.g. virtual reality), to something to be enhanced or augmented (McCullough, 2004).

Second, the importance of conceptualizing the physical world is recognized in studies of situated interaction, linked to the tradition of understanding technology use through notions such as embodiment (Dourish, 2001; Suchman, 1987) and lived experience (Agre, 1997; McCarthy and Wright, 2004): humans are deeply connected with the environment, their physical experience is at the core of interaction with the world, and their action is situated physically and spatially. From this point of view, an understanding of space, place and locality is essential in order to understand embodied interaction.

Finally, the ability of technology to create bridges between different locations and environments is important in considering space. The creation of such links usually aims to support collaboration, communication and awareness among groups of users, not necessarily co-located in the same physical space, through an architecture of media devices for sharing resources, exchanging information and monitoring remote workspaces. An actual physical environment would thus be augmented by systems that provide a link to other spaces, both physical and digital (Dourish and Bly, 1992). Important studies have explored the use, perception and sharing of virtual workspaces, and how they integrate with local physical spaces (Dourish and Bellotti, 1992). These ideas illustrate how concepts of space, place and spatiality pervade many issues surrounding digital technology research. It is crucial for designers to understand these processes; to focus on the relationship between people and the space where they are located and not only on its physical and structural characteristics.

Space and place, concepts with a long history across many disciplines, have been utilized in digital technology research for some time, but recently a more nuanced understanding of them has developed. This attention to space and place integrates work on eviscerating other crucially important notions and applying them to the context of technology research such as 'user' (Cooper and Bowers, 1995; Kuutti, 2001), 'activity' (Kaptelinin and Nardi, 2009; Leont'ev, 1978) and 'artefact' (Bannon and Bødker, 1991; Vygotsky, 1978).

When the question '*What is space?*' is asked, a possible limited answer is the one that identifies the physical context of interaction simply as a 'stage' – the setting where it occurs, something that can be mapped and/or measured as a background to activities.

However, simply by reflecting on our everyday experience, we know that the physical environments we inhabit contribute to shaping our experiences and activities in some way. We also proactively rearrange and modify the space and its elements, we organize our activities around it, encounter and interact with other people in it. We even become attached to particular environments, they *mean* something to us, and evoke feelings and emotions. Beyond the limitation of structurally orientated notions, space and place

have been thought of as culturally and socially constructed. Such conceptual frames have been more influential on digital technology research for they resonate with approaches that see human interaction with the world as essentially social and situated (Suchman, 1987).

This chapter provides an overview of theories of space and place that conceptualize the physical environment 'beyond geometry' – not simply as a structure but as the experienced physical context of interaction. The overview is presented in two parts: first, approaches that take the physical environment as a primary unit of analysis; and second, those that consider the environment as experientially, socially and culturally constructed, and have human activities and experiences as a main concern. It points out how the terms *space* and *place* themselves are conceptually loaded and carry different meanings (Casey, 1997). Finally, we review some significant examples of how different spatial notions have been used by researchers in the interactive systems design field.

FROM THE GROUND UP: THE PHYSICAL ENVIRONMENT BECOMES A THEORETICAL NOTION

Historically, the first discipline to examine how physical space is related to or influenced by human activities was environmental psychology (EP) in the 1960s. EP emerged as a result of the increasing interest in psychological research on the connections between behaviour and physical surroundings – for example, on the treatment of mental illness and the role of the physical environment in mediating complex social interactions in that context. Studies of this nature attempted to overcome the lack of psychological work regarding the environment of human behaviour. However, until relatively recently, EP was mainly concerned with the quantification of environmental stimuli and with establishing a functional relationship between such stimuli and behaviour – in other words, with linking human responses to physical settings (Proshansky et al., 1969). EP sees a one-way connection between physical settings and human reactions, but does not yet include the possibility that human agency may occur. Although this attempt to establish patterns of behaviour and their one-to-one connection to features of the space is clearly limited, interesting findings have emerged. One issue is the occurrence of discrepancies between the rational planning of physical settings and their actual use – that is, people's perceptions of and reactions to a space can be unexpected and different from the intentions of the planners. This suggests focusing not only on the mere physical structure and features of the environment, but, rather, on the connection between the space and its inhabitants, and the necessity of undertaking empirical studies of how actual spaces are used.

Proshansky et al. highlighted how 'the physical settings and the broader structures that encompass them are themselves expressions of correspondingly inclusive social systems. The physical space is an integral part of social practice' (Assumption 3, in Proshansky et al., 1969: 32). Interestingly, this assumption and its implications for planning and architecture became one object of study within cultural geography only a few years later.

As a consequence of this link between physical and social environment, the organization of the physical setting is, and can only be, dynamic. The environment is an open, dynamic system not only regarding its connection with people, but also the relationships among its elements.

New foundations for EP were laid in the 1970s, as a response to critiques of particular aspects of the early approach with a definite shift in perspective away from behaviouristic influences and towards cultural theory and anthropology. Most notably, important studies on urban perception (see Lynch and Rivkin, 1959) introduced a debate on the connection between urban structures and human use that went on to influence new streams of research in urban planning and design.

Canter and Singer (1975) highlighted how developments in psychology of perception brought attention to the physical environment and also proposed a shift in method for environmental psychology, advocating a move, whenever possible, out of the lab into the real world.

The EP vision of the environment as a dynamic process and the awareness of the complex web of connections between the nature of the physical environment, people's interaction with it and with companions and other individuals in that environment has also led to the acknowledgement that social interaction occurring within the environment needs to be studied in order to understand behaviour within that space (a clear influence from Goffman (1963)).

Canter and Singer also point out that our presence in the environment is purposeful. We actively modify, build and influence our physical surroundings; we do not simply react to a certain environmental layout or to particular variables, but modify the environment at different levels – a reference to the notion of agency that can be found in other theoretical views of space, here seen from an empirical standpoint. Therefore, attention shifts to the relationships between the complexity of the spatial environment and the wealth of human responses that it can trigger.

However, the elements that constitute the dynamic connection between people and spaces are never clearly articulated and fully investigated, and important factors such as cultural influences and personal attitudes are not considered.

Later work on EP has introduced a new and rather different set of concerns within the discipline related to the influence of the natural environment (e.g. presence of green areas, natural lighting, etc.) on humans (Cassidy, 1997).

Other influential structurally orientated views of human use of space emerged from architectural theory. Cullen (1971) focuses on human visual perception as the lens through which physical discovery of space occurs, and as the perspective from which architects should form their design process. He sees place as a substrate for human reactions that should be understood and exploited in architects' and planners' practical strategies for good design. His concern for place is mainly functional, but he goes on to consider emotional reactions that the physical/visual discovery of the built environment can engender.

Alexander (1977) went further in distilling design guidelines and abstracted a grammar describing the encounters between human movement and buildings, reframing human behaviour in terms of built environment. He argued for the need to guide design through a 'pattern language', a detailed repertoire of bodily structural relationships engendering optimal environmental experiences (Alexander, 1977).

Recent developments in architectural theory are starting to consider increasingly broad 'human-centred' issues, such as, the perspectives of 'ethic architecture' (Harries, 1997) and 'architectural experience' (Tweed, 2000).

The sociological work of Lefevbre (1992) has also influenced architectural design, despite originating from a different discipline. For Lefevbre, space is socially produced and is the result of social, political and power relationships. Particularly, the designed physicality of structures represents power relationships. He contrasts the built and natural environments by analysing how 'appropriation' by people – the modification of structure as a result of social practices – occurs. Lefevbre's theory is built 'from the ground up', focusing on the Euclidean dimensions of the environment and then reflecting on the socially and politically produced dimensions of human use.

In all of these approaches the physical environment is seen as the primary unit of analysis, albeit with a common concern for conceptualizing it not simply as a geometrical structure but also with a connection to human actions and practices. However, consideration for the breadth of practice in space

is still somewhat limited. Other theoretical approaches foreground issues of experience, emotion and values associated with space and place. An interesting case is Seamon's work on phenomenology in architecture (Seamon, 1993, 2006), at the intersection of philosophy and phenomenology with architecture and planning. His studies of the built environment demonstrate a strong concern for lived experience and for incorporating notions such as culture, values and meaning into the understanding of the physical world. Also, in approaching this body of work, we see distinctions between space and place making an appearance. In moving towards human-centred approaches to understanding the environment, 'place' emerges as a core concept, and so does a philosophical relationship with 'space'.

'BEING IN THE WORLD': SOME VIEWS ON PLACE AND SPACE

Philosophical enquiry has produced a vast range of work on the themes of space as context for action and on one's physical placement in the world. From classic Greek philosophy to Kant, Heidegger and other contemporary thinkers, space and context have been dealt with from numerous perspectives, defined in many ways and integrated into visions of human activities, consciousness and role in the world.

Casey has studied the conceptualizations of space developed by different philosophical schools and, notably, when and under which assumptions the concept of *place* was introduced into philosophical theory (Casey, 1993, 1997).

Place definition can vary slightly, but the general agreement on the concept is to see it as a lived instance of the environment, an embodied experience of space: 'A series of locales in which people find themselves, live, have experiences, interpret, understand, and find meaning' (Peet, 1998: 48). Where space normally refers to geometrical extension and location, *place* describes our experience of being in the world and investing a physical location or setting with meaning, memories and feelings. Places are entities, which 'incarnate the experience and aspirations of people' (Tuan, 1971: 181).

Here we will present a selection of perspectives particularly concerned with human experience within a physical environment.

Bachelard wrote on how poetic images are generated by real-life settings (Bachelard, 1958) and proposed a vision of space that takes into account the emotional dimension of one's experience of an environment. He explored the psychology of human experience of intimate spaces (mainly the home), thus perceiving space as a blend of experience and physical structure, and discussed how a specific space can trigger emotional responses. Bachelard's notion of *topoanalysis* is the analysis of intimate spaces according to the experiences and memories associated with them (Bachelard, 1958). Whilst he did not refer explicitly to place, he used the concept of *localized experience* in order to define the focus of his study. He recognized components such as emotions and memory associated to a physical environment and its elements as essential features of a space and, conversely he assumes that human experience is fundamentally set within and affected by one's physical environment.

Bollnow – whose work has been little known in the English-speaking world as his most important book, *Mensch und Raum* (originally published in 1963), was only translated into English in 2008 – integrated Bachelard's influence into his phenomenological work around the concept of 'lived-space' (Bollnow, 2008). He suggested an 'anthropological' concept of space (in the sense of it being focused on Man – from the Greek *anthropos*) as the one that would include human experience in the study of the physical environment. The 'anthropological' vision of space is a common trait of phenomenological approaches (Merleau-Ponty, 1945). Space is dynamically related to the perception of it by humans and to cultural influences as the boundaries of space as an entity and a concept change according to human experiences within it. Mainly movement and the possibility of movement together redefine space. Thus Bollnow's own concept of *hodological space* – from the Greek word *hodos*, meaning path or way, referring to the human experience during movement between points on a map. In contrast to the mathematical concept of space as presented on maps, plans, and the like, hodological space is based on the topological, physical, social and psychological conditions a person is faced with on the way from point A to point B, whether in an open landscape or within urban or architectural conditions. Bollnow, in other words, interprets a certain physical movement within the space as a physical, bodily dimension of a particular experience.

As mentioned above, Merleau-Ponty (1945) was the first to introduce the concept of 'anthropological space', which transcends its structural/physical dimensions to encompass human activity as constituent of the identity of the space itself: '... Besides the physical and geometrical distance which stands between myself and all things, a "lived" distance binds me to things which count and exist for me, and links them to each other ... ' (Merleau-Ponty, 1945: 333).

To a phenomenologist such as Merleau-Ponty, the most immediate and essential aspects of the lived dimension of space are sensory experience and bodily movement. Merleau-Ponty's work has also had a strong influence on contemporary authors – for example, Augé proposes a similar vision of space in his notion of 'anthropological place' as a novel methodological tool for anthropological study (Augé, 1995). If we stop considering the space as a mere shell, a container, or a location, and start looking at it as a setting for action, experiences, communication, then effective enquiry concerning human actions and activity can take place.

Another example is De Certeau's reflection on 'spatial practices', as part of his work on the practices that constitute everyday life (De Certeau, 1984). De Certeau also advocated the primacy of sensory interaction between a human being and the environment. However, De Certeau made an unusual space/place distinction – interpreting the two concepts in ways that are opposite to their most common use. He associated the term 'space' with the meaning that geography associates with 'place', so, for De Certeau, space is the notion that describes a practised, lived environment: '*space is a practiced place*' (De Certeau, 1984: 117, italics in the original text). For DeCerteau the practice of movement becomes the practice of knowing both space and others, as Crang puts it: 'Practices have no place of their own but move in the territory of the other' (Crang and Thrift, 2000: 148).

So far, we have seen how the unit of analysis here is a compound of human and environment, of 'being in the world' in Heideggerian terms. None of these positions, however, articulates fully the many dimensions of experienced space, or place. More recent developments in philosophical theory have attempted to overcome this limit.

Casey also discusses contemporary philosophical work that defines place not as a structural and abstract concept, but 'at work, part of something ongoing and dynamic, ingredient in something else' (Casey, 1993: 286). Through the vast body of work dealing with place, some philosophical studies are based entirely on this concept and show a strong influence from geographical theory. A notable example is Malpas's 'philosophical topography' (Malpas, 1999), seeing place not as something that can occur in our experience, but as the very grounding in which experience occurs. Malpas claims that, to understand human existence and experience, we can only focus on understanding place and locality. Malpas's position is based on defining place in the context of concepts related to human experience such as space, time and self. He also points out that place is bounded in terms of space and locality, and cannot be thought of outside of the structural boundaries of geometrical space.

Malpas has produced a deep analysis of Heidegger's concepts of being and dwelling,

connecting Heideggerian thought to place theory by abstracting insights from the philosopher's vast work into a topology (Malpas, 2008). The work of Malpas shows a strong geographical influence. In fact, the concept of place is analysed in the experience of geographical reality, thus considering the analysis of structural, spatial elements as fundamental to an investigation of people's experience of an environment.

While these perspectives show the emergence of notions describing an experienced space, a detailed articulation of the dimensions that constitute this experience is not apparent. Furthermore, philosophical thought does not concern itself with the ways in which we can empirically study people in places, an important aspect of much digital technology research. Therefore, now we turn to a discipline that is concerned with these issues – human geography, which, at the same time, has a strong empirical orientation to understanding human actions and experiences.

GEOGRAPHY AND THE EXPERIENCE OF PLACE: MERGING THEORY AND PRACTICE

Human geography concentrates on people as much as it concentrates on space and place. The principal goal of geography is 'to understand the Earth as the world of Man' (Broek, 1965). In geography, place implies the dynamic relationship between physical surroundings and human actors. Moreover, culture shapes place, as do social relationships. What it is to be human is linked to belonging to the physical world and making sense of it. Sense of place is about locating oneself, but also about establishing a deeper relationship to one's locale (Crang, 1998).

Historically, the vision of geography that gained prominence in the 1960s regarded the essence of the discipline to be spatial science concerned with predicting regular patterns over space. Human geographers, however, take into account influences from philosophy, sociology and anthropology to develop a more complete understanding of phenomena (see Crang and Thrift, 2000). The effects of the 'cultural turn' on the practice of geography originated a particular stream of human geography, cultural geography (Mitchell, 2000). The 'cultural turn' marked the beginning of theorization of the relationships between people and place (Entrikin, 1991), and issues such as self, embodiment and gender were brought into geographical enquiry (Massey, 1994). Political dimensions of the relationship that interconnects culture, society and environment were explored for the first time, manifesting the influence of many theoretical positions, from Wittgenstein (Curry, 2000) to Deleuze (Massey, 2005).

Cultural geography sees culture as spatial. The structures of the space are maps of political and social meaning in communities and represent their culture (Jackson, 1989), in the same way as plotting spatial structures on an actual map means representing such meanings (Propen, 2009). The main point of criticism of cultural geography is that culture is used as a sort of overarching framework and as a predictive tool for the behaviour of communities, with no open concern for personal attitudes, values and emotions. The move towards a more encompassing view of the relationship between humans and space finally happens with the *humanistic* school of geography, focusing on individuals and their experience in the world and on the notion of place itself and around empirical ways of investigating it.

Existentialism and, most of all phenomenology, are recognized as the primary philosophical influences within humanistic geography, a major concern of the discipline being the analysis of how space becomes place: how 'the abstract notion of spatiality is transformed into dimensions of meaning' (Ley and Samuels, 1978: 11).

The main philosophical influences on humanistic geography (as summarized in Crang (1998)) are as follows.

- *Intentionality* (Husserl): through intentionality we assign meanings to objects that go beyond the observable world. Places are not just a set of accumulated data but also involve human *intentions* and are shaped by the way we approach them.
- *Essences* (Bachelard and phenomenology): there is more to a place than the simple physical or sensory properties of it. People can feel an attachment to a place or the unique spirit of a place, or, *genius loci* (Relph, 1976: 48). Thus the meaning of place extends beyond the visible and includes elements belonging to the realm of emotion and memory.
- *Situatedness* (Heidegger and existentialists): the human subject only becomes able to think and act through *being-in-the-world*. Our knowledge of the world is always *em-placed*, it is always starting from, and based around, places. This is related to the concept of *dwelling* proposed by Heidegger as a constituent element of human existence (Casey, 1993).

Another epistemological property of the humanist position is holism, which adopts qualitative methods of investigation in order to conduct 'experiential fieldwork' (Rowles, 1978). Humanistic geography treats the person as constantly proactively interacting with the environment, and seeks to understand this interaction by studying it as it is represented by the individual and not as an example of some scientifically defined model of behaviour.

This is only the first step towards a fuller understanding of place. It is necessary for geographers to link individual sensory and emotional perspectives to cultural understandings of place, shaped by, for example, religion and ethnicity. For all these reasons, humanistic geography focuses on qualitative methods of data collection rather than on the quantitative methods of spatial science. The goal is to avoid generalizations and, most of all, measurements. In this respect, humanistic geographers critique quantitative geographical approaches and even cultural geography to some extent, especially when it attempts to find generalizations explained by cultural rules or conventions (Ley and Samuels, 1978).

Geographers also conduct studies of the built environment in order to critique current design and architectural practices. Relph

(1981) focuses on the changing relationships between landscapes and humanistic concerns, with particular attention to modern built environments. Associating a meaning through experience to spatial arrangements implies a dramatic change in current planning concerns and practice (Crang, 1998).

Another inheritance from philosophy is the humanistic geographers' concern for a detailed analysis of the concept of place and their attempts to understand what constitutes sense of place. For example, Relph (1976) identified what he called the 'raw materials' of the identity of places.

- *Physical appearance*: the structure and layout, the perceivable qualities of a setting.
- *Activities*: social activities, economic functions, routines, etc.
- *Meanings*: the particular significance deriving from past events and present situations. Although rooted in them, meanings are not a property of physical settings, but of human intentions and experiences.

According to Relph, 'The three fundamental components of place are irreducible one to the other, yet are inseparably interwoven in our experiences of place' (Relph, 1976: 47–48). However, in Relph's view, there is another important aspect of identity that is not as tangible, but serves to link and embrace them: what he refers to as *genius loci*, the spirit of a place (Relph, 1976). He argues that *genius loci* is perceivable when we encounter places to which a shared meaning or common symbols are attached by certain groups of individuals (Seamon, 1996).

Other researchers question whether shared universal place meanings exist at all. Indeed, a component such as the 'spirit of place' is hard to identify and recognize. Places may hold meanings unique to individuals or to particular groups. It is difficult to know whether this is connected to a general feel of the place or to aspects such as cultural heritage or social practices.

Buttimer deconstructs the notion of *genius loci* and tries to identify the various

strands from which place can be invested by meaning:

> There are many dimensions to meanings ascribed to place: symbolic, emotional, cultural, political, and biological. People have not only intellectual, imaginary, and symbolic conceptions of place, but also personal and social associations with place-based networks of interaction and affiliations. (Buttimer, 1980: 53)

Buttimer attributes the overall feel of a place to more observable elements, and points out what factors might converge, at a different degree, to create the spirit of a place.

A more recent example of a geographical framework for understanding the constituents of place is proposed by Sack (1997), who focuses on turning the concept of place into the foundation of a framework for the understanding of human values, and adds a significant component: moral action. Sack's framework is grounded on the definition of place as constituted by dynamic configurations of *nature*, *meaning* and *social relations*. For this purpose, Sack identifies nature with space in a geometrical sense. Moreover, he articulates culture as involving social relations and meaning.

> People in modern society also carve out places and create a world. Our places contain varying mixes of nature and culture, and require rules about what should and should not occur in them. These rules about place structure our lives and organise our meaning. (Sack, 1997: 8)

There are two main critiques here. First, the lack of attention to experiences, feelings and individual values that belong to the personal sphere and clearly enrich the meaning of places. Second, the inclusion of 'nature' in the framework as a given and immutable variable, without highlighting the relevance of individual perception of space and of agency into the making of a place.

A richer and more inspired investigation of what place is, and its meaning to people, has been conducted by Yi-Fu Tuan. Influenced by literature, anthropology, philosophy and psychology, Tuan refined the conceptual distinction between space and place (Tuan, 1971) and discussed extensively the existential significance of places. Tuan introduced the concept of *topophilia*, a study of 'all of the human being's affective ties with the material environment' (Tuan, 1974: 7). At this point, his definition of the concept of place was limited to that of emotional attachment, but expanded further over his next 40 years of work.

In *Space and Place*, Tuan (1977) narrows down the theme of human experience of the environment to the concepts of space and place and, more importantly, he develops this material from a single perspective, that of experience. He defines his approach from an experiential perspective, trying to define the features of human experience of space and, consequently, place. It is the continuous experience of a space or place that associates ever-changing and nuanced values to a place (Tuan, 1977). Tuan focuses on the relationships between space and place, and on how what begins as undifferentiated space *becomes* place. Inspired by anthropological studies, he also acknowledges culture as one of the most important factors in human experience, but recognizing the biological characteristics (particularly related to the human perceptual apparatus) that a person will bring into the making of a place.

He critiques approaches that focus on measuring features of the space in order to acquire spatial laws and rules, for they are dealing with only a part of the problem and need to be complemented by experiential data (Tuan, 1977: 5).

Tuan defines the experience of place as 'compounded of feeling and thought' (1977: 10). Space and place must be conceptualized with respect to the human body, as 'the human being, by his mere presence, imposes a schema on space' (1977: 36). Physical space is organized by people so that it conforms with and caters to their biological needs and social relationships.

Also human spatial abilities construct knowledge and then place. Tuan suggests a similarity between moving from space to place and learning a maze – associating meaning to space through perceptual clues establishes landmarks. What was an unknown geometrical extension is becoming familiar and meaningful, therefore bodily movement through the space is the first step towards placemaking. All the human senses then construct together in the mind the experience of a space and then of a place, as movement, sight, taste, smell, hearing and skin sensitivity all merge to make us aware of a spacious external world inhabited by objects. Subsequently, it is the association of personal and cultural values and meanings that start changing this primary idea of space.

According to Tuan, personal reflections and feelings also make places; as centres of felt value, they are the results of the experience of all the senses as well as the active and reflective mind. Therefore, there is also a direct relationship between physical environment and feeling, although we must be aware that space and spaciousness carry different sets of meaning in different cultures. The presence of others physically sharing a space and culturally sharing values and meanings also contribute to one's making of a place. People also share experiences within a space – their sense of place will have some communal elements that go beyond bodily sensations and cultural influences. The way the built environment is designed and laid out also clarifies cultural identity and social roles and relations (Tuan, 1977: 102).

Finally, Tuan discusses the relationship between time and the experience of place. He reflects on the nature of place as *pause*, as a particular occurrence within time seen as motion or flow. Place exists as it is experienced by people at a certain time, but continuously and dynamically enriched by the presence of new people, new elements, new events. Tuan also defines attachment to place as a function of time – we become more attached and emotionally tied to a place, the more time we spend within it. The third and last relationship between time and place

Tuan notes is the role that a place can take as a memorial or as time made visible. For example, environments such as museums consist wholly of displaced objects but they create a new structure of objects into place, new places.

We have seen how this body of thought has developed further the analysis of what constitutes the experience of place and refined its definition. In particular, Tuan's contribution is valuable in recognizing the importance of several spheres of experience in the making of place. It is clear in his work that space is an essential component of place – the structural, physical element where experience occurs, as humans are always physically situated.

We have thus reviewed theories 'beyond geometry' that concentrate on spatial structures and theories primarily concerned with human experiential aspects. In the following section, we will link these theoretical ideas to some notable examples of digital technology research that have been incorporating theories of space and place, in order to give a sense of how such theories can be influential, useful and applicable.

SPACE AND PLACE IN DIGITAL TECHNOLOGY RESEARCH

Since the 1990s, evolutions in technology have meant a greater interest in concepts of spatiality. For example, attempts at creating 'virtual spaces', innovations in communication between computers, and therefore a new understanding of distance, and increasing interest in the features of the physical context of interaction that can be monitored and sensed.

Erickson (1993) observed how 'real' space both structures and enriches human interaction, suggesting that spatial environments and their structures would work as effective metaphors for interface design. An even earlier paper, although not as well developed, highlighting the analogies between the design of interfaces and some principles of architectural design is by Hooper (1986).

Erickson also notes that one of the most important attributes of space is the meaning people read into it. He suggests that place, rather than space, should be used to describe those environments that people invest with actions, understandings and meanings.

Subsequently, Erickson adopts Alexander's architectural theory of patterns as a possible framework to support the design of spatially based systems (Erickson, 2000). Pemberton and Griffiths (1998) also drew on Alexander's patterns, proposing a framework for the design of interfaces based on an underlying spatial metaphor. The influence of Alexander's work has also shaped approaches to the study of collaborative systems in use. Crabtree proposes patterns as a useful approach to guide ethnographic work aimed at informing design. He applies a sociological perspective to understanding the relationship between space, place and conduct (Crabtree, 2000), where the relationship between space and technology is a practised one.

Other researchers have also been inspired by architectural theories (Lainer and Wagner, 2003; Munro et al., 1999): Lainer and Wagner reflected on the relationships between the qualities of built spaces as envisioned by architects and planners and the activities and events that take place within them. They discussed how specific qualities of social use are supported through the design of appropriate spaces and places, whether physical or digital. Wagner's more recent work carries this influence through and applies it to several design domains (Jacucci and Wagner, 2005).

More recently, attention to the importance of space and to the use of spatial metaphors for supporting social interaction has arisen also in the more technical areas of collaborative virtual environments (Benford et al., 2001), mixed reality installations and the creation of trajectories of physical discovery of the physical and the virtual (Benford and Giannachi, 2011).

In an influential paper, Harrison and Dourish (1996) propose that place, rather than space, should be the frame for understanding human behaviour and inform the design of interactive systems. They note, 'Place is a space which is *invested with understandings* of behavioural appropriateness, cultural expectations and so forth. We are *located* in "space", but we *act* in "place"' (Harrison and Dourish, 1996). More recently, Dourish applied the space/place distinction for the development of a framework for the design of collaborative technologies (1999). He later framed this vision of space and place within his phenomenologically influenced approach to design (Dourish, 2001) and reflected more deeply on the current conceptual work on spatial notions in human–computer interaction design, advocating a rediscovery of its spatial (i.e. structural) aspects (Dourish, 2006).

Similarly, Turner and Turner (2003) propose phenomenology as the perspective from which to conduct studies of place, but in order to inform the design of realistic virtual models of actual locations rather than of interactive systems located in the physical world. This is a useful approach informing research on the re-creation of engaging environments by means of digital technology.

Chalmers (2001) and Harrison and Tatar (2008) both discuss space and place, taking inspiration from post-structuralistic semiotics. For Chalmers, symbolic interpretation of a space and of the media that populate it through language is what makes of a space a place. He aims to extend Harrison and Dourish's (1996) vision through a semiologic approach concerned with how interpretation shapes our actions within a space – physical or virtual. Harrison and Tatar (2008) analyse place construction from the point of view of how people make sense of it through linguistic constructs, which are in turn socially and culturally shaped. Such an approach is useful particularly for the analysis of verbal data collected when studying interactions in place and for understanding how place characteristics become embodied in language.

Fitzpatrick proposes the concept of place as defined by Harrison and Dourish as the foundational metaphor of her 'locales

framework', a perspective for the analysis and design of collaborative systems (Fitzpatrick, 2003) whereby place represents the lived relationship with the structures and resources that support communication and interaction within a group. Fitzpatrick's work is of particular relevance for computer-supported cooperative work, showing how collaborative mechanisms are grounded in a lived context.

Brown and Perry (2001) present their analysis of the concept of place as it is proposed within geography and its relevance for the design of a different kind of technology, namely spatially orientated interactive systems such as electronic tourist guides. The application of this perspective to another case, the study of mobile workers, is discussed in Brown and O'Hara (2003). Mobile technology evaluation and design can benefit from this work, as often aspects of the experience of the environment are neglected in this area.

Rossitto's study of how mobile work becomes emplaced at several locations adopts Casey's concern for agency in place-making (Rossitto, 2008). She takes the view of place first of all as an *event* that happens through the negotiation between environment, inhabitants and their experiences. Furthermore she sees place as experienced along physical, psychological, historical and social dimensions. Agency as a crucial dimension of placemaking is important when studying how humans configure their environment by means of technology – an increasingly important concern, for example, in mobile technology design.

Ciolfi and Bannon (2005) propose a framework based on Tuan's work, that sees place as an emergent notion articulating alongside four interconnected dimensions of lived experience: physical, social, cultural and personal. The framework has been applied as a way to guide the process of designing interactive physical environments in domains such as museums (Ciolfi and Bannon, 2007) and other public spaces (Ciolfi, 2007). The framework is utilized as a way to guide both the empirical investigation of human activities in place along the four dimensions and the design of technological interventions so that the four dimensions are addressed by interactional qualities.

Tuan is also an inspiration for Greenbaum's reflection on how the process of *appropriation* of digital environments leads to the reconfiguration of physical spaces (Greenbaum, 2009). Using examples of community engagement in New York, she draws a connection between agency within digital environments and physical placemaking.

A last but notable example of how theories of space and place have been enfolded within digital technology research is Malcolm McCullough's book *Digital Ground*. Reflecting on the intersection between architecture and interaction design, McCullough highlights current and future challenges for both disciplines in the context of pervasive computing and his contribution is key in highlighting how the design of physical environments and digital products and services must be aware of each other's concerns and strategies.

CONCLUSION

In this chapter we have seen how notions of space and place are increasingly relevant to digital technology research. Moving from examples of current technological developments that link to open questions about people's interaction with technology within the physical world, we have presented two sets of theories with a common standpoint. Both sets are grounded on an understanding of the physical world 'beyond geometry', as a dynamic notion that is constructed by physicality and by human agency. The first group of theories share a focus on the physical environment as the starting point of enquiry, the second group considers human experience as the primary concern. Finally, we have provided an overview of how digital technology research carries different theoretical influences through a number of examples

of recent work, mainly within human–computer interaction.

The goal of the chapter was to introduce the reader not only to significant theorists but also to theoretical distinctions when thinking of space and place. The overall frame of the chapter is orientated to representing space and place as socially and culturally constructed notions. Considering the physical world as a continuously negotiated compound of structure and experience is essential when studying and/or designing digital technology: in the same way as technology needs to be thought of in terms of human needs and activities, so the physical setting it occupies in an increasingly pervasive way needs to be examined from a human-centred perspective.

FURTHER READING

Adams, P.C., Hoelscher, S. and Till, K.E. (eds) (2001) *Textures of Place: Exploring Humanist Geographies*. Minneapolis, MN: University of Minnesota Press.

Low, S.M. and Lawrence-Zúñiga, D. (eds) (2003) *The Anthropology of Space and Place*. Oxford: Blackwell.

McCullough, M. (2004) *Digital Ground: Architecture, Pervasive Computing and Environmental Knowing*. Cambridge, MA: MIT Press.

Moores, S. (2012) *Media, Place and Mobility*. Houndmills, Basingstoke: Palgrave Macmillan

Turner, P. and Davenport, E. (eds.) (2005), *Spaces, Spatiality and Technology*. London: Springer

REFERENCES

Alexander, C. (1977) *The Timeless Way of Building*. Oxford: Oxford University Press.

Agre, P. (1997) *Computation and Human Experience*. Cambridge: Cambridge University Press.

Augé, M. (1995) *Non-Places: Introduction to an Anthropology of Supermodernity*. London and New York: Verso.

Bachelard, G. (1958) *The Poetics of Space*. Boston, MA: Beacon Press.

Bannon, L.J. and Bødker, S. (1991) 'Beyond the Interface: Encountering Artifacts in Use'. In J.M. Carroll (ed.), *Designing Interaction: Psychology at the Human–Computer Interface*. New York: Cambridge University Press, 227–253.

Benford, S. and Giannachi, G. (2011) *Performing Mixed Reality*. Cambridge, MA: MIT Press.

Benford, S., Greenhalgh, C., Rodden, T. and Pycock, J. (2001) 'Collaborative Virtual Environments', *Communications of the ACM*, 44: 7.

Bollnow, O.F. (2008) *Human Space*. London: Hyphen Press.

Broek, J.O.M. (1965) *Geography: Its Scope and Spirit*. Columbus, OH: Charles E. Merrill Books, as reported on http://geography.about.com/od/studygeography/a/geodefinitions.htm (last accessed 7 October 2011).

Brown, B. and O'Hara, K. (2003) 'Place as a Practical Concern for Mobile Workers', *Environment and Planning A*, 35(9): 1565–1587.

Brown, B. and Perry, M. (2001) 'Of Maps and Guidebooks: Designing Geographical Technologies', *Proceedings of Designing Interactive Systems: Processes, Practices, Methods, and Techniques*, ACM, 246–254.

Buttimer, A. (1980) 'Home, Reach and Sense of Place'. In A. Buttimer and D. Seamon (eds), *The Human Experience of Space and Place*. London: Croom Helm, 166–187.

Canter, D. (1977) *The Psychology of Place*. Oxford: Architectural Press.

Canter, D. and Singer, P. (eds) (1975) *Environmental Interaction*. New York: International Universities Press.

Casey, E.S. (1993) *Getting Back Into Place: Toward a Renewed Understanding of the Place-World*. Bloomington, IN: Indiana University Press.

Casey, E.S. (1997) *The Fate of Place: A Philosophical History*. Berkeley, CA: University of California Press.

Cassidy, T. (1997) *Environmental Psychology: Behaviour and Experience in Context*. Hove: Psychology Press.

Chalmers, M. (2001) 'Place, Media and Activity', *SIGGROUP Bulletin*, 22(3): 38–42.

Ciolfi, L. (2007) 'Supporting Affective Experiences of Place Through Interaction Design', *Co-Design*, 3(S1): 183–198.

Ciolfi, L. and Bannon, L.J. (2005) 'Space, Place and the Design of Technologically Enhanced Physical Environments'. In P. Turner and E. Davenport (eds), *Spaces, Spatiality and Technology*. London: Springer, 217–232.

Ciolfi, L. and Bannon L.J. (2007) 'Designing Hybrid Places: Merging Interaction Design, Ubiquitous Technologies and Geographies of the Museum Space', *Co-Design*, 3(3): 159–180.

Cooper, G. and Bowers, J. (1995) 'Representing the User: Notes on the Disciplinary Rhetoric of Human-Computer Interaction'. In P. Thomas (ed.), *The Social and Interactional Dimensions of Human-Computer Interfaces*, Cambridge: Cambridge University Press, 50–66.

Crabtree, A. (2000) 'Remarks on the Social Organization of Space and Place', *Journal of Mundane Behaviour*, 1(1): 25–44.

Crang, M. (1998) *Cultural Geography*. London: Routledge.

Crang, M. and Thrift, N. (2000) *Thinking Space*. Abingdon, Oxon: Routledge.

Cullen, C. (1971) *The Concise Townscape*. Oxford: Architectural Press.

Curry, M.R. (1999) 'New Technologies and the Ontology of Places', *Information Studies Seminar*, University of California Los Angeles, March 1999. Available at: http://baja.sscnet.ucla.edu/~curry/Curry_Tech_Regimes.pdf (accessed 16 May 2012).

Curry, M.R. (2000) 'On Wittgenstein and the Fabric of Everyday Life'. In N. Thrift and M. Crang (eds), *Thinking Space*. Abingdon, Oxon: Routledge, 89-113.

De Certeau, M. (1984) *The Practice of Everyday Life*. Berkeley, CA: University of California Press.

Dourish, P. (1999) 'Where the Footprints Lead: Tracking Down Other Roles for Social Navigation'. In A. Munro, K. Höök and D. Benyon (eds), *Social Navigation of Information Space*. London: Springer, 15–34.

Dourish, P. (2001) *Where the Action Is: The Foundations of Embodied Interaction*. Cambridge, MA: MIT Press.

Dourish, P. (2004) 'What We Talk About When We Talk About Context', *Personal and Ubiquitous Computing*, 8(1): 19–30.

Dourish, P. (2006) 'Re-Space-ing Place: Place and Space: Place and Space Ten Years On', *Proceedings of ACM Conference on Computer-Supported Cooperative Work (CSCW '06)*, ACM, 299–308.

Dourish, P. and Bellotti, V. (1992) 'Awareness and Coordination in Shared Workspaces', *Proceedings of CSCW '92*, ACM, 107–114.

Dourish, P. and Bly, S. (1992) 'Portholes: Supporting Awareness in a Distributed Work Group', *Proceedings of the ACM Conference on Human Factors in Computing Systems (CHI '92)*. ACM, 541–547.

Dourish, P., Adler, A., Bellotti, V. and Henderson, A. (1996) 'Your Place or Mine?: Learning from Long-Term Use of Audio-Video Communication', *Computer-Supported Cooperative Work*, 5(1): 33–62.

Entrikin, J.N. (1991) *The Betweenness of Place*. Baltimore, MD: Johns Hopkins University Press.

Erickson, T. (1993) 'From Interface to Interplace: The Spatial Environment as Medium for Interaction', *Proceedings of the Conference on Spatial Information Theory*, Springer-Verlag. Available at: www.pliant.org/personal/Tom_Erickson/Interplace.html

Erickson, T. (1996) 'Design as Storytelling', *Interactions*, iii(4), July/August. Available at: www.visi.com/~snowfall/Storytelling.html

Erickson, T. (2000) 'Towards a Pattern Language for Interaction Design'. In P. Luff, J. Hindmarsh and C. Heath (eds), *Workplace Studies: Recovering Work Practice and Informing Systems Design*. Cambridge: Cambridge University Press, 252–261.

Fitzpatrick, G. (2003) *The Locales Framework: Understanding and Designing for Wicked Problems*. Dordrecht: Kluwer.

Goffman, E. (1963) *Behaviour in Public Spaces*. New York: Free Press.

Greenbaum, J. (2009) 'Appropriating Digital Environments. (Re-)constructing the Physical through the Digital'. In T. Binder, J. Löwgren and L. Malmborg (eds), *(Re)Searching the Digital Bauhaus*. London: Springer, 235–249.

Harries, K. (1997) *The Ethical Function of Architecture*. Cambridge, MA: MIT Press.

Harrison, S. and Dourish, P. (1996) 'Re-Place-ing Space: The Roles of Place and Space in Collaborative Systems', *Proceedings of CSCW '96*, ACM, 67–76.

Harrison, S. and Tatar, D. (2008) 'Places: People, Events, Loci – the Relation of Semantic Frames in the Construction of Place', *Computer-Supported Cooperative Work* 17(2–3): 97–133.

Hooper, K. (1986) 'Architectural Design: An Analogy'. In D.A. Norman and S. Draper (eds) *User Centered System Design: New Perspectives on Human-Computer Interaction*. Hillsdale, NJ: Lawrence Erlbaum, 9–23.

Jackson, P. (1989) *Maps of Meaning: An Introduction to Cultural Geography*. London: Unwin Hyman.

Jacucci, G. and Wagner, I. (2005) 'Performative Uses of Space in Mixed Media Environments'. In P. Turner and E. Davenport (eds), *Spaces, Spatiality and Technology*. London: Springer. 191–216.

Kaptelinin, V. and Nardi, B. (2009) *Acting with Technology: Activity Theory and Interaction Design*. Cambridge, MA: MIT Press.

Kuutti, K. (2001) 'Hunting for the Lost User: From Sources of Errors to Active Actors – and Beyond'. Paper written for the Cultural Usability Seminar, Media Lab, University of Art and Design Helsinki, 24 April. Available at: www.mlab.uiah.fi/culturalusability/papers/Kuutti_paper.html (accessed 7 October 2011).

Lainer, R. and Wagner, I. (2003) 'Architecture et scénographie technique', *Alliage*, 50–51; (2003) 'Le spectacle de la technique', 190–199.

Lefevbre, H. (1992) *The Production of Space*. chichester and oxford: Wiley-Blackwell.

Leont'ev, A.N. (1978) *Activity, Consciousness, and Personality*. Englewood Cliff, NJ: Prentice Hall.

Ley, D. and Samuels, M. (1978) 'Contexts of Modern Humanism in Geography'. In D. Ley and M. Samuels (eds), *Humanistic Geography: Prospects & Problems*. London: Croom Helm, 1–18.

Lynch, K. and Rivkin, M. (1959) 'A Walk Around the Block', *Landscape*, 8: 24–34.

Malpas, J. (1999) *Place and Experience: A Philosophical Topography*. Cambridge: Cambridge University Press.

Malpas, J. (2008) *Heidegger's Topology: Being, Place, World*. Cambridge, MA: MIT Press.

Massey, D. (1994) *Space, Place and Gender*. Cambridge: Polity Press.

Massey, D. (2005) *For Space*. London: Sage.

McCarthy, J. and Wright, P. (2004). *Technology as Experience*. Cambridge, MA: MIT Press.

McCullough, M. (2004) *Digital Ground: Architecture, Pervasive Computing and Environmental Knowing*. Cambridge, MA: MIT Press.

Merleau-Ponty, M. (1945) *Phenomenology of Perception*. London: Routledge.

Mitchell, D. (2000) *Cultural Geography*. Oxford: Blackwell.

Munro, A.J., Höök, K. and Benyon, D. (eds) (1999) *Social Navigation of Information Space*. Heidelberg: Springer-Verlag.

Peet, R. (1998) *Modern Geographical Thought*. Oxford: Blackwell.

Pemberton, L. and Griffiths, R. (1998) 'The Timeless Way: Making Living Cooperative Buildings with Design Patterns'. In N. Streitz, S. Konomi and H. Burkhardt (eds), *Proceedings of CoBuild98*, Springer, 142–153.

Propen, A.D. (2009) 'Cartographic Representation and the Construction of Lived Worlds: Understanding Cartographic Practice as Embodied Knowledge'. In M. Dodge, R. Kitchin and C. Perkins (eds), *Rethinking Maps: New Frontiers in Cartographic Theory*. Abingdon, Oxon: Routledge, 113–130.

Proshansky, H.M., Ittelson, W.H. and Rivlin, L.G. (1969) 'The Influence of the Physical Environment on Behavior: Some Basic Assumptions'. In H.M. Proshansky, W.H. Ittelson and L.G. Rivlin (eds), *Environmental Psychology: Man and his Physical Setting*. London: Holt, Rinehart & Winston, 27–37.

Relph, E. (1976) *Place and Placelessness*. London: Pion.

Relph, E. (1981) *Rational Landscape and Humanistic Geography*. London: Croom Helm.

Rossitto, C. (2008) 'Managing Work at Several Places: Articulating the Notion of Nomadic Work in Student Groups'. PhD dissertation, Royal Institute of Technology, Department of Computer Science and Communication, Stockholm.

Rowles, G. (1978) 'Reflections on Experiential Field Work'. In D. Ley and M. Samuels (eds), *Humanistic Geography: Prospects & Problems*. London: Croom Helm, 173–193.

Sack, R.D. (1997) *Homo Geographicus: A Framework for Action, Awareness and Moral Concern*. Baltimore, MD: Johns Hopkins University Press.

Seamon, D. (1993) *Dwelling, Seeing, and Designing: Toward a Phenomenological Ecology*. Albany, NY: State University of New York Press.

Seamon, D. (1996) 'A Singular Impact: Edward Relph's *Place and Placelessness*', *Environmental and Architectural Phenomenology*, 7(3): 5–8.

Seamon, D. (2006) 'Interconnections, Relationships, and Environmental Wholes: A Phenomenological Ecology of Natural and Built Worlds'. In M. Geib (ed.), *Phenomenology and Ecology*. Pittsburgh, PA: Simon Silverman Phenomenology Center, 53–86.

Stults, B. (ed.) (1986) *Media Spaces*. Palo Alto, CA: Xerox PARC and System Concepts Lab.

Suchman, L. (1987) *Plans and Situated Action: The Problem of Human-Machine Communication*. Cambridge: Cambridge University Press.

Tuan, Yi-Fu (1971) 'Geography, Phenomenology and the Study of Human Nature', *The Canadian Geographer*, 15: 181–192.

Tuan, Yi-Fu (1974) *Topophilia: A Study of Environmental Perception, Attitudes, and Values*. New York: Columbia University Press.

Tuan, Yi-Fu (1977) *Space and Place: The Perspective of Experience*. Minneapolis, MN: University of Minnesota Press.

Turner, P. and Turner, S. (2003) 'Two Phenomenological Studies of Place'. In E. O'Neill, P. Palanque and P. Johnson (eds), *People and Computers XVII – Designing for Society*. London: Springer, 21–36.

Tweed, C. (2000) 'A Phenomenological Framework for Describing Architectural Experience', *British Phenomenological Society Workshop on Phenomenology and Culture*, University College Cork, July.

Vygotsky, L.S. (1978) *Mind in Society*. Michael Cole, Vera John-Steiner, Sylvia Scribner and Ellen Souberman (eds). Cambridge, MA: Harvard University Press.

Weiser, M. (1994) 'The World is Not a Desktop', *Interactions*, 1(1): 7–8.

Affect and Experiential Approaches

Kristina Höök

INTRODUCTION

Experiential approaches to understanding technology and its use often start from the understanding of the individual's experience in using a particular system in the world. Whether it is a horseback rider or an artist using Photoshop, individuals have a rich embedded physical experience of the world, one that involves complex skills and knowledge. With artists each movement of their stylus can involve a huge amount of skill and judgement, just as horseback riders must remember and feel as they ride their horse. This chapter explores experiential approaches to the use and design of technology – developing an understanding of how we can design technology that in use evokes key experiences and feelings. It will put a particular focus on bodily experiences, as our bodies are always there with us, inseparable from our thinking, emotions and ways of understanding the world. Yet another reason to focus on bodily experience is the recent wave of sensors and actuators, involving users' whole bodies in digital interactions, opening up a whole new design arena.

Below we will review different approaches to experience as a design approach, alongside its philosophical background in post-Cartesianism. Design that aims to spur particular experiences is notoriously hard to achieve. What we need to do, is to develop a way forward for research into user experiences that is both empirical yet is closely tied to the design of new technical systems.

When designing for bodily experiences with digital technologies, a range of questions arise: (1) figuring out exactly what kind of experiences we are trying to design for; (2) how can we articulate those experiences so that we can share them in a design team or perform evaluations of the resulting artefact; (3) what are the aesthetic qualities of the interactive artefact that can support or spur experiences of the kind we are aiming for; and, in general (4) finding the theoretical underpinnings that may help us towards a more systematic understanding of how to design for many different interactive experiences.

Obviously, the area that addresses the design of interactive artefacts is not the first to approach design for experience. We can learn a lot from arts, music, advertising and any other field addressing experience, but what is different when producing digital interactions is their *interactive* nature. The experience we seek to spur is partly in the hands of our users and what they choose to do. The system needs to build for and sustain what the designer intended over whole interactions, sometimes repeatedly, sometimes rendering unique experiences.

FROM ERGONOMICS TO HCI AND BACK

Originally, human–computer interaction (HCI) was a spin-off from ergonomics, particularly through its focus on the cognitive aspects of interaction. With the turn to third-wave HCI and its focus on 'experiences', we are now in a sense turning back to considering the body more centrally in interaction, similarly to how ergonomics researchers have been concerned with our bodily selves. However, contrary to an ergonomics focus on documenting and altering routines, finding error-free ways of involving operators in the complex beast of man and machine as a unity, or measuring how the body has its limitations, the current HCI turn back to bodily interaction has a different goal. The renewed HCI interest lies in designing for *bodily experiences* – beyond the efficiency and task completion acts that can be achieved through body movement. The goal is to involve users bodily and thereby create for a whole range of experiences: movement-based, emotional or social.

This raises three questions. First, what kinds of experiences are we aiming to design for here? We have very few descriptions of experiences a designer may strive for and they are sometimes quite vague terms, such as *flow* (Csikszentmihalyi, 1990), describing a sense of being totally absorbed by the activity at hand, forgetting time and space or game play (Salen and Zimmerman, 2004),

describing how a game world, the interaction mechanisms and the user activities come together into a whole. Glossing over experience as being all about designing for flow or inducing a game play experience is too vague, as I have argued before (Isbister and Höök, 2009). We need to drill deeper and better understand exactly what bodily experiences we are talking about. Are we designing for pleasurable or unpleasurable ones? Are we designing for those that are subjective and unique or ones that are common and shared? Ones that deliver serendipitous experiences or ones that are evocative and emotional?

A particularly difficult issue lies in understanding how these experiences may unfold over time – both in the particular interaction with and manipulation of the artefact[1] and as parts of our everyday ongoing lives. As Löwgren (2001: 35–36) puts it, a gestalt for interactive artefacts is defined as a *dynamic gestalt* that '*we have to experience as a dynamic process*'.

A second question is, once we know what kind of experiences we are aiming to design for, we need to articulate them in a form that captures our intentions and can be shared within a design team. Here we run into many difficulties. Ways of knowing can arise from your bodily acts without any language translation in between. In fact, we sometimes entirely lack verbal expressions that can convey these experiences. The feel of the muscle tensions, the touch of the skin, the tonicities of the body, balance, posture, rhythm of movement, the symbiotic relationship with objects in our environment – these all come together in unique holistic experiences. It is not the ability to fulfil a task, but the experience of the corporeality of doing so that matters here. What are the forms of articulation we can use that capture what we aim for? These descriptions also need to be shared with the users we invite to test designs or participate in the design process.

Recent work by Hagen shows how the games industry, leading the development of bodily interactions, struggle with this issue (Hagen, 2011). Games development often involves a couple of hundred developers working together for a couple of years to bring out a first-class experience for the gamers. They need to make sure that the whole team shares the same understanding of the sought experience, thereby making the right design decisions for all the details of graphics, animations, user input and interactive behaviour. Hagen coins the term *lodestar* to describe the plethora of means used to direct the team. These lodestars can manifest themselves as PowerPoint presentations, short animations, props or artefacts, capturing aspects of the emotional/bodily feel that the game should communicate.

The articulation of the sought experience also constitutes the baseline against which we evaluate the artefact we are building. Building systems for bodily interactions is notoriously difficult and there are not many design principles we can rely on. This means that we will often bring in users repeatedly to test parts of our design. This, in turn, puts demands on user study methods. How can our users express what they experience when, for example, playing with a Wii-mote for the first time?

But before we can test our system with users, we need to create our design. As in any design situation, the design space is only limited by our imagination. Nowadays, there is a plethora of technologies and materials that can be used – interactive textiles, sensors, various actuators, graphics, many different displays and so on. Depending on what material we choose, we can draw upon different aesthetic qualities. When using gestures as input, we know that certain gestures will be more likely to spur certain experiences. When colour is used as output, we can build on perceptual and cultural understandings of different colours. How do we put these together into interesting dynamic gestalts? Much research remains to be done in this area.

Finally, the third question: what are the theoretical underpinnings we can use to understand our designs and design processes?

Where do we turn to explain bodily experiences?

Below I will go through some methods and research results for each of these three questions. As the readers will see, I will mainly draw on our own work, pointing in certain directions, rather than as a full account of all the design and research done in this area. In a sense, that chimes with my whole stance towards design in this area: it is a very personal experience. Lessons learnt are sensual, hard to articulate, hard to communicate, but hopefully it will inspire more work in this area.

STUDIES OF BODILY AND AFFECTIVE EXPERIENCES

In order to open the design space and remind ourselves of how many different bodily experiences we could be designing for, let us provide a short account of the links between emotion, body, movement and experience.

As indicated above, we are not interested in the function of the movement of a limb as the ergonomists are, but, rather, in the *experience* of doing so. Experiences are linked to many aspects of ourselves: our actual physical, corporeal abilities and experiences, the emotional experiences these spur, our memories of prior experiences and our interpretations and meaning-making processes of those experiences. My particular focus here will be on the linking of emotion to body movement.

There has been a wave of new research on emotion in diverse areas such as psychology, neurology, medicine and sociology. Neurologists have studied how the brain works and how emotion processes are a key part of cognition. Emotion processes are basically sitting in the middle of most processing going from frontal lobe processing in the brain, via brain stem to body and back (Damasio, 1994; Davidson et al., 2003). Bodily movements and emotion processes are tightly coupled.

Darwin had already discussed how certain body movements, facial expressions and emotional experiences co-occur (Darwin,

1872). An experience may start as an external stimulus, scaring us, and then move on to a bodily reaction, but it may also start as a bodily movement, stimulating our emotional processing and thinking, creating an emotional reaction. As discussed by Sheets-Johnstone, there is 'a generative as well as expressive relationship between movement and emotion' (Sheets-Johnstone, 2009). Certain movements will generate emotion processes and vice versa. Emotions are not hard-wired processes in our brains, but changeable and interesting regulating processes for our social selves. As such, they are constructed in a dialogue between ourselves and the cultural and social settings we live in (Katz, 1999; Lutz, 1986, 1988; Parkinson, 1996). Emotion is a social and dynamic communication mechanism. We learn how and when certain emotions are appropriate and we learn the appropriate expressions of emotions for different cultures, contexts and situations. The way we make sense of (and thereby experience) emotions is a combination of the experiential processes in our bodies and how emotions arise and are expressed in specific situations in the world, in interaction with others, coloured by the cultural practices that we have learnt.

Lutz, for example, shows how a particular form of anger, named song by the people on the south Pacific atoll Ifaluk, serves a very social role in their society (Lutz, 1986, 1988). Song is, according to Lutz, 'justifiable anger' and is used with children and with those who are subordinate to you, to teach them appropriate behaviour, for example in making a fair contribution to the communal meal, paying respect to elders or acting socially appropriately.

In ethnography, the work by Katz (1999) provides us with a rich account of how people individually and group-wise actively produce emotion as part of their social practices. When he, for example, discusses anger among car drivers in Los Angeles, he shows how anger is produced as a consequence of a loss of embodiment with the car (as part of our body), the road and the general experience of travelling. He connects the social situation on the road; the lack of communicative possibilities between cars and their drivers; our prejudices about others' driving skills related to their cultural background or ethnicity; etc. and shows how all of it comes together to explain why anger is produced when, for example, we are cut up by another car. He even sees anger as a graceful way to regain embodiment after, for example, having been cut up by another car.

These theories of emotion and bodily movement provide us with some backdrop to understanding the complexity of experience and what it would mean to design for it, but, to approach design, we sometimes need much more detailed accounts that we can draw upon. Let us turn to some of the studies done with the explicit purpose of feeding into design – both ethnographic and so-called autoethnographic accounts.

ETHNOGRAPHIC ACCOUNTS

While ethnographic accounts shun making statements as to what people are feeling or what is going on in their minds, traditional ethnographic accounts may also, in a sense, capture some of the inner experiences of bodily and emotional practices. For example, we did an ethnographic study of the last groups of users on earth to be exposed to mobiles, the people of the Rah island in Vanuatu (also discussed Chapter 13, by Barry Brown, in this volume). In the study, we noted how they changed their body postures and movement behaviours and shifted their focus from caring about their own bodies to caring more about their mobiles. Their care resembled caring for a fragile being, like a baby (Ferreira and Höök, 2011). We observed how they protected their mobiles dangling from a string around their neck, from dipping into water by tensing their backs into an uncomfortable position. When swimming from one island to another, they had to swim in a painful posture with one arm stretched above the head to carry and protect the mobile. On several occasions, we saw people taking a direct hit on their chest when falling to protect the mobile in their hands.

In my own Western-world culture, we have made many posture and movement adaptions a long time ago. In a sense, they are almost inaccessible to me now, as they have become totally ingrained into my everyday bodily practices. I can only mention the most obvious, like how I will never place my mobile on the kitchen sink close to water, I always worry about where to put my mobile in my handbag so that I can easily retrieve it and I keep looking about for places where I can find electricity to recharge my battery. From watching the people on Rah island and their struggles to integrate the mobiles with their practices, changes in bodily practices became much more accessible and visible to us. It became possible to describe and understand their experiences of this shift (Ferreira and Höök, 2011, 2012). Studies like these make us more sensitive as designers to how the shape and form of a technological device will interact with our bodily experiences. It may also spur ideas, such as creating forms that encourage certain ways of carrying the device or taking care of it.

Studies of bodily practices may have to go into painstaking details in order to capture the training and changes in perception needed to attend to a particular experience. For example, in a study of fly-fishing by Laurier and Brown, we learn how we might train ourselves to the particular ways of 'seeing' required to find fish (2004). Similarly, in Ingold's description of plank sawing (2006). He nicely shows that no matter how experienced you are, sawing through yet another plank will never be exactly the same as the last. There are a number of different sensations that synchronize in a particular sawing experience. Rapidly adjusting to each one of them in the dynamic unfolding of the event, will in every new instance be a unique experience. When designing digital interactions involving our bodies, we need to remember how our designs may become part of, or even

ingrained into, the everyday habits of a person – a huge responsibility for the designer.

AUTOETHNOGRAPHIC ACCOUNTS

In several of the studies on bodily/emotional experiences, first-person methodologies have been employed (Höök, 2010). There is, for example, autoethnography, a form of autobiographical personal narrative that explores the writer's experience of life.

Autoethnography (though controversial due to its subjective nature) has recently started to be used in HCI (Sengers et al., 2005). An example can be seen in my own autoethnographic study of horseback riding. Here I discussed both how impoverished digital interaction is compared to the sensory richness of riding, and also how impoverished our descriptions of those interactions are, due to the lack of an agreed upon language for describing bodily interactions (Höök, 2010). Only through describing all the details of specific riding experiences, such as the placement of my feet, arms and sitting bones, my gaze direction, mine and the horse's emotional processes and our wordless conversation, was I able to get anywhere near describing at least some aspects of my experience. Still, the actual experience must really be felt to make sense. To some extent it is *qualia* problem: you cannot ever know for sure that someone else will ever have the same experience as yourself.[2] Here is an excerpt from one of my horseback riding lessons, trying to capture the feeling of being nervous and not communicating very well with the horse, named Liberty:

> At the beginning of the lesson with Liberty I was immensely tense. Liberty has already on our way towards the riding track decided that there were ghosts everywhere and was 'spooked'. I could feel his tension and how it increased by my tensed movements. My butt-muscles and arm-muscles were extremely tense. I tried to make them relax, but failed. The tension in my arms made them straight and stiff rather than bent at the elbow

and relaxed along the sides of my body. ... During the lesson, I felt that Liberty was not listening. My feeling was that I needed to take more control, provide more signs to make him listen to me and start collaborating. I did not feel that Liberty was testing me, more like he was genuinely into his own thing, being spooked, and not that interested in me. I therefore tried to distract him, make him listen to me. I pulled at the reins to say 'hello, I am here – listen to me'. I shifted in my seat, and I made my legs 'present' in the interaction. But it felt as [if] the more I tried to get in contact, the more he closed off and did not want to listen. He just became more silly, more easily spooked. I did not feel that I was in control of the horse, something that makes any rider very nervous. ... As I was doing all of this 'give me attention'-business, I fell back into a bad habit of watching Liberty's head. By doing so, you can get some more signs of where the horse is heading, what it is thinking or looking at, and whether it is listening to you. But the rider should not be watching the horse's head. My head should be directed towards where we are going, since the balance of my head and the position of my shoulders tell the horse where we are heading. I also need to plan ahead where we should be going next, preparing the horse and myself for what is going to happen. If I look in the direction of where we are going, I will still see the horse's head in the periphery. And by looking at where we are going and planning our route, I stay in control, rather than acknowledging and giving the horse the control. On and off, I tried to stop watching Liberty's head, but because of this lack of contact or dialogue between us, I kept looking back at his head to be prepared for any action he would take. We were not 'one', but two that were not collaborating. (Höök, 2010: 232)

The muscle tensions, the movements I tried, the horse behaviour and my eye-gaze direction – all of it comes together in my experience of nervousness in this particular riding lesson. Nervousness will not always be perceived the same way and, for someone who is not a rider, this account may not even make any sense. What is interesting in the description is how it is not enough to only account for the emotional experience of being nervous, but how that experience was dependant upon and interacting with my bodily, physical position, movements, the horse's movements, tensions in my muscles and so on.

The overall idea behind these autoethnographic accounts is to provide rich descriptions of experiences that we may want to be inspired by when designing. Autoethnography may also be used as a method to guide the whole design process. By carefully observing and articulating their whole complex experiences, the designers or users may provide feedback into the design process (Sengers et al., 2005).

PAINFUL ADAPTATIONS TO TECHNOLOGY?

All of these studies point to limitations in the ways we think of today's wearable and mobile technologies and their impact on bodily behaviours and practices – often missing out on designing for much richer experiences. Some even claim that the technologies we wear today treat our bodies in a severely negative way (Longo, 2003: 25): 'Electronics, robotics, and spintronics invade and transform the body and, as a consequence of this, the body becomes an object and loses its remaining personal characteristics, those characteristics that might make us consider it as the sacred guardian of our identity'.

In that line of work, the most notable study comes from Troshynski and colleagues who studied sex offenders who had to wear a GPS-system around their ankle as part of their parole conditions (2008). Their study shows the painstaking care these offenders had to pay to the technology in order to keep both the technology itself and their own bodies intact. The technology could not take much water so they could not take a bath. It became visible to others unless they wore trousers to hide it, in effect, making them never wear shorts. They had to care extensively for the battery life and the fragility of the device itself, since they would end up in prison if it stopped working. In many ways, this piece of technology transformed their bodies in exactly the way Longo, mentioned above, fears: it invaded their bodies and their identity, changing their understanding of themselves in a negative way.

In a sense, we need to move back to the ergonomic approach, carefully considering our bodily ways of being in the world, but without putting a machine-like perspective on what the body is or could be.

ARTICULATION FOR DESIGN AND EVALUATION

However, not all bodily experiences with digital technology are impoverished, limiting or painful. There have been attempts to design for various aesthetic experiences: Schiphorst's soft(n), where design of physical interactive artefacts is done from a somaesthetic perspective (Schiphorst, 2007), Moen's design for the joy of moving in her BodyBug system (Moens, 2007), Sengers and colleagues' designs for affective presence in Affector and Mirò (Sengers et al., 2005; Boehner et al., 2003), Danielle Wilde's hipdisks that make its wearers do 'undignified' hip-torso movements (Wilde, 2011), Høbye and Löwgren's Mediated Body where users generate sound together (Höök and Löwgren, 2012), as well as systems produced by my group, such as eMoto for sending emotional messages with your mobile through gestures (Sundström et al., 2007), Affective Diary and Affective Health mirroring your everyday emotional experiences (Ståhl et al., 2009; Sanches et al., 2010), EmRoll for controlling a game bodily through breathing and moving (Zangouie et al., 2010), and the LEGA for leaving physical traces in a museum for your friends to discover (Laaksolahti et al., 2011) – all designed for emotional and bodily expressivity and the

joy of communicating with yourself or others (see Figure 12.1).

To design such systems, it is not enough to know about experiential qualities in already established bodily and emotional practices (as described in the section above). There is often no point in mimicking what we can already do in the world (even if that is often done with sports games or rock band props). Instead, we need to shift our understanding of bodily experiences into *novel* design. That shift requires recognition of the fact that the system may alter or create for entirely novel ways of experiencing and being in the world. Design researchers in this field have been stumbling around to find ways of articulating, sharing and validating their knowledge on how to design for rich emotional, bodily and social experiences. Below, we describe some of these attempts: articulation of so-called experiential qualities arising in and through the design, inventing and describing novel interaction modalities 'touching' our senses, and various design methods.

EXPERIENTIAL QUALITIES

One attempt at describing the components in a design that create a particular experience has been to isolate and describe *experiential qualities* (Löwgren and Stolterman, 2004) and how those can be achieved in an interface. The idea is that certain forms of designs will lead to certain experiences. Löwgren and Stolterman describe, for example, a quality they name *pliability* — that is, the degree to which the user perceives the interaction to have a strong

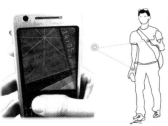

Figure 12.1   Emotional and bodily interfaces.

connection between action and outcome. He uses the example of interactive maps to explain it: when we can touch the map directly, using gestures to zoom in and out, it becomes more pliable than when we have to use a navigation toolbar on the side of the map. Other experiential qualities relating to bodily experience from my own work are, for example, *suppleness* (Isbister et al., 2007) or *affective loop* qualities (Höök, 2010).

A *supple* interaction has the following components.

- **Subtle signals**: Rich human communication and interpretation strategies (e.g. emotion, social ritual, nonverbal communication, kinesthetic engagement). Supple interfaces address users not only intellectually but also physically, typically involving their bodies in the interaction in some way. This may involve the face, gestures, biosensor data or general aspects of interaction picked up by various sensor technologies.
- **Emergent dynamics**: Taking into consideration subtle communication dynamics that require new thinking about system adaptivity and feedback. For example, increased legibility of system moves to help users actively co-construct practice and meaning and push system boundaries in interesting ways.
- **Moment-to-moment experience**: Privileging the quality of moment-to-moment experience both in terms of design and in terms of evaluation of success of design (e.g. a focus on engagement, pleasure, rapport).

We recognize suppleness in systems such as playing bowling with Wii-motes.

Systems that are designed to allow for affective loops can both be influenced by and influence users corporeally. In an affective loop experience: (i) emotions are seen as processes, constructed in the interaction, starting from everyday bodily, cognitive or social experiences; (ii) the system responds in ways that pull the user into the interaction, touching upon end users' physical experiences; and (iii) throughout the interaction the user is an active, meaning-making individual choosing how to express him- or herself. Thus the interpretation responsibility does not lie with the system.

We have built several systems that attempt to create affective loop experiences with more or less successful results. For example, eMoto lets users send text messages between mobile phones, but in addition to text, the messages also have colourful and animated shapes in the background chosen via emotion-gestures with a sensor-enabled stylus pen. Those gestures resonate with our emotional experiences through building on arousal levels – the more movement the more arousal. A negative emotion can be expressed through a more tensed gesture, while a positive emotion is expressed with looser movements or other dimensions of emotional experiences (see Figure 12.2).

These experiential qualities are not mutually exclusive. We typically combine them and put them alongside the specific requirements of the domain at hand to arrive at our design goals. The aim is that these qualities can be put up as *ideals* or strong concepts (Höök and Löwgren, 2012), for the design and the whole design team needs to strive towards making design decisions that enable such qualities in the interaction.

Figure 12.2 A user expresses her emotion using a physical interface.

HOW TO FIND EXPERIENTIAL QUALITIES?

Before entering the design phase, we first need to figure out what we aim to achieve – especially difficult as there are not that many movement-based applications (outside the gaming world) that we can be inspired by. Instead we may have to look to activities that have happened outside the realms of digital technologies.

One starting point that has been suggested is to extract the design qualities from one (extreme, unusual or particularly interesting) already existing practice and then transfer these to another. Ljungblad and Holmquist call this method *transfer scenarios* (2007). By that they mean learning from people's specific practices, using the experiential qualities of those practices and transferring them into other settings, thus creating innovative design that can attract a more general audience beyond the specific practice originally studied. They have tested this process in a study of reptile owners and transferring qualities of their relationships with their animals to the design of home robots or interactive wallpaper. What they learnt from reptile owners is that they often regard their aquariums as aesthetically appealing, living installations in their living rooms. This was transferred to people who do not own any reptiles. An interactive wallpaper was created into which users could 'inject' photos from their mobiles. At the place where those photos were injected, a flower would start growing, using the mobile photo to colour the leaves.

The difficult step in the transfer scenario method is narrowing down on (1) what the important experiential qualities are and (2) which are worth transferring, as well as (3) how to reshape and implement those in an innovative design. It is, of course, also important to find a relevant and interesting practice to study in the first place.

Some have critiqued this method as de-contextualizing the experience and thereby trivializing what we transfer.[3]

AESTHETIC QUALITIES

While we may have interesting design ideas for how technology should function, we must also deal with the realities of what the digital material lends itself to and what kinds of *aesthetic* qualities it may have. We need to respect, and cultivate, deep knowledge in a design team of the materials we are working with. Sometimes this reveals possibilities and experiences beyond what is imagined.

For example, Sundström and colleagues propose helping engineers to expose their 'material' (be it sensors, properties of wireless connectivity or algorithms) in the design process. By exploring a method they name 'inspirational bits' (Sundström et al., 2011), they are changing the engineer's role in a design team focused on designing for bodily experiences. When the whole team can *feel* the dynamic gestalt of, for example, a sensor, they can collectively come up with interesting developments. They may, for example, learn about how a wireless technology is not distributing the signal evenly in a room and how this can be used to create secret whispers in unexpected places in a museum. Schiphorst makes a similar turn to exploring the 'materials' in her work. She writes about how she exposed and interacted with the sensor and physical materials for quite some time in the design process before turning to design (Schiphorst, 2007).

In this way, the digital materials we choose will meet our physical, corporeal selves. Our abilities to move in certain ways will spur different aesthetic experiences. To describe the linkage between movement and aesthetic experience, we may draw on the theories mentioned above (from Darwin onwards), but we also need a language to describe and articulate those experiences. In my research group, we have sometimes used Laban analysis to understand body movements from an experiential perspective. Rudolf Laban was a choreographer who invented a way of describing body movements in similar ways to how a musical score describes music (Laban and

Lawrence, 1974). In a Laban analysis we get more than a description of the shape of a movement. We also get a description of the *effort* (the inner experience of what is required) involved in performing the movement. In one of our first design projects, we wanted users to express themselves by gestures when sending mobile text messages (Sundström et al., 2007). We altered the stylus pen that came with some mobile phones, adding sensors to it that could pick up on angry, happy, calm or sad gestures. In order to figure out which gestures could be relevant, we tried to stay focused on users' experience of making certain movements. We used Laban notation to write down the effort (inner experience) of the movements we wanted to portray (Figure 12.3).

Obviously, it is not only the aesthetics of the movements or bodily interactions we perform with the artefact that matters. As mentioned several times already, we also have to consider how the system responds to users' activities over time. The system may respond to, for example, breathing with interesting animations on huge screens covering whole walls, with interactive music, changing to reflect our movements, or any number of other materials. The aesthetic qualities of those materials also need to be investigated and understood.

MY AESTHETIC EXPERIENCE V. YOURS?

We may note that several of these attempts to articulate and share insights on design repeatedly return to the importance of experiencing

for *yourself* what you are designing for, but first-person methodologies do not free us from the responsibility of considering not only our own experiences but also those of our prospective users. Several interpretative methods for understanding how others interact with and understand our technology have therefore been evolving. The sensual evaluation instrument (Isbister et al., 2007) allows users to express their experience of an application in physical form. It tries to capture users' experiences beyond what they can say in words, instead making users pick up clay figures with evocative shapes expressing aspects of what they feel.

Our experiences point to the importance of moving from low-fi prototyping to high-fi – no matter which artefact is used to keep the design team on track. We have also learnt that we need to cultivate deep knowledge of the high-fi material (be it sensors, wireless communication, mobile screens, animations or sensornode programming) among all the participants in the project. Bodily and emotional interaction is extremely sensitive to the physical properties of the interaction – and everybody in the project needs to be highly aware of the possibilities and limitations of the material. Repeatedly doing build 'tech-fests' with all participants in the project where everybody gets to test the latest versions of the prototype as well as 'feel' the digital material has been a useful process to us.

A second insight is that we might not be able to verbalize or otherwise externalize the 'feel' of the interaction we are aiming for – be it suppleness, embodied self-reflection or something else. Instead, we have to repeatedly

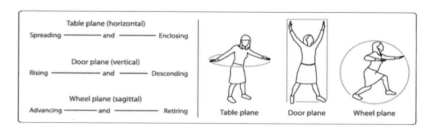

Figure 12.3   Laban notation.

expose ourselves (and other users – but most of all ourselves) to unfinished prototypes and try to seriously interact with it as if it was going to be a part of our life. Sometimes this requires that the design team use a system for a week or more, daily, together. Sometimes you have to wear bio-sensors in your everyday life and imagine the feedback on the mobile even when you know that what you see is a fake animation.

Finally, a deeper understanding of emotional and bodily interaction processes based on non-interactive as well as interactive systems is badly needed. There are very few methods for design and evaluation that we have found relevant to our work. We have invented a couple of methods to help us on the way, such as the sensual evaluation instrument mentioned above. While Laban notation has been useful to us as a means of describing movement and experience, it is a complicated way of doing so and we need a method/tool/artefact that speaks more of *interaction* than of individual movement.

THEORETICAL UNDERPINNINGS

Despite all the work we have seen on designing for embodiment (e.g. Dourish, 2001, and others), the actual corporeal, pulsating, live, felt body has been notably absent from HCI theory. Let me therefore point to some theories that have been helpful in the work we have done in the past: first, non-dualism, in the area of separating body from mind, individual from social, emotion as a separate entity; second, phenomenology and what the role of the body really is; third, ideas on how we need to approach the actual, physical body; and, finally, looking at norms for how we may want to aesthetically experience our bodies.

Non-dualism

First, while an emotion process is not enough to create an aesthetic experience, emotions will be part of the experience and *inseparable*

from the intellectual and bodily experiences. In such an holistic perspective, it does not make sense to talk about emotion processes as something separate from our embodied experience of being in the world. Gaver makes the same argument when discussing design for emotion (Gaver, 2009). Rather than isolating emotion as if it is something separate, we need to consider a broader view of interaction design, allowing for individual appropriation. Gaver pinpoints this excellently when he writes:

> Clearly, emotion is a crucial facet of experience. But saying that it is a 'facet of experience' suggests both that it is only one part of a more complex whole (the experience) and that it pertains to something beyond itself (an experience of something). It is that something – a chair, the home, the challenges of growing older – which is an appropriate object for design, and emotion is only one of many concerns that must be considered in addressing it. From this point of view, designing for emotion is like designing for blue: it makes a modifier a noun. Imagine being told to design something blue. Blue what? Whale? Sky? Suede shoes? The request seems nonsensical. (Gaver, 2009: 3598)

The Body is the Condition Through Which We Act in the World

Second, when it comes to understanding emotion and the body as basic processes for how we exist in the world, we can be inspired by phenomenology. When Merleau-Ponty writes about the body he begins by stating that the body is not an object (1962). It is instead the condition and context through which I am in the world. Our bodily experiences are integral to how we come to interpret and thus make sense of the world. This premise draws heavily on the notion of embodiment. Playing a central role in phenomenology, embodiment offers a way of explaining how we create meaning from our interactions with the everyday world we inhabit. Our experience of the world depends on our human bodies, not only in a strict physical, biological way, through our experiential body, but also through our cultural bodies. Merleau-Ponty attempts to get away from the perspective of the doctrine that treats:

> perception as a simple result of the action of external things on our body as well as against those which insist on the autonomy of consciousness. These philosophies commonly forget – in favour of a pure exteriority or of a pure interiority – the insertion of the mind in corporeality, the ambiguous relation without body, and correlatively, with perceived things. (Merleau-Ponty, 1962: 3–4)

Our Actual, Corporeal Bodies

We also need to get closer to the actual corporeal body and talk about it in concrete terms. Feminists have attempted to deal with the actual physical body in this way, highlighting in particular the differences between male and female bodies. Grosz (1994), for example, makes an interesting journey through the various philosophies of the last century (e.g. Freud's psychoanalysis, phenomenology), showing that most of them speak, in a sense, vaguely about the actual corporeal body. As a feminist, she sees very little of the female body, but instead, if anything, a normal, male body, in the theories on, for example, perception. Grosz makes the case that female bodies are different from male bodies – both corporeally and through their 'cultural completion':

> as an essential internal condition of human bodies, a consequence of perhaps their organic openness to cultural completion, bodies must take the social order as their productive nucleus. Part of their own 'nature' is an organic or ontological 'incompleteness' or lack of finality, an amenability to social completion, social ordering and organisation. (Grosz, 1994: xi)

This perspective resonates well with Merleau-Ponty's experiential and cultural bodies mentioned above, even if he, according to Grosz, never really dealt with the fact that some bodies are different from the male body, both corporeally and in terms of their cultural completion.

Relevant to our investigation here, is Grosz's emphasis on bodily completion by culture or practice. This is where our designs of digital tools come into play. Through new tools, we are in fact interfering with users' practices, with the social ordering and organization. Our bodies are shaped by the tools we surround ourselves with, not only in a metaphorical or 'cultural body' sense but also in a concrete corporeal sense. The tools we have make us experience the world in certain ways, make us use our muscles in certain ways and stimulate our nervous system in certain ways. Just like dancers, riders or runners will shape their bodies into certain forms, making them sensitive to balance, position and rhythm, computer gamers or office workers will shape their bodies into fitting with gaming or desktop activities.

Aesthetic Norms

When designing for a nondualistic stance towards body and mind, we need some way to talk about what experiences we strive to engage ourselves and our users in. While most accounts of corporeal involvement will be mainly descriptive, Shusterman's somaesthetic theory is also *normative* (Shusterman, 2008). He tells us that by engaging in certain practices, in inward listening and learning, we can know ourselves better, and thereby understand and interact with others more fully. It trains our empathy – both with ourselves and others.

Moving your body is not only a matter of performing a function, it is also an aesthetic experience. There is a plenitude of activities that we do for the pleasure of moving – dancing, sports, jogging, cycling. The pleasures of these activities are of course not only soft, flowing movements, since some of the activities involve pain, applying yourself really carefully to make your body do them, adjusting your own body in various ways, even making your body build certain muscles that you normally do not use so much, embarrassment when you do not get it right and so on.

Interaction design has perhaps been a bit too obsessed with zero learning time, an issue that will not sit well with some of the movement-based practices Shusterman is advocating. A take-away message from Shusterman is that it takes time to learn. You have to apply yourself. Getting to know yourself, your own body, changing your movements, training yourself, is not 'natural' – even if your body is 'there for you' all the time. Similarly to how you must learn to think and reason, you must learn how to listen to your body, how to improve your body knowledge.

CONCLUSION

By now is should be clear to the reader that we have opened up a huge area for investigation. Designing for pleasure or displeasure coming from our bodily interactions with digital technologies are in fact everywhere. Even an ordinary desktop interface or a mobile interaction is at the same time touching our bodies, our emotions, our experiences and our ways of being in the world. With the advent of new materials that allow sensors, wireless connectivity, actuators and all the data these will generate, we need to start to consider our bodily experiences more deeply when designing digital interactions.

In my view, we stand ill equipped for this shift. In practical terms, we are missing design methods, evaluation methods, design materials, principles and example designs. Yet, perhaps more importantly, we have still to question why we design for some experiences and not others. While we will no doubt better understand the practical issues here, it is the ethical concerns that will remain the most challenging of all.

NOTES

1. I use the term *artefact* to describe the whole system, including digital interactions, physical tools and technical implementation.

2. The *qualia* concept is used to refer to subjective experiences. Examples of qualia are the pain of a headache, the taste of wine, the experience of taking a recreational drug or the perceived redness of an evening sky (Wikipedia definition). Daniel Dennett (1988) writes that qualia is 'an unfamiliar term for something that could not be more familiar to each of us: the ways things seem to us'.

3. Oskar Juhlin, personal communication.

REFERENCES

Boehner, K., Chen, M. and Liu, Z. (2003) The Vibe Reflector: An Emergent Impression of Collective Experience. *CHI '03: Workshop on Providing Elegant Peripheral Awareness*, ACM.

Csikszentmihalyi, Mihaly (1990) *Flow: The Psychology of Optimal Experience*. New York: Harper & Row.

Damasio A.R. (1994) *Descartes' Error: Emotion, Reason and the Human Brain*. New York: Grosset/ Putnam.

Dennett, D.C. (1988) Quining qualia. In A. Marcel and E. Bisiach (eds) *Consciousness in Modern Science*, Oxford: Oxford University Press, 42–77.

Darwin, C. [1872] (1998). *The Expression of the Emotions in Man and Animals*. New York: Oxford University Press.

Davidson, R.J., Pizzagalli, D., Nitschke, J.B. and Kalin, N.H. (2003) Parsing the Subcomponents of Emotion and Disorders of Emotion: Perspectives from Affective Neuroscience. In: R.J. Davidson, K.R. Scherer and H.H. Goldsmith (eds), *Handbook of Affective Sciences*, New York: Oxford University Press, 8–24.

Dourish, P. (2001) *Where the Action Is: The Foundations of Embodied Interaction*. Cambridge, MA: MIT Press.

Ferreira, P. and Höök, K. (2011). Bodily Orientations around Mobiles: Lessons learnt in Vanuatu. In proceedings of CHI 2011: 29th ACM Conference on Human Factors in Computing Systems, Vancouver, Canada, May 2011, ACM Press.

Ferreira, P, and Höök, K. (2012) Appreciating plei-plei around mobiles: playfulness in Rah Island, In Proceedings of the 2012 ACM annual conference on Human Factors in Computing Systems, New York, NY, pp. 2015–2024.

Ferreira, Pedro, and Höök, K. (2011) "Bodily orientations around mobiles: lessons learnt in Vanuatu. *Proceedings of the 2011 Annual Conference on Human Factors in Computing Systems*, 277–286.

Gaver, William (2009) Designing for Emotion (Among Other Things). *Philosophical Transactions of the Royal Society*, 364(1535): 3597–3604.

Grosz, E. (1994) *Volatile Bodies: Toward a Corporeal Feminism.* Bloomington, IN: Indiana University Press.

Hagen, U. (2011) Designing for Player Experience: How Professional Game Developers Communicate Design Visions. *Journal of Gaming & Virtual World*, 3(3): 259–275.

Höök, K. (2010) Transferring Qualities from Horseback Riding to Design. *Proceedings of the Sixth Nordic Conference on Human–Computer Interaction*, 226–235.

Höök, K. and Löwgren, J. (2012) Strong Concepts: Intermediate-level Knowledge in Interaction Design Research. *Transactions on Computer–Human Interaction* (ToCHI).

Ingold, T. (2006) Walking the Plank: Meditations on a Process of Skill. In: J.R. Dakers (ed.), *Defining Technological Literacy: Towards an Epistemological Framework.* New York: Palgrave Macmillan, pp. 65–80.

Isbister, K. and Höök, K. (2009) On Being Supple: In Search of Rigor Without Rigidity in Meeting New Design and Evaluation Challenges for HCI practitioners. *Proceedings of the 27th International Conference on Human Factors in Computing Systems*, ACM, 2233–2242.

Isbister, K., Höök, K., Laaksolahti, J. and Sharp, M. (2007) The Sensual Evaluation Instrument: Developing a Trans-Cultural Self-Report Measure of Affect. *International Journal on Human-Computer Studies* (Special Issue: Evaluating Affective Interfaces), 65(4): 315–328.

Katz, J. (1999) *How Emotions Work.* Chicago, IL: University of Chicago Press.

Laaksolahti J., Tholander J., Lundén M., Solsona Belenguer, J., Karlsson A. and Jaensson, T. (2011) The Lega: A Device for Leaving and Finding Tactile Traces. *Proceedings of Conference on Tangible and Embedded Interactions*, (TEI '11), ACM.

Laban, R. and Lawrence, F.C. (1974) *Effort, Economy of Human Effort*, (2nd edn). London: Macdonald & Evans Ltd.

Laurier, E. and Brown, B. (2004) Cultures of Seeing: Pedagogies of the Riverbank. Manuscript, Institute of Geography, Edinburgh University. Available at: http://ericlaurier.co.uk/resources/Writings/Laurier_cultures_of_seeing3.doc.pdf

Ljungblad, S. and Holmquist, L-E. (2007) Transfer Scenarios: Grounding Innovation with Marginal Practices. *Proceedings of the SIGCHI Conference on Human Factors in Computing Systems*, ACM, 737–746.

Longo, Giuseppe O. (2003) Body and Technology: Continuity or Discontinuity? In: Leopoldina Fortunati,

James E. Katz and Raimonda Riccini (eds), *Mediating the Human Body: Technology, Communication, and Fashion.* Mahwah, NJ: Lawrence Erlbaum.

Löwgren J. (2001) From HCI to Interaction Design. In: Qiyang Chen (ed.), *Human–Computer Interaction: Issues and Challenges.* Hershey, PA: Idea Group Inc.

Löwgren, J. and Stolterman, E. (2004) *Thoughtful Interaction Design.* Cambridge, MA: MIT Press.

Lutz, C.A. (1986) Emotion, Thought, and Estrangement: Emotion as a Cultural Category. *Cultural Anthropology*, 1(3): 287–309.

Lutz, C.A. (1988) *Unnatural Emotions: Everyday Sentiments on a Micronesian Atoll and Their Challenge to Western Theory.* Chicago, IL: University of Chicago Press.

Merleau-Ponty, M. (1962) *Phenomenology of Perception.* (Trans C. Smith). London: Routledge & Kegan Paul.

Moen, J. (2007). From Hand-held to Body-worn: Embodied Experiences of the Design and Use of a Wearable Movement-based Interaction Concept. *Proceedings of the 1st International Conference on Tangible and Embedded Interaction*, ACM, 251–258). ACM.

Parkinson, B. (1996): Emotions are social. In *British Journal of Psychology*, 87 p. 663–683

Parkinson, B., Fischer, A. and Manstead, A.S.R. (2005) *Emotion in Social Relations: Cultural, Group, and Interpersonal Processes.* Philadelphia, PA: Psychology Press.

Salen, K. and Zimmerman, E. (2004) *Rules of Play: Game Design Fundamentals.* Cambridge, MA: MIT Press.

Sanches, P., Höök, K., Kosmack Vaara, E., Weymann, C., Bylund, M., Ferreira, P., Peira, N. and Sjölinder, M. (2010) Mind the Body!: Designing a Mobile Stress Management Application Encouraging Personal Reflection. *Proceedings of Designing Interactive Systems, (DIS '10)*, ACM.

Schiphorst, Thecla (2007) Really, Really Small: The Palpability of the Invisible. *Proceedings of the Conference on Creativity and Cognition*, ACM, 7–16.

Sengers, P., Boehner, K., Warner, S. and T. Jenkins (2005) Evaluating Affector: Co-Interpreting What 'Works'. *Workshop on Innovative Approaches to Evaluating Affective Systems* (CHI '05), ACM.

Sheets-Johnstone, M. (2009). *The Corporeal Turn: An Interdisciplinary Reader.* Exeter: Academic.

Shusterman, R. (2008) *Body Consciousness: A Philosophy of Mindfulness and Somaesthetics.* Cambridge: Cambridge University Press.

Ståhl, A., Höök, K., Svensson, M., Taylor, A.S. and Combetto, M. (2009) Experiencing the Affective Diary. *Personal Ubiquitous Computing*, 13(5): 365–378.

Sundström, P., Ståhl, A. and Höök, K. (2007) In Situ Informants Exploring an Emotional Mobile Messaging System in Their Everyday Practice. *International Journal on Human-Computer Studies* (Special Issue: Evaluating Affective Interfaces), 65(4): 388–403.

Sundström, P., Taylor, A., Grufberg, K., Wirström, N., Solsona Belenguer, J., and Lundén, M. (2011). Inspirational Bits: Towards a Shared Understanding of the Digital Material. *Proceedings of the 2011 Annual Conference on Human Factors in Computing Systems*, ACM, 1561–1570.

Troshynski, E., Lee, C. and Dourish, P. (2008) Accountabilities of Presence: Reframing Location-Based Systems. *Proceedings of the SIGCHI Conference on Computer–Human Interaction (CHI '08)*, ACM, 487–496.

Wilde, D. (2011) Swing That Thing: Moving to Move. The Poetics of Embodied Engagement. PhD dissertation, Monash University, Melbourne Australia, with support and supervision from CSIRO, Australia.

Zangouie, F., Gashti, M.A.B., Höök K, Tijs, T., Gert-Jan de Vries, G-J. and Westerink, J. (2010) How to Stay in the Emotional Rollercoaster: Lessons Learnt from Designing EmRoll. *Proceedings of NordiCHI '10*, ACM.

13

Ethnographic Approaches to Digital Research

Barry Brown

INTRODUCTION

Ethnography presents a certain analytic attitude to its work: it focuses on getting as close to people as possible, understanding how they see the world, working through their problems and with their ideas. This attitude has become a pervasive one in human–computer interaction (HCI) where the ethos of understanding users has moved, as the other chapters in this book demonstrate, beyond seeing users as measurable units of information processing. A recent focus of much work in HCI has thus been on understanding the details of users' experiences, understanding their motivations and their work and leisure practices. Social, cultural and bodily experiences have become much more central, at least in research, to studying users.

Yet, despite this central change in how the user is perceived and alongside it, the very job of user studies in HCI, the actual number of ethnographies conducted in HCI is relatively low. It could be said of ethnography that it lost the battle but won the war – conceptually

ethnography has been very influential, alongside the general move to more interpretivist positions, yet its time-consuming and involving nature has limited its widespread adoption.

In this chapter we give an outline of what ethnography is, and a little of its history in HCI. While we can only give the most schematic of accounts, the goal here is to inspire others as to the value of ethnography as an approach, as well as giving some idea of its history and foundations. We will skip a lengthy description of its roots in anthropology and sociology, focusing instead on questions of how ethnographic work has influenced HCI, as well as how it has contributed to new technology and design.

A SAMPLE ETHNOGRAPHY IN HCI

Perhaps the strongest contribution of ethnography to HCI is in how it can impart a strong sense of a particular situation – the struggles of those who live, work and play within that situation, and how technology is threaded into that setting. Ethnography is perhaps at its best documenting where things go wrong or right, how technology fits (or often fails to fit). Going beyond just technology, though, ethnography can also be used to document *how things are actually done*. If one is interested in questions of how such things might be done better – how technology might create positive change – this is crucial. Rather than offering correlations or measurements, in ethnography we get rich descriptions that help us understand what is going on. That is to say, it helps us to change how we think of a particular setting or a particular way in which technology comes to be used. 'Pay attention to this!' good ethnography shouts – 'you're not thinking about this!' When what developers are building is far away from what is actually needed, ethnography can gain its most valuable purchase by forcing attention on what is getting ignored.

Rather than jumping into the theory or the history of ethnography, to demonstrate this, let us start with a short description of a recent ethnography that gives something of the flavour of what ethnography does. With many thousands of ethnographic studies published in HCI over the years, it is perhaps tempting to pick a classic old paper – of which there are many (e.g. Star 1988; Harper et al. 1991, 2000; Nardi and Miller 1991; Bently et al. 1992; Bowers et al. 1995; Button and Harper 1996; Harper 1998; Bowker and Star 1999; O'Brien et al. 1999; Heath and Luff 2000; Taylor and Harper 2002; Brown and Chalmers 2003). Instead, let us pick a more recent ethnography – but one that still makes a valuable and substantive contribution and provides some insight into technology use. Published in 2011 at the main HCI conference, 'Bodily orientations around mobiles: lessons learnt in Vanuatu' by Pedro Ferreira and Kristina Höök (Ferreira and Höök 2011), this study documents the late adoption of mobile phones by the inhabitants of a set of Melanesian Islands – Vanuatu. Simultaneous with the arrival of the researchers to the island to conduct their study, the local cell phone company 'TVL' turned on a mobile phone tower providing mobile phone access to the islanders for the first time. The paper argues that, because of this 'newness', features of mobile phone use that are usually taken for granted were visible to the islanders, and to the researchers, as they had not yet become routine.

The paper is based on observations they made in the islands – participating in the life of the islanders for around three months, living with them and understanding a little of what it is to live on the islands. Since the islands are so small, islanders spend much of their day travelling between different islands on small canoes. Moreover, since much of life there involves subsistence farming they spend a lot of their time in the water, fishing or swimming. This heavy involvement in the water obviously conflicts with their new non-water-resistant mobile phones. One of the first stories the paper relates concerns how some joking between brothers almost results in a broken phone:

> One day, crossing from Motalava to Rah, Brian crossed paths with Brian's younger brother Suva who was heading in the opposite direction. Brian made some gestures to Suva, in a seemingly mocking way, as they would be often 'pulling each other's leg' in a friendly way. Suva stood up on the canoe and screamed with a challenging tone, then throwing one of his sandals in the water, close to the canoe where Brian was, probably in the hope of hitting him with some of the splash from his Croc in the water. The shoe hit a couple of meters from Brian's canoe, he immediately used the paddle to maneuver the canoe around and intercept the fallen shoe. As he was approaching the shoe, floating in the water, he stopped paddling, taking the paddle inside the canoe, just letting it 'glide' with its existing momentum. He stretched his body in the direction of the floating shoe, leaning over the side of the canoe. He held on the canoe with one hand while his other arm reached out for the shoe. At the time, his mobile phone was hanging around his neck, and was still inside the canoe, but as he stretched out, the phone was pulled higher and higher against the inner wall of the dugout and was now facing a possible dip in the water, which Brian did not seem to be realizing at the moment. Suva, who was half standing on his canoe, looking at the whole situation, yelled out something to Brian, which made him sit back in a

very sudden move, causing the boat to rock slightly. They both screamed to each other with a smile, as if saying 'that was a close one' ... As Jorege put it at one time: *'sometimes we just lean over to look at something, or to get some water out of the canoe [...] we forget that we have the phone on our chest [laughter] and then the phone goes in the water [followed by generalized laughter]'*. (Ferreira and Höök 2011: 279)

This extended extract highlights some features of ethnographic work. First, it is quite a lengthy description – ethnographic papers frequently fight with page and word limits as they are text and quote-heavy. From the extract you get a sense of what the people being studied are like (in a very general sense) – certainly they seem playful. Being (playing and working) on water seems like an everyday part of their lives, but with the new mobile phones the authors identify some 'ordinary problems' that they face, such as having to be careful about dipping the device in the water. Carrying the device around your neck seems a usual way of carrying the phone and, indeed, the rest of the paper at some length engages with how they carried their phones and the challenges of (for example) working with a machete to keep a farm plot clean, while not dropping the phone on to rocky ground.

What this paper does is focus its attention on an unusual aspect (at least to non-islanders) of the community they studied. They bring out something different from our own established understandings of mobile phone use. We spend little time thinking about our bodily orientation and mobile phones, where we can keep our mobile phones, if they might get damaged and so on. Certainly, the design of phones is much more concerned with functionality and aesthetics. Yet we still have bodies and use our phones, so more broadly the paper starts to cast light on the role of bodily orientations in mobile phone use.

In this way the ethnography demonstrates some of the key contributions that technological ethnography can make – it tells us something surprising about another place. The authors emphasize the newness of mobile

phones to those they studied – but it is actually the newness of these practices to us that is of more interest. Often with research the result is known before the investigator starts the experiment. In some ways it would be bad science often to be genuinely surprised – but what is documented here *is* a genuine surprise. Light is cast on something different, but also something the same. The paper highlights our own practices of using mobile phones, shedding light on the very practical questions of how and where we carry our mobile phone, how we keep it out of danger, how *we* orientate it to our body. The results 'come home' and help us think a different way here about Western practice, as much as about those being studied.

Reading this paper one could take an immediate finding: wouldn't these people be better off with waterproof phones? Often with ethnographic studies it is possible to draw some sort of safe, immediate design implication that can often be valuable in its own right. One often finds 'implications for design' in ethnographic papers that provide this sort of immediate 'news'. Here we could use these results to explore how to build rugged phones for the kind of environments described here. Yet it is also to undersell ethnography to take this as the payoff. Ethnography is fairly basic research – if you are doing research into the properties of materials you wouldn't expect the payoff to immediately contribute to a new type of computer. We are usually patient with basic materials research and so it is with ethnography, for one of its most powerful contributions comes in how it makes us think about technology, and the use of technology, in a new way (Button and Harper 1996; Dourish 2006).

For the Vanuatu ethnography this is in how it highlights the importance of bodily relationships with our devices, the ways in which we manage our devices and ways of carrying and holding those devices. While ergonomics has always been concerned with these issues it has tended to see them as a set of issues concerning the geometry and physiology of the body. Yet this paper highlights that while those in Vanuatu might have similar bodies to us, it is actually their *environment* and *practices* that are radically different. They live in a different place and they do different things. This ethnography nicely demonstrates some of the values of ethnographic work. While it takes as its subject strange people in a strange place, how it does what it does can be translated to places closer to home, and settings closer to home.

ETHNOGRAPHY AND PERSPECTIVES

This ethnography builds on the history of ethnography in HCI – an at times troubled enterprise. In particular, throughout its development, ethnography has struggled with serious questions about its validity, its applicability – how much it can be seen as a 'scientific' method and arguments over variations of ethnography and their validity. These arguments have in turn been echoed in disputes over the acceptability of ethnographic work in HCI.

Positivism, in its most raw variants, suggests that we should aim for the same repeatable, disengaged approach of the natural sciences. While it is rare that researchers in HCI are as explicit in stating this, often these beliefs ground critiques or positions held during disputes (although see Whittaker et al. (2000) for one presentation). Following this reasoning, researchers should be disengaged observers, they should experiment with individual identifiable phenomena, and those phenomena should be replicable by others. This leads to a focus on experimental methods and measurable results. This fits well with a focus on the *usability* of computer systems – whether someone can successfully complete a task or not, a simple definition of useable.

Max Weber was one of the first theorists to point out one of the key limitations of having social sciences follow the approach of the natural sciences. When it is people who are being studied, unlike physical phenomena, alongside any interpretation the analyst might have, the people being studied also have their own interpretations and viewpoints. Indeed, for much of what social scientists do it is those very interpretations that are their main subject matter. When it is protons or electrons being studied they do not, on the whole, hold their own interpretations, which are consequential in their actions. The order of phenomena being studied by those interested in humans, then, is radically different from the natural sciences, in that the subjective and not just the objective, takes an important role. Indeed, even the viewpoint of the analysts themselves does not escape this and often takes its place as just another interpretation of a setting. Rather than positivism, Weber argued for *Verstehen* – the understanding of the subjective meanings and purpose that individuals attach to their actions (Sharrock et al. 1995).

Weber's insight is such that it creates a clear break between investigation of the *objective* and investigation of the *subjective*. Moreover, since for human action it is the subjective that is important, if we are interested in human behaviour – of any sort – we must be interested in the subjective. The subjective can be used to suggest something is biased, hearsay – less reliable in some way. What Weber does instead is to raise the subjective to be the essential element in understanding human behaviour – all the way down to our concerns about understanding the use of computers and computer interfaces. What Weber did not do, however, was proscribe specific empirical methods for understanding what those subjective understandings might be.

This is where ethnography can find its introduction. Ethnography traces its routes to anthropology – with ethnography as something of the 'original method' for understanding different societies and cultures (Maanen 1988; Goffman 1989; Button 1992; Cooper et al. 1993; Fine 1993; Hammersley and Atkinson 1995). At its heart ethnography is based on attempts to understand the social world through the eyes of its participants, to understand the perspectives actors take, what they do, why they act the way they do, and how they talk about and communicate with others about those actions. At its core it is a focus on people's understandings, perspectives, skills, experiences, viewpoints and worldviews. Rather than seek to understand in objective terms why something might be what it is, instead there is a focus on why a particular person, or group, or culture, might see something as what it is. Importantly, the role of ethnography is not (or is not normally) about judgement or discovering what is 'really true'. Instead, it is focused on how those being studied are making sense of the situations they find themselves in.

Again, how does one document this? To this end ethnography has attracted a range of different methods, approaches, theories and disputes. At its heart is participant observation, so without some sort of participant observation it is difficult to say that something is ethnography at all. With participant observation a researcher *participates* in a particular community, living more or less competently within that community. During their participation they observe what is going on, attempting to understand the community and setting as much as possible. In a classic ethnographic study an ethnographer goes off to a different community and spends over a year there, observing, participating and hopefully becoming a member of that community. Obviously ethnographies of this style make major demands on the researcher, so this sort of deep engagement is only one form ethnography takes. With communities that are closely related to those of the researcher (say, other professional workers) it might seem excessive and certainly not necessary to spend a set amount of time with them.

Moreover, there is a range of methods that go beyond participant observation that ethnographers use. Ethnographers make extensive use of interviews – perhaps recording some video. Increasingly they make use of online resources (to the extent of conducting virtual ethnographies; Hine 2000). Yet whatever these specific methods, there is always the search for the same phenomena: an understanding of the practices that make a place *that* place, a community *that* community, an individual *that* individual.

ETHNOGRAPHY, CSCW AND HCI

While HCI is not sociology or anthropology, it is a discipline that has long borrowed from far and wide. At the beginning of its development this took the form of the heavy influence of psychology and the emergent field of cognitive science in turn. Indeed, in many ways, until after 2000, cognitive science was dominant within HCI and to a large extent suppressed divergent approaches.

However, it was not only HCI that had an interest in the human issues concerning technology – within computer supported collaborative work (CSCW), a quite distinct research tradition and focus developed, grouped around understandings of the complexities of work practice. While CSCW has never approached HCI in terms of size, perhaps its most valuable contribution has been its predominately social approach to understanding technology. Foundational books for CSCW such as *Plans and Situated Action* (Suchman 1987), and *Understanding Computers and Cognition* (Winograd and Flores 1986), helped CSCW to develop as a distinct field that distanced itself from a focus on cognition to concentrate instead on understanding the situated nature of technology use. CSCW grew as a field that focused both on building tools to support collaboration in the workplace on understanding the social nature of the workplace and technology's role in that. Ethnographic work, such as Julian Orr's (1996), was published in the first CSCW in 1986 (Orr 1986), and especially in the European variant started in 1989.

It is here that ethnography, and the ethnographic study of technology, gained one of its first homes. While fields like information science and science and technology studies were tolerant of ethnographies of technology, they did not make ethnography central like CSCW did. Moreover, within CSCW it was a particularly distinctive form of ethnography that gained a foothold. It is something of an historical accident, but the distinctive analytic perspectives of *ethnomethodology* and *conversation analysis* were most influential here. Ethnomethodology's similarity to 'ethnography' has caused some confusion for students of CSCW. However they are clearly distinctive – ethnomethodology (EM) is a sociological take on studying social order. Both conversation analysis and ethnomethodology share a common interest in 'members methods' – how it is that ordinary people do ordinary things. Ethnomethodologists developed a tradition of looking at and studying the ways in which actions are done and how those actions are seen. This meant that the practical problems of using computer systems – making sure something is working, talking to someone at the same time, balancing a phone call and the requirement to do something on the computer – are the very sorts of issues that ethnomethodologists were used to looking at (Button 1993).

This led to a flurry of work based at Lancaster University in England, developing what came to be known as the 'workplace studies' tradition (Luff et al. 2001). As a major part of this, conversation analysis work, while springing from a similar analytic tradition, took a distinctive focus on interaction, and in particular the use of video became more central (most notably in the work of Christian Heath and Paul Luff – Greatbatch et al. 1995; Heath and Luff 2000).

It was perhaps because these approaches were much more comfortable focusing on the details of interaction that they fitted better with the necessity for design implications. Indeed, one of the longstanding issues that ethnographers in CSCW dealt with was issues of relevance – how they could make the case for the relevance of their findings for how technology was used and supported. Indeed, finding connections with computer science was often a challenge – and computer scientists at times were often directly hostile to

the non-positivist approach emphasized by ethnography. Alongside ethnomethodologists, ethnographers from other perspectives (in particular distributed cognition and activity theory – Blackler 1995; Halverson and Rogers 1995; Rogers 1995; Nardi and Engestrom 1999) made important contributions, and slowly a small but well established community of researchers developed.

One of my favourite examples of a paper in this tradition is 'Workflow from within and without' (Bowers et al. 1995) – an ethnography of a computer system used at a large printing company to track the flow of work to the different printers. Due to an agreement to operate printing for the UK government, the company was contractually required to install centralized control software that logged what was getting printed on different presses. Yet this software caused no end of problems. Being a very high-volume print shop, the staff had worked out lots of different ways of managing the presses so as to use the machines most effectively. So, for example, printers would often 'jump the gun' and print out parts of a job before the job had actually been agreed upon by the client. This meant that they would be able to complete a job in time – particularly if cancelling the job was rare. Alternatively, the printers would make use of 'down time' to print parts of a routine regular job that did not change from month to month, again making effective use of time.

Yet the computer system that was installed – a 'workflow' system – enforced a simplistic model of printing, where jobs were agreed, paid for and then printed in one process. The ability to flexibly decide what to print when and where was lost – and this greatly slowed down the printers, to the point where additional staff needed to be hired. Eventually, staff found various ways of lying to the system – decoupling it from what was actually being printed, with the 'fictional correct' account being used by the system for billing, distinct from what was actually being done with respect to printing.

This type of finding is emblematic of much CSCW work – the documentation of the ways in which the complexities of work practice as actually carried out differed from how the designers of a support system conceived of that use. That computer systems when implemented often harmed or slowed down the work process would come as no surprise to those who worked with technology in the 1980s or 1990s. Indeed, it is perhaps a criticism of this work that it seldom took sufficient focus to understand how computing *did* assist and enhance work practice.

ETHNOGRAPHY MOVES FROM CSCW TO HCI

As remarked above, ethnography has not been limited to CSCW – Zuboff's study in the age of the smart machine from 1988 was one remarkable precursor of this research (Zuboff 1988), and there was longstanding use of ethnography in information systems research (see Kling 1980, 1994; Markus 1983; Kraemer and King 1988; Grint et al. 1995, amongst others), and ethnography had been a longstanding method more broadly in organizational studies. Julian Orr's *Talking About Machines*, published in 1996 but distributed widely before that, was also a very influential ethnographic study of technology – as were others (Greenbaum 1979; Suchman and Wynn 1984; Orr 1996). Yet there were few studies outside CSCW that brought together a concern for design with ethnography. At the turn of the century, however, this began to change. One influence was simply conceptual drift – CSCW research was published at the main HCI conference – and authors such as Richard Harper and Paul Dourish challenged the greater use of ethnography in computer science more broadly.

Many of these initiatives were based very much in a sociological understanding of ethnography, and a predominately ethnomethodological one at that. Yet there were strong rejections of this, too. A rather distant relative of ethnography – contextual enquiry, which involves a structured form of participant observation – gained considerable commercial popularity (Beyer and Holtzblatt 1997). Through the initiatives of researchers at Intel, a more anthropological flavour of technology research also developed (Mateas et al. 1996), one that sought to link ethnographic research more directly to broader cultural issues and concerns. In particular, the anthropological concern for culture took a much more central place. So, for example, in early work done in studies of the home, the traditions of the home take on a more central role, not simply as 'practices' but as cultural practices of home life (perhaps in part because of the ironies of corporate life, this flavour of ethnographic work came into HCI through Ubicomp, a related field).

In contemporary HCI these threads are present and there is a somewhat divergent tradition of ethnographic work. In the introduction we mentioned how ethnography had 'won the battle but lost the war'. By this we meant that even though the analytic perspective of ethnography has been accepted and is influential, the time-consuming nature of ethnographic work has perhaps hindered its further spread. There is only a handful of strictly ethnographic studies (in terms of long-term participant observation studies) published in HCI every year, even as the perspective of ethnography has had a fundamental influence. In technology trials of different sorts, and in exclusively interview-based studies, the topic of HCI now clearly encompasses the values and interpretations of users as its central phenomena. In some senses this is a victory for interpretivisim rather than strictly ethnography *per se*. Yet HCI goes beyond interpretivisim in its methodological eclecticism, something that owes much to ethnography. Moreover, the key questions of design – and the influence of ethnomethodology – remain in much of HCI with its close attention to questions of practice, of members' perspectives and so on.

DOING ETHNOGRAPHY

Having very briefly outlined some of the theoretical inspirations for ethnography, and a little of its history in HCI, let us engage a bit further with understanding how one might undergo an ethnographic study in HCI. This is of course just a schematic outline – much lengthier and more helpful accounts may be found elsewhere (Loftland 1971; Blomberg 1995; Crabtree 2003). With regard to conducting ethnography, one particular resource is the book *Analyzing Social Settings* (Loftland 1971). More specifically with regard to ethnography in HCI, we would recommend *Fieldwork for Design* (Randall et al. 2007). It emphasizes a set of key points about ethnographic work. The first is that ethnography is an analytic pursuit. It is not simply observation or documentation – 'hanging out' is not the point. Rather, it is about understanding the setting sufficiently so as to be a member of that community. At one level ethnography is being able to 'pass' – to go unnoticed amongst a community, successfully doing and speaking the way they do. Goffman, rather richly puts it as when you start to find members of that community sexually attractive (Goffman 1989). To do that is to move beyond observation, to being able to see how and why certain things are done – in the logic of the situation itself – and to take that on yourself.

It is important to realize that analysis here is not about second guessing what is going on or finding the 'hidden' truth of a setting. It is about understanding 'what is going on' at the same level as those in the setting – that is, to become a 'local expert', an expert in the same way as those you are studying are experts. At times this can mean that you might develop a different perspective from that of those being studied, but that perspective isn't necessarily superior to theirs.

PARTICIPATION/OBSERVATION

At the heart of ethnography, as mentioned above, is participant observation. As the name suggests, participant observation is based on participating in different activities and observing how they are done. Obviously, first-hand experience is a powerful method for studying different activities and lifeworlds. Participant observation can be roughly divided into two different parts – first observation and then participation. In practice one *observes* what is going on before jumping in and *participating* in the activity. Spectating on an activity is important because it both lets us see how something is seen by outsiders – and so what their reactions might be – and notice 'what anybody would see', seeing parts that would be invisible to a seasoned participant. The perspective of the observer is thus a vitally useful one and it is important to not overplay the value of *participation*. Moreover, in most settings one is seldom an expert and so listening to and observing what experts do can be more valuable than short-term participation in an activity.

Observation is also the way in which the strangeness of an activity can be brought out, by marking out its features and encouraging us to think, if only briefly, about why something is arranged in a certain way. When participating in an activity, or as we become incorporated into its rhythms, it can come to seem natural and unproblematic, as if it could not be any different. Those moments when we are an outsider are invaluable then for letting us see the unexpected and wonder about different ways that activities could be carried out. There might of course be times when we cannot recognize what is going on and even wonder what is the sense in the occasion – so we move a little closer and start to take part in the activity.

Moving on from observation, participation involves the researcher taking part in the activity to unpack how this is done. Participation is valuable in a number of ways. First, it puts us, if not at the centre of the action, then at least where what is happening is happening. Participation thus requires effort so as to find and get whatever resources or social connections are needed to be able to do that activity. If we want to take part in a football match a basic prerequisite is other people to play with, which forces us to become involved in the social life and interaction that makes the activity possible. That social life is crucial for understanding the organization of the activity, since it is in social life that the regularities of a particular activity take form. Social groups have particular expectations for the regular forms that particular activities will take, as well as a range of activities which help to form the social group as a coherent form.

Participation, in a personal sense, is also *high risk* in that it involves the research extensively in *failure* – when we attempt something our first attempts are usually unsuccessful. For some this is very uncomfortable – being placed in a situation where, so as to be able to elicit support and training from others, our failures are (to an extent) 'public'. This is valuable for how it can cast light on what 'the skills' involved in learning to do a particular activity are. As we participate we move from *gross novice* to something resembling competence. In achieving that competence we can record what it is that we have learnt – what we can do that we did not have the ability to do before. This can be difficult because one of the key aspects of skills is that, once we have mastered them, they pass from our conscious attention and become abilities – things that we 'just do'. Indeed, often during participation, it can be difficult to be clear about what exactly it is that we have learnt.

For example, say we are studying play of a particular video game. As we participate in the game we slowly become better at playing it, mastering the skills of achieving points, staying alive and so on. As we gain those skills they are briefly highlighted as we are learning them, but quickly they are pressed into the background as we go on to gain more complex skills (Sudnow 1983). For this reason it is valuable to shift from participation to observation again, to gain a chance to reflect upon what we have just done and learnt. Indeed, often the skills that one learns through participation are not particular skills of doing, but are skills of *seeing* – being able to mark out and understand what is being

done, seeing an activity as a particular case of something, and recognizing success and failure. One of the key skills in any activity is recognition as much as production – one needs to be able to see the key parts of activity as well as (in some cases) do them oneself. Take, for example, football players. To be able to play the game competently one needs to be able to read other players and see what they are doing so as to assist or compete with them. These skills come into special focus in football spectating since one is not taking part in the game *per se* but, rather, enjoying observation. To be a competent football spectator one needs to be able to read the game and to be able to see different events that are happening and what they mean for the game.

ANALYSIS

We now move on to consider the different analytic stages in participant observation. On the whole we agree with Randall et al. (2007) it is the analysis: that is particularly challenging in empirical work, much more so than fieldwork or fieldnotes. Of course, the fieldwork can seem most like *work*, particularly if you are the sort of person (as most of us are) who has some discomfort in launching into conversation with complete strangers. Moreover, in any project the fieldwork materially takes up most of the time, as well as being the most memorable part. It is during fieldwork that we approach something for the first time and its strangeness, but also its order, starts to be apparent. Analysis, however, is the *crucible* of participant observation. It is where observations and participation can be transformed into something that somebody else could understand and make sense of – losing some of their personal weight in the process, of course, but becoming potentially powerful for helping others understand what is involved in a family of different activities.

As with fieldwork, we can divide analysis loosely into two parts: a *retrospective* phase – where you attempt to document what you have done and seen – and a *prospective* phase where you try – work out what relevance that might have for anybody else to or for those attempting to understand the activity in itself. In the retrospective phase we attempt to document and note all the things we have experienced. Here the focus is on documentation, on thinking back to different parts of the experiences and in particular aspects that we did not reflect on at the time. Analysis often involves redescription of previous situations and observations, rewriting and bringing together seperate points – transforming your fieldnotes into richer more inter-threaded documents.

It can be confusing to work out what are the important parts that are worth writing down – this is particularly difficult if it is the 'first time through' at doing participant observation. Indeed, the first time we write down experiences from participant observation it is unlikely that we will be successful. Working out what is important is something that comes from the later stages of the undertaking. Essentially, what is important later impacts on what it can be useful to focus on when retrospectively reviewing one's activity. Yet there are some rough guidelines that can help the endeavour. Be careful to document all the different processes and parts of your study, including gaining access and travelling to and from the particular fieldsite. Document your own feelings and experiences and what you can grasp of others' perspectives. Be careful to write down what people actually do and where they do it. At the beginning of the fieldwork the challenge is that nothing really seems to make any sense so the details can be overwhelming. In contrast, when you have a sense of what is happening and how, it can be tempting to 'gloss' – to simply write down what an expert would see about a setting. Remember a goal of the analysis is to get at *what it is to be an expert*. So if you start to use any jargon, explain how you would know when and where to use that jargon. It is the 'taken for granted' that you

are trying to extract from your data, so the challenge is getting yourself to stop seeing what an expert would see.

It is perhaps the *prospective* analytic phase though that is the hardest of all. The prospective phase is about moving from descriptions of what happened to analytic findings that could be used to understand *future* activity. This is why we call it prospective – trying to find out what is done in a setting so that you can be quicker to understand what is going on there, how it is organized and so on. In a sense this is about finding generalities, but ones very much tied to the particular setting or activity that one is researching. In many ways these are the sorts of generalities that we commonly draw all the time – for example, 'the bakery runs out of bread after about 2 p.m.', 'the project is much smoother if the boss doesn't get involved', 'it's easier to apply for research funding if you have a track record'. Finding good prospective analysis comes from understanding what is going on in a setting to the point where one can make predictions, of a sort, about what will happen after what one has seen.

So if one is interested in something analytically, one should be able to detect what it is that will be regularly done. Moreover, if one can understand what motivates particular events or occasions one can predict what changes might take place – small changes, perhaps, but the sorts of local innovations that can enhance or detract from a particular activity. The power of prospective analysis comes from being able to learn from what goes on in one situation or activity and use it to understand other activities or settings. That is to say, to move from understanding something here, to understanding something 'there'. This can be a narrow comparison – such as drawing on lessons from looking at one sort of sports spectating to understand another – to more general comparisons – say between sport participation and video game playing. In each case there is no general or strong rule that says something in one forum should take the same form as something in another, but, rather, that contrasts and comparisons can help us learn broadly about how activities are arranged.

The challenge of prospective analysis is keeping it sufficiently open so that one does not fall into overgeneralizations or stops documenting any specific cases. Always make sure that you document actual cases and use them in trying to make any sort of prospective analysis. If something happened once, it's unlikely to be a general feature of any setting.

ONLINE ETHNOGRAPHY

One recent and interesting twist in ethnography and technology has been the growing interest in online ethnography – labelled with titles such as 'virtual ethnography' or 'netography' (Kozinets et al. 2010). This is a form of ethnography that takes as its subject communities where the predominant, or at times only, contact is online. While until only recently such communities were considered marginal, with the spread of online social networks and the like, not to mention the increasing thicket of forms of online communication, these communities have come to the fore. Indeed, until recently some were doubtful as to whether online communities were themselves possible. Online forums present a range of different social collectives of differing intensities and connections (Kraut and Resnick 2012). For example, we might think of the editors of Wikipedia, the online encyclopedia, as a community. While we will find 'legitimate peripheral participation', more central participation and the like, there is also some sort of common orientation to a set of rules, a common space and a common set of skills and abilities. It might be harder to argue that the non-authoring users of wikipedia are, similarly, a community. After all, there is little direct communication between them. Yet we might still find collaboration of a sort – the reading of different pages effecting their popularity, donating money, referencing Wikipedia in student essays or simply propagating ideas and arguments from the pages of Wikipedia.

The question of whether a virtual ethnography can be carried out would seem to depend, to an extent, on assessing the closeness of the community to determine if there is a 'what' to be studied. In terms of methods, the lack of physical contact can cause problems not so much in understanding individual informants or those being investigated, but, rather, in how one might pick up on the ambiance of a place – what the walls are covered with, who sits where, the temporal rhythms that can be obvious at a glance in situ, yet harder to see when one only has a slowly updating list of forum posts.

One of the more productive areas for virtual ethnography has been online games. Through offering a non-textual virtual environment, one where avatars move and engage, it seems that the possibility of an ethnography is much enhanced. Recent ethnographies of Everquest and Second Life, for example, have been notable (Castronova 2005). Perhaps, though, in terms of ethnographic study it has been World of Warcraft that has been the most productive source of study (Pace et al. 2010). One of the most successful online communities ever, Warcraft peaked at around 12 million subscriptions and has around 10 million currently. While these are segmented more or less by region, this seems more than enough (and with enough of a world in common) for virtual ethnographers to work in. Yet, there can be few communities that develop and change as quickly as those that are online, so one can expect methods and approaches here also to be somewhat in flux.

CONCLUSION AND A RECENT CONTROVERSY

In this chapter we have reviewed some aspects of the ethnography of technology, touching on a few notable examples, but also charting the longer-term history of ethnographic work. We can hardly give a full description of ethnography as a method, but hopefully this chapter gives a little of the flavour of the goals, theories and methods that make up the use of ethnography in HCI. Even for those who do not deploy ethnography themselves, there is much to be learnt in its broad approach: its emphasis on understanding others' worldviews, how technology is threaded into situations and what it is like in communities distant from those we are most familiar with.

In conclusion, however, it is worth noting that while controversy seldom makes the pages of textbooks of this sort, ethnography is still quite a controversial enterprise. In particular, arguments between ethnographers can cause as much ramcour as can questions about the validity of ethnography itself. At one recent CHI conference, the paper 'Ethnography considered harmful' (Crabtree et al. 2009: 890), cast a critical eye on the forms of ethnography deployed within HCI and, in particular, the growing interest in more anthropological approaches. Here we find a scathing critique of ethnography as 'cultural tourism for computer scientists':

> Ethnography thus conducted might be said to engage design in a form of tourism, providing a level of description that may be described as 'I went there and this is what I saw'. However, while attending to some of the constituent actions in a setting or settings, the danger is that this scenic form of cultural anthropology misses 'what is done in the doing' of situated action. People are not, for example, simply 'looking at screens' or engaged in 'a flurry of paper turning' in the American Megachurch. They are doing being at prayer, and doing so congregationally in the company of others. ... new approaches raise the spectre of ethnography taking a step back to general exhortation. They tell designers that they should suspend their cultural biases, adopt new values, engage in critical reflection, and so on. Instead of detailed empirical accounts of particular organized activities in particular settings relevant to systems design, new approaches offer the designer a return to a world of moral and political invective. Once again, a key question arises for designers to consider, namely, will that do the job? Can you build computer systems on that basis?

The reception for this paper can be described as heated. Taking a written example, one blog post on the paper concludes:

'I'd like to engage in debates, I think there's much to be learnt from discussion, but I'd like them to be civil and conducted around a set of principles that not only respect the products of our discourse but also the scholars that produce them'. At the conference itself the paper was given something of a harsh reception. The debate is a classically ethnographic one, in that the demands of ethnographic research are such it often pushes debates into questions *around* ethnography rather than into the studies themselves.

Yet this debate did clearly establish the centrality of ethnography to the HCI project. As mentioned at the start of this chapter, whatever the controversy around ethnography (and perhaps despite the lack of actual ethnographies in HCI), ethnography has definitely established itself as central to understanding technology, interaction and our broader lives.

REFERENCES

Bently R, Hughes JA, Randall D, et al. (1992) Ethnograpically-informed system design for air traffic control. In: Turner, J. and Kraut, R. (eds), *Proceedings of (CSCW) '92*, ACM, 123–129.

Beyer H and Holtzblatt K (1997) *Contextual Design: Defining Customer-Centered Systems*. San Francisco, CA: Morgan Kaufmann.

Blackler, F (1995) Activity Theory, CSCW and Organizations. In AF Monk and N Gilbert (eds.), *Perspectives on HCI: Diverse Approaches* London: Academic Press.

Blomberg JL (1995) Ethnography: Aligning field studies of work and system design. In AF Monk and N Gilbert (eds) *Perspectives on HCI: Diverse Approaches*. London: Academic Press.

Bowers J, Button G and Sharrock W (1995) Workflow from within and without. *Proceedings of ECSCW '95*, Kluwer Academic Press.

Bowker G and Star SL (1999) *Sorting Things Out: Classification and its Consequences*. Cambridge, MA: MIT Press.

Brown B and Chalmers M (2003) Tourism and mobile technology. In: Kuutti K, Karsten EH and Fitzpatrick G, et al. (eds) *Proceedings of ECSCW '03*, Kluwer Academic Press, 335–355.

Button G (1992) Hanging around is not the point: Calling ethnography to account. (unpublished manuscript)

Button G (1993) *Technology in Working Order: Studies of Work, Interaction and Technology*. London: Routledge.

Button G and Harper R (1996) The relevance of 'work practice' for design. *Computer Supported Cooperative Work* 4: 263–280.

Castronova E (2005) *Synthetic Worlds: The Business and Culture of Online Games*: Chicago, IL: University of Chicago Press.

Cooper G, Hine C, Low J, et al. (1993) Ethnography and Human–Computer Interaction. In: Thomas PJ (ed.), *The Social and Interactional Dimensions of Human–Computer Interaction*. Cambridge: Cambridge University Press, 11–36.

Crabtree A (2003) *Designing Collaborative Systems: A Practical Guide to Ethnography*. Heidlberg: Springer.

Crabtree A, Rodden T, Tolmie P, et al. (2009) Ethnography considered harmful. *Proceedings of CHI '09*, ACM, 879–888.

Dourish P (2006) Implications for design. *CHI '06 Proceedings of the SIGCHI Conference on Human Factors in Computing Systems*, New York: ACM Press, 541–550.

Ferreira P and Höök K (2011) Bodily orientations around mobiles: lessons learnt in Vanuatu. *Proceedings of the 2011 Annual Conference on Human Factors in Computing Systems*. ACM, 277–286.

Fine GA (1993) Ten lies of ethnography: Moral dilemmas of field research. *Journal of Contemporary Ethnography* 22: 267–294.

Goffman E (1989) On fieldwork. *Journal of Contemporary Ethnography* 18: 123–132.

Greatbatch D, Heath C, Luff P, and Campion P (1995) Conversation analysis: Human–computer interaction and the general practice consultation. In AF Monk and N Gilbert (eds), *Perspectives on HCI: Diverse Approaches*. London: Academic Press, 199–222.

Greenbaum J (1979) *In the Name of Efficiency*. Philadelphia, PA: Temple University Press.

Grint K, Case P and Willcocks L (1995) Business process reengineering reappraised: the politics and technology of forgetting. In W Orlikowski, G Walsham, M Jones, et al. (eds), *Information Technology and Changes in Organisational Work: Proceedings of IFIP WG8.2 Working Conference on Information Technology and Changes in Organizational Work*. Cambridge: Champman & Hall.

Halverson CA and Rogers Y (1995) Combining the Social and the Cognitive: Distributed Cognition, Theory and Application (tutorial). *ECSCW '95*, Stockholm, Sweden.

Hammersley M and Atkinson P (1995) *Ethnography: Principles in Practice*. London: Routledge.

Harper R (1998) *Inside the IMF*. London: Academic Press.

Harper R, Hughes J and Shapiro D (1991) Harmonious working and cscw: computer technology and air traffic control. In J Bowers and SD Benford (eds), *Studies in Computer Supported Cooperative Work: Theory, Practice and Design*. Cambridge: Cambridge University Press.

Harper R, Randall D and Rouncefield M (2000) *Organizational Change in Retail Finance: An Ethnographic Perspective*. Abingdon, Oxon: Routledge.

Heath C and Luff P (2000) *Technology in Action*. Cambridge: Cambridge University Press.

Hine C (2000) *Virtual Ethnography*. London: Sage.

Kling R (1980) Social analyses of computing: Theoretical perspectives in recent empirical research. *Computing Surveys* 12(1): 61–110.

Kling R (1994) Reading "All About" Computerization: How Genre Conventions Shape Non-Fiction Social Analysis. *The Information Society* 10(3): 147–172.

Kozinets RV, de Valck K, Wojnicki A, et al. (2010) Networked narratives: Understanding word-of-mouth marketing in online communities. *Journal of Marketing* 74: 71–89.

Kraemer K and King J (1988) Computer-based systems for cooperative work and group decision making. *ACM Computing Surveys* 20: 115–146.

Kraut RE and Resnick P (2012) *Building Successful Online Communities: Evidence-based Social Design*. Cambridge, MA: MIT Press.

Loftland J (1971) *Analyzing Social Settings: A Guide to Qualitative Observation and Analysis*. Belmont, CA: Wadsworth.

Luff P, Hindmarsh J and Heath C (2001) *Workplace Studies: Recovering Work Practice and Informing System Design*. Cambridge: Cambridge University Press.

Maanen JV (1988) *Tales from the Field: On Writing Ethnography*. Chicago IL: University of Chicago Press.

Markus M (1983) Power, politics and MIS implementation. *Communications of the ACM* 26(6): 430–444.

Mateas M, Salvador T, Scholtz J, et al. (1996) Engineering ethnography in the home. *Proceedings of CHI '96*. ACM, 283–284

Nardi B and Engestrom Y (1999) A web on the wind: The structure of invisible work. *Computer Supported Collaborative Work* 8: 1–8.

Nardi B and Miller J (1991) Twinkling lights and nested loops: Distributed problem solving and spreadsheet development. *International Journal of Man-Machine Studies* 34, 161–184.

O'Brien J Rodden T, Rouncefield M, et al. (1999) At home with the technology: An ethnographic study of a set-top-box trial. *ACM Transactions on computer-Human Interaction.* 6: 282–308.

Orr JE (1986) Narratives at work: Story telling as cooperative diagnostic activity. *Proceedings of the 1986 Conference on Computer Supported Cooperative Work.* ACM, 62–72.

Orr JE (1996) *Talking about Machines: An Ethnography of a Modern Job.* Ithaca, NY: ILR Press.

Pace T, Bardzell S and Bardzell J (2010) The rogue in the lovely black dress: Intimacy in World of Warcraft. *CHI '10, Proceedings of the SIGCHI Conference on Human Factors in Computing Systems.* New York: ACM Press, 233–242.

Randall D, Harper R and Rouncefield M (2007) *Fieldwork for Design: Theory and Practice.* New York: Springer-Verlag.

Rogers Y (1995) Combining the social and the cognitive: Distributed cognition, theory and application Tutorial presented at ECSCW '95, Stockholm, Sweden.

Sharrock W, Hughes JA and Martin PJ (1995) *Understanding Classical Sociology: Marx, Weber, Durkheim.* London: Sage.

Star SL (1988) The structure of ill-structured solutions: Boundary objects and heterogeneous distributed problem solving. In M Huhns and L Gasser (eds) *Distributed Artificial Intelligence 2.* Menlo Park, CA: Morgan Kauffman, 37–54.

Suchman L (1987) *Plans and Situated Actions: The Problem of Human–Machine Communication.* Cambridge: Cambridge University Press.

Suchman L and Wynn E. (1984) Procedures and problems in the office. *Office Technology and People* 2: 133–154.

Sudnow D (1983) *Pilgrim in the Microworld: Eye, Mind, and the Essence of Video Skill.* Oxford: Heinemann.

Taylor AS and Harper R (2002) Age-old practices in the 'new world': A study of gift-giving between teenage mobile phone users. *Proceedings of CHI '02.* ACM, 439–446.

Whittaker S, Terveen L and Nardi BA (2000) Let's stop pushing the envelope and start addressing it: A reference task agenda for HCI. *Human–Computer Interaction* 15(2): 75–106.

Winograd T and Flores F (1986) *Understanding Computers and Cognition: A New Foundation for Design.* Norwood, NJ: Ablex Publishing.

Zuboff S (1988) *In the Age of the Smart Machine: The Future of Work and Power.* New York: Basic Books.

14

The Mediational Perspective on Digital Technology: Understanding the Interplay between Technology, Mind and Action

Victor Kaptelinin

INTRODUCTION

Digital technology is a complex, multifaceted object of social science research. It can be approached from different angles, by foregrounding different facets and making different basic assumptions about the relevance of digital technology to people. The specific focus of the present chapter is on exploring the relationship between, on the one hand, technology and, on the other hand, the human mind and action. The chapter adopts a particular perspective on digital technology, namely the *mediational* perspective, according to which technology is considered a means through which human beings act in the world.[1] The underlying ideas of the mediational perspective can be briefly summarized as follows. Technologies, in a broad sense, are a special part of the world in which we live. They have been created by us, the humans, as our own projections and extensions and, therefore, are a part of 'us', as well as the outside world. At the level of an individual human being, different technologies

can occupy different places on the 'me v. the world' continuum; they can be perceived as a part of the external environment (e.g. this is how a car driver usually sees other people's cars), but can also become virtually inseparable from one's self-image (consider, e.g., glasses).

The aim of the chapter is twofold. First, it makes an attempt to explore some theoretical approaches, according to which technology is considered a mediating means that affects, and even shapes, the structure, functioning and development of the human mind and action. The analysis in the chapter is mainly based on the Vygotskian cultural–historical tradition, but some other frameworks, such as pragmatism, phenomenology and actor–network theory, are discussed as well. Second, the chapter provides an overview of some relevant existing digital technology research from the point of view of the mediational perspective. It is often believed, for instance, that digital technologies, using Don Norman's (1993) expression, are 'things that make us smart'. Is that really so and, if yes, how should the technologies be designed and used in order to actually make us smart? Do the technologies affect the human mind in some other ways, as well? The chapter discusses these and similar questions by bringing in some relevant research conducted at the intersection of psychology and technology design, including studies in the fields of human–computer interaction (HCI), interaction design, and, partly, technology-enhanced learning.

In HCI the mediational perspective was adopted rather early in the history of the field, especially by research informed by activity theory (Bødker, 1991; Kaptelinin, 1996a; Nardi, 1996; Kaptelinin and Nardi, 2006; 2012b). The concept of mediation proved to be a useful analytical tool in applied HCI research dealing with analysis and design of interactive artifacts. In particular, the notion of mediation played a key role in the instrumental interaction model (Beaudouin-Lafon, 2000), the human-artifact model (Bødker and Klokmuse, 2011) that deals with artifact ecologies and analysis of technology affordances (Kaptelinin and Nardi, 2012a).

This chapter makes an attempt to take a more theoretical stance on the concept of mediation in relation to digital technologies. Digital technologies, which offer more interactivity compared to more traditional mediational means, open up a range of additional possibilities for a deeper integration of technological artifacts into our lives, minds and identities. Digital technologies have already radically influenced the world in which we live and new developments are taking place with an incredible (and even alarming) speed. In particular, the widespread use of powerful and relatively affordable mobile devices connected to the Internet, robots and social proxies, social networks, ambient intelligence, augmented reality technologies, brain–computer interfaces, etc., is likely to have – and is already having – a significant impact on human perception, action, cognition, emotions and communication. These current trends present challenges for digital technology research. This chapter argues that

adopting the mediational perspective helps address these challenges.

The remainder of the chapter is organized as follows. The next section introduces the basic idea of mediation and how it is interpreted in some conceptual frameworks in philosophy and psychology. The section capitalizes on these frameworks to identify four aspects of mediation: material, social, cognitive and experiential. The third section gives an overview of relevant digital technology research and developments concerned with the impact of digital technology on the human mind. It starts with early visions of 'precomputer' information technology and then goes on to discuss some 'classic' HCI studies, as well as a range of new developments in personal technologies (e.g. MyLifeBits, SixthSense, and mobile-phone-based augmented reality applications). The section also explores some key conceptual challenges associated with the studies and discusses the implications of the mediational perspective for the analysis and design of interactive technologies. Finally, the fourth section concludes the chapter with a general summary and reflections.

THE CONCEPT OF MEDIATION IN PSYCHOLOGY AND PHILOSOPHY: SOME KEY IDEAS

Mediation: What Makes Humans Humans

In everyday language 'mediation' often refers to reconciling differences and contradictions between individuals, groups or institutions – for example parties in conflict. More abstractly, mediation can be defined as the act or result of connecting two (or more) entities indirectly, through an additional entity (the mediator). The concept often has a positive connotation; it is implied that mediation serves to enable, enhance or facilitate the way connected entities are related to one another. The abstract concept of mediation can be applied to any kind of interaction, including the interaction between human beings and the world. As discussed below, when applied in that way, the concept provides some important insights about the nature of human action, as well as human beings in general.

Arguably, mediation is the primary dimension along which human beings differ from other animals. The key distinctive features of humans, such as language, complex social organization, religion and the production and use of advanced tools, all involve mediation. These features represent different aspects of the same phenomenon – that is, the emergence of a complex system of objects and structures, both material and immaterial, that serve as mediating means embedded in the interaction between human beings and the world. It is mediation that has made *homo sapiens* such a successful species: while we do not have sharp claws and thick fur, we compensate for that by employing mediating artifacts, such as instruments and clothing.

Some animals also manifest instinctive or learnt behavior, characterized by a certain degree of mediation. Birds bringing straws to build nests or apes using sticks to reach for bananas are just a few examples of such mediated behavior. However, in such cases mediation is essentially limited to 'first-order mediation'. In other words, interaction between an animal and a mediating means is not mediated by other mediating means. In humans, on the contrary, a common type of mediation is complex 'higher-order mediation'. For instance, we need money to buy food at a supermarket and we also may need to get a job to earn the money, write and send an application to get the job, use a word processing program to write the application, download and install the program to be able to use it, etc. Embedded levels and the resulting complexity of mediation in humans are virtually unlimited and this is what makes us a truly unique species.

A Variety of Theoretical Approaches to Mediation

The importance of taking mediation into account to understand the nature of human

identities, cognition and action was emphasized in a number of approaches in philosophy, psychology and social sciences in general (e.g. James, 1890; Dewey, 1910, 1925; Bateson, 1972; Vygotsky, 1984; Engeström, 1987; Cole and Engeström, 1993). The most relevant to the discussion in this chapter are analyses of mediation in pragmatism, phenomenology, actor–network theory and, especially, Vygotskian cultural–historical tradition.

Important early insights into the role of material artifacts (such as clothing or tools), as well as immaterial ones (such as language), can be found in American pragmatism. For instance, William James, one of the founders of psychology as a scientific discipline, observed that the notion of Self, in a broad sense, includes certain physical artifacts:

> The body is the innermost part of the material Self in each of us; and certain parts of the body seem more intimately ours than the rest. The clothes come next. The old saying that the human person is composed of three parts – soul, body and clothes – is more than a joke. (James, 1890: 292)

John Dewey, another prominent American pragmatist, developed an instrumental view on human thought as embedded in the meaningful interaction with – and transformation of – the world. The specific function of thought in the interaction was understood as 'the experimental determinations of future consequences' (Dewey, 1925: 14), and language was considered a tool (a 'tool of tools') that made thought possible.

Within the philosophical tradition of phenomenology (e.g. Heidegger, 1962, see also Heidegger, 1993; Svanaes, 2000; Dourish, 2001) technology is understood as critically important in determining human experience. According to Heidegger (1962), the breakdowns we sometimes experience when using technology (or any other kinds of 'equipment') result in a transformation from technology being 'ready to hand' to technology being 'present at hand'. Therefore, technology is central for revealing to us the true essence of the world. The particular type of technology also matters. In a later work entitled 'The question concerning technology' Heidegger (1993) considers the 'new technology' of that time – that is, machine-powered technology – as revealing the world differently, compared to previous types of technology, such as old windmills.

More recently, a notion of mediation, inspired by both phenomenology and pragmatism, was proposed by Ihde (1990) to define the role of modern technologies, including digital technologies, in how people perceive and interpret the world. In particular, to identify the variety of ways in which technologies mediate our perception, Ihde differentiates between several types of relationships human beings have with technological artifacts. The *embodiment* relationship means that people act through technology, without being aware of the technology in question. In *hermeneutic* relationships we are aware of – and need to interpret – both the technology we are using and the world we are acting upon through the technology. In the *alterity* relationship we are only interacting with (and not through) technology. Finally, in the *background* relationship we are neither interacting through nor aware of the technology in question (Ihde, 1990; Verbeek, 2006).

In actor–network theory, an approach originating from social studies of technology (SST) and currently being widely used in a diversity of areas, mediation is described in terms of 'scripts' and 'delegation' (Latour, 1992). Implicitly or explicitly, artifacts prescribe the types of actions that need to be carried out with them. A classic example is speed bumps, which make drivers slow down. The scripts, which are associated with artifacts, may result from the deliberate attempts of designers to 'delegate' some responsibilities to the artifacts (as in the case of the speed bump), but may also be unintentional. What is important to keep in mind, according to actor–network theory, is that enlisting the help of mediators (as opposed to passive 'intermediaries') may substantially change the character of an action. A mediator, either a human being or an artifact, brings its own

agenda and becomes an actant in a new actor network.

The Concept of Mediation in the Vygotskian Cultural–historical Tradition

The concept of mediation received a thorough treatment in the Vygotskian cultural–historical tradition (Vygotsky, 1978, 1982; Leontiev, 1978, 1981; Cole and Griffin, 1983; Cole, 1996; Wertsch, 1998). Within the psychological framework developed by Vygotsky himself, mediation is, arguably, *the* most important concept of all; it serves as the cornerstone of his approach as a whole. Vygotsky proposed that the very nature of human mental processes, as opposed to those of animals, is defined by mediation. Vygotsky differentiated between 'lower' (or 'natural') mental functions, which can be observed in animals and with which every human being is born, and 'higher' mental functions, which only humans can have. According to Vygotsky, lower mental functions transform into higher mental functions in the process of development, when individuals appropriate culturally produced signs, so that the signs become mediators of their cognitive processes. Some of the main arguments provided by Vygotsky are as follows.

Animals are commonly believed to possess the same basic mental functions as human beings. Such functions include perception, memory, attention, learning and even problem solving. However, animals lack language capabilities in the sense that they cannot use signs. This limitation is more significant than it may seem, for language is not just an addition to other mental functions; its appropriation induces deep changes to all other functions as well.

For instance, as observed by Vygotsky (1984: 23–24), the process of a child solving a practical problem may look like the process of an ape problem solving. However, if a child reaches the level of development at which he or she is already capable of using language, the process of his or her problem solving is different in two significant respects. First, the process is characterized by increased freedom and flexibility. While apes are bound to the information present in the visual field – that is, their problem solving can only take place when the task at hand is perceptually present to them – children are less dependent on perception. They can temporarily move their focus of attention away from the overall goal to carry out an auxiliary task; they can also employ objects from outside of the immediate problem context. Second, children are more in control of their own problem-solving process. They can more easily initiate, stop or modify their problem-solving activities.

According to Vygotsky, the reason why human problem solving is more visual-field-independent and self-controlled (or 'voluntary'), compared to problem solving in apes, is that humans use signs. Acting on signs instead of perceptual images of the immediate environment allows for more freedom and flexibility in finding novel solutions. At the same time, if a person's behavior is expressed in signs, it can be presented to the person as a separate object, which can be acted upon. It opens up a possibility for a person to control their own mental processes. This fundamental impact of sign-mediation on mental processes is not limited to problem solving but can be extended to other mental functions as well. In particular, similar transformations were found by Vygotsky and his colleagues in their studies of memory and attention.

A key aim of theoretical explorations and empirical studies of mediation, conducted by Vygotsky and his colleagues, was to find out how exactly mediation takes place and identify the underlying laws and mechanisms of mediation. The studies foreground two main, closely related phenomena affecting mediation: the social context and internalization.

The genetic roots of mediation can be found in the social relations of a child. Initially signs are represented in a child's behavior as means embedded into the social relations of the child – that is, communication

with other people. Only then the signs are appropriated and become embedded in the individual's mental processes. In this respect sign mediation is subordinated to Vygotsky's general law of development (Vygotsky, 1978), according to which each mental function first emerges as distributed between people (inter-psychological) and only after that becomes individual (intra-psychological).

Internalization is another dimension along which signs are transformed when mediated mental processes emerge. As a means of interpersonal communication, signs are by necessity external, which makes them accessible to all participants in a communication. When signs change their status from inter-psychological to intra-psychological, they often move into the internal plane. The transition of signs from the external plane to the internal plane is not a simple 'relocation' but, rather, a complex transformation associated with qualitative changes.

The mediating role of signs is somewhat similar to that of tools. There is a complex two-way relationship between tools and signs. On the one hand, sign mediation may emerge through internalizing actions with tools as physical artifacts. On the other hand, mediation of mental functions by signs may affect the use of physical tools by supporting people in mastering their own behavior and making human actions free and independent from their immediate situations.

Vygotsky's concept of mediation made a major impact on activity theory, another socio-cultural approach in psychology, which was developed by Vygotsky's disciple, Alexey Leontiev (1978, 1981). Key aspects of Vygotsky's view of mediation – that is, social genesis and internalization – were explicitly incorporated into the conceptual framework of activity theory. However, in activity theory these ideas were placed in a somewhat different theoretical context. The overall focus of Leontiev's approach was on activity, understood as the purposeful inter-action of active subjects with the objective world (i.e. the 'S ↔ O' interaction), rather than on higher mental functions and their ontogenetic development. Accordingly, in activity theory special attention is paid to tools (that is, means that mediate an object-orientated activity as a whole) rather than signs (that is, means which mediate specific mental operations). In addition, while retaining the focus on individual development as a major object of research, Leontiev also expanded his analysis to transformations of activities and psyche in biological evolution and cultural history.

The main differences between Vygotsky's and Leontiev's views on mediation can be summarized as follows. For Vygotsky, mediation is primarily understood as sign mediation, which transforms lower mental functions into higher mental functions. It allows for freedom and control of mental operations and makes the person a true 'owner' of his or her mental functions (and, therefore, serves as a basis for free will and free action). Sign mediation first emerges in the external plane, in communication between a person and other people, and then it is translated into the internal plane through internalization. For Leontiev, mediation is primarily understood as tool mediation, which transforms human interaction with the objective world as a whole. The transformation makes it possible for a person to appropriate socially developed forms of acting in the world.[2] Tool mediation shapes the entire structure of meaningful, purposeful activities. Over time, some external components of an activity can be translated into the internal plane through internalization to ensure efficiency and, as a result, transform a person's mental processes. It should be noted that Vygotsky's and Leontiev's views on mediation do not contradict each other. They reflect differences in research foci and can be considered as complementary theoretical accounts of mediation.

The concepts of mediation, proposed by Vygotsky and Leontiev, were further developed in more recent work within the cultural–historical tradition. For instance, Wertsch (1998) argues that the unit of analysis in studies of human action should include both agent and his/her meditational means. The concepts also influenced (at least, partly) a number of more recent approaches that more specifically deal with issues briefly outlined but not thoroughly examined in the cultural–historical tradition. While Vygotsky was mostly interested in the general principles underlying the genesis of individual, subjective phenomena, *distributed cognition* frameworks (Salomon, 1993a, 1993b; Hutchins, 1995; Hollan et al., 2000), are specifically concerned with the development and structure of externally distributed cognition as it naturally takes place in everyday life. The point of departure for the *instrumental genesis* framework (Rabardel and Bourmand, 2003) is an observation that mediation rarely takes place quickly and effortlessly; it is often an outcome of an unfolding process, during which artifacts, which are initially loosely integrated into human activities, become genuine mediating means – that is, instruments.

Summary: Roles and Aspects of Mediation

The theoretical accounts of mediation, discussed above, are similar in a number of important respects (Koschmann et al., 1998; Guribye, 2005; Miettinen, 2006). They all maintain that human life is fundamentally mediated by tools and signs, which both enable and constrain human action. They all consider tools as not just simple instruments, the role of which is limited to helping people carry out specific tasks, but also rather crucial links between individuals on the one hand and culture and society on the other. Guribye (2005) observes that '… over history, the production and use of artefacts leave ideal and material traces that have an impact when artefacts are used in the present' (2005: 40).

Taken together, the conceptual analyses, discussed above, suggest that four aspects of mediation can be differentiated between: material, social, cognitive and experiential. Material mediation means interacting with the objective, material world through external objects. Social mediation refers to using artifacts to connect to other people in communication or joint action. Cognitive mediation implies involving tools and signs in perception, memory and thinking. Finally, experiential mediation means that artifacts play a role in shaping the overall human experience of being and acting in the world, including meaning making and affect.

These aspects are not independent of each other. On the contrary, they are closely intertwined, and the way they are related to one another may dynamically change over time. Consider, for instance, the mobile phone, a device almost continuously used by many people to get connected to the world. To fulfill its mediating function it should be properly handled as a *material* object – that is, operated in a certain way (e.g. recharged). Of course, its main purpose is to support *social* interactions, such as voice communication, exchange of text messages, photo sharing, etc. The device can also be a *cognitive* aid as, for instance, a built-in camera can be used to take a picture of a bus schedule and thus serve as a memory aid. Finally, the artifact may influence the overall *experience* of a person – for instance, a person who knows that he or she can call for help in case of emergency may feel more secure. All these aspects are closely interrelated. If the battery is dead (the material aspect) or if the person cannot navigate to the needed function because the language of the interface is not familiar (the cognitive aspect) or if he or she does not know who to call for help (the social aspect), the mobile phone can hardly make its owner feel safer (the experiential aspect).

A BRIEF OVERVIEW OF RELEVANT RESEARCH AND TECHNOLOGICAL DEVELOPMENTS

This section briefly discusses several threads of digital technology research and development that are particularly relevant to the relationship between technology, mind and action and in which the notion of mediation, one way or another, has played a significant role in the analysis.

The Augmentation of Human Cognition: Historical Roots

Back in 1945 Vannevar Bush presented his vision of a future information technology called 'memex'. While the device was only partly digital (the suggested information storage was analog microfilms), it has a striking resemblance to modern desktop computers:

> A memex is a device in which an individual stores all his books, records, and communications, and which is mechanized so that it may be consulted with exceeding speed and flexibility. *It is an enlarged intimate supplement to his memory.*
>
> It consists of a desk, and while it can presumably be operated from a distance, it is primarily the piece of furniture at which he works. On the top are slanting translucent screens, on which material can be projected for convenient reading. There is a keyboard, and sets of buttons and levers. (Bush, 1945, emphasis added)

As follows from the above quote, the technology was believed to advance human cognition. This intention is made clear by the very title of the paper, 'As we may think', which apparently alludes to the classic *How We Think* by John Dewey (1910). Moreover, the technology was envisioned as having an impact on human experience in general:

> Presumably man's spirit should be elevated if he can better review his shady past and analyze more completely and objectively his present problems. … His excursions may be more enjoyable if he can reacquire the privilege of forgetting the manifold things he does not need to have immediately at hand, with some assurance that he can find them again if they prove important. (Bush, 1945)

Bush's vision of technology as a 'supplement' of human cognition that might bring about large-scale changes in human experience as a whole, has been an inspiration for a number of analyses and developments – in particular, for Douglas Engelbart and his pioneering explorations into augmented intelligence. Engelbart (1963), whose work had influenced or anticipated many major developments in the fields of human–computer interaction (HCI) and computer supported cooperative work (CSCW), suggested that native human capabilities are limited and, to overcome the limitations, a larger-scale system called H-LAM/T – 'the individual [that is, Human] augmented by the language, artifacts, and methodology, in which he is trained' (Engelbart, 1963: 4) – needs to be created. The agenda for the development of augmented intelligence systems formulated by Engelbart included identification of the basic cognitive capabilities of human beings and artifacts and using these as a foundation for designing more efficient cognitive systems, in which properly designed technological artifacts would compensate for the limitations of native human capabilities.

Explorations into Computer Technologies and Human Cognition

In HCI research an account of digital technology as a tool for augmenting human cognition was proposed by Norman (1991, 1993). He suggested digital technologies be understood as *cognitive artifacts* – that is, artifacts which serve a representational function. According to Norman (1991), the performance of the individual who is employing artifacts to carry out a task represents the 'system view', which should be differentiated from the 'personal view'. The latter corresponds to the individual's perspective on the task *and* artifacts; from this perspective both the task and artifacts are components of an environment, which is external to the individual. Norman (1991) maintained that ' ... artifacts do not actually change an individual's capabilities. Rather, they change the nature of the task performed by the person' (Norman, 1991: 19). This claim, as discussed below, is contested by some contemporary research.

An influential edited collection entitled *Distributed Cognitions* (Salomon, 1993a) presented a wide range of conceptual accounts of human cognition and learning as comprising both internal processes and external means.

According to Salomon (1993b), individual and distributed cognitions:

> ... interact with one another in a spiral-like fashion whereby individuals' inputs, through their collaborative activities, affect the nature of the joint, distributed system, which in turn affects their cognitions such that their subsequent participation is altered, resulting in subsequent altered joint performances and products. (Salomon, 1993b: 122)

One of the contributing authors to the book, Perkins (1993), differentiated between the concepts of 'person solo' and 'person plus'. The former refers to the individual without any artifacts, while the latter refers to the individual taken together with available external resources. The advantages of employing external resources include providing needed knowledge, accessible representations, retrieval paths and construction arenas.

Radically new possibilities for the augmentation of human cognition are opened up by mobile and wearable technologies, since the use of such technologies can be embedded in a much wider range of everyday tasks and contexts, compared to more traditional computers. While a few decades ago Steve Mann, a pioneer of wearable technologies, had to carry a backpack full of heavy and expensive equipment (Mann, 1997) to be able to use the power of digital technology 'in the field', such technologies are currently rather compact and relatively affordable. Substantially more powerful, lightweight computing technologies, plus mobile Internet access, are now available to millions of smartphone users.

One particular possibility enabled by mobile devices (and already mentioned by Bush, 1945) is sampling information about the environment to support human memory – taking pictures, recording conversations, measuring ambient temperature, etc. In principle, data about all the events that have happened to a person can be recorded and saved in log files. This way of using technology, dubbed 'lifelogging' (Whittaker et al., 2008; Sellen and Whittaker, 2010), has been explored in recent years. A well-known project in this area is MyLifeBits, conducted at Microsoft Research (Gemmell et al., 2006). The technology used in the MyLifeBits project was SenseCam, a portable device that could be continuously worn by the user around the neck. The functionality of SenseCam made it possible to register a range of parameters, such as temperature, orientation or acceleration, and automatically take a picture if a change in the situation were detected. One of the researchers – and simultaneously participants – in the project, Gordon Bell, had been using the device for an extended period of time, making it a part of his life.

A limitation of SenseCam is that it is, essentially, an autonomous sampling device, which collects ambient data without making any direct interference into the situation. Therefore, it may well be a useful technology for information retrieval that takes place *post factum*, but it cannot help the user to deal with the task at hand. This limitation was addressed in the design of other technologies, which were intended to support the user *in situ*.

Augmented reality (AR) technologies implementing the Magic Lens concept (Rohs and Oulasvirta, 2008) are, essentially, 'augmented perception' devices. They provide their users with an enriched image of the world, e.g., presented on displays of their mobile devices, such as smartphones. For instance, a tourist pointing their phone's camera at an historical building can see a description of the building on the phone's display. An example of such technology is an AR solution developed at Layar (see www.layar.com).

Even more advanced support is provided to their users by the SixthSense system, developed at MIT (Mistry et al., 2009). The system itself is rather simple and does not contain expensive technological components. It combines a small computing device with a mobile Internet connection, a camera and projector. The camera employs image recognition to recognize objects in the environments, as well as the user's gestures. The design of the SixthSense system allows the user to bring it

directly to the task at hand; for instance, by recognizing the bar code on, say, a box of cereal, the system may find relevant information on the Web and project it directly onto the box to help the user choose the right product. The functionality of the system also includes taking pictures, controlling the image displayed by the projector through gestures recognized by the camera, etc.

Another relevant project, also at MIT, is '10x' (Roy, 2004; http://10x.media.mit.edu). The ambition of the project is to develop technologies that will amplify human cognitive capabilities at least tenfold. An example of an envisioned technology is smart glasses that help their user search for a lost item in a room. The glasses would prevent the person from looking through the same places over and over again by drawing his or her attention to the places that have not been looked at yet.

The conceptual and design explorations discussed above, from early work by Bush (1945) and Engelbart (1963) to more recent work, e.g., some projects at MIT (Roy, 2004; Mistry et al., 2009), are based on the same main assumptions, namely: (a) the natural cognitive capabilities of humans are limited; and (b) technology can be used to complement natural human cognition within an extended, more efficient cognitive system, in which the limitations of natural human cognition are compensated for by the strengths of technology. Suggestions for the most promising lines for augmenting human cognition include supporting the breaking down of complex problems into sets of more simple ones (Engelbart, 1963); supporting task performance with representational artifacts in order to help bridge 'the gulf of execution' and 'the gulf of evaluation' (Norman, 1991); enhancing short-term and long-term memory (Kozma, 1992); providing needed knowledge, accessible representations, retrieval paths and construction arenas (Perkins, 1993); and strategically targeting the weaknesses of human memory (Sellen and Whittaker, 2010).

Structural augmentation of 'native' human capabilities with digital technologies appears to be the most logical approach to enhancing human cognition, but a closer look at this approach from a meditational perspective suggests that it is not unproblematic. As follows from the socio-cultural theories, discussed above, natural human cognition already becomes mediated at an early age – as soon as the child develops language abilities. Consequently, what is considered *natural* human thinking or memory are, in fact, products of culture and, therefore, are different in different cultural contexts. Moreover, as follows, for instance, from distributed cognition analyses, in everyday problem solving humans typically employ a variety of external resources. Therefore, unaided cognitive abilities, as opposed to what is assumed in augmentation frameworks, are not 'natural' ones (Kaptelinin, 1996b; Kaptelinin and Kuutti, 1999). Instead, they are peculiar phenomena, which are typically observed only in artificial conditions. It means that extending human cognition with new types of artifacts should not be seen as *augmentation* – that is, creating new information processing systems on the basis of some native human cognitive capabilities. Instead, integration of new artifacts should be viewed as a process of development, as a 'transformation of one form of mediation into another form' (Kaptelinin, 1996b), or 'remediation' (Cole and Griffin, 1983; Bødker and Andersen, 2005).[3]

A research strategy, alternative to the analysis of augmentation, is investigation of the cognitive effect of digital technologies – that is, long-term changes caused by the use of the technologies. The effect of digital technologies on learners' cognition was a popular object of educational psychology research in the 1980s. LOGO, a programming language developed by Seymour Papert (1980), as well as the related notion of microworlds, were claimed to provide conditions not only for engaged and efficient learning of math and other subjects but also for the development of thinking skills in general. However, the evidence produced in numerous empirical assessments of the cognitive effect of LOGO,

as well as other programming languages and microworlds, has been rather inconclusive (e.g. Pea and Kurland, 1984). As suggested by Pea and Kurland (1984) it is more important to properly organize learning experiences, then choose 'the best' technology.

Technological Mediation and Human Experience

It is interesting to note that researchers' motivations behind many of the studies of the augmentation of human cognition described above have not been limited to pragmatic considerations of merely creating improved tools or achieving a better understanding of certain phenomena. Vannevar Bush (1945) presented his agenda for developing technological support for human cognition as a grand vision of the next great goal for researchers in the new, post-war era. Douglas Engelbart (1963) had deliberately chosen creating technologies for advancing human cognition as the direction of his research because he considered that the most meaningful life goal one can imagine. For Steve Mann (1997) and Gordon Bell (Gemmell et al., 2006), serving as 'test pilots' for novel wearable technologies has been literally a life-changing experience. Years, even decades, of their lives were shaped by the continuous and pervasive everyday use of personal computing devices. All this points to researchers' firm belief that personal technologies do not only serve certain practical purposes but they also make an impact on how people generally make sense of the world and themselves.

Until recently, supporting people in making sense of the world and themselves, facing important life choices and developing their personal identities have been in the blind spot of mainstream HCI and interaction design research (see Bannon, 2011). The almost exclusive emphasis of these fields has been on utility and usability. However, with the field moving from the 'second-wave HCI', focused on pragmatic aspects of technology use, to the 'third-wave HCI', concerned with experience in general (Bødker, 2006;

Hassenzahl, 2010), issues such as 'designing for the self' (Zimmerman, 2009) and 'identity construction' (Mamykina et al., 2010) increasingly find their way into HCI research.

The emerging area of 'experience design', understood as designing technology 'for all the right reasons' (Hassenzahl, 2010), is currently in the process of formulating its specific agenda and finding an appropriate theoretical foundation. One of the main challenges in the area is to define the role of technology in determining the experience as a whole. Since user experience is an integrated outcome of a variety of factors, the specific impact of a certain technology can be unpredictable, highly dependent on a particular context or even negligible. Exploring this issue is of critical importance for establishing the agenda for interaction design research and development oriented toward supporting and facilitating positive user experiences. Understanding exactly how interactive technologies influence the overall experience can help identify realistic strategies for 'experience design'. The mediational perspective places technology in the context of the purposeful human action and thus suggests a principled strategy for dealing with the above challenge.

Theoretical frameworks discussed in this chapter are contributing to forming the research with their respective insights. Dewey's emphasis on the centrality of experience makes his framework especially appropriate for understanding 'technology as experience' (McCarthy and Wright, 2004). Phenomenology that understands technology as a mediating means playing a key role in revealing the world to us (Heidegger, 1962, 1993; Svanæs, 2000; Dourish, 2001) highlights the importance of technology in how people make sense of their lives. Finally, activity theory offers conceptual apparatus for understanding how certain experiences can be supported indirectly, through providing tools, which are likely to induce changes in the structure of an activity (Kaptelinin and Nardi, 2012b).

CONCLUSION

Developments in digital technologies in the last decades have been an unparalleled success story, going far beyond the boldest visions of the past. There are massive transformations in human lives that are caused by digital technologies and associated with both challenges (e.g. coping with accelerating multitasking) and possibilities (e.g. increasingly available support for remote action). These transformations, at least some of them, are likely to have profound effects on humans and their minds. Understanding these effects presents an enormous challenge for digital technology research, which challenge is yet to be properly addressed.

As argued in this chapter, the mediational perspective on digital technology offers potentially useful insights for helping to address such a challenge. A variety of theoretical frameworks, both traditional and relatively recent ones, provide specific conceptual tools to understand the nature and properties of mediational means and their role in purposeful human action. This chapter analyzes how some of these concepts can be applied in concrete areas of digital technology research, mostly focusing on the augmentation of human cognition and the relationship between technology and human experience.

In particular, it is argued that the mediational perspective on digital technology offers an explanation of why the outcomes of analysis and design of technologies intended to make an impact on human cognition are so inconclusive. The meditational perspective maintains that technologies do not have an immediate effect on humans and their minds. There is nothing in a technology *per se* that automatically makes the technology in question 'amplify' human cognition or in some other way affect the human mind. To have such an effect a technology should first be integrated into meaningful activities and the context of a person's social relations. Only then, during the integration and as a result of subsequent internalization, may the technology transform human mental processes as well.

While four aspects of mediation – physical, cognitive, social and experiential – were identified above, the discussion in the chapter mostly deals with digital technology research, focusing on a subset of these aspects, namely, cognitive and experiential mediation. The reason why the other two aspects haven't received here the same level of attention is not because there is a lack of research into physical and social aspects of the design and use of digital technologies. On the contrary, there has been considerable interest in these issues in HCI and related areas – for instance, within the influential embodied interaction paradigm (Dourish, 2001). However, the research seldom applies the mediational perspective in a concrete and systematic way. For instance, considering human beings as actors-with-their-mediational-means (Wertsch, 1998) has direct implications for understanding physical embodiment. The mutuality between actors and their physical environments depends not only on the capabilities of the 'natural' human body but also on the meditational means, such as a knife, vaulting pole or prosthesis, available to the human. Apparently, the same applies to digital technologies (e.g., smart prostheses). However, the notion of a 'dynamic human', as either a physical or social entity, with some notable exceptions, has not yet become commonly accepted in HCI research.

A major problem with studies of the effects of technologies on the human mind is that they are typically limited in scope and thus often cannot generate sufficient evidence to properly analyze the effects. Accordingly, another direction for future digital technology research informed by the mediational perspective is to go beyond studying short-term interaction with individual artifacts and analyze complex realities of modern technology use, characterized by ecologies of artifacts, long-term developmental transformations and tight integration

of physical, cognitive, social and experiential aspects.

Some examples of existing research, including analyses of complex mediation (Bertelsen and Bødker, 2002; Bødker and Andersen, 2005) and long-term studies of technology-enabled fantasy worlds, including the 5th Dimension (Cole, 1996) and the POGO world (Rizzo et al., 2003), indicate that expanding the scope of analysis and technological support allows researchers to reveal phenomena that are impossible to reveal otherwise (such as 'webs of mediators'; Bødker and Andersen, 2005). As these studies suggest, placing technologies in the context of the activities they mediate and analyzing an interplay between all aspects of mediation, including material, social, cognitive and experiential, is an approach that has a potential to provide a more comprehensive understanding of how digital technology and the human mind are related to one another.

NOTES

1. This perspective is related to, but also different from studies of *media* – that is, channels and tools for storing and transmitting information (e.g. Bolter and Grusin, 2000). The latter are predominantly concerned with a medium as such, without necessarily considering how it is integrated into the structure of a purposeful human action.

2. An insightful historical analysis of how the experience of employing artifacts as tools in human activities is reflected in the evolution of artifacts' design is presented by Petroski (1992).

3. Not to be confused with remediation understood as representation of one medium in another (Bolter and Grusin, 2000).

REFERENCES

Bannon, L. (2011) 'Reimagining HCI: Toward a more human-centered perspective', *Interactions*, 18(4): 50–57.

Bateson, G. (1972) *Steps to an Ecology of Mind*. New York: Ballantine Books.

Beaudouin-Lafon, M. (2000) 'Instrumental interaction: An interaction model for designing post-WIMP user interfaces', *Proceedings of CHI '00*, ACM, 446–453.

Bertelsen, O. and Bødker, S. (2002) 'Interaction through clusters of artifacts', *Proceedings of ECCE '11 – Cognition, Culture and Design*, ACM, 103–111.

Bødker, S. (1991) *Through the Interface: A Human Activity Approach to User Interface Design*. Hillsdale, NJ: Lawrence Erlbaum.

Bødker, S. (2006) 'When second wave HCI meets third wave challenges', *Proceedings of NordiCHI '06*, ACM, 1–8.

Bødker, S. and Andersen, P.B. (2005) 'Complex mediation', *Human–Computer Interaction*, 20: 353–402.

Bødker, S. and Klokmose, N. (2011) 'The Human-Artifact Model: An activity theoretical approach to artifact ecologies', *Human–Computer Interaction*, 26: 315–371.

Bolter, J.D. and Grusin, R. (2000) *Remediation: Understanding New Media*. Cambridge, MA: MIT Press.

Bush, V. (1945) 'As we may think', *The Atlantic Monthly*, July. Available at: www.theatlantic.com/doc/194507/bush.

Cole, M. (1996) *Cultural Psychology: A Once and Future Discipline*. Cambridge, MA: Belknap Press of Harvard University Press.

Cole, M. and Engeström, Y. (1993) 'A cultural–historical approach to distributed cognition'. In: Salomon, G. (ed.), *Distributed Cognitions: Psychological and Educational Considerations*. Cambridge: Cambridge University Press.

Cole, M. and Griffin, P. (1983) 'A socio-historical approach to re-mediation', *The Quarterly Newsletter of LCHC*, 5(4): 69–74.

Dewey, J. (1910) *How We Think*. Boston, MA: D.C. Heath & Co Publishers.

Dewey, J. (1925) *Experience and Nature*. New York: Dover Publications.

Dourish, P. (2001) *Where the Action Is: The Foundations of Embodied Interaction*. Cambridge, MA: MIT Press.

Engelbart, D.A. (1963) 'A conceptual framework for the augmentation of man's intellect'. In: Howerton, P. (ed.), *Vistas in Information Handling*, Vol. 1. Washington, DC: Spartan Books, pp. 1–29.

Engeström, Y. (1987) *Learning by Expanding: An Activity-Theoretical Approach to Developmental Research*. Helsinki: Orienta-Konsultit Oy.

Gemmell, J., Bell, G. and Lueder, R. (2006) 'MyLifeBits: A personal database for everything', *Communications of the ACM*, 49(1): 88–95.

Guribye, F. (2005) *Infrastructures for Learning: Ethnographic Inquiries into the Social and Technical Conditions of Education and Training*. Bergen, Norway: The University of Bergen.

Hassenzahl, Marc (2010) *Experience Design: Technology for All the Right Reasons*. San Rafael, CA: Morgan & Claypool.

Heidegger, M. ([1927] 1962) *Being and Time*. English translation, 1962. New York: Harper & Row.

Heidegger, M. ([1954] 1993) 'The question concerning technology'. In: Krell, D.F. (ed.), *Martin Heidegger: Basic Writings from 'Being and Time' (1927) to 'The Task of Thinking' (1964)*. San Francisco, CA: Harper.

Hollan, J., Hutchins, E. and Kirsch, D. (2000) 'Distributed cognition: Toward a new foundation for human–computer interaction research', *ACM Transactions on Computer–Human Interaction*, 7(2): 174–196.

Hutchins, E. (1995) *Cognition in the Wild*. Cambridge, MA: MIT Press.

Ihde, D. (1990) *Technology and the Lifeworld*. Bloomington, IN: Indiana University Press.

James, W. (1890) *The Principles of Psychology*. New York: Henry Holt.

Kaptelinin, V. (1996a) 'Computer-mediated activity: Functional organs in social and developmental contexts'. In: Nardi, B. (ed.), *Context and Consciousness: Activity Theory and Human–Computer Interaction*. Cambridge, MA: MIT Press.

Kaptelinin, V. (1996b) 'Distribution of cognition between minds and artifacts: Augmentation or mediation?', *AI and Society*, 10: 15–25.

Kaptelinin, V. and Kuutti, K. (1999) 'Cognitive tools reconsidered: From augmentation to mediation'. In: Marsh, J., Gorayska, B. and Mey, J.L. (eds), *Humane Interfaces: Questions of Methods and Practice in Cognitive Technology*. Amsterdam: Elsevier/North-Holland, 145–160.

Kaptelinin, V. and Nardi, B. (2006) *Acting with Technology: Activity Theory and Interaction Design*. Cambridge, MA: MIT Press.

Kaptelinin, V. and Nardi, B. (2012a) 'Affordances in HCI: Toward a mediated action perspective'. *Proceedings of CHI '12*, ACM.

Kaptelinin, V. and Nardi, B. (2012b) *Activity Theory in HCI: Fundamentals and Reflections*. San Rafael, CA: Morgan & Claypool.

Koschmann, T., Kuutti, K. and Hickmann, L. (1998) 'The concept of breakdown in Heidegger, Leont'ev, and Dewey and its implications for education', *Mind, Culture, and Activity: An International Journal*, 5: 25–41.

Kozma, R.B. (1992) 'Constructing knowledge with learning tools'. In: Kommers, P.A.M., Jonassen, D.H. and Mayes, J.T. (eds) *Cognitive Tools For Learning*. Berlin/Heidelberg: Springer.

Latour, B. (1992) '"Where are the missing masses?": The sociology of a few mundate artifacts'. In: Bijker, W.E. and Law, J. (eds), *Shaping Technology/Building Society*. Cambridge, MA: MIT Press, 205–224.

Leont'ev, A.N. (1978) *Activity, Consciousness, and Personality*. Englewood Cliffs, NJ: Prentice Hall.

Leont'ev, A. (1981) *Problems in the Development of the Mind*. Moscow: Progress Publishers.

Mamykina, L., Miller, A.D., Mynatt, E.D., and Greenblatt, D. (2010) 'Constructing identities through storytelling in diabetes management', *Proceedings of CHI '10*, ACM, 1203–1212.

Mann, S. (1997) 'An historical account of the "WearComp" and "WearCam" inventions developed for applications in "Personal Imaging"', *The First International Symposium on Wearable Computers: Digest of Papers, IEEE Computer Society*.

McCarthy, J. and Wright, P. (2004) *Technology as Experience*. Cambridge, MA: MIT Press.

Miettinen, R. (2006) 'Epistemology of transformative material activity: John Dewey's pragmatism and cultural–historical activity theory', *Journal for the Theory of Social Behaviour*, 36(4): 389–408.

Mistry, P., Maes, P. and Chang, L. (2009) 'WUW – Wear Ur World – A Wearable Gestural Interface', *CHI '09 Extended Abstracts*, ACM.

Nardi, B. (ed.) (1996) *Context and Consciousness: Activity Theory and Human–Computer Interaction*. Cambridge, MA: MIT Press.

Norman, D. (1991) 'Cognitive artifacts'. In: Carroll, J. (ed.), *Designing Interaction: Psychology at the Human–Computer Interface*. Cambridge: Cambridge University Press.

Norman, D. (1993) *Things That Make Us Smart: Defending Human Attributes in the Age of the Machine*. Reading, MA: Addison-Wesley.

Papert, S. (1980) *Children, Computers, and Powerful Ideas*. New York: Basic Books.

Pea, R.D. and Kurland, D.M. (1984) 'On the cognitive effects of learning computer programming', *New Ideas in Psychology*, 2(2): 137–168.

Perkins, D.N. (1993) 'Person-Plus: A distributed view of thinking and learning'. In: Salomon, G. (ed.) *Distributed Cognitions: Psychological and Educational Considerations*. Cambridge: Cambridge University Press.

Petroski, H. (1992) *The Evolution of Useful Things: How Everyday Artifacts – From Forks and Pins to Paper Clips and Zippers – Came to Be as They Are*. New York: Vintage Books.

Rabardel, P. and Bourmaud, G. (2003) 'From computer to instrument system: A developmental perspective', *Interacting with Computers*, 15: 665–691.

Rizzo, A., Marti, P., Decortis, F., Moderini, C. and Rutgers, J. (2003) 'The design of POGO story world'. In: Hollnagen, E. (ed.) *Handbook of Cognitive Task Design*. Mahwah, NJ: Lawrence Erlbaum, 577–602.

Rohs, M. and Oulasvirta, A. (2008) 'Target acquisition with camera phones when used as magic lenses', *Proceedings of CHI '08*, ACM, 1409–1418.

Roy, D. (2004) '10x: Human–machine symbiosis', *BT Technology Journal*, 22(4): 1–5.

Salomon, G. (ed.) (1993a) *Distributed Cognitions: Psychological and Educational Considerations*. Cambridge: Cambridge University Press.

Salomon, G. (1993b) 'No distribution without individual's cognition: A dynamic interactional view'. In: Salomon, G. (ed.), *Distributed Cognitions: Psychological and Educational Considerations*. Cambridge: Cambridge University Press.

Sellen, A. and Whittaker, S. (2010) 'Beyond total capture: A constructive critique of lifelogging', *Communications of the ACM*, 53(5): 70–77.

Svanæs, D. (2000) 'Understanding interactivity – steps to a phenomenology of human–computer interaction'. PhD dissertation, Trondheim, Norway: NTNU. Available at: http://dag.idi.ntnu.no/interactivity.pdf.

Verbeek, P.P. (2006) 'Acting artifacts: The technological mediation of action'. In: Verbeek, P. and Slob, A. (eds), *User Behaviour and Technology Development: Shaping Sustainable Relations Between Consumers and Technologies*. Dordrecht, The Netherlands: Springer, 53–60.

Vygotsky, L.S. (1978) *Mind and Society*. Cambridge, MA: Harvard University Press.

Vygotsky, L. ([1930] 1982) 'Instrumental method in psychology' [Instrumentalnyj metod v psikhologii]. In: Luria, A. and Yaroshevsky, M. (eds), *L.S. Vygotsky: Collected Works, Vol. 1*. Moscow: Pedagogika (in Russian).

Vygotsky, L. ([1930] 1984) 'Tool and sign in child development' [Orudie I znak v razvitii rebenka]. In: Yaroshevsky, M. (ed.), *L.S. Vygotsky: Collected Works, Vol. 6*. Moscow: Pedagogika (in Russian).

Wertsch, J. (1998) *Mind as Action*. New York: Oxford University Press.

Whittaker, S., Tucker, S., Swampillai, K. and Laban, R. (2008) 'Design and evaluation of systems to support interaction capture and retrieval', *Personal and Ubiquitous Computing*, 12: 197–221.

Zimmerman, J. (2009) 'Designing for the self: Making products that help people become the person they desire to be', *Proceedings of CHI '09*, ACM, 395–404.

15

Ethnomethodology and Conversation Analysis: Empirical Approaches to the Study of Digital Technology in Action

Robert J. Moore

INTRODUCTION

Although they represent radical departures from the mainstream of their home discipline of sociology, ethnomethodology and conversation analysis (EM/CA) have played more prominent roles in the areas of human–computer interaction (HCI) and computer supported cooperative work (CSCW). EM/CA touch many of the topics in this volume including experience, taken-for-granted or tacit knowledge, embodiment, gaze, multimodal interaction, design, video analysis, ethnography, phenomenology, and more. However, despite their influence, these approaches are often poorly understood or conflated with other approaches, such as ethnography (Dourish and Button, 1998; Crabtree et al., 2009).

What unites ethnomethodology and conversation analysis is a shared concern with discovering the endogenous orderliness or local organization of concrete social activities. What divides them at times is their analytic goals and methods. Ethnomethodology arose

as an alternative to mainstream sociological theory (e.g. Parsons, 1937), and it produces empirical studies of situated human action, as well as theoretical 'respecifications' of classic topics in the human sciences (Button, 1991). Emerging at roughly the same time, conversation analysis focused on a particular area of ethnomethodological phenomena – talk-in-interaction – and produced a novel methodology with the aim of building a rigorous, empirical science (Sacks, 1984).

The goal of this chapter is to provide a practical framework for understanding the fields of ethnomethodology and conversation analysis with respect to each other and to studies of digital technologies. It will *not* provide a detailed account of EM/CA studies and their contributions to sociology and other fields, which can be found elsewhere (Garfinkel, 1967; Heritage, 1984; Suchman, 1987; Lynch, 1993; Button et al., 1995; Dourish and Button, 1998). Nor will it provide a detailed account of EM/CA research methodologies (see Jordan and Henderson, 1995; Psathas, 1995; Schegloff, 2007). Instead the goal is to provide:

(1) a brief overview of the basic theory and methods of EM/CA; (2) a brief overview of the literature, with a focus on studies of digital technologies; and (3) an example of just one of the many ways in which these fields can be adapted to the study of digital technology use.

ETHNOMETHODOLOGY

Ethnomethodology was established by Harold Garfinkel in the 1960s. Garfinkel (1968: 16–17) explains that his choice of the term 'ethnomethodology' was inspired by terms like ethnobotany, ethnophysiology or ethnophysics. 'Ethno' refers to a people's or member's body of knowledge. For example, an anthropologist might study the 'ethnobotany' of a tribe or the tribe's own knowledge and methods for dealing with botanical matters. In a similar way, Garfinkel is concerned with people's methodologies, but methodologies *for what?* The term itself does not specify. It is people's own methodologies for achieving recognizable social order with which Garfinkel is concerned. He turns a kind of anthropological eye back on his own society and makes it look strange in order to examine its mundane workings, which ordinarily go unnoticed due to their intimate familiarity. 'It is the organizational study of a member's knowledge of his ordinary affairs, of his own organized enterprises, where that knowledge is treated as part of the same setting that it also makes orderable' (Garfinkel, 1968: 18). However, Garfinkel (1968: 18) himself suggests that the term itself may be a 'mistake' because 'it has acquired a life of its own' and that it could just as well be discarded and replaced with 'neopraxiology'. This flexibility toward the label of his own field reflects a more general tendency in Garfinkel's writings to resist the reifying of his ideas into static concepts that become too familiar.

Garfinkel's early articulations of ethnomethodology were influenced in part by the social phenomenology of Alfred Schütz (1962).

Schütz (1962) examines what he calls the 'natural attitude of daily life' or the taken-for-granted expectations that underlie individuals' experience of the everyday social world. Like phenomenology, ethnomethodology is concerned with the *experience* of everyday life, while 'bracketing' questions of its objective causes or 'true states', for the sake of descriptive analysis. However, unlike phenomenology, ethnomethodology focuses on the *behavioral* aspects of human experience, while further bracketing questions of 'inner states.' Thus, ethnomethodology relies on *observation* rather than introspection as its primary mode of inquiry.

Garfinkel (1991: 13) critiques sociology, and other human sciences, for characterizing concrete social activities as 'noisy' or chaotic and therefore as places where *order* is 'not to be found'. Sociologists, such as Talcott Parsons, account for social order by proposing theoretically derived concepts that are removed from concrete activities themselves. For example, Parsons accounts for social order in part through the theorized *internalization* of values and norms (Heritage, 1984: 18–19), but Garfinkel (1967: 67–69) argues that this kind of account makes the individual into a 'judgmental dope', who merely enacts societal scripts and rules of conduct. Garfinkel's early work (1963, 1967) goes to great lengths to demonstrate that concrete, situated activities require judgment, improvization and even 'artful' work on an ongoing basis. At the beginnings of encounters there are no clear 'definitions of the situation' by which one can identify the appropriate norms or scripts to enact (Garfinkel, 1963: 200). Garfinkel (1967: 32–33) conceives of values, norms or rules, not as independent explanations of social situations, but as *features of* them. This 'reflexive' character of rules, norms and any other accounts of social activities, means that all such accounts are *resources* for participants in some situation and therefore are potential topics for ethnomethodological enquiry (Garfinkel, 1967: 1–4).

Garfinkel thus takes a radically different view of social order from that of contemporary sociologists. Garfinkel respecifies Emile Durkheim's classic aphorism, 'the objective reality of social facts is sociology's fundamental principle', for ethnomethodology as, 'the objective reality of social facts is sociology's fundamental *phenomenon*'. Where 'social facts' are further respecified as 'society's locally, endogenously produced, naturally organized, reflexively accountable, ongoing, practical achievement, being everywhere, always, only, exactly and entirely, members' work' (Garfinkel, 1991: 11). In other words, concrete social activities are orderly in themselves, and this order is achieved in and through people's own methods.

Garfinkel's respecification of sociological phenomena makes a distinctive contribution to sociological theory, of which Maynard and Clayman (1991: 387) explain:

> Ethnomethodology's 'incommensurate' ... theoretical proposal is that there is a self-generating order in concrete activities, an order whose scientific appreciation depends upon neither prior description, nor empirical generalization, nor formal specification of variable elements and their analytic relations. From an ethnomethodological standpoint, 'raw' experience is anything but chaotic, for the concrete activities of which it is composed are coeval with an intelligible organization that actors 'already' provide and that is therefore available for scientific analysis.

Garfinkel (1967) shows that the researcher does not need theories of *external* forces impinging on social settings, nor forces *internal* to the participants guiding their behavior, in order to find *order* in concrete activities. Direct observation of situated social activities reveals an endogenous orderliness or local organization. This endogenous organization further exhibits a useful feature that the researcher can exploit. Because the participants themselves must publicly display the orderly character of their actions *to each other* in order to coordinate their actions jointly, their methods are publicly available for researchers to observe as well. Garfinkel (1967: 1) writes, 'the activities whereby members produce and manage settings of organized everyday affairs are identical with members' procedures for

making those settings "account-able"'. *Accountable* here means 'observable and reportable' (Garfinkel, 1967) or recognizable. Anyone competent in the social practices in question can see *what* the participants are doing and whether they are doing it well, poorly, unusually, etc.

Garfinkel (1963, 1967) initially examined mostly ordinary activities – that is, ones any competent member of a society knows, including, playing tic-tac-toe, asking for clarification, requesting help from shop clerks, deliberating on juries, talking to counselors, doing 'being a natural, normal female', and more. However, the ethnomethodological perspective is not limited to commonsense knowledge and practices. In later work, Garfinkel and his colleagues turned to activities in more technical, professional settings: astronomy (Garfinkel et al., 1981), neuroscience (Lynch, 1985), mathematics (Livingston, 1986), logic (Coulter, 1991), software development (Button and Sharrock, 1996), and more. One concern in these so-called 'ethnomethodological studies of work' was to get to the core of the professions under study. Among his students, Garfinkel (see Lynch, 1993: 271) often referred to Sacks's critique of studies in social science that 'miss the interactional what' of particular professions. For example, Becker's (1963) classic ethnography of jazz musicians reveals a great deal about what they say and do *around* the activity of playing jazz, such as using drugs, but reveals surprisingly little about the *actual work* of 'playing jazz' itself. Garfinkel (see Lynch, 1993: 271) points instead to Sudnow's (1978) account of jazz improvization, which examines such embodied work in detail, as an alternative to Becker's (1963) ethnographic approach.

Investigating the situated work of astronomers or neuroscientists as a non-member of those communities is inherently challenging. A sociologist can perhaps see what makes astronomers' work 'talk and collaboration', but not necessarily what makes it 'astronomy', much less 'good astronomy' or 'bad astronomy'. Garfinkel and Wieder (1992: 182) describe this challenge in grasping the orderliness of professional work as the 'unique adequacy requirement of methods':

> In its weak use the unique adequacy requirement of methods is identical with the requirement that for the analyst to recognize, or identify, or follow the development of, or describe phenomena of order in local production of coherent detail the analyst must be vulgarly competent in the local production and reflexively natural accountability of the phenomenon of order he is 'studying'.

In other words, a uniquely adequate analysis of the work of technical settings requires some level of competence in the practices that constitute those settings by the researchers themselves and therefore some level of participant observation or personal participation. It is only through learning to *do the work* that a researcher or practitioner can begin to see and capture the distinctive orderly features or 'just thisness' (Garfinkel and Wieder, 1992) of that work.

Methodologically Garfinkel (1963, 1967) employed 'breaching experiments', or demonstrations, in which he intentionally violated taken-for-granted practices in order to reveal their accomplished nature through subjects' attempts to restore order. Such demonstrations included making a mark in tic-tac-toe on a line between cells, treating a fellow shopper as a clerk and ignoring his or her every attempt correct the misidentification, repeatedly asking for clarification of the same mundane utterance, providing random yes/no responses as serious counselor advice and more. In addition to eliciting such situations, Garfinkel also searched for naturally occurring 'perspicuous settings', which throw the accomplished nature of social activities into relief, such as an interview with a transgendered person about how she very self-consciously *does* 'being a normal, natural female' (Garfinkel, 1967: ch. 5) or a recording of a conversation among astronomers as they discover a new pulsar (Garfinkel et al., 1981).

Since its inception, ethnomethodology has developed into a diverse field (Maynard and Clayman, 1991). This diversity is manifested in part in the *deconstructive* and *constructive* dimensions of ethnomethodology (Clayman, 1995: 110–111). On the one hand, it critiques sociology and the human sciences for theorizing away the endogenous order of concrete social activities. This deconstructive tendency has produced 'an empirically-based form of critique whose intellectual content runs contrary to established philosophical accounts of the sciences' (Clayman, 1995: 115–116). On the other hand, ethnomethodology also contains a constructive dimension in that it reveals a new domain of sociological phenomena, ripe for empirical investigation. Harvey Sacks saw this constructive potential in ethnomethodology and from it developed a new empirical *science*.

CONVERSATION ANALYSIS

Conversation analysis (CA) was originally developed by Harvey Sacks and his colleagues, Emanuel Schegloff and Gail Jefferson, beginning in the 1960s, concurrently with ethnomethodology. CA emerged during a time in which many in the human sciences were writing about social interaction and language use. Sacks was once a student of Erving Goffman, with whom he shared a concern for the organization of *social interaction*; however, Sacks and his colleagues developed a novel methodology in part to remedy shortcomings in his former advisor's approach (Goffman, 1959). Sacks's new approach also shares concerns with the work of John Austin (1962) and John Searle (1969) in philosophy for understanding *how* social actions can be accomplished through words (Heritage and Atkinson, 1984: 5; Sacks, 1993 vol. I: 343). However, Sacks breaks with Goffman, Austin and Searle methodologically and critiques their approaches for employing *imagined examples* of isolated sentences. The problem with these kinds of approaches is not only that imagined cases may be inaccurate but also, as Sacks (1984: 25) points out, it means the researcher is 'constrained by reference to what an audience, an audience of professionals, can accept as reasonable'. Such a program therefore is *philosophical* rather than empirical. Instead Sacks aimed to study social interaction and language use empirically using actual instances of utterances rather than imagined ones. He embraced Garfinkel's notion of a 'self-generating order in concrete activities' (Maynard and Clayman, 1991) and set out to discover it. Sacks (1984: 25) writes, 'We will be using observation as a basis for theorizing. Thus we can start with things that are not currently imaginable, by showing that they happened'.

Sacks incorporated key elements of Garfinkel's emerging ethnomethodology – such as the discovery of members' methods for accomplishing social activities – into his own emerging approach. But Sacks's greatest, and most controversial (see Lynch, 1993: ch. 6), contribution to ethnomethodology was methodological. Sacks (1984: 26) admits that his subject matter of interest was chosen primarily for methodological reasons:

> It was not from any large interest in language or from some theoretical formulation of what should be studied that I started with tape-recorded conversations, but simply because I could get my hands on it and I could study it again and again, and also, consequently, because others could look at what I had studied and make of it what they could, if, for example, they wanted to be able to disagree with me.

Sacks was concerned from the start with creating a more rigorous way to study social interaction. Using tape recordings, he, along with Schegloff and Jefferson, developed a distinctive method for analyzing the endogenous orderliness of concrete social activities in conversations. With a unique transcription scheme, devised by Jefferson (2004) to capture primarily the *temporal* relationship between speakers' utterances, they demonstrated how 'talk-in-interaction' is sequentially organized by its participants on a local, turn-by-turn basis. Tapes and transcripts became materials, or 'data', shared among conversation analysts, which enabled a thriving, cumulative literature to emerge.

Sacks's approach was not only a departure from those of Goffman, Austin and Searle but, it was also a departure from that of ethnography, which traditionally employs fieldnotes as its primary source of data. Although fieldnotes may be adequate for constructing 'thick descriptions' of social activities (Geertz, 1973), they are inadequate for capturing the fine details of how those activities are produced in real time. Maynard (1989: 130–131) writes:

> With observation studies, the reader must depend on the ethnographer's skills and capacities for reliable note taking, at the very least. ... With recorded data, the audience is invited to inspect actual data along with the investigator, whose analysis can therefore be checked with what the data exhibit in relatively pristine form.

Even though ethnographers today tend to employ all of the latest recording devices when collecting observational data, their continued focus on participants' subjective *meanings* makes it difficult for fellow researchers to verify analytic claims and build on them.

Yet despite the important differences between conversation analysis and ethnography, many conversation analysts use ethnographic methods as part of their studies. As Maynard (2003: 65) argues, the proper relationship between CA and ethnography is one of 'limited affinity'. Although 'ethnographic context' cannot simply be added to a sequential analysis without first demonstrating its *relevance* and *procedural consequentiality* (Schegloff, 1991), ethnography can nonetheless inform descriptions of settings and identities, explications of special terms, phrases or courses of action, and explanations of 'curious' patterns (Maynard, 2003: 73–77). An example of the latter can be found in a study of telephone survey interviews (Moore and Maynard, 2002) in which sequential analysis of the recorded interviews reveals that, consistently, when respondents request clarification of a survey question, interviewers respond with *verbatim repeats* of the question rather than with paraphrases. While this pattern could be interpreted as evidence of a lack of competence on the part of interviewers, ethnography reveals that it is in fact the opposite. Avoiding the unscripted paraphrasing of survey questions is one method through which interviewers recognizably *do* 'standardized, unbiased interviewing'. While the merits of this practice can be debated, the ethnographic investigation of the survey center in the study (Moore and Maynard, 2002) helps explain *why* the seemingly curious sequential pattern recurs in the talk.

Since the early lectures of Sacks (1993), conversation analysis has evolved into an established field that examines the *sequential organization* of talk-in-interaction, or, how people accomplish social activities through their talk and embodied actions on a turn-by-turn basis. Recurrent topics for conversation analysts include turn-taking (Sacks et al., 1974), sequence organization (Schegloff, 2007), repair (Schegloff et al., 1977), turn design, and much more. No conversation analyst has investigated and specified the 'machinery' of ordinary conversation more than Emanuel Schegloff. After Sacks's untimely death in 1975, Schegloff emerged as the most influential figure in the nascent field. Maynard (in press) points out that one of Schegloff's 'crowning contributions' to conversation analysis is working with 'collections' of cases 'rather than just single instances where Sacks's intention was "to isolate structure in particulars"'. It is through collections that *recurrent* patterns and *generic* practices are discovered.

Like ethnomethodology, which began with ordinary social practices and expanded into the analysis of technical settings, conversation analysis likewise began with the specification of 'ordinary conversation' and expanded into studies of 'institutional talk' – that is, talk-in-interaction in professional and technical work settings (Boden and Zimmerman, 1991; Drew and Heritage, 1992). In addition, with the affordability of consumer-quality video cameras, CA expanded from the analysis of talk and audiotapes to the analysis of embodied interaction and video recordings (Goodwin, 1979; Streeck, 1996; Schegloff,

1998; Hindmarsh and Heath, 2000; Moore, 2008). Studies of embodied interaction examine how gesture, body orientation, eye gaze and artifacts are used by participants, in conjunction with their talk, in locally organizing face-to-face interactions.

Conversation analysts' adoption of a common methodology, involving mechanical recordings, transcription conventions and an analytic focus on sequential organization, has enabled the field to generate a sizable and cumulative body of findings. Not only are there numerous studies of talk-in-interaction in a variety of settings but also these studies reference each other in organic ways. Clayman (1995: 115) explains:

> Consistent with its predominantly data-driven and analytically inductive methodology, conversation analysis is devoted to explicating a progressively expanding array of interactional practices. Moreover, results in this area have been strongly cumulative in the sense that established findings have served as a foundation for subsequent investigations.

In the CA literature, generic practices have been discovered across multiple settings and adaptations of generic practices have been found in specialized settings. For example, 'repair', the redoing of a turn or component of a turn, in interaction, was first examined in ordinary (English) conversation (Schegloff et al., 1977; Schegloff, 1992) but has since been identified and examined in a variety of other forms of interaction, from 'institutional' talk (McHoul, 1990; Zimmerman, 1992; Moore and Maynard, 2002; Macbeth, 2004; Arminen et al., 2010) to adult–child interaction (Wootton, 1994; Forrester, 2008) to non-English conversation (Kim, 2001; Wu, 2006; Rosenthal, 2008; Maheux-Pelletier and Golato, 2008) and even to human–computer interaction (Suchman, 1987; Frohlich and Luff, 1990; Frohlich et al., 1993; Moore et al., 2011). These studies and many others demonstrate that 'repair' in interaction is a recurrent phenomenon across a diversity of settings. Despite the fact that CA studies individually tend to examine a relatively small number of cases, when compared with quantitative approaches, the cumulative nature of its literature provides for a distinctive kind of generality *across* studies.

In summary, Sacks and his colleagues picked up Garfinkel's concern for discovering the practices in and through which social activities are endogenously organized and locally achieved. While Garfinkel and his colleagues took ethnomethodology in the direction of 'empirically-based critique' of the human sciences (Clayman, 1995: 115), Sacks and his colleagues took it in the direction of a new inductive, empirical method, which has resulted in a thriving literature that crosses multiple disciplinary boundaries.

ETHNOMETHODOLOGY, CONVERSATION ANALYSIS AND TECHNOLOGY

From the 1980s, ethnomethodology and conversation analysis have been influential among computer scientists in the areas of human–computer interaction (HCI) and computer supported cooperative work (CSCW). With her classic book, *Plans and Situated Actions* (1987), Lucy Suchman helped introduce computer scientists to EM/CA and a concern for the situated, accomplished nature of human action. Working with computer scientists in an interdisciplinary laboratory at the famed Xerox Palo Alto Research Center (PARC), Suchman adapted the theory and methods of EM/CA to the area of human–machine interaction and technology design.

Suchman (1987) critiques the 'planning model' in cognitive science and artificial intelligence, according to which human behavior and machine behavior are much alike: both are determined by internal schemas, rules, norms or plans. Artificial intelligence attempts to create intelligent machine behavior by endowing systems with generic plans that would guide their behavior, much like the cognitive scripts and schemas of human behavior. Drawing on ethnomethodology, Suchman (1987) argues that the planning model is based on false assumptions about human behavior. Instead Suchman (1987) reviews ethnomethodological notions of *rules as resources*, rather than as causes of behavior, and highlights the emergent, contingent aspects of human action when it is situated in actual circumstances instead of imagined ones. Suchman (1987) thus invites computer scientists to take the ethnomethodological challenge of observing 'situated actions' and use *those* as a basis of intelligent system design instead of a theoretical model of human actors conceived as 'judgmental dopes'.

In addition to an ethnomethodological critique of the 'planning model', Suchman (1987) offers a novel adaptation of conversation analysis to the examination of human–machine communication or, more accurately, human–human–machine communication. Suchman (1987: ch.7) examines the use of an expert system embedded in a Xerox production photocopier. Rather than studying a solitary user's interaction with the system, she examines *pairs* of users as they interact both with the machine and with each other. In collaborating, the two users naturally work to make the sense of their actions accountable to each other. Suchman then exploits the *talk* between users to make their practical reasoning in using the *system* more visible.

Following conversation analytic concerns for data, Suchman (1987: ch.7) videotaped the interactions between users and the photocopier. In addition, she devised a novel transcription notation scheme, inspired by that of Gail Jefferson (2004), to represent the details of the users' actions and the system's responses. With mechanical recordings and detailed transcripts, Suchman (1987: ch.7) proceeds to offer a sequential analysis of breakdowns in human interactions with the machine and the machine's programmed logic. Suchman (1987: ch.7) shows how users and system reach interactional impasses. In terms of system design, a major limitation of the system is that it is insensitive to the particulars of the user's situation (Suchman, 1987: 169) and instead offers only generic instructions.

Although Suchman's (1987: ch.7) method for analyzing human–machine communication has not been widely emulated by others (with the exception of Frohlich et al., 1993), her invitation to examine *situated actions* has resonated with computer scientists. The basic notions of examining systems in real-world situations and user interactions on a fine-grained, turn-by-turn basis have led to design insights at a very concrete, actionable level. Ethnomethodological and conversation analytic studies in HCI and CSCW span a wide range of phenomena from computer-mediated communication (e.g. McIlvenny, 1990; Garcia and Jacobs, 1999; Hindmarsh et al., 2001; Moore et al., 2007a, 2007b; Szymanski et al., 2006) to cooperative work around computers (e.g. Luff and Heath, 1993b; Hindmarsh and Heath, 2000; Whalen and Vinkhuyzen, 2000) to systems design (e.g. Dourish and Button, 1998; Crabtree, 2003; O'Neill et al., 2011). A selection of these studies is highlighted below.

Beyond the Telephone

Perhaps the most natural application of ethnomethodological and conversation analytic work in the area of digital technology is the study of technologically mediated communication, a topic of interest in CA since its inception. In his classic paper and dissertation on openings in ordinary telephone calls, Schegloff (1968) analyzes how two parties in a telephone conversation coordinate the openings of calls, including mutual recognition of the other. Schegloff (1968) describes how a single 'deviant' or 'perspicuous' case out of 500, instead of being thrown out as an 'outlier', changed his analysis of the entire corpus. A single breakdown in the opening, in which a *caller* starts with 'hello' instead of the call recipient, enabled Schegloff to see that, in all of the other cases, the normal 'hello' by the call recipient is actually a 'second-pair part'. That is, the second turn in a two-turn sequence, the first turn being the mechanical ringing of the telephone, which acts as a 'summons'. Since Schegloff's (1968)

study, recordings of telephone calls have been a popular form of data for conversation analysts.

As new communication technologies emerge, conversation analysts gravitate toward examining the new forms of talk they make possible. For example, Szymanski et al. (2006) examine how the organization of push-to-talk mobile phone calls differs dramatically from traditional telephone calls, like those examined by Schegloff (1968). Mobile push-to-talk technology, which behaves more like handheld radios than traditional landline phones, enables a 'continuing state of incipient talk' (Schegloff and Sacks, 1973). Such talk is characterized by lengthy lapses in talk not found in traditional telephone conversations and characteristic of co-present parties, as well as the absence of greetings each time the conversation resumes (Szymanski et al., 2006).

Garcia and Jacobs (1999) examine turn-taking in Internet relay chat (IRC) and how it differs from that of ordinary conversation (Sacks et al., 1974). They identify the interactional consequences of a feature of the system: that turns-at-chat only become visible to the recipients after they are fully formed. Composition of the turn itself is private. As a result, the participants in IRC cannot monitor the other's turn-in-progress and therefore cannot achieve the 'one speaker at a time' feature of ordinary conversation nor the adjacency of related turns nor minimal gap and overlap by anticipating the ends of turns (Garcia and Jacobs, 1999). Moore et al. (2007a, 2007b) find the same interactional consequences of IRC-style chat systems in 3D virtual worlds (see 'Example: embodiment in avatar-mediated interaction' below).

Hindmarsh et al. (2001) examine referential practice with respect to objects in 3D virtual environments, which use voice instead of text chat. Limited fields of view mean that users are unable to take for granted what Alfred Schütz (1962) calls the 'reciprocity of perspectives'. That is, they cannot reliably assume how the spatial setting must appear to the other. In an early virtual 3D world, Hindmarsh et al. (2001) examine the referential practices of users and find that the system inadequately provides for simple acts of pointing that are easy to accomplish in face-to-face interaction. Similarly, in the domain of *physical* avatars, or robot-mediated social interaction, Kuzuoka et al. (2004) identify the 'problem of projectability'. They demonstrate the importance of providing the remote participant with resources in the robot itself for making references to objects projectable – that is, enabling recipients to *anticipate* their endpoints. Kuzuoka et al. (2004) further offer a system, GestureMan, that enables the projectability of pointing references, by relaying the remote participant's direction of eye gaze and pointing through the robot's head and arm.

In the Wild

In addition to technologically mediated communication, ethnomethodological and conversation analytic studies of technology have also examined the work settings in which technologies are situated. Combining EM/CA approaches with technical settings, this body of work, often known as 'workplace studies' (Luff et al., 2000), examines work practice in technology-heavy professional settings. In producing analyses of the local organization of situated work, such studies tend to offer recommendations for improving the design of workplace technologies and work processes so that they better support the work practices.

Whalen (1995) makes a natural extension of the analysis of telephone conversations to workplace settings by examining emergency calls phone. In addition to analyzing the sequential organization of the calls, Whalen examines other work of the call taker in the context of the call center. The call taker must record key details from the calls in a computer system in a standardized format, communicate with Dispatch and with others who are co-present and sitting nearby in the call center. Whalen takes Garfinkel and Wieder's

(1992) 'unique adequacy requirement of methods' seriously by conducting participant observation in the call center and working *as* a call taker, answering live calls over the course of the study.

Suchman (1997) categorizes studies of such call centers as those of 'centers of coordination' and groups them with ethnomethodological studies of air traffic control centers (Harper and Hughs, 1993), airport ground operations (Suchman, 1993; Goodwin and Goodwin, 1996) and line control in the London Underground (Heath and Luff, 1992). Suchman (1997: 42) proposes that such centers of coordination are characterized by 'problems of space and time, involving the deployment of people and equipment across distance, according to a canonical timetable or the emergent requirements of rapid response to a time-critical situation'. What makes such studies of interest to technologists is the fact that they examine 'technology intensive forms of practice.' They demonstrate how technology use is embedded in larger ecologies of work.

Many of these workplace studies were conducted by or in collaboration with researchers at Xerox research centers in Palo Alto, California, Cambridge, England, and Grenoble, France. The Xerox Corporation has no doubt been the single largest patron of ethnomethodological workplace studies to date. While Xerox supported studies in a wide variety of types of workplaces, many of them deal with Xerox's core business: xerographic printing. In addition to Suchman's (1987) study of photocopier interaction, Button and Sharrock (1997) examine work practices on the shop floor of a large government printing office in the United Kingdom. They offer a detailed account of how production schedules are produced and managed in practice within an individual factory and draw implications of this for a proposal to distribute print jobs across the printing office's multiple locations. In addition, Moore (2008) and Moore et al. (2010) study face-to-face interactions between customers and employees in retail print shops. They examine both how customers formulate order requirements, often with the use of gestures, and how those embodied descriptions get rendered into standardized accounts by employees with the use of paper order forms. Building on earlier workplace studies of telephone support calls between customers and copier technicians, which identify 'lack of mutual access' to the machine as a pervasive problem, O'Neill et al. (2011) demonstrate a mixed-reality support system tailored to the troubles of the work practice. The system simulates the experience of customer and remote technician standing over the machine together: a virtual 3D representation of a particular model copier is animated by sensors on an actual machine in a customer site. The system thus enables the customer to 'show' the technician the problem with the machine and the technician to 'show' the customer how to fix it. O'Neill et al. (2011) provide an exemplar of how ethnomethodological workplace studies can lead to innovative technology design.

Beyond the Social

While many ethnomethodological and conversation analytic studies have examined social interaction *through* technology and *around* technology, fewer studies have examined interaction *directly with* technology, despite Suchman's (1987) classic example. This raises the question of whether EM/CA are truly applicable to HCI or they are useful only for forms of *social* interaction. A minority of studies demonstrate the former.

The most obvious extension of ethnomethodology and conversation analysis to human–machine communication is perhaps that involving systems which attempt to simulate human conversation on the part of the machine. For example, Frohlich and Luff (1990) and Raudaskoski (1990) use findings from conversation to inform the design of dialogue interfaces between a user and a system. In addition, Yamazaki et al. (2009) use ethnomethodological analysis of embodied interactions between tour guides and museum visitors to inform the design of a museum robot that can coordinate its utterances and body movements to monitor the responses of visitors. In both cases, systems and robots are designed to *mimic* certain interactional behaviors from human conversation.

In addition to interactions with machines that simulate human behavior, a few studies examine a more pervasive form of HCI, namely single users' interactions with personal computers. Frohlich et al. (1993) examine user interaction with an X-Windows environment, an early graphical user interface (GUI) for networked workstations and with a proprietary database query system at Hewlett-Packard Labs. Like Suchman (1987), Frohlich et al. (1993) exploit the talk of a pair of users to make their work at the computer more publicly visible. They videotape the computer screen around which a user and an assistant sit and capture their conversation on the audio track. They then transcribe the talk in CA fashion and insert notations for screen and input events into the transcript. As in Suchman's (1987) transcripts, the talk between users plays a prominent role in the transcripts of Frohlich et al. (1993). Using these data, Frohlich et al. (1993) examine *repair* in the context of user interaction with the database program.

In a similar study, Luff and Heath (1993a) examine user interaction with the Macintosh's GUI and focus on the phenomena of *menu use* in the context of Microsoft PowerPoint. Like Frohlich et al. (1993), Luff and Heath (1993a) collect video recordings of the users' computer screen along with their verbal comments; however, instead of user pairs, they videotape solitary users. Consequently users' talk plays a more minimal role in Luff and Heath's (1993a) transcripts and their resulting analysis. The focus is much more on the screen-based interaction between single user and system.

More recently, Moore et al. (2011), Moore and Churchill (2011) and Moore (2013) demonstrate a novel adaptation of conversation analysis to GUI-based user interaction, which they call 'computer interaction analysis'. Like Luff and Heath (1993a), they examine solitary users' interactions with systems through screen video but *without* audio of users' talk. They use digital screen capture and eye-tracking technology to capture video of users' screens with the constantly changing targets of their eye gaze superimposed. They offer a novel transcription scheme for capturing the temporal relationships among user input events, display events and eye-gaze events. Using this approach, they examine both the phenomena of *query repair* (Moore et al. 2011) and *referential* practice (Moore, 2013) in the context of user interactions with Internet search engines (i.e. Yahoo!, Google, Bing).

Finally, Dourish and Button (1998) argue that ethnomethodology and conversation analysis are not only useful for the empirical investigation of technology use but they can also inform system *design* itself at a more fundamental, conceptual level. Dourish and Button (1998) demonstrate how concepts from ethnomethodological theory can contribute directly to computer programming. They argue that combining Garfinkel's (1967) concept of the 'accountability' of human action with the notion of 'abstraction' in system programming can create a powerful style of programming in which system behavior becomes accountably rational. Drawing inspiration from the practice of 'open implementation', Dourish and Button (1998) argue that systems' behaviors can be made more accountable to users and programmers if more of their mechanism is revealed to the user, instead of black-boxing it as is traditionally the practice. Dourish and Button (1998) call this kind of application of ethnomethodological concepts to the technology development process, 'technomethodology'.

These studies demonstrate that ethnomethodology and conversation analysis are indeed applicable to HCI, in addition to social interaction through and around technology.

EXAMPLE: EMBODIMENT IN AVATAR-MEDIATED INTERACTION

This section offers one example of an EM/CA analysis of digital technology use. The classic CA topics of technologically mediated communication and embodiment come together in a unique way in *avatar-mediated interaction*. Today the most pervasive form of avatar-mediated interaction occurs in the context of online games. Massively multiplayer online role-playing games (MMORPGs), such as World of Warcraft, or sociable virtual worlds, such as Second Life, attract millions of players from around the world. These virtual worlds contain a distinctive form of computer-mediated communication that takes face-to-face conversation as its social interaction metaphor. Such systems consist of graphical 3D environments inhabited by humanoid figures, or, 'avatars'. Avatars are puppeteered in real time by humans and serve as a medium for synchronous social interaction and play. EM/CA can be used to examine the endogenous orderliness and local organization of this social play and, in doing so, identify opportunities for improving system design.

The 3D avatars enable players to organize their interactions with other players in ways that resemble face-to-face conversation. Even with crude avatars, or 'blockies' (Hindmarsh et al., 2001), a surprisingly compelling experience of virtual co-presence can be created. Players can initiate interactions with fellow players simply by moving their avatars near one another and turning to 'face' each other. However, despite the sophistication of today's avatars, which are anatomically realistic and often animated with motion captured from *real* human bodies, they nonetheless remain *clunky* in their usability and sociality (Moore et al., 2007a, 2007b). Simple social actions, such as establishing mutual attention toward objects, communicating through avatar gestures, coordinating turns-at-chat and embodied actions with fellow players, are still difficult to achieve and in some cases impossible.

Although players can routinely coordinate their actions, it is usually not *tight coordination* (Moore et al., 2007a, 2007b). Face to face, people can fit their actions together with minimal gap and overlap (Sacks et al.,

1974: 707–708) because they can monitor turns-at-talk in real time and also observe each others' nonverbal cues. Yet it is precisely these features that are absent from current avatar-mediated interaction systems (Moore et al., 2007a, 2007b). Many player actions lack any public cues, rendering them unaccountable to fellow players. When players perform any of the following actions, their fellow players receive *no* cues from the system: composing a turn-at-chat, browsing one's inventory, consulting one's map, reading one's quest log, trading items, private messaging, and more.

For example, in Excerpt 15.1, right, a team of players is adventuring together in the superhero-themed game, City of Heroes. As with all MMORPGs, players create characters, represented visually as 3D avatars, that possess certain characteristics and abilities within the game world. By completing quests or missions, which inevitably involve battling computer-controlled opponents, players gain experience points and 'level up' their characters, unlocking more powerful abilities. In this excerpt, a team of three players has just completed one battle and has paused a moment before engaging in the next. Their characters' names are, Elizabot, Fuyuonnna and Napaul (the players are referred to using their characters' names, which is also a common practice among players).

In the brief downtime between battles, Elizabot stops (line 1) and waits for her teammates to catch up (lines 02–04). She looks toward the corner where the next group of opponents awaits (line 06) and then back, half-facing Fuyuonna (line 08). Elizabot then begins to say something by typing (line 10). She formulates a proposal regarding the tactics for the next battle, 'why don't you guys attack first' (line 18). However, while she is composing this turn-at-chat (lines 10–16), Elizabot's teammates receive *no cues* from the system that she is doing so. Consequently, they do not wait to see what she is saying. In the middle of Elizabot's turn, Fuyuonna runs off toward the next group of opponents (line 12, Figure 15.1) and she is joined by the third teammate, Napaul (line 15). Both disappear off the right side of the screen. After completing her turn-at-chat (line 18), Elizabot turns to see where her teammates went. Before receiving any response, she begins to expand her proposal with a second turn-at-chat (line 22), 'then I'll try to take aggro' which does not appear publicly (line 31) until the moment Fuyuonna initiates the next battle (line 30). Elizabot then joins the battle late (line 34). Thus, the team experiences slippage in

Figure 15.1 'Fuyuonna runs past Elizabot' (line 12) (© Robert J. Moore, 2012).

City of Heroes (standard)
```
01       ((Elizabot runs ahead of the group and stops))
02          (2.2)
03       [((Elizabot    turns    back    toward    teammates))]
04       [((Fuyuonna catches up & stops next to Elizabot))]
05          (1.7)
06       ((Elizabot turns away toward next opponents))
07          (2.5)
08       ((Elizabot turns back, half-facing Fuyuonna))
09          (0.8)
10       ((Elizabot begins typing))
11          (3.5)
12       [((Fuyuonna runs past Elizabot and off screen))]
13       [((    Elizabot    continues    typing    ))]
14          (2.1)
15       [((Napaul runs past Elizabot following Fuyuonna))]
16       [((    Elizabot    continues    typing    ))]
17          (2.1)
18 E:    why don't you guys attack first
19          (1.0)
20       ((Elizabot turns right toward teammates))
21          (0.8)
22       ((Elizabot begins typing))
23          (0.7)
24       [((Fuyuonna jumps down, approaching next opponents))]
25       [((      Elizabot    continues    typing    ))]
26          (2.0)
27       [((Napaul follows behind Fuyuonna))]
28       [(( Elizabot continues typing ))]
29          (2.4)
30       [((Fuyuonna initiates attack))]
31 E:    [ then I'll try to take aggro ]
32       ((Elizabot runs and catches up with teammates))
33          (5.5)
34       ((Elizabot joins attack late))
35          (0.9)
36 F:    I really don't think that works.
```

Excerpt 15.1

coordinating a conversation about tactics with the initiation of the next battle. Because Fuyuonna and Napaul cannot see that Elizabot is composing turns-at-chat, they cannot *wait* to hear what she has to say (lines 12, 15, 24, 27 and 30). Such slippages in coordination are common in this kind of avatar interaction system. The avatars move in real-time, but the text chat does not and, as a result, it lags behind.

While Excerpt 15.1 demonstrates the kind of slippage in coordination that results from a lack of public cues, Excerpt 15.2 (see page 231) demonstrates how public cues can enable *tighter* coordination. In this case, the researchers have augmented the system by binding visual cues, including avatar animations and chat-bubble messages, to particular game commands. For example, when a player opens the 'enhancements' window in the user interface, his or her avatar *automatically* displays a particular animation, in this case, the avatar appears to work at a laptop computer resting on a table that appears out of nowhere (see Figure 15.2, below) and emits a particular chat message, 'Adding enhancements'. Although the animation itself poorly represents the action, it nonetheless enables players, once they are told its significance, to *see* when a fellow player is performing this game action. Knowing that the enhancements window fills the entire screen, players can see at a glance that a fellow player is temporarily cut off from the 3D world and is thereby not ready to fight, travel, chat, etc. Furthermore, when the player closes the enhancements window and returns to the shared space, the laptop and table vanish as well, enabling fellow players to see that (and precisely when) he or she is done. In Excerpt 15.2, we can see how these public cues are consequential for group coordination. (The character names are Derowen, Jain Reaction and Earth Dude.)

In this case, Derowen approaches a group of opponents (line 01) to initiate the team's next battle. However, just before she enters attack range, she can see that Earth Dude has opened his enhancement window because the chat message, 'Adding enhancements' (line 03), and the 'laptop' animation on his avatar appear (line 04). In response, Derowen *retreats* from the group of opponents (line 06) and *waits*. After 2.0 seconds (line 07), Earth Dude's laptop disappears (line 08), indicating that he has closed his enhancements window, and Derowen then types, 'ready' (line 10). Jain Reaction and Derowen then discuss the state of their video recording for the study (lines 12–15). Derowen glances back and forth between her two teammates, whom she cannot view simultaneously because they are positioned too far apart (lines 16–19) and her field of view is too narrow. When she glances back at Earth Dude (line 21), she can see that he is now 'resting' to regenerate health points because his avatar is *kneeling*, one of the few public cues included by the game developers themselves. Derowen then continues to wait until she can see that Earth Dude is done resting by standing up (line 23), and only then does she initiate the next battle, when the whole team is ready to go (line 25).

The preceding analysis demonstrates how an ethnomethodological concern for the

Figure 15.2 'Adding enhancements' (line 03) (© Robert J. Moore, 2012).

City of Heroes (augmented)
```
01      ((Derowen approaches next group of opponents))
02          (1.0)
03 ED:  [          Adding           enhancements          ]
04      [((laptop appears in front of Earth Dude))]
05          (1.4)
06      ((Derowen retreats))
07          (2.0)
08      ((Earth Dude's laptop disappears))
09          (2.0)
10 D:   ready
11          (2.0)
12 JR:  running out of tape
13      ((Derowen advances and retreats))
14          (3.1)
15 D:   yeah, one min
16      ((Derowen turns toward Earth Dude))
17          (2.4)
18      ((Derowen turns her view back and forth between JR and
19      ED, approaches opponents))
20          (1.0)
21      ((Earth Dude kneels))
22      ((Derowen turns away from Earth Dude and back toward him))
23      ((Earth Dude stands up))
24          (0.8)
25      ((Derowen approaches and attacks opponents))
26          (2.6)
27      ((Jain Reaction joins attack))
```

Excerpt 15.2

situated accomplishment of social activities and a conversation analytic utilization of mechanical recordings, transcripts and sequential analysis can be applied to one form of avatar-mediated interaction. It reveals systematic features of the endogenous orderliness of these activities, namely, how public cues of player activity are used to time the initiation of the next round of team gameplay at a moment when everyone appears ready. The analysis thus demonstrates, on a turn-by-turn scale, some social interactional consequences of particular game design choices.

The preceding analysis also suggests a general design principle: in shared virtual worlds, user interface design is also always *social interaction* design. Whether intended or not, choices about how users interact with the system impact their ability to interact with fellow users simultaneously. Hiding system interactions entirely from fellow users, which are often a relevant part of the social interactional context, results in unaccountable gaps in users' public activity and lapses in social coordination (Moore et al., 2007a, 2007b). Therefore, when designing such systems, 'feedback' is not simply relevant to the individual user, as it is in single-player games, but is also relevant to other users in a fundamentally *social* environment.

CONCLUSION

Ethnomethodology and conversation analysis represent interrelated approaches to the study of a distinctive domain of sociological phenomena. Although they sometimes vary in their settings of interest, their methods and their analytic goals, EM/CA are nonetheless inextricably linked both historically and in the everyday work of many practitioners. EM/CA represent a radical departure from most other approaches in the human sciences. Human behavior is most commonly examined by abstracting variables from concrete events, aggregating them and discovering statistical relationships, or 'signals', in the numbers. One result of this is that the *mechanism* of behavior remains a black box. Independent variables go in and dependent variables come out, but exactly *how* the human participants produce the behavior locally is lost. EM/CA peer inside the black box of human conduct by observing concrete activities directly and identifying methodic patterns endogenous to the activities themselves.

Although ethnomethodology and conversation analysis began by examining ordinary social activities, they have expanded into many other kinds of human action, including institutional talk, work practice, technology use and more. Studies of human–computer interaction (Suchman, 1987; Frohlich et al., 1993; Luff and Heath, 1993a; Moore and Churchill, 2011; Moore et al., 2011) demonstrate that the basic theoretical and methodological principles of EM/CA can be adapted to very different forms of human practice. Ethnomethodological theory invites empirical investigation of *any* form of human practice, while conversation analytic concerns for objective data and reproducible results invite empirical investigation of *any* form of human practice that can be captured with mechanical recording.

Human use of digital technologies not only provides fertile topics for ethnomethodology and conversation analysis, but these approaches can also offer practical insights for the design of such technologies themselves. Because they are concerned with the local organization of concrete activities, EM/CA offer design insights at the level of sequences of activity, the same level at which technology designers work. This sets EM/CA apart from other social sciences, which tend to offer abstract, theoretical recommendations that designers often find difficult to translate into particular system features. For example, demonstrating how animated avatar cues can enable teammates in an online game to wait for each other before initiating their next battle suggests practical recommendations regarding the inclusion of more public cues for users' interactions with the system. The empirically grounded nature of EM/CA no doubt accounts in part for their relative popularity in HCI and CSCW when compared with that in their home discipline of sociology, of which they are theoretically critical.

REFERENCES

Arminen, I., P. Auvinen and H. Palukka (2010) 'Repairs as the last orderly provided defense of safety in aviation', *Journal of Pragmatics* 42(2): 443–465.

Austin, J.L. (1962) *How to Do Things with Words*. Oxford: Oxford University Press.

Becker, H.S. (1963) *Outsiders: Studies in the Sociology of Deviance*. New York: Free Press.

Boden, D. and D.H. Zimmerman (eds) (1991) *Talk and Social Structure*. Cambridge: Polity Press.

Button, G. (ed.) (1991) *Ethnomethodology and the Human Sciences*. Cambridge: Cambridge University Press.

Button, G., J. Coulter, J.R.E. Lee and W. Sharrock (1995) *Computers, Minds and Conduct*. Cambridge: Polity Press.

Button, G. and W. Sharrock (1996) 'Project work: The organisation of collaborative design and development in software engineering', *Computer Supported Cooperative Work* 5(4): 369–386.

Button, G. and W. Sharrock (1997) 'The production of order and the order of production: Possibilities for distributed organisations, work and technology in the print industry', *Proceedings of ECSCW '97*, Kluwer.

Clayman, S.E. (1995) 'The dialectic of ethnomethodology', *Semiotica* 107(1/2): 105–123.

Coulter, J. (1991) 'Logic: Ethnomethodology and the logic of language', in G. Button (ed.) *Ethnomethodology and the Human Sciences*. Cambridge: Cambridge University Press, 20–50.

Crabtree, A. (2003) *Designing Collaborative Systems: A Practical Guide to Ethnography*. London: Springer-Verlag.

Crabtree, A., T. Rodden, P. Tolmie and G. Button (2009) 'Ethnography considered harmful', *Proceedings of CHI '09*. ACM.

Dourish, P. and G. Button (1998) 'On 'technomethodology': Foundational relationships between ethnomethodology and system design', *Human-Computer Interaction* 13(4): 395–432.

Drew, P. and J. Heritage (eds) (1992) *Talk at Work*. Cambridge: Cambridge University Press.

Forrester, M.A. (2008) 'The emergence of self-repair: A case study of one child during the early preschool years', *Research on Language & Social Interaction* 41(1): 99–128.

Frohlich, D.M., P. Drew and A. Monk (1993) 'The management of repair in human computer interaction', Internal Report, Hewlett-Packard Company, Bristol, UK.

Frohlich, D.M. and P. Luff (1990) 'Applying the technology of conversation to the technology for conversation', in P. Luff, G.N. Gilbert and D.M. Frohlich (eds), *Computers and Conversation*. London: Academic Press, 187–220.

Garcia, A. and J. Jacobs (1999) 'The eyes of the beholder: Understanding the turn-taking system in quasi-synchronous CMC', *Research on Language and Social Interaction* 32(4): 337–367.

Garfinkel, H. (1963) 'A conception of, and experiments with "trust" as a condition of stable concerted actions', in O.J. Harvey (ed.), *Motivation and Social Interaction*. New York: Ronald Press, 187–238.

Garfinkel, H. (1967) *Studies in Ethnomethodology*. Englewood Cliffs, NJ: Prentice Hall.

Garfinkel, H. (1968) 'The origins of the term "Ethnomethodology"', in R.J. Hill and K.S. Crittenden (eds), *Proceedings of the Purdue Symposium on Ethnomethodology*, Institute Monograph Series no. 1, Institute for the Study of Social Change, Purdue University, 5–11.

Garfinkel, H. (1991) 'Respecification: Evidence for locally produced, naturally accountable phenomena of order*, logic, reason, meaning, method, etc. in and as of the essential haecceity of immortal ordinary society (I) an announcement of studies', in G. Button (ed.), *Ethnomethodology and the Human Sciences*. Cambridge: Cambridge University Press, 10–19.

Garfinkel, H., M. Lynch and E. Livingston (1981) 'The work of discovering science construed with materials from the optically discovered pulsar', *Philosophy of the Social Sciences* 11: 131–158.

Garfinkel, H. and D.L. Wieder (1992) 'Evidence for locally produced, naturally accountable phenomena of order*, logic, reason, meaning, method, etc., in and as of the essentially unavoidable and irremediable haecceity of immortal ordinary society: IV two incommensurable, asymmetrically alternate technologies of social analysis', in G. Watson and R.M. Seiler (eds), *Text in Context*. London: Sage, 175–206.

Geertz, C. (1973) *The Interpretation of Cultures*. New York: Basic Books.

Goffman, E. (1959) *The Presentation of Self in Everyday Life*. New York: Doubleday.

Goodwin, C. (1979) 'The interactive construction of a sentence in natural conversation', in G. Psathas (ed.), *Everyday Language: Studies in Ethnomethodology*. New York: Irvington, 97–121.

Goodwin, C. and M. Goodwin (1996) 'Formulating planes: Seeing as a situated activity', in Y. Engestrom and D. Middleton (eds), *Cognition and Communication at Work*. New York: Cambridge University Press.

Harper, R. and J. Hughes (1993) '"What a F-ing System!: Send 'em all to the same place and then expect us to stop 'em hitting": Making technology work in air traffic control', in G. Button (ed.), *Technology in Working Order: Studies in Work, Interaction and Technology*. London: Routledge, 127–144.

Heath, C. and P. Luff (1992) 'Collaboration and control: Crisis management and multimedia technology in London Underground line control rooms', *Computer Supported Cooperative Work*, 1: 69–94.

Heritage, J. (1984) *Garfinkel and Ethnomethodology*. Cambridge: Polity Press.

Heritage, J. and J.M. Atkinson (1984) *Structures of Social Action: Studies in Conversation Analysis*. Cambridge: Cambridge University Press.

Hindmarsh, J., M. Fraser, C. Heath and S. Benford (2001) 'Virtually missing the point: Configuring CVEs for object-focused interaction', in E.F. Churchill, D.N. Snowdon and A.J. Munro (eds), *Collaborative Virtual Environments*. London: Springer, 115–133.

Hindmarsh, J. and C. Heath (2000) 'Embodied reference: A study of deixis in workplace interaction', *Journal of Pragmatics* 32: 1855–1878.

Jefferson, G. (2004) 'Glossary of transcript symbols with an introduction', in G. Lerner (ed.), *Conversation*

Analysis: Studies from the First Generation. Amsterdam: John Benjamins Publishing, 13–31.

Jordan, B. and A. Henderson (1995) 'Interaction analysis: Foundations and practice', *The Journal of the Learning Sciences*, 4(1): 39–103.

Kim, K. (2001) 'Confirming intersubjectivity through retroactive elaboration: Organization of phrasal units in other-initiated repair sequences in Korean conversation', in M. Selting and E. Couper-kuhlen (eds), *Studies in Interactional Linguistics*. Amsterdam: John Benjamins Publishing.

Kuzuoka, J., K. Yamazaki, A. Yamazaki, J. Kosaka, Y. Suga and C. Heath (2004), 'Dual ecologies of robot as communication media: Thoughts on coordinating orientations and projectability', *Proceedings of CHI '04*, ACM.

Livingston, E. (1986) *The Ethnomethodological Foundations of Mathematics*. London: Routledge & Kegan Paul.

Luff, P. and C. Heath (1993a) 'The practicalities of menu use: Improvisation in a screen-based activity', *Journal of Intelligent Systems* 3: 251–296.

Luff, P. and C. Heath (1993b) 'System use and social organisation: Observations on human–computer interaction in an architectural practice', in G. Button (ed.), *Technology in Working Order: Studies of Work, Interaction and Technology*. London: Routledge, 184–210.

Luff, P., J. Hindmarsh and C. Heath (eds) (2000) *Workplace Studies: Recovering Work Practice and Informing System Design*. Cambridge: Cambridge University Press.

Lynch, M. (1985) *Art and Artifact in Laboratory Science: A Study of Shop Work and Shop Talk in a Research Laboratory*. London: Routledge & Kegan Paul.

Lynch, M. (1993) *Scientific Practice and Ordinary Action: Ethnomethodology and Social Studies of Science*. New York: Cambridge University Press.

Macbeth, D. (2004) 'The relevance of repair for classroom correction', *Language in Society*: 33: 703–736.

Maheux-Pelletier, G. and A. Golato (2008) 'Repair in membership categorization in French', *Language in Society* 37(5): 689–712.

Maynard, D.W. (1989) 'On the ethnography and analysis of discourse in institutional settings', *Perspectives on Social Problems*. 1: 127–146.

Maynard, D.W. (2003) *Bad News, Good News: Conversational Order in Everyday Talk and Clinical Settings*. Chicago, IL: University of Chicago Press.

Maynard, D.W. (in press) 'Everyone and no one to turn to: Intellectual roots and contexts for conversation analysis', in J. Sidnell and T. Stivers (eds), *Handbook of Conversation Analysis*. New York: Blackwell-Wiley.

Maynard, D.W. and S.E. Clayman (1991) 'The diversity of ethnomethodology', *Annual Review of Sociology* 17: 385–418.

McHoul, A. (1990) 'The organization of repair in classroom talk', *Language in Society* 19: 349–77.

McIlvenny, P. (1990) 'Communicative action and computers: Re-embodying conversation analysis?', in P. Luff, N. Gilbert and D. Frohlich (eds), *Computers and Conversation*. London: Academic Press, 91–132.

Moore, R.J. (2008) 'When names fail: Referential practice in face-to-face service encounters', *Language in Society* 37(3): 385–413.

Moore, R.J. (2013) 'A Name Is Worth a Thousand Pictures: Referential Practice in Human Interactions with Internet Search Engines', in A. Neustein and J.A. Markowitz (eds) *Mobile Speech and Advanced Natural Language Solutions*. New York: Springer, 259–286.

Moore, R.J. and E.F. Churchill (2011) 'Computer interaction analysis: Toward an empirical approach to understanding user practice and eye gaze in GUI-based interaction', *Computer Supported Cooperative Work* 20: 497–528.

Moore, R.J., E.F. Churchill and R.G.P. Kantamneni (2011) 'Three sequential positions of query repair in interactions with internet search engines', *Proceedings of CSCW '11*. ACM, 415–424.

Moore, R.J., N. Ducheneaut and E. Nickell (2007a) 'Doing virtually nothing: Awareness and accountability in massively multiplayer online worlds', *Computer Supported Cooperative Work* 16: 265–305.

Moore, R.J., C.H. Gathman, N. Ducheneaut and E. Nickell (2007b) 'Coordinating joint activity in avatar-mediated interaction', *Proceedings of CHI '07*. New York: ACM: 21–30.

Moore, R.J. and D.W. Maynard (2002) 'Achieving understanding in the standardized survey interview: Repair sequences', in D.W. Maynard, H. Houtkoop-Steenstra, N.C. Schaeffer and J. van der Zouwen (eds), *Standardization & Tacit Knowledge: Interaction and Practice in the Survey Interview*. New York: John Wiley, 281–311.

Moore, R.J., J. Whalen and C.H. Gathman (2010) 'The work of the work order: Document practices in face-to-face service encounters', in N. Llewellyn and J. Hindmarsh (eds), *Organisation, Interaction and Practice: Studies in Ethnomethodology and Conversation Analysis*. Cambridge: Cambridge University Press, 172–197.

O'Neill, J., S. Castellani, F. Roulland, C. Juliano, L. Dai and N. Hairon (2011) 'From ethnographic study to mixed reality: A remote collaborative troubleshooting system', *Proceedings of CSCW '11*, ACM, 225–234.

Parsons, T. (1937) *The Structure of Social Action*. New York: McGraw-Hill.

Psathas, G. (1995) 'Conversation analysis: The study of talk-in-interaction', *Qualitative Research Methods Series* 35. London: Sage.

Raudaskoski, P. (1990) 'Repair work in human–computer interaction – a conversation analytic perspective', in P. Luff, N. Gilbert and D. Frohlich (eds), *Computers and Conversation Analysis*. London: Academic Press, 151–171.

Rosenthal, B.M. (2008) 'A resource for repair in Japanese talk-in-interaction: The phrase TTE YUU KA', *Research on Language & Social Interaction* 41(2): 227–240.

Sacks, H. (1984) 'Notes on methodology', in J.M. Atkinson and J.C. Heritage (eds), *Structures in Social Action: Studies in Conversation Analysis*. Cambridge: Cambridge University Press, 21–27.

Sacks, H. (1993) *Lectures on Conversation*. Oxford: Blackwell.

Sacks, H., E.A. Schegloff and G. Jefferson (1974) 'A simplest systematics for the organization of turn-taking for conversation', *Language* 50: 696–735.

Schegloff, E.A. (1968) 'Sequencing in conversational openings', *American Anthropologist* 70: 1075–1095.

Schegloff, E.A. (1991) 'Reflections on talk and social structure', in D. Boden and D.H. Zimmerman (eds), *Talk and Social Structure*. Cambridge: Polity Press, 44–70.

Schegloff, E.A. (1992) 'Repair after next turn: The last structurally provided defense of intersubjectivity in conversation', *American Journal of Sociology* 98: 1295–1345.

Schegloff, E.A. (1998) 'Body torque', *Social Research* 65: 535–586.

Schegloff, E.A. (2007) *Sequence Organization: A Primer in Conversation Analysis*. Cambridge: Cambridge University Press.

Schegloff, E.A., G. Jefferson and H. Sacks (1977) 'The preference for self-correction in the organization of repair in conversation', *Language*, 53(2): 361–382.

Schegloff, E.A. and H. Sacks (1973) 'Opening up closings', *Semiotica* 7: 289–327.

Schütz, A. (1962) *Collected Papers I: The Problem of Social Reality*. M.A. Natanson and H.L. van Breda (eds). Dordrecht: Martinus Nijhoff.

Searle, J.R. (1969) *Speech Acts*. Cambridge: Cambridge University Press.

Streeck, J. (1996) 'How to do things with things: Objets trouvés and symbolization', *Human Studies* 19: 365–84.

Suchman, L. (1987) *Plans and Situated Actions: The Problem of Human–Machine Communication*. Cambridge: Cambridge University Press.

Suchman, L. (1993) 'Technologies of accountability', in G. Button (ed.), *Technology in Working Order: Studies of Work, Interaction and Technology*. London: Routledge, 113–126.

Suchman, L. (1997) 'Centers of coordination: A case and some themes', in L.B. Resnick, R. Sälijö, C. Pontecorvo and B. Burge (eds), *Discourse, Tools, and Reasoning: Essays on Situated Cognition*. Berlin: Springer-Verlag, 41–62.

Sudnow, D. (1978) *Ways of the Hand: The Organization of Improvised Conduct*. Cambridge, MA: Harvard University Press.

Szymanski, M.H., P.M. Aoki, E. Vinkhuyzen and A.Woodruff (2006) 'Organizing a remote state of incipient talk: Push-to-talk mobile radio interaction', *Language in Society* 35: 393–418.

Whalen, J. (1995) 'A technology of order production: Computer-aided dispatch in public safety communications', in P. ten Have and G. Psathas (eds), *Situated Order: Studies in the Social Organization of Talk and Embodied Activities*. Washington, DC: University Press of America.

Whalen, J. and E. Vinkhuyzen (2000) 'Expert systems in (inter)action: Diagnosing document machine problems over the telephone', in P. Luff, J. Hindmarsh and C. Heath (eds), *Workplace Studies: Recovering Work Practice and Informing Systems Design*. Cambridge: Cambridge University Press, 92–140.

Wootton, A.J. (1994) 'Object transfer, intersubjectivity and third position repair: Early developmental observations of one child', *Journal of Child Language* 21: 543–64.

Wu, R.J.R. (2006) 'Initiating repair and beyond: The use of two repeat-formatted repair initiations in Mandarin conversation', *Discourse Processes* 41: 67–109.

Yamazaki, K., A. Yamazaki, M. Okada, Y. Kuno, Y. Kobayashi, Y. Hoshi, K. Pitsch, P. Luff, D. Lehn and C. Heath. (2009) 'Revealing Gaugin: Engaging visitors in robot guide's explanation in an art museum', *Proceedings of CHI '09*, ACM.

Zimmerman, D.H. (1992) 'The interactional organization of calls for emergency assistance', in P. Drew and J. Heritage (eds) *Talk at Work*. Cambridge: Cambridge University Press, 418–469.

16

Behavioral Trace Data for Analyzing Online Communities

Cliff Lampe

INTRODUCTION

Recently, there has been a dramatic increase in the number of people using social media, including social network sites, on a regular basis (Hampton et al., 2011). This increase in use has led to new opportunities in understanding social processes by studying the behavioral traces – that is, records stored as people use online communities. In this chapter, we will examine the pros and cons of using behavioral trace data from online communities for research and relate that to other ways these social processes are studied. In understanding the effects of this new paradigm for online interaction, work in social media should be associated with the more general research on online communities, as both areas involve the research of many-to-many interactions mediated by information technology. Research on online communities has a rich history of findings, but also highlights a number of research techniques that can be considered in light of the current online environments. In particular, online communities have given access to behavioral data of users as they interact with the system, which provides invaluable insights into large-scale social behavior.

While terms have shifted over the years, the idea of the online community is comparable to more recent terms like 'Web 2.0' and 'social media' when the basic characteristics of these systems are considered. At a high level, all of these terms involve interactions where many people communicate directly with each other through some type of computer mediation. In general, these people are creating the content of the system collaboratively, whether that content is in the form of articles, photos or comments in a discussion. This is a broad class of online interactions and can cover an incredibly diverse set of topics and users. However, there are three common characteristics that define whether a site is social media: (1) users are engaged in direct, relatively unmediated interaction with one another; (2) users are creating the content of the site, in some form or another; and (3) user interactions are mainly supported by networked computing through a collection of individual applications that are bundled together as a system. In other words, online communities are places where people can come together to share content supported by multiple computer tools and applications.

Given that online communities are social systems at their core, there have been as many research methods used to study them as have been applied to offline social systems. Studies of online communities have included surveys (Ellison, Steinfield, and Lampe, 2007; Lampe, Ellison, and Steinfield, 2008; Papacharissi and Mendelson, 2011; Tufekci and Wilson, 2012), ethnographies (Barkhuus and Tashiro, 2010; boyd, 2004; Toma, Hancock, and Ellison, 2008) and experiments (Chen et al., 2009; Lampe and Garrett, 2007; Ling et al., 2005; Tong et al., 2008).

However, the distinguishing characteristic of online communities that separates analysis of them from other social systems is that, because they are mediated by computer technology, many of these systems contain records of how people use the site and interact with one another. Previous scholars have identified several ways in which the tools available to users of online communities allow for different types of interactions from those in offline groups (Hollan and Stornetta, 1992; Resnick, 2001; Wellman, 1997a), including the ability to broadcast to large groups, allow for rapid feedback channels, and support interaction with computational tools, among others. A key feature of technology-mediated group interactions is the ability to keep a history of the interactions that take place. Online communities use underlying databases to enable their interactive features. For example, a site that allows users to post comments on a news story must keep some archive of those comments so the next user can read them. For sites that allow real identities or pseudonyms for persistent users, some record of those profiles is captured in a table in the underlying database. These databases underpinning online communities are the source of this rich set of user interaction data. The more users are in a system, and the longer they use that system, the more valuable the database grows. For example, Twitter is a social network site that has hundreds of millions of users and hundreds if not thousands of interactions every second. Each of those 'tweets' posted to the site is stored by Twitter and includes information like the time it was posted, who posted it, who reposted it and more. Each Twitter user also is recorded in a database, with records of their activities stored by the site and of potential value to researchers.

The records of the choices users make as they interact with and through a system can include the choices of an individual user, like profile fields they have filled out or pages they have visited, but can also include interactions between users, such as records of messaging between users or voting on each others' content. These records of user activity are sometimes called 'digital traces' or 'behavioral traces' as they are like trails of the behaviors in which people have engaged in digital environments.

In this chapter, we present a brief definition and history of research in online communities and look at how that genre of online interaction is comparable to social media generally and social network sites specifically. We then consider insights gained from analysis of user data archived in online communities and discuss ways in which that data is typically attained and analyzed. Finally, we examine caveats regarding the use of user archives for understanding interactions in online communities.

RESEARCH IN ONLINE COMMUNITIES

Communication has been supported by computing systems for as long as computers have been networked together (Licklider and Taylor, 1968). Very early networks supported social interactions around science fiction fans (Abbate, 2000) and that sociality was carried into the Usenet discussion groups (Pfaffenberger, 2002), Multi-User

Dungeon (MUD) games (Curtis, 1992), and e-mail systems (Hiltz and Turoff, 1978). In each of these systems, the tools and the infrastructures differed, but the general idea of using computer networks to support communication was consistent.

Rheingold (1993) framed early scholarship in mediated group interactions by describing the WELL – a discussion forum related to a range of topics – as an 'online community'. The WELL was based on a bulletin board system (BBS), a protean example of computer-mediated communication dependent on dial-up phone lines for connection. In the second edition of his book, he declares that he would have used the term 'online social networks' instead of online community if he had known about work in that area (Rheingold, 2000). A main proponent of the concept of online social networks has been Barry Wellman, who argued social networks that are expressed through computer technologies have many of the same principles as offline social networks (Wellman, 2001), though they often have characteristics which enable new types of interactions, such as archive and broadcast features (Wellman and Gulia, 1999).

Another area of research related to studies of online communities was work in groupware systems used in organizations (Grudin, 1988). This work pioneered many of the studies of how distributed individuals were affected by mediated communication, including early work in how misunderstanding could be exacerbated in online messages (Sproull and Kiesler, 1991). Insights from this work include awareness that social systems predict groupware adoption, exceptions are difficult to handle in groupware environments and computing systems are inflexible compared to the nuances of social systems (Ackerman, 2001). A classic example from this literature is the difficulty in establishing shared calendaring systems. How and why people schedule meetings is a complex process that involves issues of authority, social norms and organizational culture (among other things) but calendaring systems are often blind to those underlying forces. In another example, Facebook allows users to place people in different groups in order to help preserve social context as the user posts content to the site. However, who belongs in which group of friends is a complex and dynamic process for humans and the system designed to do this has not been flexible enough for most people (Marwick and boyd, 2010; McLaughlin and Vitak, 2012). This area has also provided insights into how channels of communication (for example body language, shared references like a whiteboard, tone of voice, etc.) affect different types of social processes, with more types of information being needed in communication around complex types of interactions (Olson and Olson, 2001). For example, text environments where you may not have rapid feedback from the group or access to their body language may work well for a 'loosely coupled' task like scheduling a meeting, but 'tightly coupled' tasks like deliberating about strategy or creative design may require more types of information provided by those richer communication channels. However, there is also evidence that repeated interactions over time in text-based environments can also lead to deep understanding of other people in an electronic group (Walther, 1992; Walther and Parks, 2002).

Another type of online community that was important in the history of research in this area was Usenet. Usenet is an online discussion system based on protocols that allow researchers to easily access the content of the system. Usenet is a 'newsgroup' system, with text-based discussions that are threaded and grouped by topic (Quarterman and Hoskins, 1986). This access led to a wealth of research regarding social interactions in Usenet, including perceptions of author value (Fiore, Tiernan and Smith, 2002), how 'lurkers' play a role in these systems (Nonnecke, Preece and Andrews, 2004), the size of groups that are sustainable (Butler, 2001) and how information is diffused between different groups in Usenet (Whittaker et al., 1998). The data in

Usenet was important in establishing early versions of recommender systems (Resnick et al., 1994) and the importance of visualization as a technique for understanding social patterns (Smith, 1999).

As this brief history shows, the past 30 years of research in online communities has been active. The current environment of user engagement in computer-mediated communication has many of the same characteristics as these studies, but there are two major changes that are still being explored in research in this area. First, the scale of participation in online communities and social media has grown precipitously in the past few years (Madden and Zickuhr, 2011), bringing more diverse types of users into online communities. This brings issues of motivation, literacy, efficacy and agency into sharper focus than in previous years of research. This increased use has been matched with changes in how systems are designed and developed as well, with a greater push to commercialize systems, which in turn leads to less access to systems that are privately owned and simpler interfaces as online communities vie for users in a competitive market. The second change in online communities research is that the sites themselves have grown more complicated and diverse. Rather than being simple text chats, social media now encompass a broad range of media, activities and tools.

SOCIAL MEDIA AND SOCIAL NETWORK SITES ARE ONLINE COMMUNITIES

Social network sites (SNSs) are a specific type of online community where the content being collaboratively created is information about the users, including profile information and status updates. Social network sites, as defined by boyd and Ellison (2007) are defined by three characteristics: (1) a profile of some type defining the user; (2) the ability to articulate links with other users; and (3) the ability to see the links that have been made by others on the site. What constitutes a profile can range from rich, contextual information about a person, to a brief description, to pseudonymous information. For example, the site WebKinz[1] allows children to engage in a social network site, but the profile is a virtual pet. What constitutes a connection between users can also vary by SNS. On Facebook, users need to agree to share a connection (synchronous link), whereas on Twitter one can follow a person without that person reciprocating the connection (asynchronous link). None of these features that are specifically characteristics of SNS are in conflict with the definition above of online communities and social media. Although SNS may have particular features that are not shared by all online communities, the necessary conditions of being an online community are shared by all SNS.

ATTAINING BEHAVIORAL TRACE DATA

As described above, behavioral trace data has two great advantages: scale and validity. In terms of scale, many of these social media sites log interactions ranging from the thousands to the trillions, making a large amount of data potentially available to researchers. Interactions of individual users and groups have ecological validity, as they show actual behaviors of people (as opposed to recall of behaviors common in surveys and interviews). There are multiple methods for getting access to the behavioral trace data of users in these sites. Each of the acquisition methods has benefits and deficits, which are detailed below.

Scraping Data

Researchers can use programs to access the site and 'scrape' the data into an external database. This method can be an effective way to collect data when the researcher does not have access to the databases underlying the online community, but there are many limitations to this method. Due to concerns about user rights, intellectual property and server efficiency, many sites have 'terms of use' that prohibit people from using automated scripts to download user data. This prohibition raises an ethical concern for researchers regarding accessing data 'owned' by a private entity. Most online communities in the modern environment are commercial entities privately owned. Does the researcher have a right to make a copy of data owned by that third party? Ethical issues aside, there are legal issues, as well as technical issues since sites often block computers that use automated scraping from accessing their servers.

An additional limitation of screen scraping is that not all behaviors of interest to the researcher may be posted to the site, limiting which data can be automatically retrieved with this method. For example, individual user visits are usually not displayed on a page, so would not be accessible. Many sites require users to authenticate and on social network sites many users have privacy setting options that make them invisible to anyone but people they've 'friended'. This makes them effectively invisible to this method, affecting overall sampling.

Research that has used this method has helped to study sites from which behavioral data was not readily available. Lampe, Ellison and Steinfield (2007) automatically downloaded Facebook profile data with the permission of the site in order to show how certain profile fields were more useful for establishing identity than others. Adamic and Glance (2005) used a data-scraping program to download links on political blogs to show, in an influential article, how only a small percentage of politically ideological blogs connect to content from differing ideologies (Adamic and Glance, 2005).

Accessing Data from an API

Another method used to retrieve site use data is to create an application in those online communities that make an application performance interface (API) available for writing third-party applications. For example, Facebook allows third-party developers to create applications and, when users adopt those applications, many fields in their profile are made available to the application. Consequently, if users adopt the application, automatic retrieval of such fields as number of friends, posted demographic information and more, are possible. This method can be effective for retrieving information that is behind user authentication walls and is consistent with site terms of use. However, one must convince users to load the app, which can affect the breadth of sampling.

Several researchers have used this method to conduct research on social network sites. Gilbert and Karahalios (2009) used an application to download key characteristics of how users were interacting with each other to compute a measure of how strong the ties were between those people, which they validated with direct user assessments of tie strength. Hogan developed a Facebook application called 'namegen' that researchers can use to have Facebook users download their social networks and profile information for research use (Hogan, 2010). He and his colleagues have used this tool to study how socio-economic status affects the generation of social capital through Facebook in university students (Brooks, Hogan, and Titsworth, 2011).

Accessing Server Logs

Another way of accessing network data is to receive data directly from the site hosting the social interactions, either through a deal with the organization that owns the site or, in many cases, by working for the organization as a researcher. The benefits of this data include its completeness in that it includes users who would be hidden behind privacy settings in the scraping method and often includes fields that are not visible through the interface. For example, individual time stamps for actions on the site might be captured by the server, but not reported on pages. The largest weakness of this method is the difficulty in receiving data from organizations that own the sites where the activity is taking place. Another potential weakness of this method is that the site designers and owners may not log all of the activity that would be of interest to the researcher. For example, it is common in online community databases to aggregate small user actions like content voting, rather than capture each vote separately. Studying the separate votes may tell us about first mover bias in voting or show how voting develops over time, which could help improve overall content voting processes. For the engineer, aggregating the votes in the database is parsimonious and efficient, but for the researcher it represents a loss of possibly useful data. Researchers can ask for additional information to be logged, but site owners have to worry about server and database operations when considering such requests.

In our own work, we've looked at social network sites using behavioral trace data (Lampe, Ellison and Steinfield 2007), as well as other sites, including an early news aggregation and discussion online community called Slashdot, which was one of the first online discussion sites to allow users to rate comments by other users. In examining Slashdot, we used behavioral trace data to study how people were rating the value of comments to news stories in the site (Lampe and Resnick, 2004), how those ratings affected the way new users learned how to participate in the site (Lampe and Johnston, 2005) and how people used comment ratings to guide which content they wanted to read. (Lampe, Johnston, and Resnick, 2007). Ganley and Lampe (2009) studied how a social network embedded in Slashdot that included both positive links ('Friends') and negative links ('Foes') was used to effect content voting behavior on the site. The data for these studies resulted from a private deal between the researchers and the owners of Slashdot.

Many large social media companies have research teams who have direct access to server data and use that to conduct research on social interactions that is subsequently published in the academic literature. Moira Burke worked with researchers at Facebook to examine server logs of user behavior and study how picture posting was related to new user adoption of Facebook (Burke, Marlow, and Lento, 2009), how social network ties on Facebook are related to outcomes like social capital and loneliness (Burke, Marlow, and Lento, 2010) and how different types of Facebook activity led to different outcomes regarding social capital (Burke, Kraut, and Marlow, 2011). Another Facebook researcher has published an analysis of how happy people are in different nations by the words they use in their status updates (Kramer, 2010).

Other researchers have made special arrangements with a site or organization to receive a dataset that can be used for research purposes. Lampe and colleagues created an agreement with the online user-generated content site Everything2 to have access to server log data on participants in order to conduct research on user patterns on the site (Lampe, Wash, Velasquez, and Ozkaya, 2010). Golder and colleagues received a dataset of early interactions on Facebook in order to test what types of communications like 'pokes' and direct messages were occurring between participants (Golder, Wilkinson, and Huberman, 2007). Some researchers have received sole access to information about large online games, like Everquest II (Keegan et al., 2011) in order to study social interaction in those large-scale online communities.

In these cases, it is not simply a matter of asking for the data and then loading it into a stats program for analysis. Organizations need to trust that the researchers will not do harm to their users, so appropriate privacy, anonymity and data security protocols need to be established. The data being provided is usually very large, so the researchers need to have a computing environment and personnel capacity that is able to store and analyze data that can be on a massive scale. Often, researchers will have an arrangement with a member of an organization who acts as an internal proponent of their research agenda, which can be problematic when the organization changes staff and that proponent is gone, leaving the researcher without data. While unusual access to online community data can be rewarding for research, the costs of doing such work should not be underestimated.

Datasets Made Available by Social Media Organizations

Sometimes, an organization that operates a large online organization which would be of interest to users makes data available through large exports that anyone can use. The most widely used of these is the Wikipedia dataset, which is released for download by anyone who wants it.[2] Occasionally, smaller sets of Wikipedia data are made available, as in the case of a recent competition sponsored by the Wikimedia Foundation, which operates Wikipedia along with other wiki-based peer-production sites. In this competition,[3] contestants worked from a shared set of data to better predict which new editors would become regular contributors. This question is of importance to Wikipedia, as new member contributions have fallen sharply since 2007 (Suh et al., 2009).

Another site that has made large sets of data available for researchers is the open source software-hosting site Sourceforge. Through a deal with the University of Notre Dame,[4] Sourceforge made anonymized data available to a wide set of researchers, with Notre Dame acting as an intermediary hosting service (VanAntwerp and Madey, 2008). Several papers have been published with that data by the Notre Dame team (Gao et al., 2007; Gao and Madey, 2007) that examine how open source projects, where participants are distributed and don't know much about each other, are organized in order to be successful.

As with data received as described above, the scale of these data can be such that only researchers with access to computational tools and methods are fully able to take advantage of them. This requires that research teams have such capacity in place, either through

collaborations with data scientists or through special training to work with this type of data. As with screen-scraped data, the datasets that are provided are usually 'as is', meaning not all variables of interest to the researcher may be included in the data. For example, in work on Slashdot (Lampe, Johnston, and Resnick, 2007) the site only captured aggregate hits per day from users, whereas it was important for the research to tell when specific users were accessing the site. Negotiations with the site administrators had to be engaged in order to collect the logs necessary, which required additional effort for all parties.

Creating Social Network Sites and Online Communities

If a researcher does not have access to one of the commercial social network sites or if they have questions that are not appropriate to those sites, they can try to create their own social media sites and study the interactions on those sites. This has the advantage of better control over the interface and the behavioral data of users as they make requests to the server. Other methods of receiving behavioral data from websites do not allow for this level of control over the tools that shape how users behave. Creating a site from scratch also allows the researcher to instrument the site in such a way that they can capture all user behavior they feel is relevant to their question. However, the downside of this method is the cost and effort required to create, market and sustain a site in order to get the number of people needed to conduct social analyses.

Ed Chi, who has created many online communities to research the intersection of social interaction and technical features, has argued researchers can test systems that have been built either in the lab or 'in the wild' (Chi, 2009). In the former case, potential users interact with the online community in controlled lab settings. This method offers strong internal validity, but sacrifices external validity. The latter method increases external validity, but sacrifices control over variables, and requires more effort to build a natural base of users.

Because of this cost, many of the online communities that have been created for research purposes have come from industry research centers. Industry researchers have the advantage of access to non-research resources that can be essential for the creation of online communities, including programming support and in some cases marketing staff. Some industry laboratories choose to implement social software in their own organizations, which provides access to a naturally bounded population. Disadvantages of building sites within an industry context is that organizational priorities may shift, causing the site to be abandoned due to business, rather than research, decisions. In addition, data from these sites can be subject to a variety of employment laws or intellectual property policies, making it difficult to share data from these sites with other researchers not in the organization.

IBM research has created many online communities to examine how people interact online. An example of an IBM online community that was used for academic research was Beehive, a social network site open only to IBM workers. The site had more than 70,000 users at its peak and was used to test such questions as why people use social network sites at work (DiMicco et al., 2008), how people use social network sites in the organization to access social capital and knowledge in radically dispersed organizations like IBM (Steinfield et al., 2009) and how people might be introduced to one another through the system (Chen, et al., 2009). The research on Beehive used multiple methods, including access to behavioral trace data, as well as interviews, surveys and experiments through interface changes.

An example of online community creation for research outside of industry research laboratories is the University of Minnesota Grouplens research laboratory. This group has created several sustainable online communities, including MovieLens[5] and Cyclopath.[6]

Using MovieLens, the group tested basic research in how technical systems like ratings interacted with social processes. For example, this group studied how social psychological principles of social loafing could be overcome using interface changes (Ling et al., 2005), how people responded to different types of rating systems in an online community (Cosley et al., 2003) and how to use system feedback to increase participation of users (Ludford et al., 2004). With the Cyclopath research, the group has been conducting fundamental research in how users adopt different tasks in online communities and how users progress from casual consumers of content of the online community into participating members (Panciera, 2011; Panciera et al., 2010). These studies looked at similar research questions about newcomers in online communities as had been asked using the Wikipedia common dataset (Panciera, Halfaker, and Terveen, 2009), but added depth to the argument through the ability to track users through their IP addresses before they make accounts, creating a fuller view of user behavior – another benefit of hosting an online community.

HOW BEHAVIORAL TRACES HAVE BEEN ANALYZED

Social network analysis (SNA) is one method that has been heavily applied to the analysis of behavioral trace data. SNA is a method that examines the structure of social connections and how that structure relates to social outcomes (Easley and Kleinberg, 2010). Researchers using SNA as a method might understand the overall topology of social network structures, including how groups are connected or how different connections within a network might lead to different social outcomes. Other SNA methods include examining the strength of the connection (or 'ties') between people (Granovetter, 1982) and how those differences might affect social outcomes. An example of SNA used to study behavioral trace data is Hogan (2010), who used an application on Facebook to download user networks and show how people are interconnected.

While SNA has been used for a long time in social sciences (Wellman, 1997b), the availability of communication data embedded in an online community has made a natural match between this analytical method and type of data. Adamic and Glance (2005) used social network analysis to examine links between political blogs to see how much cross-ideological posting was occurring between blogs of different viewpoints. Welser et al. (2011) looked at how people who edit Wikipedia communicate to each other differently based on what types of edits they make to the site. In these examples, researchers are able to impute social networks from connections like links between blogs or shared participation in Wikipedia pages. Social network sites are even richer sources of the data needed for SNA, since they are designed around users articulating their social networks, often mirroring the social networks they have in their offline lives (Lampe, Ellison, and Steinfield, 2006).

A good example of a study that uses SNA and behavioral trace data to understand social processes is that by Bakshy et al. (2012), who looked at the role of information diffusion in Facebook networks by using social network analysis to determine how different people in the network are key to the transmission of information throughout the network (Bakshy et al., 2012). These researchers were interested in how information was spread by members of a Facebook user's network. They designed an experimental condition applied to over 230 million Facebook users where one condition saw information about friends sharing a URL and others did not. The dependent variable in this case was whether the sharing person's friend was likely to also post the URL to be seen in their 'Newsfeed', a sequential list of posts by a user. Tie strength was used as an independent variable and was measured by examining how often the user interacted on the site with

the person who had reposted their content. They found that, overall, weak ties (people with whom the poster did not interact with more regularly) were more influential in propagating URLs on the site. This example shows how social network measurements drawn from behavioral trace data can be used as an independent variable to explain social outcomes, in this case the role of social influence on information diffusion. The behavioral data in this case was drawn from the population of Facebook users and shows the power of having access to such a large network. Before this type of access to behavioral trace data, SNA was limited in how much network data could be obtained by other methods. Even with large amounts of data, SNA must be carefully matched to research questions. In the example above, the researchers can only describe patterns that are propagated on Facebook, so would not account for the same information being shared from Facebook to other channels, like Twitter, e-mail or face-to-face conversation.

The use of SNA to understand behavioral trace data has become so prevalent that several recent textbooks have been made available to help researchers apply SNA techniques to behavioral trace data (Easley and Kleinberg, 2010; Hansen, Shneiderman, and Smith, 2011; Monge and Contractor, 2003). However, others have pointed out that researchers need to be aware of validity concerns in applying SNA to behavioral trace data, including imputing the completeness and meaning of links between actors in those networks (Howison, Wiggins and Crowston, 2011).

LIMITATIONS TO BEHAVIORAL TRACE DATA

Access to the large amounts of social behavioral data available in online communities is a great advantage to understanding both social systems and technology. However, these data are also constrained in what they can tell researchers. Behavioral trace data can often give remarkable detail about what people do in online communities, but they rarely provide information about why people make the choices they do in those systems. For example, two people may be connected in a site like Twitter through the linking feature, but the existence of the link does not show why that link was made, what value the people get from their connection or how often they attend to that link absent of an explicit action in the system. Several studies have attempted to validate online behaviors with offline beliefs and opinions regarding use (Burke, et al., 2010; Gilbert and Karahalios, 2009), but more multi-methodological work needs to be conducted in order to triangulate behavioral data with psychosocial mechanisms that underlie the behavior. An example of multi-methodological work that combines behavioral trace data with other methods to enable a fuller exploration of the subject is that of Kairam et al. (2012), who were interested in the ways users of the social network site Google+ were using features of the site to selectively share information with parts of their network. They took a random sample of 100,000 Google+ users and employed behavioral trace data to show how people were using the 'Circles' feature, which allows a person to divide their online social network into different categories and then easily choose which of those categories to share posts with. The authors used content analysis to determine which types of groups Google+ users had created, finding that people largely used life facets (work, school, personal) and tie strength (close friends, acquaintances) to segment their networks. In order to add more explanatory power to the behavioral trace data, the researchers also conducted surveys and interviews with smaller subsets of users, allowing them to learn more about the motivations users had for sharing information with different groups.

An additional limitation on behavioral trace data is that it tends to track behavior in one 'channel' of communication and rarely looks at user behaviors in multiple contexts.

For example, a person might use and contribute to Twitter, Wikipedia, Facebook and Pinterest, but since each site maintains separate databases, it is not feasible to track that person's behavior across those sites. Indeed, some sites allow for anonymous or pseudonymous posting, making it exceedingly difficult to compare individual users across multiple sites. This limits the types of claims researchers can make from behavioral trace data taken from individual sites. However, there are still insights about behaviors within individual sites that generalize to overall online communities by connecting these insights to broader theories of mediated interaction.

CONCLUSION

Understanding the social interactions in online communities requires a wide variety of research methods and empirical perspectives. From ethnographies to lab experiments, we have employed many data-gathering techniques to online communities and social media. However, one advantage of studying online communities is that they contain a record of the behaviors of people who use them.

Online communities are special in their ability to provide detailed records of user behavior, which we have called behavioral trace data here. These data can provide extremely granular records of what people do in online systems, allowing for new ways of understanding human interactions on a large scale. This data has been analyzed in many ways, but has been especially useful in the application of social network analysis to understanding large-scale social interactions. Although there are caveats to the use of behavioral trace data in this manner, the opportunities in this area to understand not just mediated human interaction but also social processes more generally are greatly enhanced by the availability of the data. Using behavioral trace data in conjunction with other types of data-collection methods allows us access to exciting new opportunities in social science.

NOTES

1. www.webkinz.com
2. dumps.wikimedia.org
3. www.kaggle.com/c/wikichallenge
4. www.nd.edu/~oss/Data/data.html
5. www.movielens.org/login
6. cyclopath.org

REFERENCES

Abbate, J. (2000). *Inventing the Internet*. Cambridge: MIT Press.

Ackerman, M. (2001). The Intellectual Challenge of CSCW: The Gap between Social Requirements and Technical Feasibility. In J. Carroll (ed.), *Human–Computer Interaction in the New Millennium*. New York: ACM Press.

Adamic, L., & Glance, N. (2005). The Political Blogosphere and the 2004 U.S. Election: Divided They Blog. Paper presented at the 2nd Annual Workshop on the Weblogging Ecosystem: Aggregation, Analysis and Dynamics (WWW2005), Japan.

Bakshy, E., Rosenn, I., Marlow, C., & Adamic, L. (2012). The Role of Social Networks in Information Diffusion. Paper presented at the World Wide Web (WWW) 2012, Lyon, France.

Barkhuus, L., & Tashiro, J. (2010). Student Socialization in the Age of Facebook. *Proceedings of the 28th International Conference on Human Factors in Computing Systems*, Atlanta, Georgia, USA.

boyd, d. (2004). Friendster and Publicly Articulated Social Networks. Paper presented at the Conference on Human Factors and Computing Systems, Vienna, Austria.

boyd, d.m., & Ellison, N. (2007). Social Network Sites: Definition, History, and Scholarship. *Journal of Computer Mediated Communication*, 13(1), article 11.

Brooks, B., Hogan, B., & Titsworth, S. (2011). Socioeconomic Status Updates: Family SES and Emergent Social Capital in College Student Facebook Networks. *Information, Communication & Society*, 14(4), 529–549.

Burke, M., Kraut, R., & Marlow, C. (2011). Social Capital on Facebook: Differentiating Uses and Users. *Proceedings of the 2011 Annual Conference on Human Factors in Computing Systems*, Vancouver, BC, Canada.

Burke, M., Marlow, C., & Lento, T. (2009). Feed Me: Motivating Newcomer Contribution in Social Network Sites. *ACM Conference on Human Factors in Technical Systems*, Boston, MA.

Burke, M., Marlow, C., & Lento, T. (2010). Social Network Activity and Social Well-Being. *Proceedings of the 28th International Conference on Human Factors in Computing Systems*, Atlanta, Georgia, USA.

Butler, B.S. (2001). Membership Size, Communication Activity, and Sustainability: A Resource-Based Model of Online Social Structures. *Information Systems Research*, 12(4), 346–362.

Chen, J., Geyer, W., Dugan, C., Muller, M., & Guy, I. (2009). Make New Friends, but Keep the Old: Recommending People on Social Networking Sites. *Proceedings of the 27th International Conference on Human Factors in Computing Systems*, Boston, MA, USA.

Chi, E.H. (2009). A Position Paper on 'Living Laboratories': Rethinking Ecological Designs and Experimentation in Human–Computer Interaction. *Proceedings of the 13th International Conference on Human–Computer Interaction. Part I: New Trends*, San Diego, CA.

Cosley, D., Lam, S.K., Albert, I., Konstan, J.A., & Riedl, J. (2003). Is Seeing Believing?: How Recommender Interfaces Affect Users' Opinions. *Computer–Human Interaction (ACM-CHI)*, Minneapolis, MN.

Curtis, P. (1992). Mudding: Social Phenomena in Text-based Virtual Realities. *Conference on Directions and Implications of Advanced Computing*, Berkeley, CA.

DiMicco, J., Millen, D.R., Geyer, W., Dugan, C., Brownholtz, B., & Muller, M. (2008). Motivations for Social Networking at Work. *Proceedings of the ACM 2008 Conference on Computer Supported Cooperative Work*, San Diego, CA, USA.

Easley, D., & Kleinberg, J. (2010). *Networks, Crowds, and Markets: Reasoning About a Highly Connected World*. New York: Cambridge University Press.

Ellison, N., Steinfield, C., & Lampe, C. (2007). The Benefits of Facebook 'Friends': Social Capital and College Students' Use of Online Social Network Sites. *Journal of Computer Mediated Communication*, 12(4), article 1.

Fiore, A.T., Tiernan, S.L., & Smith, M.A. (2002). Observed Behavior and Perceived Value of Authors in Usenet Newsgroups: Bridging the Gap. *CHI '02*, Minneapolis, MN.

Ganley, D., & Lampe, C. (2009). The Ties that Bind: Social Network Principles in Online Communities. *Decision Support Systems*, 47(3), 266–274.

Gao, Y., & Madey, G. (2007). Network Analysis of the SourceForge.net Community. *Third International*

Conference on Open Source Systems (OSS '07), Limerick, Ireland.

Gao, Y., VanAntwerp, M., Christley, S., & Madey, G. (2007). A Research Collaboratory for Open Source Software Research. *29th International Conference on Software Engineering + Workshops*, Minneapolis, MN.

Gilbert, E., & Karahalios, K. (2009). Predicting Tie Strength with Social Media. *Proceedings of the 27th International Conference on Human Factors in Computing Systems*, Boston, MA, USA.

Golder, S., Wilkinson, D., & Huberman, B.A. (2007). Rhythms of Social Interaction: Messaging within a Massive Online Network. *3rd International Conference on Communities and Technologies (CT '07)*, East Lansing, MI.

Granovetter, M.S. (1982). The Strength of Weak Ties: A Network Theory Revisited. In P.V.M.N. Lin (ed.), *Social Structure and Network Analysis*. Thousand Oaks, CA: Sage.

Grudin, J. (1988). Why CSCW Applications Fail: Problems in the Design and Evaluation of Organizational Interfaces. *ACM conference on Computer-supported Cooperative Work*, Portland, OR.

Hampton, K.N., Goulet, L.S., Rainie, L., & Purcell, K. (2011). *Social Networking Sites and our Lives*. Washington, DC: PEW Research Center's Internet & American Life Project.

Hansen, D.L., Shneiderman, B., & Smith, M.A. (2011). *Analyzing Social Media Networks with NodeXL: Insights from a Connected World*. Burlington, MA: Morgan Kaufmann.

Hiltz, S. R., & Turoff, M. (1978). *The Network Nation: Human Communication via Computer*. London: Addison-Wesley.

Hogan, B. (2010). Visualizing and Interpreting Facebook Netoworks. In D. Hansen, M. A. Smith & B. Shneiderman (eds), *Analyzing Social Media Networks with NodeXL*. Burlington, MA: Morgan Kaufman pp. 165–180.

Hollan, J., & Stornetta, S. (1992). Beyond Being There. *Conference on Human Factors in Computing Systems (CHI)*, Monterey CA.

Howison, J., Wiggins, A., & Crowston, K. (2011). Validity Issues in the Use of Social Network Analysis with Digital Trace Data. *Journal of the Association for Information Systems*, 12(12).

Kairam, S., Brzozowski, M.J., Huffaker, D., & Chi, E.H. (2012). Talking in Circles: Selective Sharing in Google+. *Conference on Human Factors in Computing Systems (CHI)*, Austin, TX.

Keegan, B., Ahmad, M.A., Williams, D., Srivastava, J., & Contractor, N. (2011). What Can Gold Farmers Teach

Us about Criminal Networks? *XRDS*, 17(3), 11–15. doi: 10.1145/1925041.1925043

Kramer, A.D.I. (2010). An Unobtrusive Behavioral Model of 'Gross National Happiness'. *Proceedings of the 28th International Conference on Human Factors in Computing Systems*, Atlanta, Georgia, USA.

Lampe, C., Ellison, N., & Steinfield, C. (2006). A Face(book) in the Crowd: Social Searching vs. Social Browsing. *ACM Special Interest Group on Computer-Supported Cooperative Work*, Banff, Canada.

Lampe, C., Ellison, N., & Steinfield, C. (2007). Profile Elements as Signals in an Online Social Network. *ACM Conference on Human Factors in Computing Systems (CHI)*, San Jose, CA.

Lampe, C., Ellison, N., & Steinfield, C. (2008). Changes in Use and Perception of Facebook. *ACM Conference on Computer-Supported Cooperative Work (CSCW)*, San Diego, CA.

Lampe, C., & Garrett, R. K. (2007). It's All News to Me: The Effect of Instruments on Ratings Provision. *Proceedings of the 40th Annual Hawaii International Conference on System Sciences*.

Lampe, C., & Johnston, E. (2005). Follow the (Slash) dot: Effects of Feedback on New Members in an Online Community. *International Conference on Supporting Group Work, (GROUP '05)*, Sanibel Island, FL.

Lampe, C., Johnston, E., & Resnick, P. (2007). Follow the Reader: Filtering Comments on Slashdot. *ACM Conference on Human Factors in Computing Systems (CHI '07)*, San Jose, CA.

Lampe, C., & Resnick, P. (2004). Slash(dot) and Burn: Distributed Moderation in a Large Online Conversation Space. *Conference on Human Factors in Computing Systems (CHI)*, Vienna, Austria.

Lampe, C., Wash, R., Velasquez, A., & Ozkaya, E. (2010). Motivations to Participate in Online Communities. *Proceedings of the 28th International Conference on Human Factors in Computing Systems*, Atlanta, Georgia, USA.

Licklider, J.C.R., & Taylor, R. (1968). The Computer as a Communication Device. *Science and Technology*, April 1968.

Ling, K., et al. (2005). Using Social Psychology to Motivate Contributions to Online Communities. *Journal of Computer-Mediated Communication*, 10(4).

Ludford, P.J., Cosley, D., Frankowski, D., & Terveen, L. (2004). Think Different: Increasing Online Community Participation using Uniqueness and Group Dissimilarity. *Conference on Human Factors in Computing Systems*, Vienna, Austria.

Madden, M., & Zickuhr, K. (2011). 65% of Online Adults Use Social Networking Sites. Washington, DC: PEW Internet & American Life Project.

Marwick, A.E., & boyd, d. (2010). I Tweet Honestly, I Tweet Passionately: Twitter Users, Context Collapse, and the Imagined Audience. *New Media & Society*.

McLaughlin, C., & Vitak, J. (2012). Norm Evolution and Violation on Facebook. *New Media & Society*, 14(2), 299–315.

Monge, P.R., & Contractor, N. (2003). *Theories of Communication Networks*. Oxford: Oxford University Press.

Nonnecke, B., Preece, J., & Andrews, D. (2004). What Lurkers and Posters Think of Each Other. *37th Hawaii International Conference on System Sciences*, Hawaii.

Olson, G.M., & Olson, J.S. (2001). Distance Matters. In J. Carroll (Ed.), *HCI in the New Millennium*. New York: Addison-Wesley.

Panciera, K. (2011). User Lifecycles in Cyclopath: A Survey of Users. *Proceedings of the 2011 iConference*, Seattle, Washington.

Panciera, K., Halfaker, A., & Terveen, L. (2009). Wikipedians are Born, Not Made: A Study of Power Editors on Wikipedia. *Proceedings of the ACM 2009 International Conference on Supporting Group Work*, Sanibel Island, Florida, USA.

Panciera, K., Priedhorsky, R., Erickson, T., & Terveen, L. (2010). Lurking? Cyclopaths?: A Quantitative Lifecycle Analysis of User Behavior in a Geowiki. *Proceedings of the 28th International Conference on Human Factors in Computing Systems*, Atlanta, Georgia, USA.

Papacharissi, Z., & Mendelson, A. (2011). Toward a New(er) Sociability: Uses, Gratifications and Social Capital on Facebook. In S. Papathanassopoulos (ed.), *Media Perspectives for the 21st Century* New York: Routledge. pp. 212–230.

Pfaffenberger, B. (2002). A Standing Wave in the Web of Our Communications: Usenet and the Socio-Technical Construction of Cyberspace Values. In C. Lueg & D. Fisher (eds), *From Usenet to CoWebs: Interacting with Social Information Spaces*. New York: Springer-Verlag.

Quarterman, J.S., & Hoskins, J.C. (1986). Notable Computer Networks. *Communications of the ACM*, 29(10), 932–971.

Resnick, P. (2001). Beyond Bowling Together: SocioTechnical Capital. In J. Carroll (ed.), *HCI in the New Millenium*. Harlow: Addison-wesley.

Resnick, P., Iacovou, N., Suchak, M., Bergstrom, P., & Reidl, J. (1994). GroupLens: An Open Architecture for Collaborative Filtering of Netnews. *ACM Conference on Computer Supported Cooperative Work*, Chapel Hill, NC.

Rheingold, H. (1993). *The Virtual Community: Finding Connection in a Computerized World*. Boston, MA: Addison-Wesley.

Rheingold, H. (2000). *The Virtual Community: Homesteading on the Electronic Frontier* (2nd edn). Cambridge, MA: MIT Press.

Smith, M.A. (1999). Invisible Crowds in Cyberspace: Mapping the Social Structure of the Usenet. In P. Kollock & M.A. Smith (eds), *Communities in Cyberspace*. Abingdon, Oxon: Routledge. pp. 195–219.

Sproull, L., & Kiesler, S. (1991). *Connections: New Ways of Working in the Networked Organization*. Cambridge, MA: MIT Press.

Steinfield, C., DiMicco, J.M., Ellison, N.B., & Lampe, C. (2009). Bowling Online: Social Networking and Social Capital within the Organization. *Proceedings of the Fourth International Conference on Communities and Technologies*, University Park, PA.

Suh, B., Convertino, G., Chi, E.H., & Pirolli, P. (2009). The Singularity is Not Near: Slowing Growth of Wikipedia. *Proceedings of the Fifth International Symposium on Wikis and Open Collaboration*, Orlando, FL.

Toma, C., Hancock, J., & Ellison, N. (2008). Separating Fact from Fiction: An Examination of Deceptive Self-Presentation in Online Dating Profiles. *Personality and Social Psychology Bulletin*, 34, 1023–1036.

Tong, S.T., Heide, B.V.D., Langwell, L., & Walther., J.B. (2008). Too Much of a Good Thing?: The Relationship between Number of Friends and Interpersonal Impressions on Facebook. *Journal of Computer-Mediated Communication*, 13(3), 531–549.

Tufekci, Z., & Wilson, C. (2012). Social Media and the Decision to Participate in Political Protest: Observations From Tahrir Square. *Journal of Communication*, 62(2), 363–379. doi: 10.1111/j.1460-2466.2012.01629.x

VanAntwerp, M., & Madey, G. (2008). Advances in the SourceForge Research Data Archive (SRDA). *Fourth International Conference on Open Source Systems (WoPDaSD '08)*, Milan, Italy.

Walther, J.B. (1992). Interpersonal Effects in Computer-mediated Communication: A Relational Perspective. *Communication Research*, 19, 52–90.

Walther, J.B., & Parks, M.R. (2002). Cues Filtered Out, Cues Filtered In: Computer-mediated Communication and Relationships. In M.L. Knapp & J.A. Daly (eds), *Handbook of Interpersonal Communication* (3rd edn). Thousand Oaks, CA: Sage.

Wellman, B. (1997a). An Electronic Group is Virtually a Social Network. In S. Kiesler (ed.), *Culture of the Internet*. Hillsdale, NJ: Lawrence Erlbaum.

Wellman, B. (1997b). Structural Analysis: From Method and Metaphor to Theory and Substance. In B. Wellman & S.D. Berkowitz (eds), *Social Structures: A Network Approach*. Cambridge, MA: Cambridge University Press.

Wellman, B. (2001). Computer Networks as Social Networks. *Science*, 293(14), 2031–2034.

Wellman, B., & Gulia, M. (1999). Net Surfers Don't Ride Alone: Virtual Community as Community. In B. Wellman (ed.), *Networks in the Global Village*. Boulder, CO: Westview Press.

Welser, H.T., Cosley, D., Kossinets, G., Lin, A., Dokshin, F., Gay, G., & Smith, M. (2011). Finding Social Roles in Wikipedia. *Proceedings of the 2011 iConference*, Seattle, Washington.

Whittaker, S., Terveen, L., Hill, W., & Cherny, L. (1998). The Dynamics of Mass Interaction. *Computer-Supported Cooperative Work*, Seattle, WA.

17

Multimodal Methods for Researching Digital Technologies

Carey Jewitt

This chapter provides an introduction to the field of multimodality and discusses its potential application for researching digital data and environments. It begins by outlining what multimodality is, its theoretical origins in social semiotics and its underlying assumptions. A number of concepts central to multimodality are introduced: these include mode, semiotic resource, materiality, modal affordance, multimodal ensemble and meaning functions. The scope and potential of multimodality for researching digital technologies are then discussed. The chapter sets out an illustrative example of multimodal research. It concludes with a discussion of the limitations and challenges of a multimodal approach for researching digital technologies.

WHAT IS MULTIMODALITY?

Multimodality is an interdisciplinary approach drawn from social semiotics that understands communication and representation as more than language and attends systematically to the social interpretation of a range of forms of making meaning. It provide concepts, methods and a framework for the collection and analysis of visual, aural, embodied and spatial aspects of interaction and environments (Jewitt, 2013; Kress, 2010). While other modes of communication, such as gesture, have been recognized and studied extensively (e.g. McNeil, 1992), multimodality investigates the interaction between communicational means and challenges the prior predominance of spoken and written language in research (Scollon and Scollon, 2009). Speech and writing continue to be understood as significant but are seen as parts of a multimodal ensemble. Multimodality emphasizes situated action – that is, the importance of the social context and the resources available for meaning making, with attention to people's situated choice of resources, rather than emphasizing the system of available resources. Thus it opens up possibilities for recognizing, analyzing and theorizing the different ways in which people make meaning and how those meanings are interrelated.

Multimodality provides resources to support a complex fine-grained analysis of artifacts and interactions in which meaning is understood as being realized in the iterative connection between the meaning potential of a material semiotic artifact, the meaning potential of the social and cultural environment it is encountered in and the resources, intentions and knowledge that people bring to that encounter. That is, it strives to connect the material semiotic resources available to people with what they mean to signify in social contexts. Changes to these resources and how they are configured are therefore understood as significant for communication. Digital technologies are of particular interest to multimodality because they make a wide range of modes available, often in new inter-semiotic relationships with one another, and unsettle and remake genres in ways that reshape practices and interaction. Digital technologies are thus a key site for multi-modal investigation.

Underlying this approach is the idea that language, and other systems or modes of communication (e.g. gesture, gaze), is shaped through the things that it has been used to accomplish socially in everyday instantiations, not because of a fixed set of rules and structures. This view of language as a situated resource encompasses the principle that modes of communication offer historically specific and socially/culturally shared options (or 'semiotic resources') for communicating. With this emphasis, a key question for multi-modality is how people make meaning in context to achieve specific aims.

Three interconnected theoretical assumptions underpin multimodality. These are briefly introduced and discussed below.

The first assumption underlying multi-modality is that, while language is widely taken to be the most significant mode of communication, speech or writing are a part of a multimodal ensemble. Multimodality 'steps away from the notion that language always plays the central role in interaction, without denying that it often does' (Norris, 2004: 3) and proceeds on the assumption that all modes have the potential to contribute equally to meaning. From a multimodal perspective, language is therefore only ever one mode nestled among a multimodal ensemble of modes. While others have analyzed 'non-verbal' modes, multimodality differs in that language is not its starting point, nor does it provide a prototypical model of all modes of communication. The starting point is that all modes which are a part of a multimodal ensemble – a representation and/or an interaction – need to be studied with a view to the underlying choices available to communicators, the meaning potentials of resources and the purposes for which they are chosen.

The second assumption central to multi-modal research is that all modes have, like language, been shaped through their cultural, historical and social uses to realize social functions as required by different communities. Therefore each mode is understood as having different meaning potentials or semiotic resources and to realize different kinds of communicative work. Multimodality takes all communicational acts to be constituted of and through the social. This also draws attention to the ways in which communication is constrained and produced in relation to social context and points to how modes come into spaces in particular ways.

This connects with the third assumption underpinning multimodality – that people orchestrate meaning through their selection and configuration of modes. Thus the interaction between modes is significant for meaning making. Multimodal communication is not in and of itself new, communication has always been multimodal, however, digital media has led to an increased interest in the multimodal character of communication as it foregrounds the need to consider the particular characteristics of modes, multimodal configurations, and their semiotic function in contemporary discourse worlds (Ventola et al., 2004). The meanings in any mode are always interwoven with the meanings made with those of other modes cooperating in the communicative ensemble. The interaction between modes is itself a part of the production of meaning.

A BRIEF BACKGROUND

Multimodality was developed in the early 2000s (see Kress and van Leeuwen, 2001; Kress et al., 2001, 2005; van Leeuwen, 2005; Jewitt, 2009). It originated from linguistic ideas of communication, in particular the work of Michael Halliday on language as a social semiotic system. Halliday's work shifted attention from language as a static linguistic system to language as a social system – how language is shaped by the ways that people use it and the social functions that the resources of language are put to in particular settings. In *Language as Social Semiotic* (1978) Halliday sets out a theory of language built on a social functional perspective of meaning and a framework for understanding language as a system of options and meaning potentials: in summary the idea of meaning as choice.

Hodge and Kress in *Social Semiotics* (1998), and later Kress and van Leeuwen in *Reading Images* (2006), expanded attention from language to other semiotic systems (or modes), laying the groundwork for extending and adapting social semiotics across a range of modes and opening the door for multimodality. Kress and van Leewen extended principles developed in relation to language to the visual. They examined visual texts to identify a range of semiotic resources, meaning potentials, available choices and the organizing principles underpinning their configuration to visually communicate ideologies and discourses. Multimodality has taken ideas from linguistics that are theoretically transportable to other modes, such as turn taking, coherence, composition, and it has explored the currency of these in relation to the particularities of other modes. In doing so it has extended and adapted Halliday's conception of meaning across a range of modes by taking the specific resources and

organizing principles of spoken and written language as a starting point and extending their essence to other modes in ways recognize and the resources of gesture, gaze, and image differ in significant ways. As multimodality has developed it has also looked beyond linguistics for resources to assist with analysis and to further explore the situated character of meaning making, including sociolinguistics, film theory, art history and iconography and musicology.

Multimodality foregrounds the modal choices people make and the social effect of these choices on meaning. There is therefore a strong emphasis on the notion of context within social semiotic multimodal analysis. The context shapes the resources available for meaning making and how these are selected and designed. Signs, modes and meaning making are treated as relatively fluid, dynamic and open systems, intimately connected to the social context of use. From this perspective analytical interest in the modal system (its resources and principles) is strongly located in (and regulated through) the social and cultural. When making signs, people bring together and connect the available form that is most apt to convey the meaning they want to express at a given moment.

Kress introduced a strong emphasis on the social character of meaning and developed the concept of the motivated sign (Kress, 1997). This served to foreground the agency of the sign maker and the process of sign making. In *Before Writing* (Kress, 1997) he offers a detailed account of the materiality and processes of young children's engagement with texts and how they interpret, transform and redesign the semiotic resources and signs available to them – what has been described as chains of semiosis. From this perspective, signs (e.g. talk, gesture and textual artifacts) are analyzed as material residues of a sign maker's interests. The analytical focus is on understanding their interpretative and design patterns and the broader discourses, histories and social factors which shape that. In a sense, then, the text is seen as a window on to its maker. Viewing signs as motivated and constantly being remade draws attention to the interests and intentions that motivate a person's choice of one semiotic resource over another (Kress, 1993). This 'interest' connects a person's choice of one resource over another with the social context of sign production – returning to the importance of meaning as choice within social semiotic theories of communication. The modal resources that are available to the person are an integral part of that context – hence the importance of multimodality to understanding the process of meaning making.

Multimodality can, at least in part, be understood as a response to the demands to look beyond language in a rapidly changing social and technological landscape. It is curious to understand how the use of digital technologies extends the range of resources for communication, reshapes the relationship between resources such as image and writing, and has the potential to significantly reconfigure notions of spatiality and embodiment as well as genre conventions, all of which can lead to adapted and some new types of texts and interactions.

KEY CONCEPTS

This section outlines in more detail six concepts introduced above that are key for multimodality: mode, semiotic resource, materiality, modal affordance, multimodal ensembles and meaning functions.

Mode

This term refers to a set of socially and culturally shaped resources for making meaning: a 'channel' of representation or communication (Kress and van Leeuwen, 2001). One definition of a mode is that it has to comprise a set of elements/resources and organizing principles/ norms that realize well-acknowledged regularities within any one community. That is something which can only be recognized as a mode when it is a known/usable system of communication within a community. The ability for the 'grammar' of the modal system to be broken is seen as a 'test' that it exists. Another 'test' for whether or not a set of resources can count as a mode is if it is possible for it to articulate all three of Halliday's (1978) meaning functions – that is, can a set of resources be used to articulate 'content' matter (ideational meaning), construct social relations (interpersonal meaning) and create coherence (textual meaning). Accepted examples of modes include writing, image, moving image, sound, speech, gesture, gaze and posture in embodied interactions. What constitutes a mode is the subject of debate. For instance, van Leeuwen (1999) has explored when sound and music can be thought of as modes, while Bezemer and Kress (2008) have discussed whether color and layout can be considered as modes. As these examples suggest, modes are created through social processes, fluid and subject to change – not autonomous and fixed. For example, the meanings of words and gestures change over time. Modes are also particular to a community/culture where there is a shared understanding of their semiotic, rather than universal, characteristics.

Semiotic Resource

This term is used to refer to a means of meaning making that is simultaneously a material, social and cultural resource. In other words a semiotic resource can be thought of as the connection between representational resources and what people do with them:

> Semiotic resources are the actions, materials and artifacts we use for communicative purposes, whether produced physiologically – for example, with our vocal apparatus, the muscles we use to make facial expressions and gestures – or technologically – for example, with pen and ink, or computer hardware and software – together with the ways in which these resources can be organized. Semiotic resources have a meaning potential, based on their past uses, and a set of affordances based on their possible uses, and these will be actualized in concrete social contexts where their use is subject to some form of semiotic regime. (van Leeuwen, 2005: 285)

This definition highlights the historical development of connections between form and meaning, aligned with Bakhtin's notion of intertextuality. Kress (2010) emphasizes that these resources are constantly transformed. This theoretical stance presents people as agentive sign-makers who shape and combine semiotic resources to reflect their interests.

Materiality

Materiality refers to how modes are taken to be the product of the work of social agents shaping material, physical 'stuff' into cultural semiotic resources. This materiality has important semiotic potentials in itself: sound has different affordances from written inscription, while gesture offers different material potentials from colour, and so on. All modes, on the basis both of their materiality and of the work that societies have done with the material (e.g. working sound to become speech or music) offer specific potentials and constraints for making meaning. The materiality of modes also connects with the body and its senses, that in turn place the physical and sensory at the heart of meaning.

Modal Affordance

The term modal affordance is contested and continuously debated within multimodal research. It originated from the psychologist James Gibson's (1979) work on perception and agent–situation interaction, which defined affordances as the 'action possibilities' latent in an environment, and in which the potential uses of any object arise from its perceivable properties in relation to how it is perceived by an actor's capabilities and interests. Donald Norman later took up this term in relation to the design of artifacts, with an emphasis on both the material and social dimensions of materiality (1990).

Adapted by Kress (e.g. 2010), the term 'modal affordance' refers to the potentialities and constraints of different modes – what it

is possible to express and represent or communicate easily with the resources of a mode and what is less straightforward or even impossible – and this is subject to constant social work. From this perspective, the term 'affordance' is not a matter of perception, but, rather, is a complex concept connected to both the material *and* the cultural, social and historical use of a mode. Modal affordance is shaped by how a mode has been used, what it has been repeatedly used to mean and do and the social conventions that inform its use in context. As indicated by van Leeuwen's definition of semiotic resource, where a mode originates, its history of cultural work, its provenance, shapes the meaning potential of a semiotic resource. These affordances contribute to the different communicational and representational potentials or modal logics of modes (although it is important to note these are open to change and disruption). The affordances of the sounds of speech, for instance, usually happen across time and this sequence in time shapes what can be done with (speech) sounds. The logic of sequence in time is difficult to avoid for speech: one sound is uttered after another, one word after another, one syntactic and textual element after another. This sequence becomes an affordance or meaning potential: it produces the possibilities for putting things first or last or somewhere else in a sequence. The mode of speech is therefore strongly governed by the logic of time. Like all governing principles they do not hold in all contexts and are realized through the complex interactions of the social as material and vice versa – in this sense the material constitutes the social and vice versa. Modal affordance suggests all modes are partial in making meaning, so that the designed selection of modes, into multimodal ensembles, allows this partiality to be managed.

Multimodal Ensembles

Representations or interactions that consist of more than one mode can be referred to as a multimodal ensemble. The term draws attention to the agency of the sign maker – who pulls together the ensemble within the social and material constraints of a specific context of meaning making. Multimodal ensembles can therefore be seen as a material outcome or trace of the social context, available modes and modal affordances, the technology available and the agency of an individual. When several modes are involved in a communicative event (e.g. a text, a website, a spoken interchange) all of the modes combine to represent a message's meaning (e.g. Kress et al., 2001, 2005). The meaning of any message is, however, distributed across all of these modes and not necessarily evenly. The different aspects of meaning are carried in different ways by each of the modes in the ensemble. Any one mode in that ensemble is carrying a part of the message only: each mode is therefore partial in relation to the whole of the meaning, and speech and writing are no exception (Jewitt and Kress, 2003). Multimodal research attends to the interplay between modes to look at the specific work of each mode and how each mode interacts with and contributes to the others in the multimodal ensemble. This raises analytical questions, such as which modes have been included or excluded, the function of each mode, how meanings have been distributed across modes and what the communicative effect of a different choice would be. At times the meaning realized by two modes can be 'aligned', at other times they may be complementary and at other times each mode may be used to refer to distinct aspects of meaning and be contradictory or in tension. Lemke noted (2002: 303) 'No [written] text is an image. No text or visual representation means in all and only the same ways that text can mean. It is this essential incommensurability that enables genuine new meanings to be made from the combinations of modalities'. Modal affordance in the context of multimodal ensembles raises the question of what image is 'best' for and what words, as well as what other modes and their arrangements are 'best' for in a particular context. The

relationships between modes as they are orchestrated in interactions (and texts) may themselves realize meanings through particular modal combinations, different weightings of modes (Martinec and Salway, 2005) or modal density in an ensemble (Norris, 2009). The structure of hyperlinks, for example, realizes connections and disconnections between elements that may contribute to the expansion of meaning relations between elements. The question of what to attend to, what to 'make meaningful' is a significant aspect of the work of making meaning and is foregrounded by a multimodal focus. Further, as meaning makers decide on modal 'best fit' and how to combine modes for a particular purpose, analysis of the moment-by-moment processes of constructing multimodal ensembles can enable the analyst to unpack how meanings are brought together.

Meaning Functions

As noted earlier, multimodality is built on a functional theory of meaning, an idea of meaning as social action realized through people's situated modal choices and the way they combine and organize these resources into multimodal ensembles. It distinguishes between three different but interconnected categories of meaning choices (also called meta-functions) that are simultaneously made when people communicate.

1. Choices related to how people realize content meanings (known as Ideational meaning) – that is, the resources people choose to represent the world and their experience of it, for example, what is depicted about processes, relations, events, participants and circumstances;
2. Choices related to how people articulate Interpersonal meanings – that is, the resources people choose to represent the social relations between themselves and those they are communicating with – either directly via interaction or via a text or artifact. For example, the visual or spatial depiction of elements as near and far, direct or oblique, are resources used to orientate viewers or inter-actors to a text or one another;

3. Choices concerned with textual or organizational meaning – for example, the choice of resources such as space, layout, pace and rhythm for realizing the cohesion, composition and structure of a text or interaction.

Multimodality applies these meaning functions to all modes to better understand their meaning potential – 'what can be meant' or 'what can be done' with a particular set of semiotic resources – and to explore how these three interconnected kinds of meaning potentials are actualized through the grammar and elements of their different modal systems.

A key point to draw attention to here is that the concepts outlined in this section can be applied across any kind of representation or interaction – be it a printed or digital text (Jewitt, 2002), a classroom with or without technology (Jewitt et al., 2011) or a complex interaction in a digitally mediated environment such as a surgical operating theatre (Bezemer et al., 2011). Thus, a researcher can employ multimodality to investigate the modal meaning potentials of a resource (e.g. mobile application, tangible environment) as well as how people make use of these resources in interaction.

THE POTENTIAL OF MULTIMODALITY FOR RESEARCHING DIGITAL TECHNOLOGIES

This section gives a sense of the scope and potential of multimodality for researching digital technologies: how it has been used to date, the kinds of questions it can be used to address and what research insights it can provide to inform the evaluation of technology design and use. The following four potentials of multimodal research are discussed in this section.

1. The systematic description of modes and their semiotic resources.
2. Multimodal investigation of interpretation and interaction with specific digital environments.

3. Identification and development of new digital semiotic resources and new uses of existing resources in digital environments.
4. Contribution to research methods for the collection and analysis of digital data and environments within social research.

The Systematic Description of Modes and their Semiotic Resources

A multimodal approach can be used to create an inventory of the meaning potentials available to people when using a technology in a particular context. This may be done through a systematic description of the modes and their semiotic resources, materiality and modal affordances and the organizing principles of a device and/or application. Building on the notion of meaning as choice and the concept of the meta-functions, some multimodal researchers use a style of diagramming called system networks to map the meaning potentials of a mode. This is a diagrammatic taxonomy of the systematic, semiotic options that are possible within a semiotic or lexico-grammatical system. This maps the potential of modal resources to articulate content, interpersonal and textual or organizational meanings – in an artifact or interaction. The options should preferably be of the either/or type. As described by Kress and van Leewen (2006), for instance, a visual image may either be a 'demand for information' (a kind of visual question) or an 'offer of information' (a kind of statement) – it cannot be both. A 'demand for information', in turn may be either 'polar' (yes/no question), or open, and so on. When analyzing other modes than language, some semiotic relations are better described as scaled along a continuum – for example the semiotic dimensions of color have been mapped as a set of continuum scales concerning hue, brightness, luminosity and so on (Kress and van Leeuwen, 2002). System networks provide an analytical tool for mapping the range of semiotic resources and options made available by a mode in a given context. In this way system networks provide a way to push the formal analysis of a mode (or a semiotic resource) to a logical limit.

To date system networks have been used to describe the semiotic options available within a range of modes including language (Halliday and Matthiessen, 2004), visual communication and color (Kress and van Leeuwen, 2002, 2006), action (Martinec, 2000) and sound, voice and music (van Leeuwen, 1999), as well as three-dimensional objects (e.g. tables, Bjorkvall, 2009). Networks have been used to explore multimodal genres and multimodal ensembles including online newspapers (Knox, 2007; Caple and Knox, 2012), film and media texts (Bateman, 2008) and interactive media texts (White, 2012).

In the case of digital texts, mediated interaction and environments, multimodal inventories can be of use in both understanding the meaning-making potentials and constraints that different technologies place on representation, communication and interaction, and how users of those technologies notice and take up those resources in different ways. This can inform both the redesign of technological artifacts and environments as well as their introduction into a set of practices, for example for learning or work.

Multimodal Investigation of Interaction in Specific Digital Environments

Multimodal researchers have also used system networks to focus on how modal resources are taken up and used in a specific context. They map and compare people's choice of mode, semiotic resources in specific contexts and some examine how these modal choices are shaped by the materiality and affordances of a mode and the research subjects' knowledge and experience. Some multimodal researchers, particularly those who are focused on meaning making as a process and are thus perhaps less concerned with mapping the

resources of the mode itself, use system networks as a much looser heuristic tool to explore meanings. Multimodal studies investigate how these resources are used in specific contexts and how people talk about them, justify them and critique them in order to understand how semiotic resources are used to articulate discourses across a variety of contexts and media, such as school, workplaces, online environments, textbooks and advertisements.

The import of the body and spatiality in the contemporary digital landscape is evident in emergent bodily interaction-based technologies (Price et al., 2009). Much work has been done on the classroom as a multimodal environment of learning and the role of position, posture, gesture and gaze has been shown to be key to learning and teaching in the production of school English and science (e.g. Kress et al., 2001, 2005). Multimodal attention to how bodily modes and space feature in interaction – their semiotic resources and affordance has potential for researching digital technologies. For instance, Wii games serve to reconfigure the relationships between players' physical (and therefore social relationship) bodies, now with digital sensory feedback via wristbands and body straps, virtual avatars, and the screen in ways that require physical digital mapping in interesting ways for what it means to collaborate and 'play together'. Multimodality provides a set of resources to describe and interrogate these remappings – for example to get at the interaction across the 'physical' and the 'virtual' body. This type of digital remapping and extending of the physical is paramount in a range of digitally remediated contexts. The question of how screens and digital technologies remediate the role of the body is also relevant for understanding online multimodal interaction. Jones in his analysis of how people construct and consume multimodal displays of their selves in social networking environments examines 'how the different digital technologies available for producing and consuming displays affects the kinds of relationships that are possible between users of these sites and the kinds of social actions that these displays allow them to take' (Jones, 2009: 82). A focus on mode, semiotic resources, materiality and modal affordance provides a descriptive language for examining interaction in these complex sites. For instance, multimodal research in a surgical operating theatre shows the interactional impact of digital technologies being inserted into older established social environments (Bezemer et al., 2011). Surgeons undertaking keyhole surgery work in screen-based digital environments that, like the Wii, reorient their gaze, body posture, team configurations and require them to engage in physical–digital mapping. A multimodal approach also asks if the use of blended physical–digital tools of applications like those discussed here generate new forms of interaction and enable new action, as well as physical, perceptual and bodily experiences.

Multimodality has been applied to a range of multimodal digital genres to explore questions of digital identities and literacy, notably in the field of education (Alvermann, 2002; Jewitt and Kress, 2003; Marsh, 2006). It has also has been used to analyze the orchestration of music, filmic shots and editing features in video productions, digital animation and games (e.g. Walton, 2004; Burn, 2009), as well as online environments, (Jones, 2009) and more recently interactions with mobile and Geographic Information System (GIS) technologies (Hollett and Leander, this volume).

The relationships across and between modes in multimodal texts and interaction are a central area of multimodal research, and multimodal research often investigates the relationship between a given context and the configuration of modes in a text or situated interactions – both to better understand the modal resources in use and to address substantive questions. The textual or organizational meta-function has been a focus of this work, for instance understanding how multimodal cohesion (van Leeuwen, 2005) is realized (or not) through the integration of different semiotic resources in multimodal

texts and communicative events via rhythm, composition, information linking and modal density or intensity (Norris, 2004).

The ways in which contemporary digital texts are organized via textual features such as digital layering and hyperlinking and the impact of this on how people navigate multimodal digital texts has also been examined (Lemke, 2002; Zammit, 2007). This work is potentially useful when thinking about the take up of designed resources (e.g. Jewitt, 2008). There is a large body of multimodal research that explores the dynamics of the interaction between image and language. This includes the early work of Kress and van Leeuwen (2006) on the visual articulation of meaning, Lemke's (1998) work on the role of image and writing in science textbooks, work by Martinec and Salaway (2005) rethinking Barthes's classification of image–text relations, and Bezemer and Kress's (2008) development of a framework for the analysis of image, writing, typography, color and layout in school textbooks. Focusing on multimodal texts, for instance, Kress and Bezemer investigated the learning gains and losses of different multimodal ensembles of learning resources in science, mathematics and English from the 1930s to the first decade of the twenty-first century, including digitally represented and online learning resources. They provide a multimodal account of the changes to the design of learning resources and their epistemological and social/pedagogic significance. They conclude that image and layout are increasingly meshed in the construction of content and color so that layout and typography can increasingly be seen as communicative modes. With a focus on multimodal interaction Jewitt, in her book *Technology, Literacy and Learning* (2008), explores the fundamental connection between a range of modal resources (including color, image, sound, movement and gesture and gaze), digital technologies, knowledge, literacy and learning. In this and other work she shows how teacher and student engagement with the modal resources made available by technologies reshapes practices such as reading and writing and the particular ways in which students and teachers interact in school science and English classes and looks at the impact of this on learning. These studies show how digital technologies stretch, foreground and in some cases remake modes, semiotic resources and their configurations in contemporary materiality and modal affordances, as well as the intersemiotic relations possible in multimodal ensembles.

Identification and Development of New Digital Semiotic Resources and New Uses of Existing Resources

In addition to creating inventories of modes and semiotic resources, and analyzing how these are used in a range of specific contexts, multimodality contributes to the discovery and development of new semiotic resources and new ways of using existing semiotic resources.

> Studying the semiotic potential of a given semiotic resource is studying how that resource has been, is, and can be used for purposes of communication, it is drawing up an inventory of past and present and maybe also future resources and their uses. By such inventories are never complete, because they tend to be made for nature specific purposes. (van Leeuwen, 2005: 17)

The discovery and development of new modal resources is linked to social change and society's need for new semiotic resources and new ways of using existing semiotic resources as the communicational landscape changes. Two factors central to this are the potentials of digital technology and the importing of semiotic resources in a global society. Digital synthesizers and other digital technologies, for example, have reshaped the possibilities of the 'human' voice to create new semiotic resources and contexts for the use of 'human' voices – in digital artifacts, public announcements, music, and so on (van Leeuwen, 2005). This digital reshaping of voice has in turn impacted the non-digital use of voice – for example, providing different

tonal or rhythmic uses of the non-digital voice not previously imagined. Modal semiotic resources common to print-based texts, such as textual linking, layering, layout and the organization of time, are also foregrounded and reconfigured in significant ways by digital technologies. Knox (2007), for example, has explored how online newspapers have reshaped newspaper layout, genres and the relationship of image, writing and video, and has mapped the 'wash-back' influence from online to print-based newspapers, as well as reading pathways (Knox, 2007; Caple and Knox, 2012). Adami (2009) has examined the multimodal patterns of coherence and turn-taking on the social networking site YouTube. Multimodal tools also have the potential to identify and describe the reconfigurations of space, time and embodiment that digital technologies (e.g. mobile and GIS) make available and address questions about how these technologies influence people's interaction and experiences.

Multimodality moves beyond intuitive ideas about what a technology can do, to provide detailed analysis of the way semiotic resources of digital technologies work, what they can and cannot do. It enables the construction of explicit understandings of a form of communication and thus makes it possible for these to be discussed, taught and evaluated. Multimodality can also help to design and implement new uses for semiotic innovations.

Contribution to Research Methods

Researchers increasingly need to look beyond language to better understand how people communicate and interact in digital environments. This places new demands on research methods with respect to digital texts and environments where conventional concepts and analytical tools may need rethinking. Multimodality makes a significant contribution to existing research methods for the collection and analysis of data and environments within social research. It provides methods for the collection and analysis of types of visual digital data including screen capture data and eye-tracking data (e.g. see Holsanova, 2012) and researcher-generated and naturally occurring digital video data (e.g. Kress et al., 2001, 2005; Norris, 2004; Bezemer and Jewitt, 2010). The use of digital video technology and a multimodal focus pose what has become a key challenge for social research, namely how to transcribe or *re*-present multimodal data (e.g. movement in time and space). Increasingly, the topic of transcription is subject to innovation and experimentation in multimodal approaches. This might range from the inclusion of line drawings and stills from video footage to the use of software such as Comic Life and Transana (e.g. Baldry and Tibault, 2005; Plowman and Stephen, 2008; Flewitt et al., 2009; Bezemer and Mavers, 2011). As already discussed, multimodality provides tools for mapping and analyzing the visual, embodied and spatial features of interaction with digital technologies as well as the analysis of music, film, digital animation, games, adverts and other new media (e.g. Knox, 2007; Burn, 2009; Jones, 2009; Adami, 2009; Caple and Knox, 2012).

Having outlined the scope and potential of a multimodal approach for researching digital technologies in general terms, the following section illustrates its application.

ILLUSTRATIVE CASE STUDY

This short case study concerns the learning of mathematical concepts in a digital programming game environment and is focused on the interaction of two students (aged seven years) with the resources of Playground, an object-orientated programming tool (Jewitt and Adamson, 2003). The excerpt discussed here focuses on how the students' emergent conception of 'bounce' was shaped through their selection and use of the modal resources available to them (the full case study is reported in Jewitt, 2008).

In the students' original design using paper and pen, the game concerned a small creature

being chased by an alien that fired bombs to catch it. The movement of their characters (a creature and an alien) and bounce of the bullets were realized using modes and semiotic resources drawn from a static image, writing and cartoon-visual genres (e.g. a time-lapse drawing and wiggly lines to signify vibration and the sound of an explosion).

Programming the game in Playground offered the students additional modes and semiotic resources for their design, notably ready-made visual elements and backgrounds, color, movement and sound and the removal of the written mode. Detailed analysis of the students' game as a product as well as video data of the process of production shows that these modal resources demanded different kinds of representational commitment, design decisions and thinking on the part of the students. The move from the page to screen underpinned changes in ideational, interpersonal and textual meaning, resulting, for instance, in increasing the stakes for the little creature: now it will be killed instead of being caught, suggesting a shift in the students' understanding of the affordance (social rules and expectations) of genre from board game to adventure/action game on the screen. The students' digital redesign of the multimodal frame of the game redefined the game narrative and the necessity to consider the movement of the elements. In addition, they needed to specify the spatial and dynamic relationships between the elements in the game.

In the students' written description of the game, bounce is represented as 'the bombs go sideways by arrows and then if [the bomb] touches the bars it goes different ways'. That is, bounce is represented as a matter of movement and change of direction when something is touched. The semiotic resources and affordance of writing as a mode do not require the students to make explicit the 'cause' of this change in movement – the player, the bomb or the bars.

The digital environment of Playground represents the idea of bounce in three modes and each provides different semiotic resources for the students' construction of the entity

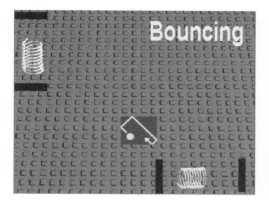

Figure 17.1   The program for 'bounce'.

Figure 17.2   The students' game design in Playground.

'bounce'. Figure 17.1 shows the program for 'bounce'. It uses the mode of writing – the word 'Bouncing' – to name and classify the movement in everyday terms. It uses the mode of still image – two images of a spring and an image of a ball – to specify particular potentials of bounce as a mechanical and regularly ordered entity rather than an organic, unpredictable bouncing (e.g. a rabbit). Third, it uses the mode of animated movement – three repeated animated sequences, one of a spring moving up and down between two bars, another of a spring moving sideways between two bars and a third sequence of a ball moving at angles within a square. The animated sequences work to give meaning to the entity 'bounce' in the context of the Playground program.

These modal resources work together as a multimodal ensemble to associate the (ideational) meaning of 'bouncing' within the mathematical paradigm of the system. This introduction of movement as a design resource raised a key question for the students in their design, 'What is it that produces bounce?' and 'What it is that bounces?'

Initially, the students in their game design (shown in Figure 17.2) programmed the sticks to bounce (that is, they added the behaviour of bouncing to the sticks), placed them on the game and then played the game. The sticks bounced off. It was the visual experience of playing the game that led the

students to realize their mistake and how to rectify it.

Through their engagement with the Playground environment the students worked out their ambiguities about agency – ambiguities that the affordances of writing and static image in their own paper design masked.

The students used gaze and gesture as a resource to address these questions and the process of programming bounce in their game. They students created different kinds of spaces on the screen through their gesture and gaze with/at the screen itself and their interaction with and organization of the elements displayed on the screen. These spaces marked distinctions between the different kinds of practices that the students were engaged with. In their creation and use of these spaces the students set up a rhythm and distinction between game planning, game design and construction and game playing. The students gestured 'on' the screen to produce a plan of the game: an 'imagined-space', overlaying the screen in which they gesturally placed elements and imagined their movement, and used gesture and gaze to connect their imagined (idealized) game with the resources of the application as it ran the program. The temporary and ephemeral character of gesture and gaze as modes enabled their plans of the game to remain fluid and ambiguous.

The role of gesture was central in their unfolding programming of the bouncing behaviour in three ways.

1. Gestures gave a way into understanding how the students are thinking about the concept 'bounce'. Initially the two students' talk and gesture is strongly coordinated and suggestive of a shared vision of how they imagine the bullet moving (from the alien to the left stick, then to the top-right stick). When the students stop acting in unison, however, two alternative versions of the movement of the bullet emerge (Figure 17.3). Student 1 traces the bullet moving in a *vertical* line down to the bottom-right stick. She then traces it in a *horizontal* line to the dog, wiggles the pen to indicate somewhere in that area. She is working with the entity 'bounce' as a generalized concept of movement, as going from one place to another. Student 2 works with the entity 'bounce' as a more specialized kind of movement. She indicates that a bullet would not move in a perpendicular line from the top-right to the bottom-right stick (as gestured by student 1). Holding her finger on the top-right stick she then gesturally traces an 'imagined' stick to the right of the alien before slowly trailing her finger off the edge of the screen. This 'gestural overlay' adds another stick to the visual design of the game, which in turn enables her to imagine the movement of the bullet bouncing from the top-right stick to the bottom-right stick, then off past the dog.

2. Examining the students' use of gesture in this way helped to identify areas of difficulty. The two students' accounts both end with a faltering tone of voice and lexical (e.g. 'whatever', 'ends somewhere') and gestural vagueness of wiggles and trailing off. These gestures are material signs of uncertainty of how the movement of a bullet would come to an end if it did not hit the dog. Would the ball keep bouncing or would it go off screen? This is itself an uncertainty of what is *producing the bounce*, is it the ball or the something that is hit by the ball.

3. The students' use of gesture can be analyzed to explore their hypothesis. Student 2 used a gestural overlay to 'estimate' where the ball would bounce, which in turn led to the amendment of the game – student 2's suggestion that they need to place some horizontal sticks on the planet.

The *invisibility*, the visual absence, of the bullets at this stage of the design is what proves to be problematic for the students. They prioritized the meaning of the visual within the multimodal ensemble of the game and, modally, at that point in the game-design, the students were working visually and not multimodally. The students were looking at the game to decide where to 'attach' the bounce. The 'sticks' (bars) were visible on screen but the bullets are 'within the alien' and are only visible when the game is being played. In this visual mode of working the system does not make the bullets available as something that the students could specify as the object that the 'I bounce' refers to. In short, when working visually, the notion of agency depends on visual presence. In sum, what was made visible on the

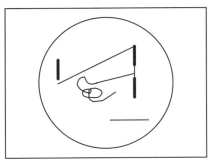

Student 1

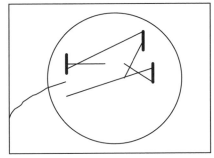

Student 2

Figure 17.3   The students' gestures with the screen.

screen proved to be particularly important in the students' design process. The students appeared to associate visual presence with agency: 'If it couldn't be seen it couldn't be acting' seemed to stand behind the students' programming process.

This example shows how the availability of multimodal resources changes the representations that students are working with, as well as the work of interpreting them, particularly what it is that the students need to attend to and what they need to specify. Finally, it highlights the potential of examining multimodal interaction and the range of representational resources available on screen to understand technology-mediated learning.

LIMITATIONS AND CHALLENGES

Although multimodal research has much to offer, it also has several limitations. A criticism sometimes made of multimodality is that it can seem rather impressionistic in its analysis. How do you know that this gesture means this or this image means that? In part this is an issue of the linguistic heritage of multimodality – that is, how do you get from linguistics to all modes. In part it is the view of semiotic resources as contextual, fluid and flexible – which makes the task of building 'stable analytical inventories' of multimodal semiotic resources complex. It is perhaps useful to note that this problem exists for speech or writing. The principles for establishing the 'security' of a meaning or a category are the same for multimodality as for linguistics and other disciplines. It is resolved by linking the meanings people make (whatever the mode) to context and social function. Increasingly, multimodal research looks across a range of data (combining textual/video analysis with interviews, for example) and towards participant involvement to explore analytical meanings as one response to this potential problem.

Linked with the above problem of interpretation is the criticism that multimodality is a kind of 'linguistic imperialism' that imports and imposes linguistic terms on everything. These critics overlook the fact that much of the work on multimodality has its origins in social semiotic theory of communication and the social component of this perspective sets it apart from narrower concerns with syntactic structures, language and mind and language universals that have long dominated the discipline. This view of communication can be applied (in different ways) to all modes.

Multimodal analysis is an intensive research process, both in relation to time and labor. Multimodal research can be applied to take a detailed look at 'big' issues and questions through specific instances. Nonetheless, the scale of multimodal research *can* restrict the potential of multimodality to comment beyond the specific to the general. The development of multimodal corpora may help to overcome some of these limitations, as might the potential to combine multimodal analysis with quantitative analysis in innovative ways.

CONCLUSION

This chapter has provided an introduction to the field of multimodality. It has discussed what multimodality is, sketched its theoretical origins and presented its underlying assumptions. Throughout the chapter the key concepts central to this approach have been introduced, discussed and illustrated through their application within the literature and in the case study example presented above. In this way the chapter has set out the scope and potential of multimodality for researching digital technologies with reference to four significant areas: (1) the systematic description of modes to research meaning making in complex digitally mediated environments and the evaluation and design of multimodal digital artifacts, interactions and experiences; (2) the investigation of interpretation and interaction in specific digital environments; (3) the identification and development of new digital semiotic resources and new uses; and (4) a contribution to research methods. Finally, the chapter points to some of the limitations and challenges of a multimodal approach for digital technologies.

REFERENCES

Adami, E. (2009) 'We/YouTube': exploring sign-making in video-interaction', *Visual Communication*, 8(4): 379–399.

Alvermann, D. (ed.) (2002) *Adolescents and Literacies in a Digital World*. New York: Peter Lang.

Baldry, A. and Tibault, P. (2005) *Multimodal Transcription and Text Analysis*. Harlow: Equinox.

Bateman, J. (2008) *Multimodality and Genre: A Foundation for the Systematic Analysis of Multimodal Documents*. Harlow: Palgrave Macmillan.

Bezemer, J. and Jewitt, C. (2010) 'Multimodal analysis', in L. Litosseliti (ed.), *Research Methods in Linguistics*. London: Continuum. pp. 180–197.

Bezemer, J. and Kress, G. (2008) 'Writing in multimodal texts: a social semiotic account of designs for learning', *Written Communication*, 25(2): 166–195.

Bezemer, J. and Mavers, D. (2011) 'Multimodal transcription as academic practice: a social semiotic perspective', *International Journal of Social Research Methodology*, 14(3): 191–207.

Bezemer, J., Murtagh, G., Cope, A., Kress, G. and Kneebone, R. (2011) '"Scissors, please": the practical accomplishment of surgical work in the operating theatre', *Symbolic Interaction*, 34(3): 398–414.

Bjorkvall, A. (2009) 'Practical function and meaning: a case study of IKEA tables', in C. Jewitt (ed.), *The Routledge Handbook of Multimodal Analysis*. Abingdon, Oxon: Routledge.

Burn, A. (2009) *Making New Media: Creative Production and Digital Literacies*. New York: Peter Lang.

Caple, H. and Knox, J. (2012) 'Online news galleries, photojournalism and the photo essay', *Visual Communication*, 11(2): 207–236.

Flewitt, R., Hampel, R., Hauck, M. and Lancaster, L. (2009) 'What are multimodal data and transcription?', in C. Jewitt (ed.), *Routledge Handbook of Multimodal Analysis*. Abingdon, Oxon: Routledge. pp. 40–53.

Gibson, J. (1979) *The Ecological Approach to Visual Perception*. Hillsdale, NJ: Lawrence Erlbaum.

Halliday, M.A.K. (1978) *Language as Social Semiotic: The Social Interpretation of Language and Meaning*. London: Edward Arnold.

Halliday, M.A.K. and C. (2004), *An Introduction to Functional Grammar*. London: Hodder Education

Hodge, R. and Kress, G. (1988) *Social Semiotics*. Cambridge: Polity Press.

Holsanova, J. (2012) 'Methodologies for multimodal research', *Visual Communication* (Special Issue), 11(3).

Jewitt, C. (2002) 'The move from page to screen: the multimodal reshaping of school English', *Visual Communication*, 1(2): 171–196.

Jewitt, C. (2008) *Technology, Literacy and Learning: A Multimodal Approach*. Abingdon, Oxon: Routledge.

Jewitt, C. (ed.) (2013) *The Routledge Handbook of Multimodal Analysis* (2nd edition). Abingdon, Oxon: Routledge.

Jewitt, C. and Adamson, R. (2003) 'The multimodal construction of rule in computer programming applications', *Education, Communication and Information*, 3(3): 361–382.

Jewitt, C., Bezemer, J. and Kress, G. (2011) 'Annotation in school English: a social semiotic historical account', *National Society for the Study of Education Yearbook*, 110(1): 129–152.

Jewitt, C. and Kress, G. (eds) (2003) *Multimodal Literacy*. New York: Peter Lang.

Jones, R. (2009) 'Technology and sites of display', in C. Jewitt (ed.), *Routledge Handbook of Multimodal Analysis*. Abingdon, Oxon: Routledge. pp. 114–126.

Knox, J. (2007) 'Visual-verbal communication on online newspaper home pages', *Visual Communication*, 6(1): 19–36.

Kress, G. (1993) 'Against arbitrariness: The social production of the sign as a foundational issue in critical discourse analysis', *Discourse and Society*, 4(2): 169–191.

Kress, G. (1997) *Before Writing: Rethinking Paths to Literacy*. London: Routledge.

Kress, G. (2010) *Multimodality: A Social Semiotic Approach to Contemporary Communication*. Abingdon, Oxon: Routledge.

Kress, G., Jewitt, C., Bourne, J., Franks, A., Hardcastle, J., Jones, K. and Reid, E. (2005) *English in Urban Classrooms: A Multimodal Perspective on Teaching and Learning*. Abingdon, Oxon: Routledge.

Kress, G., Jewitt, C., Ogborn, J. and Tsatsarelis, C. (2001) *Multimodal Teaching and Learning*. London: Continuum.

Kress, G.R. and van Leeuwen, T. (2001) *Multimodal Discourse: The Modes and Media of Contemporary Communication*. London: Edward Arnold.

Kress, G. and van Leeuwen, T. (2002) 'Colour as a semiotic mode: Notes for a grammar of colour', *Visual Communication*, 1(3): 343–368.

Kress, G. and van Leeuwen, T. (2006) *Reading Images: The Grammar of Visual Design*. (2nd revised edition). Abingdon, Oxon: Routledge.

Lemke, J. (1998) 'Metamedia literacy: Transforming meanings and media', in D. Reinking, M. McKenna, L. Labbo and R. Kieffer (eds), *Handbook of Literacy and Technology: Transformations in a Post-Typographic World*. Hillsdale, NJ: Erlbaum. pp. 283–302.

Lemke, J. (2002) 'Travels in hypermodality', *Visual Communication*, 1(3): 299–325.

Marsh, J. (2006) 'Global, local/public, private: Young children's engagement in digital literacy practices in the home', in K. Pahl and J.Rowsell (eds), *Travel Notes from the New Literacy Studies*. Clevedon, North Someset: Multilingual Matters. pp. 19–39.

Martinec, R. (2000) 'Types of processes in action', *Semiotica*, 130 (3/4): 243–268.

Martinec, R. and Salway, A. (2005) 'A system for image–text relations in new (and old) media', *Visual Communication*, 4(3): 337–371.

McNeil, D. (1992) *Hand and Mind: What Gestures Reveal about Thought*, Chicago, IL: University of Chicago Press.

Norman, D. (1990) *The Design of Everyday Things*. New York: Doubleday.

Norris, S. (2004) *Analyzing Multimodal Interaction*. Abingdon, Oxon: RoutledgeFalmer.

Norris, S. (2009) 'Modal density, modal configurations: Multimodal actions', in C. Jewitt (ed.) *Routledge Handbook of Multimodal Analysis*. Abingdon, Oxon: Routledge. pp. 78–90.

Plowman, L. and Stephen, C. (2008) 'The big picture? Video and the representation of interaction', *British Educational Research Journal*, 34(4): 541–565.

Price, S., Roussos, G., Falcão, T.P. and Sheridan, J.G. (2009) 'Technology and embodiment: relationships and implications for knowledge, creativity and communication', www.beyondcurrenthorizons.org.uk (accessed 24 August 2012).

Scollon, R. and Scollon, S. (2009) 'Multimodality and language: a retrospective and prospective view', in C. Jewitt (ed.), *The Routledge Handbook of Multimodal Analysis*. Abingdon, Oxon: Routledge. pp.170–180.

Van Leeuwen, T. (1999) *Speech, Music, Sound*. London: Macmillan.

Van Leeuwen, T. (2005) *Introducing Social Semiotics*. Abingdon, Oxon: Routlegde.

Ventola, E., Charles, C. and Kaltenbacher, M. (2004) *Perspectives on Multimodality*. Amsterdam: John Benjamins Publishing.

Walton, M. (2004) 'Behind the screen: the language of web design', in I. Snyder and C. Beavis (eds), *Rewriting Literacy in the Network Society*. Hampton, Australia: New Dimensions.

White, P. (2012) 'Reception as social action: the case of marketing', in S. Norris (ed.), *Multimodality in Practice* (Chapter 11). Abingdon, Oxon: Routledge.

Zammit, K. (2007) 'The construction of student pathways during information-seeking sessions using hypermedia programs: A social semiotic perspective'. Unpublished PhD thesis, University of Western Sydney.

18

Projection, Place, and Point-of-view in Research through Design

Steven Dow, Wendy Ju and Wendy Mackay

The paradigm of research through design (RtD) combines the forward-thinking, artifact-generating practices of design with the knowledge-generating goals of research (Fallman 2003; Nelson & Stolterman 2003; Zimmerman et al. 2007; Zimmerman et al. 2010; Koskinen et al. 2011). This new epistemology has emerged as a complement to traditional lab experimentation and field observation research, particularly in design-orientated disciplines such as human–computer interaction (HCI) where researchers often create artifacts to understand people in a new light (Fallman 2003). In RtD, artifacts are not the end goal, but a means for framing an alternative future and uncovering human needs, desires, emotions and aspirations (Carroll & Kellogg 1989; Gaver et al. 1999; Gaver et al. 2004; Dunne 2011).

As more researchers adopt this approach, there is an increasing need to explore the rationale for and methodological implications of RtD and understand how to judge the validity of its research contributions. This chapter has several goals: (1) to define RtD and differentiate it from traditional epistemologies in the human sciences; (2) to describe key aspects of this research orientation with respect to how it *projects* a future state, where it *places* design artifacts to gather data and how researchers impose their philosophical *points-of-view*; (3) and, finally, to discuss some of the standards the research community has adopted to evaluate RtD contributions. We hope this treatment of RtD helps to expand its definition and provides a useful framework for researchers and students in the design research community.

WHAT IS RESEARCH THROUGH DESIGN?

In 2004, Bill Gaver's design studio at the Royal College of Art created a rather peculiar artifact – an electronic coffee table with

Figure 18.1   Gaver's Drift Table exemplifies RtD.

a tiny display of aerial photography that moves slowly based on the weight distribution on its surface (Gaver et al. 2004; see Figure 18.1). Gaver and his colleagues were not interested in whether people wanted to buy the Drift Table. It was never intended to be widely adopted or commercially viable, but, rather, to serve as an object of enquiry. This provocative artifact helped Gaver and his colleagues to investigate non-routine household activities and 'ludic' emotions, such as curiosity and reflection.

In RtD, researchers develop and deploy novel artifacts – digital or physical – as a tactic to learn about specific aspects of the human experience (Frayling 1993). While traditional design practice focuses primarily on producing artifacts, systems or services to make some cultural or economic impact (Kelley 2002; Buxton 2007; Kolko 2007), RtD uses designed artifacts to produce theory (Carroll & Kellogg 1989; Fallman 2003; Zimmerman et al. 2007; Zimmerman et al.

2010). This design-orientated research method goes beyond the artifact to make insights about people, culture or interactions. Moreover, RtD is distinct from 'research on design', the study of what people do when they do design, and 'research for design', which focuses more on how to improve design practice (Forlizzi et al. 2009; Dow 2011).

RtD leverages design as a fundamental research method and, hence, shares many characteristics with professional design practice. Design researchers employ techniques such as sketching and prototyping (McCloud 1993; Schrage 1996; Davidoff et al. 2007). They gather feedback through user studies and design crits (Tohidi et al. 2006; Dow et al. 2011). Much like practitioners, design researchers discover critical dimensions of a design space through a process of iteration and experimentation (Kelley 2002; Dow et al. 2009; Dow et al. 2010). RtD also strives to project a future state of the world and to engage real-world human response (Simon 1996; Gaver et al. 2003). The key distinction between RtD and design practice is that the primary activity of design researchers is to theorize and formalize knowledge about how and why people interact with design artifacts (Fallman 2003; Nelson & Stolterman 2003; Zimmerman et al. 2007; Zimmerman et al. 2010; Koskinen et al. 2011).

From a research perspective, design allows researchers to investigate how people think, behave or interact in scenarios that have yet to be or may never be. It is this orientation towards the future – towards 'what might be' (Peirce 1998; Martin 2009) – that makes RtD powerfully different from traditional research paradigms. By thinking forward, RtD acknowledges the reality of change in the world and the role of human agency in that change (Nelson & Stolterman 2003). Design researchers prototype their visions of the future, seek to understand the implications of those visions and work to communicate and disseminate their insights to a broader audience.

RtD is a knowledge-building endeavor that falls outside the boundaries of traditional human research (Nelson & Stolterman 2003; Gaver 2012), such as ethnography and psychology experiments. In anthropological ethnography, researchers describe people and activities in a specific time and place (Malinowski 1984; Dourish 2006). Designed artifacts often reside in such environments, but the ethnographer does not place them there. Ethnographers often take care not to contaminate the local culture through their very presence or by introducing new artifacts (LeCompte 1999). Some ethnographers advocate reflexivity, to describe how the researcher gained access to their site and how personal circumstances might affect their data (Geertz 1977; Rode 2011). While RtD frequently employs ethnographic-style methods for data gathering, its practitioners take a more liberal orientation towards intervention. Design researchers often take the liberty of explicitly intervening by designing and introducing artifacts and observing the subsequent effects in some context (Gaver et al. 2004). For example, to better understand how to improve work practices in air traffic control towers, Mackay et al. performed detailed ethnographic observations of how controllers use paper flight strips (Mackay, 1999), but then conducted RtD by developing a system to track and graphically augment the physical appearance of the strips with digital information (Mackay et al. 1998).

Research through traditional laboratory experimentation provides another point of contrast. Lab experimentation values repeatability, generalizablity and internal validity (Creswell & Clark 2006). Lab researchers carefully isolate key factors and create measures to understand the causal relationships of specific stimuli (Stangor 2010). RtD also borrows from laboratory-based methods to study how people react to new artifacts (Frens 2006). The formal lab setting helps to control the multitude of factors under observation. While controlled experiments manipulate one variable at a time, design-based lab studies may explore a wide variety of variables (Frens 2006; Dow et al. 2007; Ju & Takayama 2009). For example, Dow and colleagues contrasted augmented reality with desktop-based interaction using the same interactive narrative (Dow et al. 2007; see Figure 18.2). This broad-sweeping experiment helped the researchers study how a range of variables affected player engagement. In the lab, design researchers use artifacts to explore multiple dimensions of a design space.

RtD represents a new epistemology. It not only introduces new artifacts but also employs a very different kind of logic. With field observation, researchers use *inductive* reasoning to draw logical conclusions from data (Copi & Cohen 2005). With lab experiments, researchers utilize *deductive* reasoning to predict outcomes

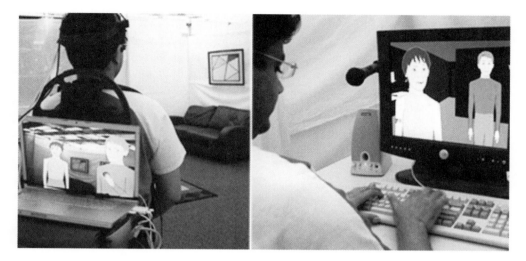

Figure 18.2   Dow and colleagues conducted semi-controlled lab studies with different interfaces to the interactive drama Façade.

based on theory (Creswell & Clark 2006). Design researchers, on the other hand, use *abductive* reasoning to conjecture a probable future based on an incomplete set of observations (Peirce 1998; Kolko 2011). They ask 'What if?' questions based on a set of ideals and agendas (Fallman 2003). By focusing on the possibilities that arise from some future design (Moggridge 2007), design researchers are bound to different epistemological commitments than pure ethnographic field studies, traditional psychology experiments or other research approaches, such as simulation or historical perspectives (Axelrod 1997; Wyche et al. 2006).

Runkel and McGrath argue that no scientific research method stands alone, since each approach involves tradeoffs and thus they urge researchers to triangulate methods (Runkel & McGrath 1972). Mackay extends this reasoning to HCI, demonstrating how researchers can include design methods as part of this process of triangulation (Mackay & Fayard 1997). RtD brings to the table several features that other methods lack: an orientation towards abductive

logic, an exploration of yet-to-be-defined variables and the invention of artifacts that serve as blueprints and provide implications for future designs.

HOW RESEARCH THROUGH DESIGN TAKES SHAPE

Research in the natural and human sciences may start from a theoretical, empirical or design perspective, but, in order to qualify as research, it must contribute to theory. Research *through* design contributes theory by reflecting on tangible projections of the future. These research artifacts serve diverse agendas and take on many shapes and forms (Carroll & Kellogg 1989; Gaver 2012). In the field of HCI, RtD addresses topics as diverse as sustainability, family communication health, spirituality, entertainment and productivity. HCI researchers often produce various artifacts, ranging from paper-based rapid prototypes to full-scale physical prototypes that leverage emerging technologies. This chapter acknowledges this diversity and outlines a framework to discuss the myriad of ways research through design takes shape.

We introduce a framework to map RtD along three key dimensions:

- **projection**: how far the design looks into the future
- **place**: where and how design artifacts gather knowledge
- **point-of-view**: the rhetorical stance of the design researcher.

Design researchers must choose how far forward in the future to investigate, how and where to observe their designs in action and what philosophical stance to take in conducting and communicating their design research. The next sections explore each of these in turn, with examples from the HCI community to demonstrate the choices available to design researchers.

Projection

Design is always about the future, exploring what could be, beyond what already is (Nelson & Stolterman 2003). Design researchers seeking to generate theory can choose how far into the future to project their designs, setting their sights on the long-, mid- or near-term. The timelines of these different design trajectories deeply affect the character and quality of RtD. We characterize three stages of projection.

Design breakthroughs	Introduce a novel idea that fundamentally changes a field, opening possibilities that inspire new designs and new theory.
Point designs	Populate the design space with examples that define, explore and extend the theoretical dimensions of the design space.
Design principles	Identify key theoretical elements that can be taught and employed by design practitioners.

Although these stages discretize a continuum, the categories help to indicate how design researchers generate different kinds of knowledge based on their degree of projection. Rare innovative ideas spark a sense of wonderment and often push the limits of current technology, but are often too expensive or impractical for current products. For design practitioners, these design breakthroughs serve as inspiration. As the design space becomes populated with point designs, the underlying technologies often become cheaper and more accessible, which often leads to early commercial products. Eventually, within a problem space, principles emerge that clearly articulate why certain design choices are preferred (Alexander et al. 1977). Design professionals and educators use these principles to inform their work (Duyne et al. 2006).

Design Breakthroughs

The most future-orientated, and thus most rare, design projections involve radical reconceptualizations of a design problem. Rather than exploring an existing design space, these design projections inspire new design spaces. Long-term design projections are often preceded by an 'a-ha' moment (Parnes 1975), an insight that enables the designer to conceive of an entirely new set of design possibilities. Design breakthroughs often take the form of a video, such as Apple's 1987 Knowledge Navigator Video, or a diorama, such as Norman Bel Geddes's Futurama exhibit at the 1968 World's Fair, or a technical demo that can serve as a 'proof of concept'. Design breakthroughs are often based on technology that does not yet exist, at least not in an easily incorporated form, and may use techniques such as storyboards, exhibits or 'Wizard of Oz' enactments to simulate the experience (McCloud 1993; Dow et al. 2005; Truong et al. 2006).

Successful breakthroughs produce knowledge by offering designers, users and other researchers a glimmer of what is possible.

Figure 18.3   Wellner's DigitalDesk demonstrated the idea of mixing virtual documents into a physical space.

Often, a demonstration of key technologies or interfaces help distinguish these research contributions from science fiction. A classic example is Wellner's DigitalDesk (Wellner 1993), which seamlessly combined physical and virtual documents in a shared physical environment (see Figure 18.3). Similarly, Bishop's Marble Answering Machine offered a novel way to use physical marbles to interact with digital audio messages (Bishop 1992). Bishop communicated the concept by creating an animated sketch. When other researchers focused on digitizing everything paper, the DigitalDesk and the Marble Answering Machine helped usher in a revolution in tangible computing and encouraged designers to explore novel ways of integrating physical and virtual objects (Ishii & Ullmer 1997; Klemmer et al. 2001). They challenged

the then-prevailing belief that the coming of the digital age marked an end of the material world.

Point Designs

Upon realizing the potential of a new design space, other researchers populate it with novel 'point designs' (Card 1996; Gaver 2012). Point designs help define and clarify the dimensions that comprise a design space. Rather than exploring each dimension individually, each point design embodies the intersection of multiple design dimensions, extends existing dimensions and identifies previously undetected holes within the design space. Over time, as many researchers contribute point designs, it extends to accommodate new user groups, new contexts and new technologies. Point designs are typically more developed than design breakthroughs and take the form of working prototypes that can be tested by users in limited settings. For example, PapierCraft extended Wellner's DigitalDesk work by enabling authors to annotate paper copies of a document and implement the changes online (Liao et al. 2008). Going beyond DigitalDesk, PapierCraft featured fully functional prototypes and addressed more pragmatic issues facing office workers.

Similarly, Ishii's musicBottles (see Figure 18.4), which play musical pieces when bottle stoppers are removed, act as a poetic counterpoint to Bishop's Marble Answering Machine and provide another example of using a tangible interface to control digital audio content (Ishii et al. 2001). Each new design extends and fleshes out a concept and pushes the boundaries of a design space. The resulting portfolio of point designs, created by the community of designers, serves as the foundation for the creation of design principles.

Design Principles

As point designs populate a design space, researchers explicate the underlying principles that govern the domain and capture

Figure 18.4   Ishii's musicBottles extended Bishop's concept of using tangible objects to control digital audio (© Hiroshi Ishii).

this knowledge as design patterns, guidelines and theories. *Design patterns* are collections of designs that work best in different scenarios for different purposes (Alexander et al. 1977; Duyne et al. 2006). For instance, Landay and Borriello championed the creation of design patterns in ubiquitous computing. They argue, 'by communicating solutions to common problems, design patterns make it easier to focus efforts on unique issues' (Landay & Borriello 2003). Design patterns help researchers and practitioners avoid reinventing solutions to known problems and provide 'framing constructs' for further designs (Zimmerman et al. 2010).

Design guidelines are organizing constructs that facilitate designing or experimenting in a space. For example, Nielson's ten heuristics for UI design (Nielsen 1990) and Spool's usability engineering guide (Spool et al. 1998) capture best practices for HCI. Critics argue that guidelines oversimplify design problems and provide more value towards the review of existing designs, rather than the creation

of new designs. Guidelines rarely explain how to manage conflicts between rules, nor do they encourage truly innovative designs (Greenberg & Buxton 2008).

Design theories are more elaborate and general-purpose: they attempt to speak to deeper 'truths' in a design space. For example, Beaudoin-Lafon and Mackay theorize generative principles for interaction techniques (e.g. reification, polymorphism and reuse) and demonstrate how they can be used individually and collectively to address a wide set of commands with a simpler interface (Beaudouin-Lafon & Mackay 2002). Similarly, Suchman's theory about situated action and Hutchin's concept of distributed cognition have deeply influenced several decades of interaction design (Suchman 1987; Hollan et al. 2000).

Place

Design researchers gain insights by placing designs into situations with people. Just as researchers come from different scholarly and philosophical traditions, they deploy their design experiments differently. 'Place' refers not just to the deployment location, but to data-gathering methods and trade-offs associated with each setting.

Laboratory settings	Create controlled settings and carefully monitor interactions around new designs.
Field deployments	Place artifacts into real-life settings where people actually work and play.
Exhibitions	Present designs to a broader public, inviting interaction and discourse around the work's conceptual underpinnings.

Each setting incurs important advantages and limitations. For better triangulation, many design researchers place their designs in several settings to better understand the theoretical implications. A small-scale field deployment may yield different, yet complementary, insights from a large-scale lab study. This section describes the trade-offs of different placement models with examples of each.

Laboratory

In a laboratory setting, potential users are invited to interact with a new artifact or service. The design researcher can explore variations within a specified design space while controlling or excluding irrelevant factors. New designs are often compared to alternative or existing designs using a variety of quantitative and qualitative measures, including performance metrics and subjective opinion.

Jonsson et al.'s experiments with in-car voice provide a simple illustration of the benefits of lab-based RtD (Jonsson et al.'s 2005). While designers of automotive speech systems have numerous variables to consider – the gender of the car voice, the volume, the content of the messages, the urgency, the effects on product safety, cost, likability, etc. – Jonsson et al.'s experiment

narrowed the focus to examine how the emotion and energy level of the in-car voice system affected drivers in a driving simulation. The experiment found that drivers who interacted with voices that matched their own emotional state drove better and communicated more than drivers interacting with mismatched emotions. This type of experiment produced implications for the design of new in-car voice services, a feat that would be difficult, unsafe and expensive to recreate in field settings. It also provided benefits outside of the product design process by contributing to theory about emotional state in voice response systems.

Unlike conventional lab experimentation, design researchers are not strictly confined to isolating specific variables. Because of the exploratory nature of design research, researchers typically modify a number of design features prior to narrowing down to a critical few. In RtD, designs may vary by a large or small degree. This degree of variation determines how precisely a researcher can claim causal factors. For example, if a new design includes seven new distinct design concepts and performs 10 per cent better than the prior design, one cannot know exactly which features are most salient. However, this bucket approach manages to both explore broad possibilities and demonstrate an effect. Through further iteration and experimentation, the design researcher can narrow the variation between prototypes and make stronger claims about the casual factors.

Field

In field-based RtD, researchers place designs into an everyday context to study how people interpret, create meaning and behave around these new artifacts. Placing prototypes into the field helps the researcher contextualize how individuals and groups live with and make sense of new artifacts in ways that would be difficult to do in lab settings. As Gaver et al. articulated, 'we don't emphasize

Figure 18.5   Ju and Sirkin's 'waving hand' kiosk explored greeting dynamics in public settings.

precise analyses or carefully controlled methodologies; instead, we concentrate on aesthetic control, the cultural implications of our designs, and ways to open new spaces for design' (Gaver et al. 1999). By placing designs in context, design researchers can generate theory by reflecting on observations of use.

The deployment of novel prototypes into real-world environments allows researchers to understand how people naturally react to artifacts in a particular environment. For example, Ju and Sirkin placed an interactive kiosk with a waving hand into various quasi-public locations, like the entranceway of a bookstore or the lobby of a computer science building (Ju & Sirkin 2010; see Figure 18.5). The contextual nature of the interaction makes the field a natural choice for the study's location. The experiment had to be designed around the vagaries of uncontrolled settings – for example, the fact that people do not just walk around alone, but often in dyads and groups. It also allowed researchers to see how people in dyads and larger groups interact with a design object differently from individuals alone.

Field-based RtD bears resemblance to action research (also called participatory action research) where a group of stakeholders make deliberate changes to a system in order to simultaneously address a problem situation and further the goals of social science (Gilmore et al. 1986). Kurt Lewin first described action research as 'comparative research on the conditions and effects of various forms of social action' (Lewin 1946). Both design researchers and action researchers can iteratively create changes and observe the effects of those changes, towards the goal of social transformation.

Exhibition

Exhibition-based research follows from traditions in art and design, rather than from natural, physical or social science. The place for conducting this style of RtD is not the laboratory or a field site, but a venue for public engagement, such as museums, showroom floors, or even well-trafficked websites. Here, the designer researcher's goal is to use designed artifacts to express ideas and to instigate public discourse. Such design researchers often draw intellectual inspiration from historically important artistic movements like constructivism, surrealism, romanticism, classicism, expressionism or minimalism (Arnason & Mansfield 2009).

Tony Dunne and Fiona Raby from the Royal College of Art explicitly use design to provoke discussion about the implications of emerging technologies (Dunne 2011). In their 2005 project, 'Evidence Dolls', Dunne and Raby created a set of special dolls to explore perspectives on genetics (see Figure 18.6). These hypothetical products were intended to spark public debate on issues such as designer babies, desirable genes, mating logic and DNA theft (Dunne & Raby 2005). Their concepts were not tested empirically per se, but were intended to trigger thoughts and discussions among their audience. As Dunne articulated, 'new ideas are tried out in the imagination of visitors, who are encouraged to draw on their already well-developed skills as window-shopper and high-street showroom-frequenter' (Dunne 2006). By placing their tangible designs in a public exhibition setting, design researchers tap into the conscious, reflective and critical public eye.

Research intended for public exhibition is judged by different standards from RtD in the lab or field. In order to be successful, exhibited artifacts need to engage an audience and effectively communicate an idea. In other words, the community at large judges the validity and impact of exhibition-based RtD. This impact can be measured, for example, through the size of the audience, the number of public comments, the amount of traffic to a website or the aggregate qualitative sentiment of critics. Further, the possibility of reaching a large audience serves as one motivator for exhibition-based researchers (Hustwit 2009).

Our concept of 'place' in RtD design does not imply a particular philosophical stance. We intentionally separate the methodological concerns of the research setting from the philosophical stance imposed on design artifacts (see 'Point-of-view' below). This distinction is notably different from Koskinen and colleague's concept of showroom-based RtD, where designers apply critical theory and pragmatist philosophy and introduce artifacts to galleries and public settings in

Figure 18.6   Dunne and Raby's 'Evidence Dolls' challenged exhibition visitors to think about the future of genetic engineering (photo: Kristof Vrancken).

order to make a cultural or political statement (Koskinen et al. 2011). To give our framework more flexibility, we make a deliberate distinction between placement and the philosophical stance. In our view, research through design has the potential to adopt any philosophical stance in any research setting.

Point-of-view

A third dimension of RtD is *point-of-view* – the design researcher's philosophical perspective on a subject matter. While traditional scientists are expected to be as objective as possible, design researchers explicitly put forth a point-of-view, a subjective argument about the future. A novel perspective can add valuable new insights into the interpretation of a design space. Although this is not intended to be a comprehensive list, we discuss several prevailing points-of-view within HCI design research.

Pragmatic perspectives	Focus on the identification and resolution of common and everyday needs encountered by people in their day-to-day lives.
Utopian perspectives	Present an idealized vision of a possible future centered around some trend, artifact or service.
Critical perspectives	Call attention to the complications or possible dystopian outcomes of possible futures.

Through point-of-view, design researchers impose a philosophy, which in turn, affects the holistic user experience, not only towards the interface, but towards the perception of the whole project. Differing points-of-view can often result in significantly different prototypes of similar technologies. For example, in the domain of robot design, some researchers and designers focus on pragmatic issues, such as understanding how people read and react to various humanoid robots so that they can make them more pleasing to people (Nomura et al. 2007). Others take a more critical perspective, for example examining the basic assumption that robots could or should look

like people (Hoffman 2007). Both of these are different from the utopian perspective – often promulgated by TV series like *Star Trek: The Next Generation* – that robots will look and act just like other human beings. Opposing viewpoints serve to calibrate the community to address societal needs and to avoid undesired consequences.

These three perspectives are not comprehensive, as there are many other points-of-view (e.g. political, ethical, anarchistic) that designers could impose on novel artifacts. In this section, we explain how designers influence research outcomes (sometimes unconsciously) by projecting their philosophy on designed objects. Hence, in RtD, research outcomes are as much a reflection of the designer as an indication of how people will react.

Pragmatic Perspectives

The pragmatic perspective in design research is characterized by a focus on addressing everyday problems or needs. This perspective has roots in human factors and industrial design, where designers and researchers look for ways to fix problems imposed by newly engineered systems. Sometimes these problems were psychological – for instance, the well-known industrial designer Henry Dreyfuss explored different ways of using design to make passengers feel comfortable in commercial airplanes, testing the use of plush upholstered seats and curtained windows so that people might feel like they were in a living room rather than hurtling through the air in a steel tube at thirty thousand feet (Dreyfuss 1955). Pragmatists may focus on resolving physical limitations, such as Mountford and North's (1980) work on voice input to reduce pilot workload. Human-centered research methods – such as lab experiments and field studies – often help to either pinpoint problems or to evaluate possible solutions.

Some RtD focuses on exploring the pragmatic issues that will arise with impending technologies. For instance, Takayama et al. used online surveys to investigate how people view the future role of robots (Takayama et al. 2011). The research team created video prototypes using animated robots that illustrated the designed interactions, allowing the team to probe factors – such as 'performed' forethought and reaction – that influence robot design before having to build any robots. Video prototypes may result in different reactions when realized in physical form, but they help reduce the number of possible designs to a reasonable subset that can be tested with higher-fidelity iterations.

Utopian Perspectives

Some design researchers make the case for new technologies or design directions by idealizing the broader implications of these directions. One classic HCI example is Doug Englebart's oNLine System (also known as 'The Mother of All Demos'; Englebart 1968). Engelbart and his colleagues at SRI developed a series of interconnected technologies with the purpose of 'augmenting human intellect'. This radical demo included prototypical designs for the computer mouse, video collaboration, hypertext and copy–paste (see Figure 18.7). This utopian perspective promoted the then-unusual idea that computers should be used to help humans to perform work, rather than to perform work for them.

Similarly, Apple's Knowledge Navigator video also embodies this utopian vision of technology (Dubberly 2007). This video depicts a college professor in his office, interacting with a tablet device with an intelligent agent that helps him manage his schedule, research information and communicate with other researchers (see Figure 18.8). The work presents a utopian vision of a future in which computing and communication technologies allow people to interact rapidly with people and data. Great vision designs are like movies and novels – they need to achieve resonance with a fairly wide audience to make a difference. With this broad impact comes critique and discussion. For instance, many people critique Knowledge Navigator's magical portrayal of technology. The system's speech recognition, artificial intelligence and video networking work seamlessly and, as a result, it gave some future users and designers a false impression of what is possible.

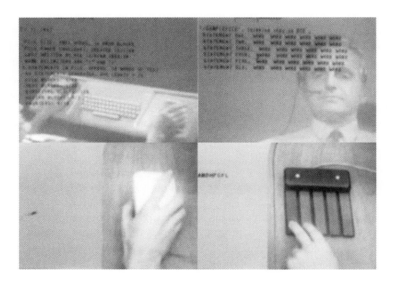

Figure 18.7 Englebart's oNLine System demonstrated a vision of 'augmenting human intellect' that had a profound influence on the field of human–computer interaction (attributed to SRI International).

Figure 18.8   Apple's Knowledge Navigator video presented a utopian vision of how people could one day interact with intelligent agents.

Utopian perspectives get traction when accompanied by a concrete prototype (Schrage 1996). For example, the Tivoli/Liveboard project at Xerox PARC created a utopian vision of workplace meetings where a smart whiteboard not only enables people from remote sites to share a view of a common written space but also enables groups to get rid of the mildly subservient 'notetaker' role (Pedersen et al. 1993; Moran et al. 1998). The systems sought to clarify the human-orientated value proposition well before the underlying technologies permitted these ideas to be tractable in the consumer market.

Critical Perspectives

Some design researchers take a critical perspective in order to instigate debate and reflection around design's unintentional and inadvertent consequences. Dunne and Raby, for example, use provocative point designs as a way of questioning assumptions promulgated by consumer culture (Dunne 2011). They created the *Technological Dream Series: No. 1, Robots* to question the commonly held belief that someday robots will do everything for us. Their series featured 'robot' artifacts and videos of projected

interactions that explicitly rejected traditional notions of the robot, as portrayed by popular media, such as the Jetsons™. Their robots had little quirks and neuroses – human dependencies that reflected Dunne and Raby's skepticism over the capability or desirability of mainstream robots.

Similarly, Purpura and colleagues proposed a tongue-in-cheek system called *Fit4Life* as a critical response to the view that technology alone can persuade healthier behavior (Purpura et al. 2011). The authors describe a conglomeration of real technologies – including a food consumption sensor and a 'thinsert' to measure body weight through footwear – to monitor and persuade users to live healthier lives. The system is comprised entirely of plausible technology currently in development within the Ubicomp research community. However, when put together into a hyper sense-and-react system, the intervention seems absurd. Throughout most of the paper, the authors lead the reader to believe the system is real, only to later provoke reflection about values and politics of design in persuasive computing. Other critical approaches to design focus on how technology affects marginalized groups, such as LeDantec's study of urban homeless people (Le Dantec & Edwards 2008) and diSalvo's efforts to engage local communities in policy discourse (Di Salvo & Lukens, 2009).

Because the goals of critical design are to provoke thought and reflection, design researchers often create new methods to capture these diverse interpretations. For instance, Sengers and Gaver employed the traditional lab-based method of Likert-scale ratings to evaluate their designs, but they encouraged study participants to label the anchors on each scale (Sengers & Gaver 2006). This unconventional tactic makes it difficult to compare study results. However, as they point out, the goal was not to show that their designs outperform reference designs, but to discover the values important to users and to question whether designers should be the arbiters of which values or metrics are used to judge designs.

DISCUSSION

Our framework for RtD illustrates the wide range of activities undertaken by researchers that utilize design, and illuminates how such diverse goals lead to different approaches. We hope our framework invites diversity and expands the definition of RtD, with an implicit goal of allowing more researchers to find a home under the RtD umbrella. Some design researchers may specialize within a particular combination of projection, place and point-of-view. Others may try different styles of RtD from project to project. However, it takes a community of researchers to explore all aspects of a design space and to construct time-honored theory.

How Communities of Design Researchers Explore Projection, Place and Point-of-View

In the dimension of *projection*, the trajectory of the research is not uniform or smooth, but has the tendency to evolve in specific ways.

Research artifacts tend to progress from breakthroughs to point designs to patterns, and, in so doing, help to discover, colonize and then systemize new arenas of design. Individuals may explore a particular design space through a portfolio of designs (Gaver 2012), but typically it takes an entire research community a long period of time to evolve from breakthroughs to principles.

In choosing a *place* for research through design, individual researchers can and often do use a multifaceted strategy by undertaking different research methods (e.g. they start in the lab and then move to the field; or maybe they start on an exhibit floor and then test key variables in the lab). Design researchers often traverse locations in order to triangulate insights that are not evident from any one method or perspective alone. While the adoption of multiple disparate methods can open design researchers to charges of dilettantism, this strategy also allows researchers to find the methods that answer the questions they have, instead of answering just the questions that a particular method affords.

The fact that design researchers often depart from very different *points-of-view* performs a tremendous service for the design and research community as a whole. Within any domain, the ongoing debates between the utopians and critics, between the positivists and constructivists, between the pragmatists and visionaries and so on, not only raise the intellectual bar but they also serve to calibrate what kind of future we collectively desire. While each group often argues that their perspective warrants primacy, it is actually the dialectic effect of contrasting approaches that helps the community as a whole.

Assessing RtD

RtD emphasizes future possibilities and results in theory that can guide design practice and reveal insights about people, culture or interactions. As the community of design researchers grows, there has been increasing call to clarify expectations about the outcomes of RtD (Zimmerman et al. 2010; Gaver 2012). Gaver has highlighted how RtD is intended to be 'provisional, contingent and aspirational' and, hence, does not necessarily need to abide by standards, protocols and specific guidelines (Gaver 2012). Reflecting on our framework and the growing body of literature on RtD, we have compiled a list of qualities to consider when conducting RtD and when judging papers that conduct RtD. In our view, 'good' RtD strives to embody one or more of these qualities:

- **predictive**: the design researcher makes some kind of conjecture about the future through the creation of design artifacts
- **relevant**: the designed artifacts and theory-generation speak to larger social concerns that are either pertinent now or will be in the future
- **novel**: the designed artifacts are not facsimiles, but original and unique designs
- **fruitful**: the designed artifacts yield valuable data and provide fodder for theory-building
- **suitable**: the design research chooses data gathering methods that are appropriate and consistent for the chosen setting
- **reflexive**: the design researcher acknowledges his or her philosophical stance and reflects on how the point-of-view affects the audience reaction
- **aesthetic**: the design researcher creates artifacts with an intentional appearance and form.

Research *on* design and *for* design can help design researchers hone their abilities. For example, design researchers can improve their craft through deliberate reflection (Schön, 1995), iterative prototyping (Dow et al. 2009) and parallel design (Dow et al. 2010) – practices currently advocated by professional designers. The framework outlined in this chapter seeks to help design researchers contemplate the range of methods and viewpoints and choose an approach that best helps them contribute to a body of knowledge.

CONCLUSION

In this chapter, we discussed an array of research activities that use *design* as a key tool for developing theories about people, cultures and interactions. By characterizing how RtD varies with regard to *projection* into the future, *places* where the research occurs and *points-of-view* adopted by the design researchers, we hope to help rationalize and justify the variations and to highlight the commonalities across different RtD activities. RtD sometimes borrows methods and techniques from traditional social science research or the fine arts. However, since it is distinct in purpose from traditional epistemologies, it needs to be evaluated on its own merits. Within our framework, we provided exemplars of different kinds of RtD and discussed how each serves its intended goals.

Case Study: A RtD Perspective on the Evolution of Telepresence

Every novel design space begins with a few inspiring examples. These early examples lead to fast followers who continue to populate the space of ideas and discover unforeseen constraints and opportunities. Once enough designs enter the space, design researchers can reflect on them, describe key principles, and articulate a language for that particular design domain. This evolutionary process is evident in a number of interactive systems.

Telepresence exemplifies how research through design traverses our framework. In 1980, Minsky coined the term to describe technologies that let people feel present at a remote location (Minsky 1980). However, the original idea appeared much earlier, as a dystopian vision of the future: Chaplin's 1934 film *Modern Times* shows the factory boss peering into the workers' bathroom and Radford's 1964 film *Nineteen Eighty-Four* presents a chilling depiction of 'big brother' constantly monitoring citizens' daily lives. These films offered a critical perspective on telepresence without explaining how such technology might actually come to be.

The public's first contact with a working telepresence technology was Bell Laboratories' Picturephone at the 1964 New York World's Fair (see Figure 18.9). Callers could sit in telephone booths with small monitors and have a video-based conversation with a stranger in another booth. The Picturephone represented a breakthrough design intended to transform communication in daily life. Ahead of its time, the Picturephone was discontinued by 1970. In 1968, Douglas Englebart's oNLine System demonstrated remote collaboration in a utopian view of interactive computing and also represented an early breakthrough in the evolution of telepresence (Englebart 1968).

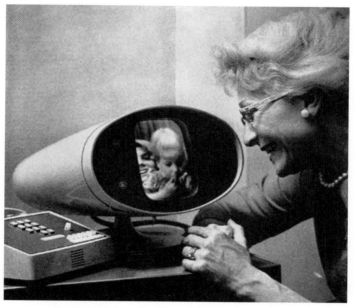

Figure 18.9   An advertisement for the Bell Picturephone (developed at Bell Laboratories, a subsidiary of Alcatel-Lucent).

Another early inspiration for telepresence came from the art world, with Galloway and Rabinowitz 's 1980 project, 'Hole in Space', a 'public communication sculpture' where two wall-sized video displays connected public sites in New York and Los Angeles (Galloway & Rabinowitz 1980). People on the street could see life-sized images of strangers in the other city, as well as hear and speak with them. The exhibition was a point design that expanded the design space to crowd-to-crowd, rather than only person-to-person interaction. By the late 1980s, researchers Xerox PARC, EuroPARC and the University of Toronto began a systematic exploration of telepresence with each lab creating multiple point designs to address specific remote communication issues, such as always-on casual video connections between offices (Gaver et al. 1992; Buxton 1992) and time-shifted telepresence between locations in the US and Europe (Dourish & Bly 1992). Similarly, Paulos's 'Personal Roving Presence' project experimented with embedding telepresence technologies into a physical robot form (Paulos & Canny 1998), and provided an early precedent to modern research on embodied social proxies (Venolia et al. 2010) (see Figure 18.10). The community's portfolio of designs highlighted diverse design dimensions, exposed design opportunities, and explored the benefits and disadvantages of telepresence.

(Continued)

(Continued)

Figure 18.10   Paulos's 'Personal Roving Presence' (PROP) embedded telepresence technology atop a physical rover.

With sufficient point design exploration, research through design within this domain could focus on understanding how key variables affect the human experience of telepresence. Olson and Olson examined face-to-face collaboration to enumerate key characteristics that an ideal telepresence system would have to support communication, such as rapid feedback, immediate corrections, multiple communication channels and shared local context (Olson & Olson 2000). Hollan and Stronetta push back against the always-on face-to-face inevitability of telepresence to suggest there are objectives 'beyond being there' that would suggest lower-cost, more ephemeral, more anonymous forms of telepresence (Hollan & Stornetta 1992). Extending this notion, Carroll et al. (2006) explored how to support activity awareness that transcends moment-to-moment social interaction, providing both co-located and remote collaborators a 'common ground' to enhance understanding of each other's activities.

Through diverse projections, places, and points-of-view, design researchers have shaped what telepresence technology has become today. Some of the earliest projections into the future, such as videophones and large collaborative wall displays (e.g., Cisco, Polycom, Skype, WebMeeting, etc.), are becoming commonplace, while other visions, like the *Star Trek* Holodeck, are starting to be explored in research labs (Dow et al. 2007). Because these new telepresence technologies are ubiquitous, researchers can more easily evaluate design alternatives in lab and field settings, and designers can more readily incorporate telepresence technologies as part of their public exhibitions. Finally, we continue to see contrasting points-of-view, from utopian visions that appear in marketing campaigns, to dystopian warnings in feature films, to pragmatic analyses that lead to more carefully refined and useful telepresence products.

REFERENCES

Alexander, C., Ishikawa, S. & Silverstein, M. (1977) *A Pattern Language: Towns, Buildings, Construction*. Oxford: Oxford University Press.

Arnason, H.H. & Mansfield, E.C. (2009) *History of Modern Art* (6th ed.). Harlow: Pearson.

Axelrod, R. (1997) Advancing the art of simulation in the social sciences. *Complexity*, 3(2): 16–22.

Beaudouin-lafon, M. & Mackay, W. (2002) Prototyping tools and techniques. In Sears, A. and Jacko, J.A. (eds), *Human–Computer Interaction Handbook*. New York: Erlbaum. pp. 1006–1031.

Bishop, D. (1992) *Marble Answer Machine*. London: Royal College of Art.

Buxton, B. (2007) *Sketching User Experiences: Getting the Design. Right and the Right Design*. SanFrancisco, CA: Morgan Kaufmann.

Buxton, W.A.S. (1992) Telepresence: integrating shared task and person spaces. *Proceedings of the conference on Graphics interface 92*, (1992), 123–129.

Card, S. (1996) Pioneers and settlers: methods used in successful user interface design. In M. Rudisill, C. Lewis, R. Poison & T. McKay (eds.), *Human–Computer Interaction Design: Success Stories, Emerging Methods and Real-World Context*. San Fransico, CA: Morgan Kauffmann.

Carroll, J.M. & Kellogg, W.A. (1989) Artifact as theory-nexus: hermeneutics meets theory-based design. *SIGCHI Bulletin.*, 20(SI), pp. 7–14.

Copi, I.M. & Cohen, C. (2005) *Introduction to Logic* (12th edn). Upper Saddle River, NJ: Pearson.

Creswell, J.W. & Clark, D.V.L.P. (2006) *Designing and Conducting Mixed Methods Research* (1st edn). Thousand Oaks, CA: Sage.

Le Dantec, C.A. & Edwards, W.K. (2008) Designs on dignity. In *Proceedings of the twenty-sixth annual CHI conference on Human factors in computing systems*, Florence, Italy.

Davidoff, S. et al. (2007) Rapidly exploring application design through speed dating. *Proceedings of Ubicomp*.

DiSalvo, C. & Lukens,, J. (2009) Towards a critical technological fluency: the confluence of speculative design and community technology programs. In *Proceedings of Digital Arts and Culture Conference*.

Dourish, P. (2006) Implications for design. In *Proceedings of the SIGCHI conference on Human Factors in computing systems*, Montre'al, Quebec, Canada.

Dourish, P. & Bly, S. (1992) Portholes: Supporting awareness in a distributed group work. *Computing*, 541–547.

Dow, S. (2011) How prototyping practices affect design results. *interactions*, 18(3): 54.

Dow, S. et al. (2010) Parallel prototyping leads to better design results, more divergence, and increased self-efficacy. *Transactions on Computer–Human Interaction*, 4.

Dow, S. et al. (2005) Wizard of Oz support throughout an iterative design process. *IEEE Pervasive Computing*, 4(4): 18–26.

Dow, S. et al. (2007) Presence and engagement in an interactive drama. In *Proceedings of the ACM SIGCHI Conference on Human Factors in Computing Systems*. San Jose, CA, 1475–1484.

Dow, S.P. et al. (2011) Prototyping dynamics: sharing multiple designs improves exploration, group rapport, and results. In *Conference on Human Factors in Computing Systems*.

Dow, S.P., Heddleston, K. & Klemmer, S.R. (2009) The efficacy of prototyping under time constraints. In *Proceedings of ACM Conference on Creativity and Cognition*, ACM, 165–174.

Dreyfuss, H. (1955) *Designing for People*. New York: Simon & Schuster.

Dubberly, H. (2007) The making of knowledge navigator. In B. Buxton, *Sketching User Experiences*. San Francisco, CA: Morgan Kaufmann.

Dunne, A. (2006) *Hertzian Tales: Electronic Products, Aesthetic Experience, and Critical Design*. Cambridge, MA: MIT Press.

Dunne, T. (2011) Dunne & Raby papers. Available at: www.dunneandraby.co.uk/content/bydandr

Dunne, T. & Raby, F. (2005) Evidence Dolls. Available at: www.dunneandraby.co.uk/content/projects/69/0.

Duyne, D.K. van, Landay, J.A. & Hong, J.I. (2006) *The Design of Sites: Patterns for Creating Winning Web Sites* (2nd edn). Upper Saddle River, NJ: Prentice Hall.

Englebart, D.C. (1968) oN-Line System. Available at: http://en.wikipedia.org/wiki/NLS_(computer_system)

Fallman, D. (2003). Design-oriented human–computer interaction. *Proceedings of the Conference on Human Factors in Computing Systems*, Fort Lauderdale, FL, 225.

Forlizzi, J., Zimmerman, J. & Stolterman, E. (2009) From design research to theory: evidence of a maturing-field. *Proceedings of the International Association of Societies of Design Research*.

Frayling, C. (1993). Research in art and design. *Royal College of Art Research Papers*, 1: 1–5.

Frens, J.W. (2006). *Designing for Rich Interaction: Integrating Form, Interaction, and Function*. Eindhoven, The Netherlands: Department of Industrial Design.

Galloway, K. & Rabinowitz, S. (1980). *Hole in Space*. Santa Monica, CA: Mobile image video-tape.

Gaver, W. (2012) What should we expect from research through design? *Proceedings of the Annual Conference on Human Factors in Computing Systems* (CHI '12), ACM, 937–946.

Gaver, W., Beaver, J. & Benford, S. (2003) Ambiguity as a resource for design. *Proceedings of the SIGCHI Conference on Human Factors in Computing Systems*, ACM, 233–240.

Gaver, W., Dunne, T. & Pacenti, E. (1999) Design: cultural probes. *Interactions*, 6(1): 21–29.

Gaver, W. et al. (1992). Realizing a video environment: EuroPARC's RAVE system. *Proceedings of the Conference on Human Factors in Computing Systems*, ACM, 27.

Gaver, W. et al. (2004) The drift table: designing for ludic engagement. *Extended Abstracts on Human Factors in Computing Systems* (CHI '04), ACM, 885–900.

Geertz, C. (1977) *The Interpretation Of Cultures*. New York: Basic Books.

Gilmore, T., Ramirez, R. & Krantz, J. (1986) Action-based modes of inquiry and the host-researcher relationship. *Consultation*, 5(3).

Greenberg, S. & Buxton, B. (2008) Usability evaluation considered harmful (some of the time). *Proceedings of the Twenty-Sixth Annual Conference on Human Factors in Computing Systems* (CHI '08), ACM, 111.

Hoffman, G. (2007) *On Non-Humanoid Expressive Robots*. Ensemble: Fluency and Embodiment for Robots Acting with Humans. Thesis, MIT, Cambridge, MA.

Hollan, J., Hutchins, E. & Kirsh, D. (2000) Distributed cognition: toward a new foundation for human–computer interaction research. *Transactions on Computer–Human Interaction*, 7(2): 174–196.

Hollan, J. & Stornetta, S. (1992) Beyond being there. *Proceedings of the SIGCHI Conference on Human Factors in Computing Systems,* Monterey, CA.

Hustwit, G. (2009) *Objectified* (documentary film).

Ishii, H., Mazalek, A. & Lee, J. (2001) Bottles as a minimal interface to access digital information. *CHI Extended Abstracts on Human Factors in Computing Systems* (CHI '01), Seattle, WA.

Ishii, H. & Ullmer, B. (1997) Tangible bits. *Proceedings of the SIGCHI Conference on Human Factors in Computing Systems*, Atlanta, GA.

Jonsson, I., Nass, C., Harris, H., Takayama, L. (2005) *Matching in-Car Voice with Driver State: Impact on Attitude and Driving Performance*. Stanford, CA: Stanford University Press.

Ju, W. & Sirkin, D. (2010) Animate objects: how physical motion encourages public interaction. *Persuasive Technology*: 40–51.

Ju, W. & Takayama, L. (2009). Approachability: how people interpret automatic door movement as gesture. *International Journal of Design*, 3(2).

Kelley, T. (2002) *The Art of Innovation*. New York: Broadway Business.

Klemmer, S.R. et al. (2001) The designers' outpost: a tangible interface for collaborative web site. *Proceedings of the 14th Annual Symposium on User Interface Software and Technology,* ACM, 1–10.

Kolko, J. (2011) Abductive thinking and sensemaking: the drivers of design synthesis. *Design Issues*, 26(1): 15–28.

Kolko, J. (2007) *Thoughts on Interaction Design*. Oxford: Elsevier.

Koskinen, I. et al. (2011) *Design Research Through Practice: From the Lab, Field, and Showroom*. Oxford: Elsevier.

Landay, J.A. & Borriello, G. (2003) Design patterns for ubiquitous computing. *Computer*, 36(8): 93–95.

LeCompte, M.D. (1999) *Ethnographer's Toolkit* (7-volume paperback boxed set, 1st edn). WalCreek, CA: AltaMira Press.

Lewin, K. (1946) Action research and minority problems. *Journal of Social Issues*, 2(4): 34–46.

Liao, C. et al. (2008) Papiercraft. *ACM Transactions on Computer-Human Interaction*, 14(4): 1–27.

MacKay, W.E. (1999) Is paper safer?: The role of paper flight strips in air traffic control. *Transactions on Computer–Human Interactions*, 6(4): 311–340.

Mackay, W.E. & Fayard, A.-L. (1997) HCI, natural science and design. *Proceedings of the Conference on Designing Interactive Systems Processes, Practices, Methods, and Techniques,* Amsterdam, The Netherlands.

Mackay, W.E. et al. (1998) Reinventing the familiar. *Proceedings of the Conference on Human Factors in Computing Systems* (SIGCHI '08), ACM.

Malinowski, B. (1984) *Argonauts of the Western Pacific*. Waveland long Grove, IL: Press.

Martin, R.L. (2009) T*he design of business: Why design thinking is the next competitive advantage*. Boston, MA: Harvard Business School Press.

McCloud, S. (1993) *Understanding Comics: The Invisible Art*. New York: Harper.

Minsky, M. (1980) Telepresence. *Omni magazine* (June).

Moggridge, B. (2007) *Designing Interactions* (1st edn). Cambridge, MA: MIT Press.

Moran, T.P., van Melle, W. & Chiu, P. (1998) Tailorable domain objects as meeting tools for an electronic whiteboard. *Proceedings of the 1998 Conference on Computer Supported Cooperative Work*, ACM.

Nelson, H.G. & Stolterman, E. (2003) The design way: intentional change in an unpredictable world: foundations and fundamentals of design competence, Educational Technology.

Nielsen, J. (1990) Ten Usabilities Heuristics. Available at: www.useit.com/papers/heuristic/heuristic_list.html

Nomura, T. et al. (2007) Implications on humanoid robots in pedagogical applications from cross-cultural analysis between Japan, Korea, and the USA. *The 16th IEEE International Symposium on Robot and Human interactive Communication*, 1052–1057.

Olson, G.M. & Olson, J.S. (2000) Distance matters. *Human–Computer Interaction*, 139–179.

Parnes, S.J. (1975) *Aha!: Insights Into Creative Behavior* (1st edn). Buffalo, NY: D.O.K. Publishers.

Paulos, E. & Canny, J. (1998) PRoP. *Proceedings of the Conference on Human Factors in Computing Systems*, (SIGCHI '98), ACM, 296–303.

Pedersen, E.R. et al. (1993) Tivoli. *Proceedings of the Conference on Human Factors in Computing Systems*, (SIGCHI '93), ACM.

Peirce, C.S. (1998) *Collected Papers of Charles Sanders Peirce*. Bristok: Thoemmes Continuum.

Purpura, S. et al. (2011) Fit4life. *Proceedings of the Conference on Human Factors in Computing Systems* (SIGCHI '98), ACM.

Rode, J.A. (2011) Reflexivity in digital anthropology. *Proceedings of the Conference on Human Factors in Computing Systems* (SIGCHI '11), ACM.

Runkel, P.J. & McGrath, J. (1972) *Research on human behavior: A systematic guide to method*. New York: Holt, Rinehart & Winston.

Schön, D.A. (1995) *The Reflective Practitioner: How Professionals Think in Action*. Aldershot. Ashgate Publishing.

Schrage, M. (1996) Cultures of prototyping. In T. Winograd (ed.), *Bringing Design to Software*. New York: ACM Press. pp. 191–213.

Sengers, P. & Gaver, B. (2006) Staying open to interpretation. *Proceedings of the 6th Conference on Designing Interactive systems*, ACM.

Simon, H.A. (1996) *The Sciences of the Artificial* (3rd edn). Cambridge, MA: MIT Press.

Spool, J. et al. (1998) *Web Site Usability: A Designer's Guide* (1st edn). San Francisco, CA: Morgan Kaufmann.

Stangor, C. (2010) *Research Methods for the Behavioral Sciences* (4th edn). Belmont, CA: Wadsworth Publishing.

Suchman, L. (1987) *Plans and Situated Actions*. Cambridge: Cambridge University Press.

Takayama, L., Dooley, D. & Ju, W. (2011) Expressing thought. *Proceedings of the 6th International Conference on Human-Robot Interaction*, Lausanne, Switzerland, 69.

Tohidi, M. et al. (2006) Getting the right design and the design right. *Proceedings of the Conference on Human Factors in Computing Systems* (SIGCHI '06), ACM.

Truong, K.N., Hayes, G.R. & Abowd, G.D. (2006) Storyboarding. *Proceedings of the 6th Conference on Designing Interactive Systems*, ACM.

Venolia, G. et al. (2010) Embodied social proxy. *Proceedings of the 28th International Conference on Human Factors in Computing Systems*, (SIGCHI '10), ACM.

Wellner, P. (1993) Interacting with paper on the DigitalDesk. *Communications of the ACM*, 36(7): 87–96.

Wyche, S., Sengers, P. & Grinter, R.E. (2006) Historical analysis: using the past to design the future. *Proceedings of Ubiquitous Computing*: 35–51.

Zimmerman, J., Forlizzi, J. & Evenson, S. (2007) Research through design as a method for interaction design research in HCI. *Proceedings of the Conference on Human Factors in Computing Systems* (SIGCHI '07), ACM.

Zimmerman, J., Stolterman, E. & Forlizzi, J. (2010). An analysis and critique of research through design. *Proceedings of the 8th Conference on Designing Interactive Systems*, ACM, 310.

Design Research: Observing Critical Design

Laurel Swan and Kirsten Boehner

INTRODUCTION

Design research is a vast area encompassing research into topics as diverse as design process, the cultural impacts of design and the use of design as a form of enquiry and intervention. In this chapter, we will look at how one provocative, inspirational, and at times contested, body of design research has emerged and what its potential lessons for digital technology and research practice might be. In particular, we present our observations on critical design, a form of design research first articulated through the work and writings of Anthony Dunne and Fiony Raby of the Royal College of Art (RCA) in London.

For digital research communities, where usability and utility are key criteria of success, critical design challenges expectations. An iconic critical design project, 'Evidence Dolls' (2005) created small dolls to store a lover's collected genetic material for later DNA extraction. Another more recent project, 'The Cloud Project' (2009), used a functioning ice cream truck to explore nanotechnology,

weather manipulation and snow cones. What could digital research possibly learn from such examples?

Critical design is fundamentally difficult to define, as it is somewhat deliberately amorphous to accommodate its evolving nature. Even the rhetoric around critical design is subject to shifts; Dunne notes that the ideas associated with critical design come in many guises (FAQs[1]), and many practitioners favour the term 'speculative design' over 'critical design'. This flexibility in nomenclature is illustrative of critical design as a whole, marking it as overlapping with sympathetic practices and responding to changing contexts. Our aim in this chapter, then, is to present critical design ideas and projects in an approachable way, but also as much as possible to present critical design on its own terms, without glossing over its complexity, nuances and malleability.

We approach critical design from our own backgrounds as researchers in the field of human–computer interaction (HCI) with a focus on disciplinary intersections, working within design departments where critical design is practised, referenced, and at times reimagined. This is important to consider as it colours our prejudices, interpretations, and assessments. This outsider status, as non-designers, gives us the vantage point of observing and hopefully shedding light on the unfolding spectacle of critical design. At the same time, our insider access allows us to speak not for the designers but in a manner that honours the design voice.

We begin with a description of the origins and trajectory of critical design followed by a selection of illustrative design projects. For each project, we provide a short commentary, highlighting some of the salient features of critical design. After the project review, we look at critical design as an evolving set of practices, perspectives and values. Finally, we discuss what critical design has to contribute to and reveal about HCI and the larger field of design research.

ORIGINS OF CRITICAL DESIGN

In 1999, Anthony Dunne introduced the foundations of critical design in his book *Hertzian Tales*. Through essays and conceptual projects, Dunne described critical design as an alternative to design that reinforces the status quo, or, 'affirmative design'. He writes:

> The primary purpose of this book is to set the scene for relocating the electronic product beyond a culture of relentless innovation for its own sake, based simply on what is technologically possible and semiologically consumable, to a broader context of critical thinking about its aesthetic role in everyday life. (Dunne, 1999: xv)

Dunne's use of design as a form of critical thinking is manifested in projects such as 'The Faraday Chair' (1995),[2] a piece that plays with perceptions of electromagnetism. Taking its name from a nineteenth-century electromagnetic pioneer, 'The Faraday Chair' looks like a cross between a large aquarium, a sterile womb and perhaps a fragile coffin, and its purpose is to provide a refuge from the electrical emissions of everyday appliances. It requires the sitter to lie down, curled up in foetal position, head resting on a pillow, with oxygen provided through a snorkel-like hose to the outside. Although the chair does not actually block electromagnetic forces, it points the viewer towards thinking about whether such a sanctuary is required and how one might use it.

In developing this notion of the critical, Dunne and Raby articulate critical design as an approach rather than a strict manifesto, expecting that the practice will evolve through their work, the work of their students at the RCA and the work of others. *Design Noir* (Dunne and Raby, 2001) compiles many of these exemplary projects and marks another step in the development of critical design concepts. Some of the examples in this book reframe existing products and patents for potential products in a critical light. A non-lethal gun prototype that shoots both cayenne pepper at and video footage of the target prompts Dunne and Raby to think about weapons that have a metaphysical memory. In a similar fashion, they reimagine the 'Truth Phone', a voice stress analyser intended to assess levels of deception from a caller as useful, not for detecting truth and falsehood, but for setting the stage for storytelling and moral dilemmas.

Another type of exploratory work described in *Design Noir* plays with 'value fictions' for complicated pleasures – arguing that existing products only design for non-controversial and anaemic versions of our true selves. A 'value fiction' turns science fiction on its head by pairing contemporary (i.e. existing or feasible) technology with futuristic (i.e. imagined or extrapolated) values. For example, the 'Life Counter' (2001) by Ippei Matsumoto counts down the remainder of a person's expected lifespan in years, days, hours or seconds. Depending on which face the user chooses to view, he or she will have a different sense of the passage and value of time and his or her mortality.

The final set of projects in *Design Noir*, known as the 'Placebo Project', represents another important trajectory in the development of critical design, namely an attempt to move it work 'out of the gallery and into everyday life' (Dunne and Raby, 2001: 75). 'The Nipple Chair', for instance, looks and functions like a straight-backed chair that might be used at a desk or dining table. The added component is a set of nodules, or nipples, on the chair's back that vibrate when radiation passes through the sitter's upper body. Volunteers lived with these prototype systems in their homes and described how they used the objects, what sense they made of them and whether it caused them to think about electromagnetic fields. Whereas it might be difficult to imagine 'The Faraday Chair' in a context other than a museum, 'The Nipple Chair' pushes the concept of critical design into domestic and lived spaces.

Projects such as these described in *Design Noir* continued to develop both within and outside the RCA. However, five years on from this publication, in 2006, competing tensions emerged, on the one hand calling for a more definitive voice (e.g. 'What exactly is critical design beyond a series of ideas and projects?') and, on the other hand calling for re-evaluating its tenets (e.g. 'Is it too "dark"?'). Dunne and Raby address many of these concerns in their 'FAQs for critical design'.[3] The format of the FAQ itself is telling; when asked to create an entry for critical design for a design encyclopedia, Dunne and Raby instead offered the FAQs, a simple and direct articulation of their original intentions and subsequently emerging ideas, rather than a more formal and pedantic definition.

To the question 'What is critical design?', Dunne and Raby respond: 'critical design uses speculative design proposals to challenge narrow assumptions, preconceptions and givens about the role products play in everyday life …'[4], an imperative motive because 'society has moved on but design has not'.[5] They describe how critical design is not new or even particularly novel: it has historical roots in Italian radical design of the 1970s and overlaps with practices such as activism and interrogative design, and many

designers practise something similar in spirit if not in name. Christening it 'critical design' did not bring the ideas into being so much as add to ongoing discussion and debate. In response to the question, 'What is it for?', they suggest, 'Mainly to make us think. But also raising awareness, exposing assumptions, provoking action, sparking debate, even entertaining in an intellectual sort of way, like literature or film'.[6] This conjunctive approach to outlining the intent of critical design functions to open it up rather than pin it down.

Dunne and Raby also use the FAQs to articulate what critical design is not, at least from their perspective. They identify several misconceptions, such as the idea that critical design is a movement rather than an attitude, it is against mass-production, too jokey to take seriously, too negative and dark and more art than design. Two of these charges, those of being 'dark' and 'art', they address in more detail. Regarding its frequent focus on dark scenarios, Dunne and Raby argue this is a counterbalance to the oversimplification of human nature in commercial design briefs. Instead of being negative for negativity's sake, it uses dark matter as a means to explore authentic human experiences. As for critical design being more art than design, Dunne and Raby acknowledge borrowing methods and approaches from art, yet take pains to remain within the realm of design: 'too weird and it will be dismissed as art. … If it is regarded as art it is easier to deal with, but if it remains as design it is more disturbing, it suggests that the everyday as we know it could be different, that things could change'.[7]

In discussing misconceptions of critical design, Dunne and Raby acknowledge that the practice could stray in these directions, and may have been perceived as doing so by some commentators. One of the defining characteristics of critical design, its openness and malleability, is therefore not without its challenges. Reading between the FAQ's cautionary lines paints a picture of what bad critical design could be: erring on the side of jokey at the expense of serious critique, being too esoteric or preachy to cause personal dissonance and reflection or simply slipping into parody or farce. This awareness that critical design will be misunderstood and misappropriated is a tacit recognition that it will change, in both favourable and unfavourable ways. The conversational tone of the FAQ underscores this sense that the practice will have different viewpoints, some in agreement and some in discord, and as such, does not provide the final word but suggests with answers come more questions.

Thus, the development of the ideas and practices underlying critical design can be traced through *Hertzian Tales*, *Design Noir*, and the critical design FAQs. These reflections punctuate the array of projects coalescing under the critical design umbrella, and further delineate what ties the work together, what sets some work apart, as well as working to identify fertile tensions in values and foci. However, as important as these works may be for charting an historical trajectory, critical design as a practice is furthered by design projects as opposed to textual explanation or analysis. In the following section, then, we will look at emblematic projects represented by the designers' own images and associated content, albeit lightly edited. As befitting a design approach, the text illustrates the images rather than the other way around. Although we provide analytic commentary in response to each project, our intent is to let the designs speak for themselves and leave room for alternative interpretations.

EXAMPLES OF CRITICAL DESIGN

In order to select emblematic projects of critical design, we surveyed a wide variety of projects identified as critical design from 1999 to 2011.[8] We also considered projects that seemed similar to, but did not identify themselves explicitly as examples of critical design, to get a sense of its boundaries. From a substantial body of critical design work, we narrowed our selection to three emblematic projects, chosen after discussions with

various people practising critical design. These projects were chosen because they generated considerable interest during exhibitions, gained a degree of media recognition and/or had maintained a lifespan beyond their initial presentation. Thus, this set is not meant as a complete or authoritative view of critical design, nor as wholly representational of all the facets of critical design, but, rather, as a means for allowing the reader to examine and experience a few critical design projects more closely.

Example Project 1: 'Life Support'

'Life Support' (Figure 19.1) was first exhibited in 2008, as the final student project of Revital Cohen for her MA at the RCA in London. It went on to be exhibited in a number of other venues including the ISEA 2009 exhibition, BRIT Insurance Designs of the Year 2009, Foundation for Art and Creative Technology and the Designhuis, Eindhoven, as well as being included in numerous design publications.

Excerpts From the Designers[9]

Assistance animals – from guide dogs to psychiatric service cats – unlike computerised machines, can establish a natural symbiosis with the patients who rely on them. Could animals be transformed into medical devices? This project proposes using animals bred commercially for consumption or entertainment as companions and providers of external organ replacement. The use of transgenic farm animals, or retired working dogs, as life support 'devices' for renal and respiratory patients offers an alternative to inhumane medical therapies [Figure 19.2].

Could a transgenic animal function as a whole mechanism and not simply supply the parts? Could humans become parasites and live off another organism's bodily functions?

A patient suffering from kidney failure gives a blood sample to the lab, the scientists cut from the patients' genome the regions that code for blood production (bone marrow tissues), and immune response (the major histocompatibility complex). They then extract the genome from the nucleus of a somatic cell taken from a sheep and substitute the corresponding regions of the

Figure 19.1 '**Life Support**' (© Cohen 2009).

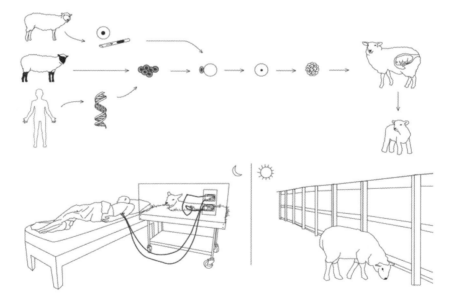

Figure 19.2 '**Dialysis Sheep**' (© Cohen 2009).

sheep's genome with the DNA cut from the patients' genome.

This recombinant DNA is then inserted into the nucleus of a pre-prepared sheep egg cell. Cell division in the egg is initiated and after a few divisions implanted into the receptive ewe.

The surrogate ewe gives birth to the transgenic lamb, which is given to the donor patient. During the day, the dialysis sheep is free to roam in the patient's back garden, graze to cleanse its kidneys, and drink water containing salt minerals, calcium and glucose.

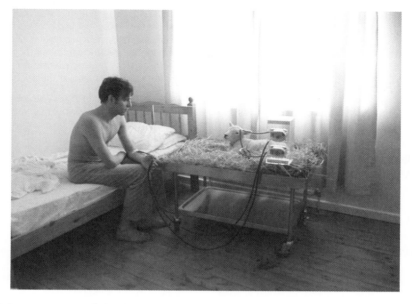

Figure 19.3   Enactment of the 'Dialysis Sheep' (© Cohen 2009).

At night, the sheep is placed on a special platform at the patient's bedside. The transgenic sheep's kidneys are connected via blood lines to the patient's fistula (a surgically enlarged vein). During the night, peristaltic pumps remove waste products from the patient's blood by pumping it out of the body, through the sheep's kidney (a natural, organic filtering system) and returning it, cleaned, to the patient [Figure 19.3].

This happens over and over again throughout the night. Each time the 'clean' blood is returned to the body, it picks up more waste products from the cells it circulates through, and brings these newly collected toxins back to the sheep's kidney to be removed. The sheep then urinates the toxins.

Commentary on 'Life Support'

'Life Support' works particularly well as an example of the speculative nature of critical design. Using a near-future technological speculation, the project is believable on a functional level, tapping into the history of utilizing animals in scientific and medical research. Using animals for medical exploration is a familiar practice, with bovine insulin for the treatment of diabetes and pediatric heart valves made from pig valves being two examples. Although potentially unsettling or problematic, these practices have over time become accepted, in part by repeated exposure; that is, because we have become accustomed to thinking about the associated issues.

'Life Support' however, raises these questions again for us and forces us to reconsider the role animals should play in relation to human welfare. In this particular context, what exactly would a relationship like this be? Would it be medical, personal or, as seems most likely, some combination of both? Would a person become attached to a sheep that cleans its blood? It does seem distinctly possible, given that the person's well-being and life depend on the animal performing a vital function for them, making it a more personal and intimate relationship than most.

This raises an interesting feature of critical design, that of changing the parameters of a familiar relationship. By shifting orientations, critical design projects can work to unseat determined taxonomies, which can be deeply unsettling. This in turn can raise issues we thought we had come to terms with and force us to re-examine if we are as comfortably resolved about them as we had thought. Thus 'Life Support' manages to be believable, realistic, yet unsettling, verging on the disturbing, and in that sense is emblematic of several criteria of critical design.

Example Project 2: 'Afterlife'

'Afterlife' Figure 19.4 was created by designers James Auger and Jimmy Loizeau. The pair met whilst students at the RCA and have been collaborating on projects together since 2000. 'Afterlife' has been part of several exhibitions, including 'Design and Elastic Mind' at the Museum of Modern Art, New York and the St Etienne Design Biennale.

Excerpts From the Designer[10]

There are many perspectives and beliefs on what happens to us after our lives on this planet come to an end. When faced with our own mortality or that of a loved one, notions of what the afterlife may hold; whether it be in a spirit world such as heaven or reincarnated into another body or form, spiritual faith can offer great comfort and reassurance.

Science and reason though have started to undermine these traditional belief systems as we strive to find logic and meaning in our existence … [I]n terms of comfort and reassurance what then is there for the grieving atheist?

Under normal circumstances after death, the human body would be assimilated back into this natural system [of the universe]. The 'Afterlife'

device intervenes during this process to harness the chemical potential and convert it into usable electrical energy via a microbial fuel cell – a device that uses an electrochemical reaction to generate electricity from organic matter. This electricity is contained within a familiar dry cell battery.

Phase II took place in 2009. We asked 15 people to propose what they would do with an afterlife battery charged either by themselves or their partner/family [Figure 19.5]. [One participant chose to power an electric toothbrush with her afterlife battery writing:] 'After 10 years of marriage one takes the occasional comment of "… your breath … perhaps you should brush your teeth" with mock horror, laughingly taken, with another kiss dispatched as punishment. I therefore feel it appropriate that on the passing away of my beloved Jimmy, the only place he would be truly happy (being a man of action) would be in my mouth busily keeping my teeth and breath fresh as a daisy morning, noon and night in a high-quality electric toothbrush.'

Commentary on 'Afterlife'

'Afterlife' works well as a critical design project in part because of its simplicity. Taking an emerging technology, microbial fuel cells, the project explores the possibilities of how that technology could be deployed. Microbial fuel cells rely on micro-organisms for fuel, and micro-organisms are available from any living being. Usually associated with micro-organisms such as e-coli, microbial fuel cells become a much different concept

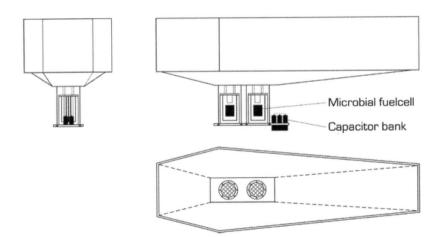

Figure 19.4   Speculative sketch for 'Afterlife' fuel harvesting (© Auger and Loizeau 2009).

Figure 19.5 'Afterlife' battery (© Auger and Loizeau 2009).

when harvesting energy from a more sentient source, such as a pet or a person. Taking it that one step further, and using the chemical decomposition of a newly dead person to power a fuel cell, introduces a completely different take on the technology and on our reactions to it.

'Afterlife' sits at the speculative end of the critical design spectrum because it focuses on the idea and possibility of a design without necessarily intending to produce an artefact. Because it resides on the speculative side, its potential for realization is ambiguous to the audience, leaving the viewer to wonder, 'Is this real?' and, 'If it is, how do I feel about it?' The reality of microbial fuel cells, still in nascent development stages but functional under very specific conditions, further complicates the reaction, in that it makes it feasible. The battery of dad with the invective to 'shine on' (Figure 19.5) plants the idea of utility and purpose – something that might be functional, and even perhaps comforting. The addition of Phase II adds a further complexity, tying in notions of regret and unfinished business on earth, as well as ideas of how couples could continue to connect and help one another after one partner has gone. As such, an emerging technology aimed at providing alternative fuel sources becomes an opportunity to provide a meaningful postscript to one's life.

Example Project 3: Biojewellery

Biojewellery began in 2003 at the RCA in response to a design brief on 'consuming monsters: big, perfect, infectious' for exploring issues around biotechnology. Design researchers Nikki Stott and Tobie Kerridge, from the RCA, teamed up with Ian Thompson, a bioengineer at King's College London to develop a proposal for making jewellery from extracted bone tissue. In 2005, the project kicked off a second time with a new round of funding from the Engineering and Physical Science Research Council's 'Partnerships for Public Engagement'.

Excerpts From the Designers[11]

The purpose of extracting and culturing tissue is to put it back into the patient's body to heal them. We imagined however that this material could also be used outside the body – but to do what? The answer to this question may be in another more profound question: what value would people place on a product that has been manufactured with materials derived from their own body?

Using chips of bone donated by couples, bone tissue was cultured ... and then combined with precious metals to create rings. The result: each person wears the body of their lover on their hand [Figures 19.6 and 19.7].

'Ye are Blood of my Blood, and Bone of my Bone. I give ye my Body, that we Two might be One.

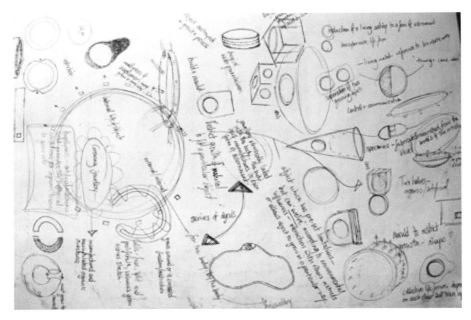

**Figure 19.6   Biojewellery design sketches for rings made from cultured bone (©
Biojewellery 2003).**

**Figure 19.7   Biojewellery participant
sketch for ring design (© Biojewellery
2003).**

**Figure 19.8   Biojewellery specimens,
clockwise from top bioactive ceramic
scaffold on which tissue growth takes
place; a sample of cow marrow; a model
ring of cow marrow and etched silver
(© Biojewellery).**

I give you my Spirit 'til our Life shall be Done.'
(Celtic wedding vows)

We needed to find – or rather, to woo – couples
for whom the idea of giving a physical part of
themselves to each other appealed. Over 200
couples sent us e-mails describing how they met,
what the rings would mean to them and why they
would like to donate their cells.

In order to obtain permission from [Central
Office of Research Ethics Committee] for tissue

engineering research, we had to submit a fifty
page application [to explain] why the bone
cells would be collected, what scientific tests
would be conducted on the cells and how the
scientific results would be disseminated
[Figure 19.8].

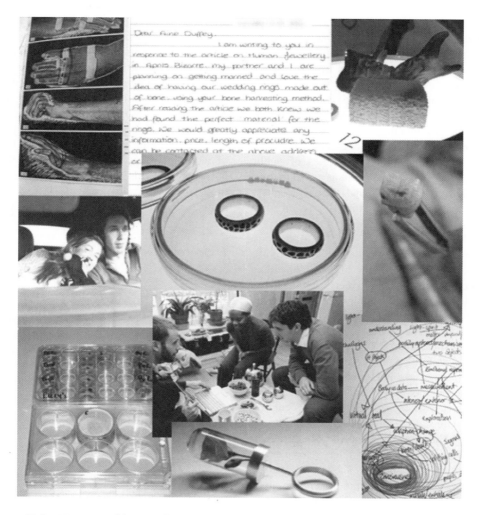

Figure 19.9   Montage of images from the Biojewellery website (© Biojewellery).

[After permission from the Central Office of Research was secured, the project moved on to the phase of extracting and then culturing the bone tissue.] Once the resuscitated osteoblast cells had expanded in sufficient numbers, they were detached from the culture dish ... re-suspended in medium and seeded on to the test scaffold hydroxyapatite (HA) rings ... the completed rings were removed ... fixed using a mix of chemicals ... then washed and left to air-dry prior to being infiltrated with epoxy resin and fashioned into rings [Figure 19.9].

Commentary on Biojewellery

Biojewellery as a project incorporates several components that often make up successful critical design projects. The first is an accessible front story with sympathetic characters and artefacts: couples, engagement rings and marriage – all topics of immediate interest and understanding to the vast majority of a public audience. The idea of creating a ring for eternal commitment out of one's tissue has a certain resonance, along the lines of the Victorian practice of wearing lockets containing fragments of loved ones' hair. Furthermore, Biojewellery takes an emerging technology, that of bioengineering and tissue culture, and applies it to an unusual and unexpected area; romance and matrimony.

Juxtaposing the fundamentally personal nature of biotissue with the emotive idea of engagement rings and matrimony is both evocative and disconcerting, both hallmarks of good critical design projects.

Biojewellery is notable for its focus on dialogue. By going beyond speculation with this scenario and inviting couples' participation in the process, Biojewellery adds a further dimension as well as the complexities of public engagement. Sparking debate is a goal of critical design in general, and the Biojewellery team made a point of inviting discussion about biotissue culturing at various forums. The proposal to make jewellery out of a loved one's reconstituted cells provided an entry for the public to discuss a range of issues regarding bioengineering. As the project continued and the designers sought additional public funding, it became imperative to rethink how the project was framed overall to make it more accountable and accessible to a larger audience (Kerridge, forthcoming). The designers gradually became less interested in the design as artefacts in themselves, and more interested in the process of exploring the public's engagement with science through the design process.

We believe these three projects – Life Support, AfterLife and Biojewellery – to be exemplar critical design projects, showcasing features we have found to be common across a much larger survey. Although limited in number, the projects presented here demonstrate some of the complexity and scope of Critical Design work, with a range of topics, forms of engagement, and the role of participants and audiences. Having outlined the origins of critical design and introduced the reader to some emblematic critical design projects, we now turn to discussing some of the salient features of Critical Design, and how it is evolving, before concluding with a discussion of its intersections with HCI.

FEATURES OF CRITICAL DESIGN

In the course of surveying materials associated with critical design – historical texts and the FAQs, numerous projects developed over the past 15 years, conversations with critical design practitioners, plus scholarly and curatorial takes on critical design – we became aware of characteristics shared by many critical design projects. The following features appeared to us to have some recurring resonance and we thus offer them as guidelines for thinking about what ties this body of work together. We wish to make clear, however, that this is not meant as a reductionist taxonomy of critical design, but, rather, as a lens to view a diverse and disparate body of work.

1. **Speculative v. real**: critical design projects are often set in the future, or the near future, and, as such, are more about what could be as opposed to what is. Often, they will be built around an existing but relatively new discipline such as bioengineering, or a newly emerging technology, such as a recent medical advance. Using these new technologies as a starting point, they develop and extrapolate from them, taking them into unexpected trajectories as a way of exploring unusual spaces.

2. **Layered**: Many critical design projects have seemingly simple front stories that belie their underlying complexity. These simple front stories tend to be innately catchy – that is, they have a hook, in order to capture the audience's interest. This simple front end works to hide the layers of research and effort that lie beneath the surface, as much of the work that has gone into a project remains virtually invisible in the finished project, accessible only through careful perusal of informational material associated with the project.

3. **Tension**: critical design projects generally aim to create a tension around how they are approached, leaving the viewer to determine, 'Is this real?' and, subsequently, 'How do I feel about it?' By questioning implicit values, the projects often elicit strong and/or diametrically opposed responses: wheareas some people may find a project stimulating and profound, others may find it distasteful or irrelevant, whilst still others may find it a combination of all of the above.

4. **Aesthetic**: As designed efforts, the look, feel and experience of critical design projects and their associated artefacts are highly intentional and purposefully designed. This means that the

projects have a particular form of a developed aesthetic, one that is communicated through all aspects of the project. This can result in a diverse array of design aesthetics, ranging from slick and highly stylized to whimsical to quite rudimentary and unpolished. In all cases, this is a deliberate decision, with the aesthetic intended to convey a particular message or elicit a particular response.

5. **Provoking thought and dialogue**: Ultimately, critical design projects are meant to raise questions and encourage dialogue. They are intended to be provocative, but importantly not solely for provocation's sake. Rather, they offer an alternative means of posing questions that should be addressed, such as why the current designed landscape looks the way it does, and what might it look like if it were tweaked in one direction or another, and what questions does this raise in terms of our own beliefs and values?.

These five features suggest ways of approaching a critical design project for understanding it on its own terms. For example, in what ways does it destabilize a mainstream take on a particular design space? Where does the front-end hook give way to a more complicated analysis or exploration? Projects that do not embody these criteria may be pushed in a critical design direction, if desirable, by attending to this feature set – for example, exploring how to provoke thought or deepening the tension that animates the project's ideas. Although these features may be helpful in understanding what critical design is, we do not propose that they sum up the diversity of the space. Instead, they point to important markers, not only about what critical design has looked like to date but also about important areas of critical design that are evolving.

EVOLUTION OF CRITICAL DESIGN

Throughout this chapter, we have alluded to critical design's amorphous quality and changing nature. Although it is possible to summarize periods of critical design work, for example along *Hertzian Tales* or *Design Noir* themes, it is too early to write a definitive

retrospective of the practice as a whole. One of the important ways critical design has expanded over the years is in moving outside the boundaries of the Design Interactions department at the RCA. Although still very much associated with Dunne and Raby and their students at the RCA, critical design has also been adopted by designers outside this institution, as well as being further developed by RCA graduates in their own studio practice. This move has increased the scope of critical design and extended its range as different designers meld the practice into their own in interesting ways.

Pushing the range of critical design further than institutional boundaries is the move across disciplines. A recent forum by the Design Research Network brought together designers and researchers interested in combining critical design with participatory design, critical theory, science and technology studies, social activism, HCI and other fields.[12] Entitled 'Before and After Critical Design', the forum focused on other forms of critical practice while questioning the limits and new directions of critical design. They discussed critical design in terms of what is being critiqued, who is doing the critiquing, how the critiquing is resourced and where the critiquing happens. This reading of critical design on the one hand considers the limitations, for instance whether the gallery format reduces the chance for dialogue, while at the same time wondering how these limitations might suggest possibilities for extending existing principles or evolving into something completely different. For example, several authors within this forum are working to extend critical design into a more involved public engagement (Bowen, 2009; DiSalvo and Lukens, 2009; Kerridge, forthcoming). As such, this boundary is interesting in that it suggests not a restraining point, but a potentially directional one.

It is perhaps not surprising that a practice will change once outside its foundations, such as its institutional home and founding discipline. This suggests the practice has legs outside of its origins. Yet it is also essential for a practice to evolve internally as well. One of the most noticeable ways that critical design has changed is in its tone. although Dunne and Raby articulated in their 2006 FAQ[13] that critical design was not meant to be 'dark for the sake of being dark', and that it aimed for 'the positive use of negativity', nonetheless it began to be identified with having a dystopic overtone and an expectation of subversive elements. In this sense, the 'critical' of critical design was narrowed in the minds of some to mean a kind of negative judgement.

However, in recent years, a new generation of critical design projects have emerged with different sensibilities. The 'critical' is more readily recognized as a questioning of given values, stories, assumptions, trajectories and resulting products. Rather than focusing on dystopian scenarios with sinister overtones, these projects surprise us not for being dark but simply for being unexpected. The 'Toaster Project' (2009), for example, set out to make a toaster from scratch – including sourcing the iron ore and smelting it in a personal microwave into steel. Toasters themselves are not strange – in fact they are so commonplace they are available for less than £5 – and as a result are rarely thought about beyond the choice of colour and number of slots. The 'Toaster Project' unravels what amounts to a complex web of parts, relationships, know ledge and power in the making of the benign, ordinary toaster. Far from an overtly dystopian viewpoint, the 'Toaster Project' encourages reflection on topics as varied as mining, commodification and the push and pull between convenience and control.

As another recent example, 'Back, Here Below, Formidable' (Humeau, 2011) appears whimsical or perhaps otherworldly as opposed to disturbing or dystopic. In this project, the designer recreated the lost sound of a woolly mammoth by building a speculative larynx for the species based on collaborations with paleontologists, animal scientists and speech experts. The project ticks the critical design feature set outlined in the earlier sections: it is speculative, involves layers of meaning,

employs a particular aesthetic, creates tension in the viewer and provokes thought and dialogue. It does all of this in a surprising and quirky manner as opposed to an uncomfortable one. The emergence of projects such as 'Back, Here Below, Formidable' and the 'Toaster Project' does not suggest that dark undertones have given way to the whimsical and weird in critical design. What it does suggest, however, is that the features of critical design are open enough to allow for the creativity of individual practitioners and their bespoke, and at times unusual, interpretations of the world. The designers operate from a specific context – orientated to an ever-growing past (including past critical design projects) and directed towards an ever-changing future. What constitutes 'affirmative design'–that is, design for the status quo moves as our design context continuously changes and therefore changes the field for critical design. As Dunne and Raby noted, whatever it is called, there will always be a charge to design for the unexpected, ignored and undervalued.

CRITICAL DESIGN AND HCI

Following these representative examples and overview of critical design, we now consider how digital technologists might draw from this varied practice. As noted earlier, we will focus on lessons for and from the field of HCI. Although a varied practice itself, HCI's historical emphasis on science-based criteria for producing knowledge and usability-based criteria for defining the success of technology systems makes for an incompatible fit with the features and values outlined for critical design. The speculative nature of critical design runs counter to the here-and-now and utility focus of HCI.

Yet, even if seemingly at odds, critical design is not unknown to mainstream HCI. Recent frameworks of design perspectives in HCI have taken note of critical design, albeit as an outlier for its design-led (as opposed to user-led or participatory-based) approach (Fallman, 2008; Forlizzi et al., 2008; Sanders and Stappers, 2008). In some of these frameworks, critical design is associated with another design-led approach that has received relatively more attention in HCI: cultural probes (Gaver and Dunne, 1999; Gaver et al., 1999, 2001). Cultural probes embody a playful and provocative attitude towards imagining the hopes, dreams, fears, wants, needs, etc. of unfamiliar cultures and spaces in order to inspire design. Bill Gaver and colleagues first used the cultural probe methodology in a project using digital technology to increase the presence of elderly people in their communities. The form and approach of the cultural probes were a novelty to the HCI audience when first introduced in 1999.

Although the work was originally developed at the RCA, both Gaver and Dunne resist conflating cultural probes with critical design. According to gaver (forthcoming), his intersection with critical design materializes through the remit to design for all of life, messy and otherwise, and the caution to recognize how designs embody preferred forms of interactions, scripting certain values over others. Critical design turns this power of commodification on its head, by scripting alternative values instead of dominant ones. Gaver contrasts his work with critical design because of his emphasis on open-endedness and provisional meaning, allowing for unexpected values to surface. Furthermore, Gaver's work tends towards more positive alternatives to dominant consumer culture as opposed to the darker versions often viewed as hallmarks of early critical design.

It is not surprising that from the outside the two design practices may seem quite similar, whereas from the inside they are markedly different. It is instructive to unpack this conflation, however, for what it tells us about the perceptions and potential adoption of critical design. A shorthand account of critical design could be as something 'edgy', design-based and different. Cultural probes certainly fit this bill; they stand in marked contrast to many conventional HCI approaches and surprise not only participants but also the

community of HCI practitioners. However, although cultural probes are provocative, they are not disturbing in the way some critical design projects are. For the cultural probes to work, accessibility is a key factor. That some within the HCI community equate cultural probes with critical design speaks to the murky area of alternative approaches to research and design. There is a standard, recognized approach to design research and anything that does not fit this mould may be lumped into a category of 'other'.

The uptake of cultural probes within the HCI community is instructive for the uptake of the similar, yet different, practice of critical design. Although striking a balance between the unfamiliar and accessible, cultural probes have generated mixed reactions within the HCI community: some researchers have boxed them off as too alternative some researchers have embraced their alternative approach as a needed antidote to a predominant scientific approach and other researchers have attempted to mine the methods but divorce the methodology. What many took away from the idea of cultural probes was a recipe for design research, a toolkit of props, instead of adopting the design sensibility underlying the methods (Boehner et al., 2007). This emphasis on methods suggests that the path from the edge to the core of HCI is a documented, understandable, verified method. The 'what' that is designed for may be alternative as long as the 'how' is not.

The appeal of standard methods anticipates that, for critical design to be taken into the HCI community, it will be through an attempt to appropriate the methods. The enigmatic nature of the ideas and sensibility of critical design, however, will resist this attempt to blunt its edges. It may be possible to deconstruct critical design into a philosophical essence and a formalized method (e.g. step 1: take an existing technology, step 2: extrapolate implications to the extreme, step 3: stylize it with design noir), yet, the results of such a deconstructive exercise or a formulaic approach would be hollow, in the same manner that formalized approaches to cultural probes also fall short.

Given these mismatches between critical design and HCI, it is understandable that for the most part meaningful intersections are few, with neither field finding much purchase in the other. However, as both critical design and HCI evolve, potential intersections seem promising. Whereas user studies and built systems still remain the hallmark of HCI, there are growing examples of work that could fall under the rubric of speculative work (Aoki and Woodruff, 2005; Gaver and Martin, 2000; Sengers and Gaver, 2006 DiSalvo et al., 2009). Critical Design's growing emphasis on dialogue and engagement also resonates with work in HCI on activism and participatory engagement beyond typical participatory design (Hirsch, 2009; Paulos et al., 2008; Parker et al., 2012). Finally, a sympathetic, as opposed to derivative, movement is the expanding focus in HCI beyond usability to 'experience' (McCarthy and Wright, 2004; Bødker, 2006). The call from HCI researchers and practitioners to examine a larger range of values and design for more complex and integrated experiences such as emotion, intimacy and spirituality echoes the invective in critical design to look beyond the status quo.

We will conclude with three observations of how critical design provides a potentially rich resource for HCI and digital technology in general. First, with the call to design for all of experience, critical design reminds us that experience runs a gamut of emotions and values. We cannot neatly demarcate experience into dark and light, odd or mainstream, yet it would appear there is a greater emphasis in designing for positive and popular experiences. Designing for some experiences over others not only ignores the full complexity of lived experience, but also institutionalizes design choices. All design is about making choices but when these choices become reified, for example when we neglect a particular range of emotions, these choices become invisible. In other words, it becomes a 'nonchoice' in terms of why, how and for whom

we design. Finding ways to uncover these tracks is the work of practices like critical design.

Second, the process of doing critical design as a designer-led approach is a potentially rich resource for HCI. Rather than positioning designer-led approaches as somewhat suspect outliers, we can use it to understand how certain features of critical design are sustained and supported. critical design challenges HCI research because the rigour of its practice is not well documented and therefore often missed. Outsiders to critical design mainly encounter it through its product – for example, a speculative proposal or a designed object. What is missing is a sense of how the concept unfolded, changed and mutated through internal critiques, through push-back of the materials, through discovery of new facts or resources and through contributions from other collaborators. How do critical designers achieve this kind of layered meaning in their designs? What are their strategies for navigating the challenges of interdisciplinary work? How does the aesthetic of projects develop? How is iterative design different in critical design than in what we know of as iterative design in HCI? How does the visual imperative in critical design both enhance and constrain the work? Thinking of design as a conversation (Schön, 1994) promotes understanding how design-led and user-led approaches play off each other. From this perspective, critical design and HCI both stand to gain, in terms of how to foster authentic, engaged dialogue.

Finally, critical design is organic. It is not codified into a set approach, nor is it a manifesto. Part of trying to understand critical design is recognizing that it has foundations but is also evolving and the boundary between what is and what could be is moving. A challenge for HCI, and for design research in general, is how to share and discuss ideas, practices, patterns, synergies and salient features without needing to codify it. One recognized and valued contribution within HCI for understanding design research is outlining a framework: defining separate parts and their relationships. What we have identified in this chapter as salient features could be marked as a 'framework for critical design', yet it is not a framework in the sense of prescribing all the set features and their relationships. As discussed earlier, the feature set outlines broad sensibilities that change in their implementation. Our purpose here is not to try and nail critical design down but to identify salient points in order to watch them evolve.

CONCLUSION

This chapter aimed to provide a close look at critical design in terms of its history, recent work and issues for the future. For some readers, this may be an introduction to critical design, whereas for others it may be an opportunity to gain a better understanding of what the practice entails, laying the groundwork for further discussion and exploration. We present this work not as a conclusive account of critical design but as our observations as 'inside outsiders' to the practice. As HCI practitioners, we have tried to bridge these very different communities by presenting critical design through the designers' own work and words, while also anticipating the challenges and obstacles between the two different perspectives. We have only touched on many of these issues, however, as the body of work critiquing critical design is growing both internally and externally.

Ultimately, this chapter points out a range of open issues with critical design; issues that are interesting because they are open. What we find simultaneously useful and challenging for digital technology research is the amorphous and slippery notion of what critical design is and could be, navigating the same dichotomy critical design itself proposes to traverse in design. Trying to conclusively define critical design moves from an exercise of questioning it to defending it. Examining the boundaries and flux of critical design, or any design research practice, is essential for keeping the practice relevant.

NOTES

1. http://www.dunneandraby.co.uk/content/bydandr/13/0
2. http://collections.vam.ac.uk/item/O63805/chair-faraday-chair/
3. www.dunneandraby.co.uk/content/bydandr/13/0
4. Ibid.
5. Ibid.
6. Ibid.
7. Ibid.
8. To review this large sample, please see the projects of: (1) *Hertzian Tales* (Dunne, 1999); (2) *Design Noir* (Dunne and Raby, 2001); (3) RCA shows from Computer Related Design and Design Interactions departments 1999–2011; (4) the 2011 'What If' exhibit at the Science Gallery in Dublin (www.sciencegallery.com/whatif); and (5) the 2011 Talk to Me exhibit at the Museum of Modern Art in New York City (http://moma.org/interactives/exhibitions/2011/talktome/objects/#category=all&tag=critical-design).
9. Source for images and text www.revitalcohen.com/project/life-support
10. Source for images and text www.auger-loizeau.com/index.php?id=9
11. The following images and text are taken from the project website (www.biojewellery.com) and the exhibition brochure for Biojewellery (exhibited in various forms at the RCA, the Dana Centre, Guy's Hospital, and other public institutions and conferences).
12. Design Research Network, Feature Discussion, 2 Sept. 2011. www.designresearchnetwork.org/drn/content/feature-discussion%3A-and-after-critical-design
13. www.dunneandraby.co.uk/content/bydandr/13/0

REFERENCES

Aoki, P. and Woodruff, A. (2005) Making Space for Stories: Ambiguity in the Design of Personal Communication Systems. *Proceedings of the Conference on Human Factors in Computing Systems (SIGCHI '05)*, ACM 181–190.

Bødker, S. (2006) When Second Wave HCI Meets Third Wave Challenges. *Proceedings of the 4th Nordic Conference on HCI (NordiCHI '06)*, ACM.

Boehner, K., Vertesi, J., Sengers, P. and Dourish, P. (2007) How HCI Interprets The Probes. *Proceedings of the Conference on Human Factors in Computing Systems (CHI '07)*, ACM, 1077–1086.

Bowen, S.J. (2009) A Critical Artefact Methodology: Using Provocative Conceptual Designs to Foster Human-Centred Innovation PhD thesis. Sheffield Hallam university, Sheffield.

DiSalvo, C., Boehner, K., Knouf, N. and Sengers, P. (2009) Nourishing the Ground for Sustainable HCI: Considerations from Ecologically Engaged Art.

Proceedings of the Conference on Human Factors in Computing Systems (CHI '09), ACM, 385–394.

DiSalvo, C. and Lukens, J. (2009) Towards a Critical Technological Fluency: The Confluence of Speculative Design and Community Technology Programs. *Proceedings of the 2009 Digital and Arts and Culture Conference (DAC '09)*, 1–5.

Dourish, P. (2006) Implications for Design. *Proceedings of the Conference on Human Factors in Computing Systems (SIGCHI '06)*, ACM, 541–550.

Dunne, A. (1999) *Hertzian Tales*. Cambridge, MA: MIT Press.

Dunne, A. and Raby, F. (2001) *Design Noir: The Secret Life of Electronic Objects*. Basel: Birkhauser.

Fallman, D. (2003) Design-Oriented Human–Computer Interaction. *Proceedings of the Conference on Human Factors in Computing Systems (SIGCHI '03)*, ACM, 225–232.

Fallman, D. (2008) The Interaction Design Research Triangle of Design Practice, Design Studies, and Design Exploration. *Design Issues*, 24(3): 4–18.

Forlizzi, J., Zimmerman, J. and Evenson, S. (2008) Crafting a Place for Interaction Design Research in HCI. *Design Issues*, 24(3): 19–29.

Gaver, W. (forthcoming) Speculative Design (manuscript).

Gaver, W. and Dunne, A. (1999) Projected realities. *Proceedings of the Conference on Human Factors in Computing Systems (SIGCHI '99)*, ACM, 600–607.

Gaver, W., Dunne, T. and Pacenti, E. (1999) Cultural Probes. *Interactions*, 6(1): 21–29.

Gaver, W., Hooker, B. and Dunne, A. (2001) *The Presence Project*. London: Royal College of Art.

Gaver, W. and Martin, H. (2000) Alternatives: Exploring Information Appliances through Conceptual Design Proposals. *Proceedings of the Conference on Human Factors in Computing Systems (SIGCHI '00)*, ACM.

Hirsch, T. (2009) Learning from Activists: Lessons from Designers. *Interactions*, 16(3): 31–33.

Kerridge, T. (forthcoming) Speculative Design and Public Engagement (thesis).

McCarthy, J. and Wright, P. (2004) *Technology as Experience*. Cambridge, MA: MIT Press.

Parker, A., Kantroo, V., Lee, H., Osornio, M., Sharma, M. and Grinter, R. (2012) Health Promotion as Activism: Building Community Capacity to Effect Social Change. *Proceedings of the Conference on Human Factors in Computing Systems (SIGCHI '12)*, ACM.

Paulos, E. Foth, M., Satchell, C., Kim, Y., Dourish, P. and Choi, J. (2008) Ubiquitous Sustainablity: Citizen Science and Activism. Human-Computer Interaction Institute. Paper 200. Available at:http://repository.cmu.edu/hcii/200

Sanders, E. and Stappers, P. (2008) Co-creation and the New Landscape of Design. *CoDesign*, 4(1): 5–18.

Schön, D. (1994) *The Reflective Practioner*. London: Ashgate.

Sengers, P. and Gaver, W. (2006). Staying Open to Interpretation: Engaging Multiple Meanings in Design and Evaluation. *Proceedings of DIS 2006*, ACM.

Wodiczko, K. (1999) *Critical Vehicles*. Cambridge, MA: MIT Press.

Wolf, T., Rode, J., Sussman, J. and Kellogg, K. (2006) Dispelling Design as the 'Black Art' of CHI. *Proceedings of the Conference on Human Factors in Computing Systems (SGICHI '06)*, ACM.

Projects Cited

Afterlife (2009), James Auger and Jimmy Loizeau

Back, Here Below, Formidable (2011), Marguerite Humeau

Biojewellery (2003), Nikki Stott, Tobi Kerridge and Ian Thompson

Evidence Dolls (2005), Tony Dunne and Fiona Raby

Faraday Chair (1995), Tony Dunne

Life Counter (2001), Ippei Matsumoto

Life Support (2008), Revital Cohen

Nipple Chair (2001), Tony Dunne and Fiona Raby

The Cloud Project (2009), Zoe Papadopoulou and Cathrine Kramer

Toaster Project (2009), Thomas Thwaites

PART 4

Environments and Tools for Digital Research

While a variety of theoretical approaches were covered in the previous section, developments in digital interfaces also provide us with a broad range of technologies and digital environments. This section presents a selection of research undertaken with a number of different digital technologies, which draw on a variety of sites of practice. Each chapter offers a review of its specific area of digital technology research, drawing on key texts and studies, and a research example to demonstrate the particular considerations, potentials and constraints for researching this specific type of digital technology.

In Chapter 20, Sara Price looks at tangible technologies – technologies that are graspable, where computational power is embedded in everyday artefacts or the environment, which can be wirelessly linked to various forms of digital representation and where interaction depends on the manipulation of physical artefacts and/or bodily action. This chapter explores tangible research in the context of education, introducing key research approaches in the field with an illustrative research example and outlining challenges for research design and evaluation, particularly key research issues around notions of physical–digital mappings, concepts of engagement, different design parameters, and collaboration.

In Chapter 21, Leah Buechley also explores 'material' aspects of computing, but the focus here is on how new materials can enable interaction design researchers to design and build new technologies. Drawing on developments in materials science and engineering it discusses how materials like paper, fabric, paint and wood can be used to create exciting and new forms of electronics and how material-based connectors, sensors and actuators can be constructed and employed by interaction designers, computer scientists and engineers, with a particular emphasis on electronic textiles, wearable computing and educational technology.

Chapter 22 (by Eve Hoggan) continues the theme of the material and tangible through its focus on haptic interfaces. It highlights the crucial nature of our sense of touch in our everyday interactions with the world, bringing 'touch' back into focus through technological advances that enable interactive systems to be enhanced with haptic modality. It describes the basic concepts of haptic interaction, different actuator technologies and methods of designing haptic interfaces. Its review of the most recent novel research in the field focuses on a variety of application areas and highlights the key research challenges in the field of haptic interface design.

Chapter 23 (by Yvonne Rogers, Nicola Yuill and Paul Marshall) extends from specific technologies to look more broadly at multi-user displays, particularly with respect to evaluating their benefits for collaboration. A primary theme of this chapter is a comparison of lab versus in-the-wild studies. The authors argue that each approach can further the development of the other: in-the-wild studies provide a 'contextual backdrop', sensitizing researchers to how this is most likely (rather than should) be used in practice. While this chapter may focus on multi-user displays, the discussion and issues can be applied more broadly across the majority of novel technology-based research.

Chapter 24 (by Yoosoo Oh and Woontack woo) looks at combined physical–digital environments based on ubiquitous virtual reality (ubiquitous VR) systems. These systems enable the superimposing of virtual reality aspects on to the physical space. This chapter defines what is meant by ubiquitous VR and describes its primary features. A research example is used to describe a ubiquitous VR environment, which is capable of simulating a smart space by enabling the addition of new virtual entities to existing real entities and vice versa.

Chapter 25 (by Tyttollett and kevin Leander) moves into the realm of location-based technologies and environments, such as geographic information systems (GIS) and global positioning systems (GPS), WiFi, mobile devices and sensors. The chapter considers the general problem of tracking and understanding

users and provides an in-depth consideration of sensing technologies, suggesting how they open up opportunities to embed rich environmental and human data within location data. The chapter offers a description of two exemplary contexts – health research and research on education and game-based learning – before presenting the issues and challenges in conducting research with location-based technologies and within location-based environments.

In Chapter 26, Niall Winters continues the 'mobile technology' theme, but specifically in the context of international development. This chapter highlights the slow progress mobile technology has made in this area as well as the challenges that persist in understanding the contribution mobile learning can make to development goals. It reviews recent reports from international bodies on mobile learning to establish how these technologies are understood and instantiated in practice. Recognizing the challenges faced when working in developing contexts, and drawing on the latest literature, this chapter suggests the importance of directly focusing on how technologies mediate learning practices in a culturally appropriate manner in order to understand the role of mobile learning in developing contexts.

Chapters 27 and 28 (by Catherine Bearis and Kirsty Young, respectively) shift the focus to examining research in online contexts. Chapter 27 centres around digital games-based research in educational contexts. Games-based learning via online and mobile Internet-based technologies is seen as providing great potential for innovative, effective and accessible contemporary teaching and learning, yet research focusing on these areas is fragmented and provides a mixed picture of their use. This chapter highlights the interdisciplinary nature of the field, which is informed by quite different research approaches and trajectories, each with its own epistemological framework and assumptions. Through an example in Second Life it explores these approaches, discussing the ongoing tensions between research traditions and providing an outline of issues for further development and insights for the future.

Chapter 28 examines social networking sites (looking at functions; education; generations; gender and cultures; identity and social ties/ networks) and the subsequent issues arising for social research. It examines the rise of second-generation Internet platforms, particularly how online content created through social media provides insights into individuals' lives and the transformation of communities and society, both locally and globally. The chapter concludes by identifying both the challenges and potential for social science researchers using SNS as a context and source of research data.

Chapter 29 (by Kas'ka Porayska-Pomsta and Sara Bernardini) turns to learner modelled environments (LMEs) as a method through which learning can be both supported and studied. Learner modelling is presented as the core defining characteristic of any LME and as a necessary prerequisite for a technology-enhanced educational tool capable of adapting the interaction and the pedagogical support to the individual learners' needs in real time. The chapter draws on diverse approaches to offer a coherent description of the learner modelling field, while discussing predominant approaches to learner modelling and highlighting key research questions related to the design of LMEs.

The final chapter in this section concerns methods for 'innovation' in real-world product design, highlighting the need to balance *invention* – coming up with new ideas – and *enquiry* – understanding how the world works. Here Larstrik Holmquist presents the innovation framework, which aims to combine these two key features in order to foster 'true' innovation. The chapter provides illustrative examples of this innovation process, offering pointers for the future of digital technology research.

20

Tangibles: Technologies and Interaction for Learning

Sara Price

INTRODUCTION

With recent developments in computing and networking, new kinds of interfaces, such as tangible interfaces, and consequently new forms of interaction with technology, have emerged. 'Tangibles' generally refer to interfaces where computational power is embedded in everyday artefacts or customized objects, which can be wirelessly networked or linked to various forms of digital representation. The emergence of increasingly small microchips and digital sensing technologies means that embedding technology in both artefacts and the environment is becoming more commonplace.

In the field of human–computer interaction (HCI) this group of technologies may be described as graspable interfaces (e.g. Fitzmaurice et al., 1995), tangible interaction (e.g. Ullmer and Ishii, 2001) and tangible bits (e.g. Ishii and Ullmer, 1997). Shaer and Hornecker's (2010: 4) definition offers a useful description for the purposes of this chapter: 'Interfaces that are concerned with providing tangible representations to digital information and controls, allowing users to quite literally grasp data with their hands' and thus physically manipulate associated representations. There are three key categories of systems that sit under this umbrella term: constructive assembly kits, token and constraint systems and interactive surfaces.

Interaction with tangibles depends on the manipulation of physical artefacts and/or physical forms of action, offering the opportunity to build on our everyday interaction and experience with the world, exploiting senses of touch and physicality. A key feature of these technologies is the high level of flexibility in design and the degree to which the design space is extended. This applies to the objects themselves – for example, their shape, size, colour, weight and texture; to the actions that can be placed upon them – for example, they can be impactive, requiring physical contact with an artefact (e.g. grasp and grip) or non-impactive (e.g. gesture); and to the associated digital information. Digital information, in the form of sound, narration, images, text or animation, can be flexibly combined with artefacts (e.g. Zuckerman et al., 2006), the environment (e.g. Klopfer and Squire, 2007; Price at al., 2010) or action (e.g. Price and Rogers, 2003; Raffle et al., 2006) to provide contextually relevant information based on abstract concepts or on enhancing key components of the task or concept with which the user is engaging. This potential to link to a wide variety and mix of representational media offers new possibilities and challenges for designing information artefacts and representations for learning. At the same time it demands particular considerations when researching these technologies, environments and interaction with them.

In terms of research methods they offer a complex domain for research. One key factor is the number of variables to take into consideration when studying tangible environments. Another is the choice of research approach, given the unique, novel and *not* off-the-shelf technology that it entails, and the different disciplinary perspectives involved, including computer science, art and design, psychology and social science more broadly. This chapter aims to outline these challenges for research design and evaluation, its focus being on research in the context of tangible learning environments and learning interaction (which might inform design of user interfaces), but not on the building or development process of tangible interfaces. It begins with a review of related research and an introduction to key research approaches in the field. This provides the context for an illustrative research example investigating the use of tangibles in an education context. Through this example, the chapter will explore some key research issues – for example, notions of physical–digital mappings, concepts of engagement, the effect of different design parameters. Finally, the chapter will outline critical future research directions and related challenges.

LITERATURE REVIEW

This section offers a review of how these technologies have been used in research to date and provides the context for an illustrative research example investigating the use of tangibles in a science-learning context. Research with tangible technologies can involve a number of different aspects, including the design and building of the system or environment (which often takes an iterative participant design approach); observing and analysing user interaction (in the wild or in the lab); measuring specific features of interaction that are of interest to the research question, such as the design of physical–digital mappings, learning outcomes, engagement. Tangibles have been designed for use in a variety of contexts, from museum exhibits (e.g. Horn et al., 2008; Wall and Wang, 2009) and interactive music installations (e.g. Jorda, 2003), to tools that support planning and decision-making (e.g. Underkoffler and Ishii, 1999). A number of tangible technology-based projects specifically explore applications in the learning domain, with emphasis on various aspects of interaction, from how design influences interaction, the learning process and social interaction to engagement and edutainment, together with a strand of work that focuses on special needs learners. A number of tangible systems for learning in different contexts have been developed during the last decade. Studies of such systems primarily inform us about levels of engagement and enjoyment, the technical achievements of mapping to learning activities that may be promoted through tangible interfaces, but with increasing insights into collaborative forms of interaction and a developing interest in the role of embodied forms of interaction through digital environments for learning.

Early examples of tangibles used popular, familiar toys, such as balls and blocks, digitally embedding them with, for example, light-emitting diodes (LEDs) or accelerometers. Bitball is a transparent sphere that records and transmits information about its own movement through the use of accelerometers (Resnick et al, 1996); Stackables and Programmable Beads comprise assembling blocks that allow children to explore

Figure 20.1 From left to right: Bitball and Programmable Beads (© Mitch Resnick MIT).

dynamic behaviour patterns (Resnick et al., 1998; see Figure 20.1); while SystemBlocks and FlowBlocks generate visual representations of behaviour according to the way the objects are combined (Resnick et al., 1998; Zuckerman et al., 2006; see Figure 20.2). In other work blocks are used as tangible programming elements to ease programming tasks for children by arranging blocks with different functions (e.g. Wyeth and Purchase, 2002; Schweikardt and Gross, 2008).

Other kinds of assembly or constructive kits allow children to build their own, personalized models, stimulating their creativity and imagination. For example, Topobo (Raffle et al., 2004) enables children to build creatures out of digitally embedded pieces that can record and play back physical motion to facilitate children's learning about movement and locomotion (Figure 20.3). This process of creating models is thought to foster a greater understanding of the functioning of things (Klopfer and Squire, 2007) and provide opportunities for children to produce knowledge by expressing themselves through the representations they create (Marshall et al., 2003) – that is, the artefact embodies the children's activity and thoughts.

While familiarity may engage children, the linking to ambiguous or less familiar representations in tangible systems has been shown to promote curiosity and exploration (Rogers et al., 2002). Chromarium, a system to explore colour mixing through physical and digital tools, suggested that children engaged in more experimentation and reflection when objects were linked to less familiar (digital) representations. Subsequent work also suggested that some level of ambiguity provokes children's interest, curiosity and reflection (Price et al., 2003; Randell et al., 2004). In contrast to Topobo, knowledge here is produced through exploration (leading to conclusions), rather than expressivity. There is less space for creativity,

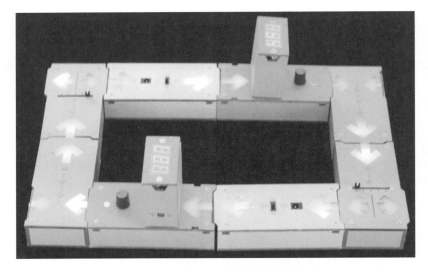

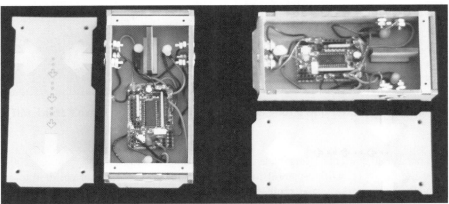

Figure 20.2 FlowBlocks (designed and developed by Oren Zuckerman during his PhD at MIT Media Lab).

suggesting expressive and exploratory systems lend themselves to different learning activities and processes.

Tangible environments have also been shown to encourage collaborative interaction. Tangibles combined with tabletop environments (e.g. Reactable: Jordà, 2003; Sensetable: Patten et al., 2001; LightTable: Price and Pontual Falcão, 2009) show increased collaboration through features of shareable interfaces that accommodate face-to-face interaction and multiple, simultaneous users, thus encouraging communication. Recent work illustrates

how these interactive properties support productive collaborative knowledge-building (e.g. Fleck et al., 2009; Pontual Falcão and Price, 2010).

Another strand of work centres around the physically active nature of interaction with tangibles. Antle and colleagues have explored this through notions of metaphor in tangible environments, especially those that relate to 'embodied' interaction (e.g. Antle, 2009; Macaranas et al., 2012) – in particular, understanding how the design of metaphorical mappings between schematic action and

Figure 20.3 Topobo (Hayes Raffle, MIT).

system response improves learning performance. For example, with Springboard, learners explore abstract concepts of 'balance', such as 'social justice', through varying degrees of their own physical bodily balance that triggers visual displays of balance related to a number of social justice issues (Antle et al., 2011).

Other empirical work on tangibles and learning investigates their value in supporting children with special educational needs. Early work suggests that tangible systems positively encourage social activity, fostering social interaction and skill development. A tangible application developed for one- to four-year-olds was found to offer more opportunities for facial, gestural and verbal interaction, as well as slowing down interaction, which was thought to allow more control over the interface (Hengeveld et al., 2009). Research with children on the autistic spectrum found that using Topobo engendered more onlooking, cooperative and parallel play than traditional Lego (Farr et al., 2010a). Furthermore, a digitally enhanced Playmobil set, the Augmented Knights Castle, was found

to encourage more collaborative play and less solitary play in the same community of children (Farr et al., 2010b). Current work is investigating how tangible environments might foster more independent exploration in children with learning disabilities (e.g. Pontual Falcão and Price, 2012). This work currently seeks to inform educators about features of tangibles that may be useful for students with learning disabilities and to inform design of artefacts that are accessible across different learning communities. Findings to date suggest important design factors include immediate system feedback (as soon as action is performed); clear mappings between action and effect, both at a physical and a conceptual level; and the use of visual representations and spatial configurations, which are more effective than audio (Pontual Falcão and Price, 2012).

Collectively, this work is beginning to indicate the value of different designs for different kinds of learning processes, learning activities and learning outcomes, as well as for different learner communities, providing the grounding for continued research.

APPROACHES TO RESEARCH

Since the nature of the research field draws on various academic fields, such as computer science, education and design, a single research approach is not usually taken, but often combined motivations underpin the research. While this chapter is centrally concerned with perspectives from social science (psychology, education, design), it begins by emphasizing the primarily interdisciplinary nature of tangible interaction research. It then outlines the central theoretical and conceptual approaches that commonly underpin the research and discusses both technology-driven and non-technology-driven research approaches.

Interdisciplinary Approaches

Tangible interaction research is often driven from different disciplinary perspectives and theoretical bases: computer science, where developments in new techniques in computing and technologies are of central concern; design, where understanding design processes and practices is of central interest; psychology, where research commonly looks at aspects of interaction related to cognition (e.g. perception, action, reasoning, social interaction); and education, where interest lies in how new technologies can support different aspects of learning (process or outcome). The importance of the interaction between these disciplines, or a subset of them, has resulted in a large body of interdisciplinary research, which intersects with and sits under the umbrella of HCI, a community comprising experts from these disciplines that typically work together.

Interdisciplinarity is central to research on tangibles in general, but for specific communities like education, it also demands domain experts, such as teachers or educators. Since tangible technologies are not 'off-the-shelf' they require new design and development. This means that computer science plays an important role in informing the development of the technical application (both what is currently possible and researching new ways of developing devices that work in desired ways); psychologists and education theorists are important in informing design that supports effective learning strategies; designers are central to designing and developing digital media that is linked to the tangible artefacts; and domain experts are instrumental in focusing the tangible as a tool to be effective for the learning domain or in the educational community. Such interdisciplinary teams, therefore, aim to deliver research that would be implausible for a single discipline alone and work towards being culturally embedded, that is, taking a broader, more real-world perspective.

This interdisciplinary nature of the work can be both creative and challenging. Bringing together teams of researchers demands the integration of different research perspectives, requiring the establishment of common ground, particularly around shared understanding (e.g. terminology, perspectives) and fulfilment of research agendas or directions. However, collective perspectives offer a broader range of ideas, commonly pushing and extending the boundaries of research and development.

Theoretical Approaches

Theories of learning and cognition offer a compelling rationale for the value of tangible interaction for supporting learning (e.g. see also O'Malley and Stanton Fraser, 2004), being compatible with constructivist theoretical concepts including hands-on engagement (e.g. Tobin, 1990); experiential and discovery theories of learning (Bruner, 1973); construction of models (e.g. Papert, 1980; Resnick et al., 1996); collaborative activity, transformative communication (Cohen, 1994; Pea, 1994; Webb and Palincsar, 1996) and embodied forms of interaction (Antle, 2009). Increasingly work draws on theoretical ideas around embodiment – a much-debated term that broadly refers to relationships between the body and mind, how bodily interactive processes, such as perception and action, aid, enhance or constrain social and cultural development Several research projects draw on these theoretical notions, but focus on different aspects of learning activity – for example, narrative construction (Annany and Cassell, 2001), exploration and construction (e.g. Raffle et al., 2006), models of phenomena (Moher et al., 2005; Price et al, 2009), pattern-based interaction (Yonemoto et al., 2006), collaboration (e.g. Farr and Yuill, 2010); and metaphorical concepts (e.g. Antle et al., 2011).

Conceptual Approaches

Other approaches that shape research are offered through the development of frameworks, which may focus on descriptive taxonomies, research guidance or analytical perspectives (also see Mazalek and Hoven, 2009). Early frameworks provide descriptive taxonomies, which specify technical configurations of different systems, but say little about the relative strengths and weaknesses of different designs in terms of interaction (e.g. Ullmer and Ishii, 2001; Koleva et al., 2003; Fishkin, 2004). More recent frameworks focus on human interaction and the relationship between design and interaction experience. For example, Hornecker and Burr's (2006) framework encompasses analytical approaches to design, interaction and bodily movement, highlighting the need to design physical tools and their interrelations as well as digital representations.

Other frameworks provide the basis for informing design, for example, Antle (2007), drawing on literature from cognitive psychology, identifies five properties of tangible systems for designers to consider. These primarily concern physical–digital mappings: perceptual (the mapping between the perceptual (often appearance) properties of the physical and digital aspects of the system), behavioral (the mapping between the input behaviors and output effect of the physical and digital aspects of the system) and semantic mappings (the mapping between the information carried in the physical and digital

aspects of the system); but also specify designing 'space for action' (space for control through physical action) and 'space for friends' (the ways in which the system supports collaboration).

Frameworks specifying the importance of empirical research approaches for learning (e.g. Marshall, 2007; Price et al, 2008) have also been proposed. Marshall (2007) proposes an analytical framework with six perspectives intended to guide tangible interface empirical research and development. The perspectives draw on a research review and focus on properties or dimensions of tangible systems that relate to learning: learning activity; learning domains; learning benefits; integration of representations; concreteness and sensoridirectness; and effects of physicality. Price's (Price et al., 2008) framework specifies the artefact–action–representation relationships in tangible systems with a view to framing empirical research around the representational properties of tangible environments. The framework has four primary parameters: location, dynamics, correspondence and modality (detailed in the section 'Example research study' below).

Frameworks, such as these, provide a structure for designing and framing research or offer perspectives for analysing research and can be used in conjunction with other theoretical approaches to learning, cognition and interaction.

Technology and Non-technology-driven Approaches

Some research studies into tangibles for learning take a technology-driven approach, while others are driven initially from a non-technology perspective. Technology-driven – or 'technology- inspired' (Rogers et al., 2002) – approaches are claimed to be effective where interactive experiences with unknown (or novel) technologies are largely unexplored: 'a mix of serendipity and invention where creative experimentation is what drives the research' (Rogers et al., 2002: 373). In contrast, work that takes a 'non-technology' approach is equally informative. For example Manches and O'Malley (2012) investigated the effect of physical manipulation on children's reasoning to inform design and evaluation of novel forms of interaction like tangible interaction.

However, much research combines elements of these two approaches to consider the technological opportunities in conjunction with our understanding of learning and cognition and current educational practice. While these approaches might help to steer the research, the majority of work in this area has been exploratory, in the sense of not being systematic or context-dependent. While this is very fruitful in gaining some insight into tangible interaction, and particularly in identifying important areas for future research, the need for systematic and focused research remains.

EXAMPLE RESEARCH STUDY

This section provides an illustrative example of how 'tangible technology' research has been undertaken, identifying the particular features of the technology for the research questions and outlining the research approaches, methods and findings.

Motivation for Research: Representation Framework

Tangible technologies that enable physical, hands-on forms of interaction and flexible linking of digital representations to physical objects offer opportunities for engaging with invisible scientific phenomena in new ways. While this may be assumed to offer learning benefits, the specific advantages and limitations for learning need to be demonstrated. Since research in this area is complex, not least because of the number of variables to take into consideration (e.g. hands-on learning interaction; physical–digital combinations; representation design), the need for more structured research is apparent. To address this, a research framework was developed

that focuses on one of the unique properties of tangible environments – the facility to flexibly link artefacts with digital representation, promising greater representational power. The flexibility of such coupling brings an exponential number of parameters for linking together representation, object or environment and action. The proposed research framework therefore focuses on the relationship between different artefact–representation combinations and the role that they play in shaping cognition.

The framework (Price et al., 2008) has four primary parameters, which specify different dimensions for empirical research with respect to learning interactions.

1. *Location* refers to the different spatial locations of digital representations in relation to the object or action triggering the effect. For example in a 'discrete' design, input and output are located separately – that is, a manipulated object triggers a digital representation on an adjacent, but separate, screen (e.g. Chromarium used an adjacent digital display to show the effects of mixing colours on cubes embedded with RFID technology; Gabrielli et al., 2001); in a 'co-located' design, input and output are contiguous—that is, the digital effect is directly adjacent to the artefact (e.g. Urp, a model urban planning environment, displays shadows or wind patterns of architectural structures on a surrounding horizontal table surface ;Underkoffler and Ishii, 1999); an 'embedded' design comprises a digital effect within an object (e.g. FlowBlocks are sensor-embedded blocks that when connected together send light signals through the blocks to help children explore different causal structures ;Zuckerman et al., 2006).

2. *Dynamics* is concerned with the flow of information during interaction. For example, digital effects or feedback can be immediate or delayed or may be dependent on multiple objects or interactions to be triggered. The resultant causal relationships can be quite complex, requiring better understanding of the impact of such flow of information on cognition.

3. *Correspondence* refers to the metaphors involved in the nature of representations of artefacts and the actions placed upon them. 'Physical correspondence' refers to the degree to which the physical properties of the objects are closely mapped to the learning concepts, the emphasis

being on the degree of correspondence to the metaphor of the learning domain. 'Symbolic correspondence' defines objects that act as common signifiers – for example, blocks used to represent various entities, where the object may have few or no characteristics of the entity it represents. For example, a block could represent a book or abstract entities, like chromosomes or circuit components. 'Literal correspondence' defines objects the physical properties of which are closely mapped to the metaphor of the domain it is representing. For example, a rigid block representing chromosomes reveals none of the fragility or separation that is inherent in the process of genetic changes, whereas loosely magnetically connected 'strips' could convey underlying 'fragile' features of the learning concept.

4. *Representational correspondence* encompasses design considerations of the representations themselves and how this corresponds to the artefact and action within the context or subject domain of use. Meaning mappings between physical and digital representations can be designed with different levels of association (direct to ambiguous) between symbol and symbolized according to the concept being displayed or indeed the desired interaction/reflection. For example, research suggests that ambiguous mappings between sound and environment engender different levels of reflection about meaning in context than direct mappings (Randell et al., 2004).

5. *Modality* of representation impacts on different aspects of the whole interaction and can be considered in parallel to all other categories. Although the visual mode is often a predominant form of representation, the potential for audio and tactile modes in tangible computing requires a broader understanding of their role for learning.

While a framework approach offers the basis for structuring research, and the potential for examining different design parameters, there are a number of limitations. It requires a substantial amount of different studies to provide a comprehensive view of learning with tangibles together with specification of the design and development of the tangible environments that enable this level of detailed investigation. Also, issues of systematic, reductionist approaches to research versus in-the-wild studies are raised.

Context of the Research

This example is situated within a science-learning context, the learning experience being designed to fit within the UK science curriculum. Specifically, a tangible environment was designed to investigate the role of tangible technologies in supporting learning about the behaviour of light, particularly basic concepts of reflection, transmission, absorption and refraction of light, and derived concepts of colour. These phenomena are invisible, hard to show in a classroom context (beyond visual illustration) and exploit the physical properties of objects. Of particular interest here was, first, to understand how the physical properties of objects that are central to the scientific idea, and their linking to digital representations, might shape interpretation. For example, green reflects green, while red reflects red; rough objects reflect in a diffuse manner, while smooth ones do not. Second, we wanted to examine the differential effects of representation location on interaction and cognition; and third to explore the role of hands-on manipulation in shaping action and contributing to scientific understanding.

The Environment

A purpose-built tangible tabletop environment was developed. The system consisted of a table with a frosted glass surface, which was illuminated from underneath by infrared LEDs. This enabled an infrared camera under the table to track objects placed on the table surface, using reacTIVision software for object recognition (Kaltenbrunner and Bencina, 2007). In order for the camera to track the objects, each object was tagged with a paper marker called a 'fiducial'. Thus, each object could be individually identified, together with its location and orientation. When distinct objects were recognized by the system, different digital effects were projected on to the tabletop (see Sheridan et al., 2009 for more technical details) and (Figure 20.4).

The digital effects were designed to illustrate light behaviour. Thus, a torch acted as a

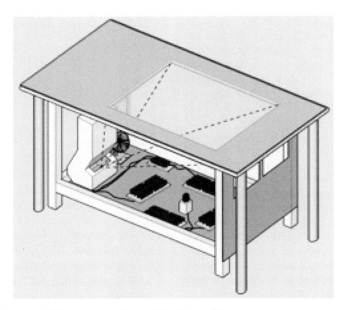

Figure 20.4   Schematic illustration of tangible lighttable.

light source (causing a digital white light beam to be displayed when placed on the surface) and objects that were placed in the beam reflected, refracted and/or absorbed the digital light beams, according to their physical properties (shape, material and colour). For example, pointing the torch at a green block caused a green beam to be reflected (Figure 20.5 left).

The torch, when placed on the surface, was 'always on', while the other objects only produced digital effects if they were placed in the pathway of the digital light beam. The digital effects changed when someone directly manipulated the objects – either by taking them off the table or altering their position on the table – which caused the light beam to be interrupted or redirected.

Scenario Design

Designing scenarios for purpose-built systems offers another research challenge. The physically based nature of the environment that uses real-world objects requires consideration in terms of design, particularly where levels of realism (i.e. objects) are combined

with schematic ideas (digital representation). Initially distinct phases for learning about each concept (reflection, refraction and absorption) were proposed. However, as the scientific concepts being explored are interrelated, sequencing removed important overall coherence of the phenomena. In addition, rather than leading children towards solving well-defined tasks, one aim of the application was to encourage free collective exploration and promote discovery learning. By experimenting with the different types of objects and the torch, children would have the opportunity to explore how light behaved with different combinations of objects and draw conclusions about the different phenomena involving light. The elements of the system were designed to encourage children's reasoning and thinking about light behaviour and the expected outcomes of the interactive sessions were dialogues between the children about the learning topic and collective knowledge building through conceptual conclusions drawn from their interaction with the interface.

The design and choice of the kind of digital representations to be used when learning

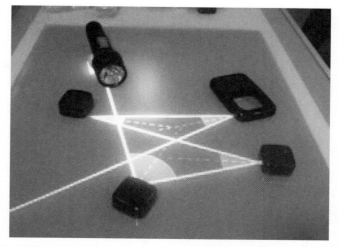

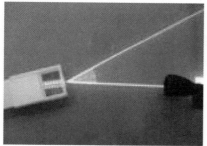 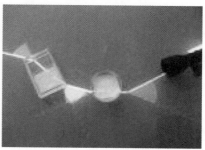

Figure 20.5 Reflection, refraction and absorption displayed on tangible lighttable.

about light in a tangible environment was also complex, with technical limitations having to be taken into account. Informal interviews with the teachers, the piloting of different designs with children and adults and input from domain experts and different academic disciplines were all instrumental in informing the design. Choices included showing absorbed colours inside or next to the object that shows the light beam as white or as the spectrum of colours and illustrating reflection through ripples, arrows or straight lines.

Study Design

A number of challenges around study design emerged. First, an appropriate task needs to be designed. Here an explorative task was chosen

in order to study interaction at a general level – to see what children intuitively did and intuitively inferred through their interaction. Other work might choose well-defined tasks to study particular learning concepts or particular forms of physical interaction.

Second, the location of studies needs to be considered – for example lab-based or 'in the wild'. In this work, lab-based studies were the only realistic option, since moving the table and situating it in a school or museum proved impractical. This creates subsequent challenges of bringing groups of students into the lab and has implications for analytical interpretation, particularly if this focuses on learning outcomes or teacher interaction. Since the focus of these studies was to examine aspects of representation design and related interaction and interpretation, a lab-based environment sufficed.

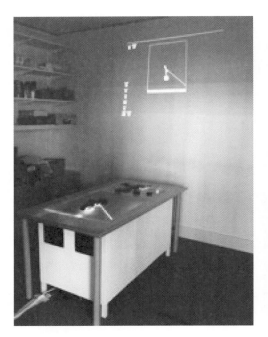

Figure 20.6   Discrete and co-located arrangements.

For the studies discussed here, 21 children from Year 7, aged 11–12 years (11 female and 10 male), and 22 children from Year 9 (10 female and 12 male) aged 13–14 years, from two schools in the UK took part. Children worked with the tangible table in groups of three and were selected by the teacher on the basis of being able to work well together. One study focused on comparing interaction and interpretation in a 'discrete' representation design with a 'co-located' design (Figure 20.6).

Each session lasted 35–45 minutes. Children were asked to freely explore the interface (by moving the objects on the tabletop) to find out about light behaviour. During the interaction, a researcher facilitator prompted the group with general questions like 'What's happening here?' and 'Why do you think this is happening?' to guide students through the exploration of the concepts towards making inferences and drawing conclusions. All sessions were video recorded. After engaging with the tangible system, children were interviewed in their groups to obtain information

on their understanding of key concepts of light behaviour, feedback on the system as a whole and their general experience.

Analytical Approaches

A thematic analysis approach was taken with all video data, the specific themes being related to aspects of the framework and to different studies undertaken. To develop coding schemes based on themes, group and paired analysis with researchers took place. One challenge here was selecting video focus – on the tabletop surface providing detailed views of manipulation and hands-on interaction or taking more global views of the 'whole' view of interaction. In this example, data were analysed from a tabletop focus of interaction, together with verbal interaction, which enabled examination of key aspects of the representation framework. In contrast, more recent work looking at 'embodied' forms of interaction took multiple video data views of interaction to access aspects of gaze and body posture as well as manipulation data.

Summary of Findings

Collectively the studies generated a number of key research findings, which feed into the research framework and offer insights into design, as well as indicating important future directions of research.

Representation Location (Discrete v. Co-located)

Findings indicate that interaction differences in the two location modes have some key implications for learning. First, they were found to have different attention demands. In the discrete mode learners tended to look at the screen, in similar ways to mouse-based interaction, using the objects as input devices, following the effect of their actions on a separate screen. Here the collaborative focus becomes the screen. On the other hand, when learners' attention was directed to the table surface, they could see each other's actions while looking at the table surface for the system's feedback. The opportunity this provides for learners to give opinions on others' actions and events changes the nature of the collaboration: explicit awareness of others' hand actions facilitated exploration and increased collaborative forms of construction and interpretation. When looking at a separate screen, users may more easily lose track of each other and tended to work by themselves.

Second, the co-located approach fostered more rapid dynamic interaction, which enabled access to increased exemplary instances of scientific phenomena and enhanced explorative activity. On the other hand, slower interaction in the discrete mode allowed more 'time' for thinking. This raises questions about the value and realization of different forms of reflection – reflection in action and reflection on action – for learning with co-located shared interfaces and highlights the need to specifically design learning activities that slow down interaction and promote opportunities for reflection to occur during 'calm' periods at various points in the learning task. Overall these findings build on previous work (Sensetable: Patten et al., 2001) showing that users

preferred information displayed on the sensing surface rather than on a separate screen, precluding the need to divide their attention between the input (sensing surface) and the output (separate screen display), by illustrating the interactive and cognitive effect of such different designs.

Physical Digital Mappings

The behavioural mappings in the environment were of a tight coupling design (Antle, 2007) and children had little difficulty in understanding the cause and effect relationships (i.e. physical action input and digital output). As well as the design of representations themselves, a key factor that underlies interpretation in tangible environments is the design of the physical–digital mappings. However, findings here suggest that children's interpretation of scientific phenomena resulted from an interaction between different design choices for physical objects and associated representations, preconceptions and previous real-world experience.

In terms of physical correspondence, issues were raised around mappings of real-world objects to virtual, artificial environments, in which the object behaves as itself. Although the torch was actually representing a torch, it could not be turned on or used in the 3D space in the same way as in the real world. Thus, the system constraints on objects or actions do not always map to familiar interaction in the real world. This highlights issues around the design of tangible interfaces and the potential impact on learning of mixed metaphors or requirements to shift from one metaphor to another The mapping between physical objects and their meaning and function within the environment was not always literally interpreted by children, who sometimes perceived objects to have a symbolic correspondence. The torch, being an object taken directly from the 'real world' with familiar affordances of interaction, was intuitively manipulated within a 3D space (lifting, switching on), rather than within the constraints of the 2D surface. However, such technical constraints were rapidly accommodated and the meaning (source of light) and purpose (shining light on objects) of the torch in the environment were unambiguous and comprehensible. On the other hand, the coloured blocks (although representing themselves) were perceived as being representative of something else, giving rise to a variety of interpretations. For example, the spectrum of absorbed colours shown inside the objects evoked the common experiment of decomposing white light through a prism and induced the perception that the block represented a prism. Furthermore, the notion of reflection, being mostly associated with concepts of optics, led to the interpretation of blocks as mirrors or lenses and never as regular opaque objects (Price and Pontual Falcão, 2009).

Interpretations of the digital effects were also affected by real-world experience and familiar representations. For instance, the representation of absorbed colours as a colour spectrum was immediately associated with a rainbow (Figure 20.5, centre). Although children were excited by the representation, the representation itself did not appear to facilitate their understanding of the phenomenon of absorption. In fact, children described it as light going through (the object) in the form of a rainbow, the word rainbow being often repeated, which was not the intention of the design. This raises issues about using representations that evoke a distinct familiar phenomenon, with other purposes, and again about the ability of children to transfer across domains (Price and Pontual Falcão, 2009).

Findings here suggest that while designers may have underlying rationales for choices of literal or symbolic correspondences (see 'Motivation for research: representation framework' earlier this chapter), learners do not necessarily infer the same correspondence metaphor. Using physical blocks or real blocks in conjunction with theoretical scientific models, which are represented symbolically, blurs the boundaries between what is real and what is symbolic. Studies here suggest that this may have an impact on at least

two things, but warrants further research. First, interpretation of phenomena in relation to objects was intuitively based on previous experience (e.g. seeing blocks as mirrors rather than as opaque blocks), hindering their tendency to attend to, for example, the physical properties of the blocks, and constraining any extension to their reasoning. Second, despite using real objects, their ability to generalize to other objects was limited.

Given that tangible systems do not just exploit the physicality of the real world but also aggregate digital models enabling access to phenomena 'invisible' in everyday interaction and manipulation of symbolic models, a key issue is how to effectively mesh together an accurate model of reality with artificial scenarios.

With physical environments the constraints of forms of representation may impact on the utility of certain illustrations of phenomena. Let's take the concept of absorption and ways of illustrating absorption of the different light waves in combination with reflection. On the one hand, with a physical object (red), a digital representation depicting a red beam being reflected off the object makes an effective combination for physical–digital representation of invisible phenomena. On the other hand, depicting absorption of the remaining light waves (thus making us 'see' red) is not so easy in the physical object itself. In the studies described here, such absorption was shown 'inside' the object – but as a 'fixed' representation rather than, for example, illustrating a dynamic process of the absorption taking place. This may have contributed to students' classification of this as a 'rainbow', thus distracting them from the key point of the representation. Now let's think about this in a purely digitally represented environment, where an object is illustrated on a screen, a light source is shone on the object and a red beam reflects off the object, while at the same time a dynamic depiction of the other light waves gradually being absorbed in the object is illustrated. The point here is that the constraints of the physical object used in the tangible system must be taken into account when considering the most effective form of illustrating invisible phenomena. It could be argued that this is a technical constraint and in the future technology will have advanced to a point that would enable the depiction of such absorption processes inside the physical objects themselves.

Collaboration

Overall the studies provided insight into the role of tangibles and interactive surfaces in collaborative interaction (Pontual Falcão and Price, 2009, 2010). First, the co-located design promoted a high level of awareness of others and of action–effect relationships and provided a common and unique focus of attention. Everyone's actions and the consequent digital effects were visible to all participants on the shared surface, which facilitated collective exploration and collaborative knowledge construction. In contrast, with the discrete mode, the physically separated input–output coupling made the action–effect relationships less clear and being aware of others' actions harder.

Second, the design invited parallel actions and the dynamics provoked rapid changes in configurations. This sparked unexpected events in the dynamic display, which in turn aroused curiosity, drew attention to relevant instances of the phenomena, engendered further exploratory and enquiry activity and promoted the need for verbal negotiation (about what was happening) and synchronization of actions to 'build' a particular configuration. More interestingly, the parallel actions and rapid dynamic changes resulted in many instances where one child's actions 'interfered' with another's current, or planned, configurations. High levels of (accidental or intentional) interference were highly successful in provoking curiosity, drawing attention to relevant instances of the phenomena, engendering exploratory and enquiry activity and promoting verbal negotiation and synchronization of actions. Overall this facilitated effective forms of collaborative interaction.

Shared Resources and Representations

Third, the shared resources within the tangible environment were fundamental in promoting interference and fostering particular ways of sharing. The potential for interference was dependent on the children actively sharing some kind of resource that allowed them some control or influence on the physical or digital resource – this could be objects/artefacts or digital representation. The design meant that although the digital effect from the torch was key to interaction, it did not preclude others from controlling the interaction. Several blocks enabled simultaneous possession of objects for manipulating the configuration and even shy children could gain access to the digital light beam using the objects they were manipulating and, in so doing, were forced to get involved with the group activity. Thus, this design was useful in encouraging all children to be actively included in the collaborative activity.

Awareness of others' actions enabled sharing of resources through gesture or 'physical asking' and they were shared through an implicit protocol of handing resources over. The physicality and availability of the devices contributed to balanced levels of participation. The digital representations were collective – that is, everyone's input fed into the same common digital representation, which contributed to collective knowledge building.

Along with this work, a growing body of evidence suggests the key role that tangible technologies may play in supporting collaborative interaction and exploratory forms of interaction (e.g. Ha et al., 2006; Hornecker et al., 2008; Do-Lenh et al., 2009; Fleck et al., 2009). In learning contexts this is of significant interest, with a general trend to promote both a collaborative nature of learning and student-led learning – or at the very least more student-centred learning.

FUTURE RESEARCH: CHALLENGES/ DIRECTIONS

In this section a number of key research challenges and important research directions are outlined.

Research Challenges

One significant challenge is accounting for rapid changes in technical development and the availability of off-the-shelf technologies. Developments in technology generate increasingly new ways of interaction. Designing studies that can continue to inform such developments is challenging, but fundamental to their sustained value. One advantage of adopting a framework approach based on key properties of the environments, such as representation, is that it offers insight into design implications across technologies.

Another technology-related challenge is choosing whether to develop purpose-built systems to test particular design features or to employ off-the-shelf technology. Such decisions are commonly driven by the research question and research aims. For example, designing for deployment in current educational contexts would likely involve readily available technology; whereas examining theoretical possibilities would likely involve developing purposely designed applications.

A second consideration is whether to carry out research with tangibles *in situ*, for example in a school classroom or museum or the informal contexts of home, or undertake lab-based studies.

A third issue is that of novelty: technology environments such as these are inherently novel to learners, impacting on interaction, and generally heightening engagement and enjoyment. Novelty factors demand longitudinal data, with larger sample sizes. Yet this is problematic when emergent technologies are not yet commonplace or embedded into classroom practice.

Last, but not least, approaches to evaluation, specifically in relation to learning, remains under-researched. Much evaluation focuses on 'usability' and ensuring that learners can easily master the interface (designing for intuitiveness) and on engagement in terms of levels of fun and enjoyment.

Research Directions

Two key research directions are worthy of note here. First, research with tangible learning environments has begun to move from more open-ended exploratory research to more in-depth research on specific forms of interaction. In particular, interest in the role of tangible technologies in fostering collaboration is growing and investigation into the role of embodied interaction in learning and its relationship with tangible learning environments is developing (e.g. Antle et al, 2011; Price and Jewitt, 2013). Current work is developing notions of what embodiment means for learning in digital environments (MODE, n.d.). Digital technologies provide new opportunities to explore and study how the body and embodiment contribute to communication and learning. The mainstreaming of tangible, mobile and sensor-based technologies places embodiment well beyond a question of physical–digital augmentation and opens up new research directions to gain insight into the role of 'embodiment' in technology-learning environments – for example, how the body mediates interaction and experiences and the relationships between context and situatedness and environment–interaction–cognition.

A further significant area of research is tackling the challenge of how to foster the embedding of technologies into classroom education. While research has established the learning opportunities tangibles may provide for students, for such technologies and their accompanying applications to be successfully integrated into educational contexts also requires a focus on teachers and teachers' use of technology. Previous work highlights a number of concerns, including that technology does not reflect pedagogic approaches (Major, 1995), there is a lack of training or familiarity with computers and the time involved in learning a new tool (Mueller et al., 2008). Brown and Green (2008) suggest the need to consider new ways for teachers to use technologies that support modification, creativity and tailoring to student age, ability and subject domain. Such a move requires tools that enable teachers and educators to design and customize their own learning activities with these tools with relative ease. In addition, changing teachers' beliefs about the value of their students learning with technology is a major catalyst for the adoption of new forms of teaching. A critical approach here is engaging them in the design and development of new technologies and approaches to teaching.

ACKNOWLEDGEMENTS

This work was supported by the EPSRC: EP/F018436. Thanks to George Roussos, Taciana Pontual Falcão and Jennifer Sheridan, for their invaluable contributions to the project, and students and teachers from Woodlands and Sweyne Park schools for their participation in the research studies.

REFERENCES

Annany, M. and Cassell, J. (2001) 'Telling tales: A new toy for encouraging written literacy through oral storytelling', *Presentation at Society for Research in Child Development*, Minneapolis, MN.

Antle, A. (2007) 'Tangibles: Five properties to consider for children', *Workshop on Tangibles, Conference on Human Factors in Computing Systems* (CHI '07), San Jose, CA.

Antle, A. (2009) 'Embodied child computer interaction – why embodiment matters', *Interactions*.

Antle, A., Corness, G. and Bevans, A. (2011) 'Springboard: Designing image schema based embodied interaction for an abstract domain', *Human Computer Interaction Series*, Springer, 7–18

Brown, A., & Green, T. (2008) 'Issues and trends in instructional technology: Making the most of mobility and ubiquity'. In M. Orey, V. J. McClendon, & R. Branch (Eds.), *Educational media and technology yearbook (Vol. 33)*. Westport, CT: Greenwood Publishing Group Libraries Unlimited.

Bruner, J. (1973) *Going Beyond the Information Given*. New York: Norton.

Cohen, E.G. (1994) 'Restructuring the classroom: Conditions for productive small groups', *Review of Educational Research*, 64 (1): 1–35.

Do-Lenh, S., Kaplan, F. and Dillenbourg, P. (2009) 'Paper-based concept map: The effects of tabletop on an expressive collaborative learning task', *British Proceedings of the 23rd HCI Group Annual Conference on People and Computers* (BCS–HCI '09), ACM, 149–158.. Cambridge, UK: ACM Press.

Farr, W. and Yuill, N. (2010) 'Tangible technology: Collaborative benefits for autism', *Autism: The International Journal of Theory and Practice*, March.

Farr, W., Yuill, N. and Raffle, H. (2010a) 'Social benefits of a tangible user interface for children with autistic spectrum conditions', *Autism*, 14(3): 237–252.

Farr, W., Yuill, N., Harris, E. and Hinske, S. (2010b) 'In my own words: Configuration of tangibles, object interaction and children with autism', *Proceedings of the 9th International Conference on Interaction Design and Children*. New York, 30–38.

Fishkin, K.P. (2004). A taxonomy for and analysis of tangible interfaces. *Personal and Ubiquitous Computing*, 8, (5), 347–358

Fitzmaurice, G., Ishii, H. and Buxton, W. (1995) 'Bricks: Laying the foundations for graspable user interfaces', *Proceedings of Conference on Human Factors in Computing Systems* CHI '95, ACM, 442–449.

Fleck, R., Rogers, Y., Yuill, N., Marshall, P., Carr, A., Rick, J. and Bonnett, V. (2009) 'Actions speak loudly with words: Unpacking collaboration around the table', *Interactive Tabletops and Surfaces*, ACM.

Gabrielli, S., Harris, E., Rogers, Y., Scaife, M. and Smith, H. (2001) 'How many ways can you mix colour?: Young children's explorations of mixed reality environments', *CIRCUS* 2001.

Ha, V., Inkpen, K.M., Whalen, T. and Mandryk, R.L. (2006) 'Direct intentions: The effects of input devices on collaboration around a tabletop display', *Proceedings of Tabletop 2006*, IEEE.

Hengeveld, B., Hummels, C., Overbeeke, K., Voort, R., van Balkom, H. and de Moor, J. (2009) 'Tangibles for toddlers learning language', *Proceedings of 3rd International Conference on Tangible and Embedded Interaction*, ACM, 161–168.

Horn, M., Solovey, E. and Jacob, R. (2008) 'Tangible programming and informal science learning: Making TUIs work for museums', *Proceedings of the 7th International Conference on Interaction Design and Children*, ACM, 194–201.

Hornecker E, and Buur, J. (2006) 'Getting a grip on tangible interaction: A framework on physical space and social interaction', *Proceedings of CHI 2006*, ACM, 437–446.

Hornecker, E., Marshall, P., Dalton, S. and Rogers, Y. (2008) 'Collaboration and interference: Awareness with mice or touch input', *Proceedings of the Conference on Computer Supported Cooperative Work*, ACM, 167–176.

Ishii, H. and Ullmer, B. (1997) 'Tangible bits: Towards seamless interfaces between people, bits and atoms', *Proceedings of the Conference on Human Factors in Computing (SIGCHI '97)*, ACM, 234–241.

Jorda, S (2003) 'Interactive music systems for everyone: Exploring visual feedback as a way for creating more intuitive, efficient and learnable instruments', *Proceedings of the Stockholm Music Acoustics Conference*, (SMAC 03), Stockholm, Sweden

Kaltenbrunner, M. and Bencina, R. (2007) 'reacTIVision: A computer-vision framework for table-based tangible interaction', *Proceedings of Tangible and Embodied Interaction*, ACM, 69-74.

Klopfer, E. and Squire, K. (2007) 'Case study analysis of augmented reality simulations on handheld computers', *Journal of the Learning Sciences*, 16(3): 371–413.

Koleva, B., Benford, S., Hui Ng, K. and Rodden, T. (2003) 'A framework for tangible user interfaces', *Physical Interaction Workshop on Real World User Interfaces. Mobile HCI Conference*, Udine, Italy.

Macarana, A., Antle, A. and Reicke, B. (2012) 'Bridging the gap: Attribute and spatial metaphors for tangible interface design', *Proceedings of Conference on Tangible Embedded and Embodied Interaction*. ACM, 161–168.

Major, N. (1995) 'Modelling teaching strategies', *Journal of AI in Education*, 6(2): 117–152.

Manches, A. and O'Malley, C. (2012) 'Tangibles for learning: A representational analysis of physical manipulation', *Personal and Ubiquitous Computing*, 16(4): 405–419.

Marshall, P. (2007) 'Do tangible interfaces enhance learning?', *Proceedings of the First International Conference on Tangible and Embedded Interaction*, ACM, 163–170.

Marshall, P., Price, S., and Rogers, Y. (2003) 'Conceptualising tangibles to support learning', *Proceedings of Interaction Design and Children*, ACM, 101–110

Mazalek, A. and Hoven, E. van den (2009) 'Framing tangible interaction frameworks', *Tangible Interaction for Design Special Issue of AIEDAM*, 23: 225–235.

MODE (n.d.) Multimodal methodologies for researching digital data. Available at: http://mode.ioe.ac.uk/research/researchproject2

Moher, T., Hussain, S., Halter, T. and Kilb, D. (2005) 'RoomQuake: Embedding dynamic phenomena within the physical space of an elementary school classroom', *Proceedings of the Conference on Human Factors in Computing Systems*, ACM, 1655–1668.

Mueller, J., Wood, E., Willoughby, T., Ross, C. and Specht, J. (2008) 'Identifying discriminating variables between teachers who fully integrate computers and teachers with limited integration', *Computers & Education*, 51: 1523–1537.

O'Malley, C. and Stanton Fraser, D. (2004) 'Literature review in learning with tangible technologies', NESTA *Nesta Futurelab*, 12.

Papert, S. (1980) *Mindstorms: Children, Computers and Powerful Ideas*. New York: Basic Books.

Patten, J., Ishii, H., Hines, J. and Pangaro, G. (2001) 'Sensetable: A wireless object tracking platform for tangible user interfaces', *Proceedings of the Conference on Human Factors in Computing Systems* (SIGCHI '01), ACM, 253–260.

Pea, R.D. (1994) 'Seeing what we build together: Distributed multimedia learning environments for transformative communications', *Journal of the Learning Sciences*, 3(3): 285–299

Pontual Falcão, T. and Price, S. (2009) 'What have you done!: The role of "interference" in tangible environments for supporting collaborative learning', *Proceedings of the 8th International Conference on Computer Supported Collaborative Learning*, Rhodes, Greece.

Pontual Falcão, T. and Price, S. (2010) 'Interfering and resolving: How tabletop interaction facilitates co-construction of argumentative knowledge', *International Journal of Computer-Supported Collaborative Learning*. New York: Springer, 23–29.

Pontual Falcão, T. and Price, S. (2012) 'Independent exploration with tangibles for students with intellectual disabilities', *Proceedings of the 11th International Conference on Interaction Design and Children*. Bremen, Germany.

Price, S. (2008) 'A representation approach to conceptualising tangible learning environments', *Proceedings of the Second International Conference on Tangible and Embedded Interaction*, Bonn, Germany.

Price, S. and Jewitt, C. (2013) 'A multimodal approach to examining "embodiment" in tangible learning environments', *Proceedings of 7th International Conference on Tangible Embedded and Embodied Interaction*. Barcelona, Spain, February 2013.

Price, S. and Pontual Falcão, T. (2009) 'Designing for physical–digital correspondence in tangible learning environments', *Proceedings of the 8th International*

Conference on Interaction Design and Children, Como, Italy

Price, S. and Rogers, Y. (2003) 'Let's get physical: The learning benefits of interacting in digitally augmented physical spaces', J. Underwood and J. Gardner (eds), *Computers & Education. Special Issue: 21st Century Learning*, 43: 137–151

Price, S., Pontual Falcão, T., Sheridan, J.G. and Roussos, G. (2009) 'The effect of representation location on interaction in a tangible learning environment', *Proceedings of the Third International Conference on Tangible and Embedded Interaction*, Cambridge, UK, 85–92.

Price, S., Rogers, Y., Scaife, M., Stanton, D. and Neale, H. (2003) 'Using 'tangibles' to promote novel forms of playful learning', *Interacting with Computers*, 15/2, May 2003, 169–185.

Price, S., Sheridan, J. and Pontual Falcão, T. (2010) 'Action and representation in tangible systems: Implications for design of learning interactions', *Proceedings of the Fourth International Conference on Tangible, Embedded and Embodied Interaction*, Cambridge, MA.

Price, S., Sheridan, J.G., Pontual Falcão, T. and Roussos, G. (2008) 'Towards a framework for investigating tangible environments for learning', *International Journal of Arts and Technology* 1(3/4): 351–368.

Raffle, H. (2004) 'Topobo: A 3-D Constructive Assembly System with Kinetic Memory'. Master's thesis, School of Architecture and Planning, MIT, Cambridge MA.

Raffle, H., Parkes, A., Ishii, H. and Lifton, J. (2006) 'Beyond record and play: Backpacks: Tangible modulators for kinetic behavior', *Proceedings of the Conference on Human Factors in Computing Systems* (SIGCHI '06), ACM, 681–690.

Randell, C., Price, S., Rogers, Y., Harris. E. and Fitzpatrick, G. (2004) 'The ambient horn: Designing a novel audio-based learning experience', *Personal and Ubiquitous Computing*, 8(3): 144–161.

Resnick, M., Martin, F., Sargent, R. and Silverman, B. (1996) 'Programmable bricks: Toys to think with', *IBM Systems Journal* 35: 3–4.

Resnick, M., Martin, F., Berg, R., Boovoy, R., Colella, V., Kramer, K., et al. (1998) 'Digital manipulatives: New toys to think with', *Proceedings of the Conference on Human Factors in Computing Systems* (SIGCHI '98), ACM, 281–287.

Rogers, Y., Scaife, M., Harris, E., Phelps, T., Price, S., Smith, H., Muller, H., Randall, C., Moss, A., Taylor, I., Stanton, D., O'Malley, C., Corke, G. and Gabrielli, S.

(2002) 'Things aren't what they seem to be: Innovation through technology inspiration', *Designing Interactive Systems Conference*, ACM, 373–379.

Schweikardt, E. and Gross, M. (2008) 'The robot is the program: Interacting with roBlocks', *Proceedings of the Second International Conference on Tangible and Embedded Interaction*, (TEI '08).

Shaer, O. and Hornecker, E. (2010) 'Tangible user interfaces: Past, present and future directions', *Foundations and Trends in Human–Computer Interaction*, 3(1–2): 1–138.

Sheridan, J.G., Tompkin, J., Maciel, A. and Roussos, G. (2009) 'DIY design process for interactive surfaces', *Proceedings of the 23rd Conference on Human–Computer Interaction, (HCI '09).*

Tobin, K. (1990) 'Social constructivist perspectives on the reform of science education', *The Australian Science Teachers Journal*, 36(4): 29–35.

Ullmer, B. and Ishii, H. (2001) 'Emerging frameworks for tangible user interfaces', in John M. Carroll (ed.), *Human–Computer Interaction in the New Millenium*. Bosto, MA: Addison-Wesley, 579–601.

Underkoffler, J. and Ishii, H. (1999) 'Urp: A luminous-tangible workbench for urban planning and design', *Proceedings of the Conference on Human Factors in Computing Systems* (SIGCHI '99). ACM, 386–393.

Wall, C. and Wang, X. (2009) 'InterANTARCTICA: Tangible user interface for museum-based interaction', *The International Journal of Virtual Reality*, 8(3): 19–24.

Webb, N. and Palincsar, A.S. (1996) 'Group processes in the classroom', in R. Calfee and C. Berliner (eds), *Handbook of Educational Psychology*. New York: Prentice Hall, 841–873.

Wyeth, P. and Purchase, H. C. (2002) 'Tangible programming elements for young children', *Proceedings of the 2nd International Conference on Human Factors in Computing Systems* (SIGCHI' 02), ACM, 774–775.

Yonemoto, S., Yotsumoto, T. and Taniguchi, R. (2006) 'A tangible interface for hands-on learning', *Tenth International Conference on Information Visualisation*, IEEE, iv, 535–538.

Zuckerman, O. Grotzer, T. and Leahy, K. (2006) 'Flow Blocks as a conceptual bridge between understanding the structure and behavior of a complex causal system', *Proceedings of the Seventh International Conference of the Learning Sciences*, 880–886.

21

Material Computing: Integrating Technology into the Material World

Leah Buechley

INTRODUCTION

A survey of my apartment yields a rich and varied collection of electronic devices. These range from the simple (non-computational) power strips and coffee grinders to complex devices like my laptop and smartphone. In between these two extremes are things like my printer, wireless router, refrigerator, blender, washing machine and microwave. All of these devices are quite different from one another – each is designed for a different unique task; they vary widely in size; and each has a distinct set of inputs (buttons, knobs and switches) and outputs (screens, lights, motors and sounds).

Yet for all of their diversity, the electronics in my apartment have a few striking things in common. They're all made almost entirely from plastic and metal – wood, ceramic and fabric, for example, are nowhere to be found. They're all hard and smooth. They are all either black, grey or white in color. They are almost all rectangular in shape – from my slim laptop to my hulking refrigerator. My primary means of interacting with almost all of them involves either pressing buttons, flipping switches or turning knobs.

Researchers have long argued we should build interfaces that involve more than buttons, keyboards and knobs – interfaces which engage the entire body and leverage our intuitive understanding of the physical world (Ishii and Ullmer, 1997; Weiser, 1999). Many elegant prototypes have explored these ideas; however, designers have been limited by the electronic materials they had access to – namely circuit boards, wires and standard electronic components. A growing body of research in materials science and materials engineering has dramatically expanded the materials available to technology designers.

This chapter surveys those developments, exploring the new material palette that is available to technology designers, along with projects that are using it to challenge the look, feel and functionality of technology. I examine a range of materials that can be employed in electronics design, including fabric, paint, paper and wood, and discuss the practical and cultural implications of incorporating these materials into new technologies.

More specifically, I discuss how diverse materials can be used as: (1) *connectors* (or wires) that connect one component to another; (2) *inputs* (or sensors) that capture information from people or environments; and (3) *outputs* (or actuators) that display information or act on the world. The heart of this chapter is organized into three sections that explore each of these areas.

MATERIAL COMPUTING

Before beginning a survey of connectors, inputs and outputs, it is worth taking a moment to more closely examine the concept of material computing. There are different ways that we might increase the diversity of materials used in interaction design. We could, for instance, use different materials to build cases for electronics.

This chapter is primarily concerned with electronics that are *constructed from* unusual materials. I will examine materials that can constitute the core components of electronics themselves. Along the way, I will survey materials from a range of different categories, including plastics, metals, textiles and paper.

An exhaustive survey of the topic is impossible in the allotted space, so the intent of this chapter is to provide researchers with pointers to relevant developments and examples of imaginative or provocative projects. I will focus on material-based connectors, sensors and actuators that can be constructed and employed by interaction designers, computer scientists and engineers – spending less time examining materials or devices that can only be constructed in materials science labs – and, because I have a background in electronic textiles, wearable computing and educational technology, I will emphasize these application areas.

MATERIAL CONNECTORS (CONDUCTORS)

Silver, copper and gold are the three most conductive metals. Non-metal materials can also be highly conductive, among them carbon and intrinsically conductive polymers (Inzelt, 2010). In the past 75 years, these materials have been combined with or applied to other materials to create conductive papers (Hu et al., 2009), textiles (Shim et al., 2011), ink (Russo et al., 2011), elastomers (Ruschau et al., 1992), clay (Johnson and Thomas, 2010) and glass (Fraser, 1973). Many of these materials are now commercially available and interaction designers are employing them in an increasing number of projects. Figure 21.1 shows three simple circuits constructed from different materials. One is sewn together with conductive thread, one is painted on to paper with conductive ink and one is sculpted out of conductive modelling clay.

Ongoing materials research is continuously adding to the library of conductors. Carbon nanotubes (CNTs), for example, are being used to, among other things, create conductive inks, papers and fabrics (Baughman et al., 2002; Paradise and Goswami, 2007; Shim et al., 2011). CNTs are extremely strong, highly conductive and elastic (Paradise and Goswami, 2007). There are, however, considerable health risks associated with exposure to these materials (Lam et al., 2006) and they are not yet widely used in consumer applications.

While the newest materials are often confined to the laboratory, researchers are also refining technologies to make materials more accessible and user-friendly. For example Russo et al. recently developed a pen that functions like a traditional 'gel' pen, but writes in conductive silver traces (Russo et al., 2011). The paper-based circuit shown in Figure 21.1 is an example of the kind of drawing/circuit that can be quickly created with this pen.

In some cases, a new class of conductive material has given rise to a new discipline – for example, the domain of e-textiles (electronic

 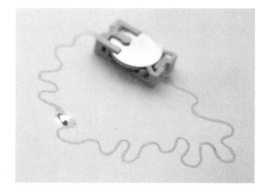

Figure 21.1   Circuits on different substrates. Top 2 left, Top right, a circuit on fabric constructed from conductive thread (© Leah Buechley), a circuit on paper painted with conductive ink (© Leah Buechley), above a circuit made with conductive modelling clay (© AnnMarie Thomas).

textiles) has been enabled by conductive threads and fabrics. E-textile practitioners use conductive threads and fabrics to seamlessly integrate traditional electronic components into textiles (Post et al., 2000; Berzowska, 2005a). Essentially, they replace wires and circuit boards with textile materials.

Figure 21.2 shows examples of designs in this tradition. In the gown on the left, by Diffus, conductive threads connect chemical sensors, microcontrollers and LEDs. The dress senses pollution levels and then visualizes them in animated patterns of light. The dress on the right, 'Masai Dress' by Studio 5050, functions as a whimsical musical instrument. Metal beads brush up against stitching in conductive thread on the front of the dress, triggering melodic bells. This seemingly simple replacement of wires with thread

and fabric has not only led to new designs but also sparked new communities of practitioners (Buechley and Hill, 2010).

Replacing wires with conductive paints also leads to new approaches. Figure 21.3 shows a project from my research group called LivingWall that explores some of these possibilities (Buechley et al., 2010). We built a series of interactive wallpapers from paper, conductive paint, traditional paint and electronics. These papers – essentially very large printed circuit boards – can monitor their environments, serve as ambient information displays and light-emitting surfaces and communicate with other networked devices. Yet – flexible, roll-able, and decorative – they also function like traditional wallpaper.

We have also been using conductive paints to introduce novices to electronics (Buechley,

Figure 21.2   Electronic textiles. Left: a pollution-sensing gown by Diffus (© Diffus Design). Right: 'Masai Dress' by Studio 5050 (© Despina Popodoupulos).

Figure 21.3   LivingWall. Left: a completed installation. Right: painting the wallpaper. (© Leah Buechley).

et al., 2009; Buechley and Perner-Wilson, 2012). Paints have unique affordances that make them particularly suitable for this purpose. People have an intuitive understanding of how to work with paper and paint and can leverage these understandings to make sense of electricity. Since electricity flows through drawn lines, the act of drawing provides people with a physical understanding of the path taken by electrical current. Moreover, people receive immediate and complete visual feedback about their constructions. There are no hidden wires or connections as there are in breadboards. Figure 21.4 shows images from some of our paintable electronics workshops.

Figure 21.4   Projects from paintable electronics workshops. From left to right: a simple circuit; a painted handmade bowl that glows when it is picked up. Paint functions as a capacitive sensor and a conductive connector in this third project (© Leah Buechley).

Figure 21.5   VooDooIO. Left: the deformable backing surface. Right: the push-pin components (© Nicholas Villar).

AnnMarie Thomas and her team have employed conductive modelling clay in a similar fashion, using it as an accessible, appealing and intuitive material for introducing electronics (Johnson and Thomas, 2010). One of their constructions is shown above in Figure 21.1. Their recipe relies on the salts and water in the dough to conduct electricity. Instructions for making conductive and insulating dough, as well as a portfolio of example projects, can be found in Thomas (2011).

In a different vein, researchers have used conductive elastomers (rubber materials) and fabrics to create bulletin board-like substrates capable of hosting electronic sensors and actuators (Lifton et al., 2005; Villar et al., 2006; Villar and Gellersen, 2007). For example, Villar et al. developed a system they

call VooDooIO – shown in Figure 21.5 – that consists of a set of input devices like switches, knobs and joysticks that can be pressed into a silicone surface rather like thumbtacks. The flexible backing provides the components with power and a communication bus and can be bent around or applied to other surfaces. The collection of inputs can then be used by users, who are able to build their own custom user interfaces. Systems explored by the researchers included personal gaming consoles, rapidly reconfigurable music mixing stations and large-scale geographical explorations (Villar and Gellersen, 2007). In this last example, map information was projected on to a large VooDooIO surface.

MATERIAL INPUTS

The previous section described how different materials can be used to electrically connect standard electronic components. In these examples, the carrying substrate can be made from paper, fabric or modelling clay, but the rest of the system is made from traditional electronic devices. Materials, however, can do much more than conduct electricity. Many materials have electrical properties – capacitance, magnetism, inductance and resistance – that change in response to stimuli like light, temperature or pressure. Others, like piezoelectric crystals, produce electricity in response to stimuli (Fraden, 2010). We can use these dynamic materials to build inputs on a wide range of substrates.

To begin with simple examples, we can examine the conductive materials I surveyed in the previous section, which can all be used to build switches. Surprisingly sophisticated devices can be constructed from simple switches (digital inputs). Figure 21.6 shows examples of switches my student Hannah Perner-Wilson constructed from thread, fabric, wood and paint that function as tilt sensors, stroke sensors and knobs (Perner-Wilson et al., 2011).

Capacitive and electromagnetic field sensors can also be constructed from any highly conductive material. To construct a simple capacitive touch/proximity sensor, one can measure the amount of electrical charge in a conductive material over time and use this information to determine if another conductive object (like a human body) is nearby (Fraden, 2010). Figure 21.6 also shows examples of capacitive touch sensors constructed

from textiles and paints (Orth, 2007; Buechley et al., 2010).

These examples illustrate the way that material-based electronics can be used as decorative and even sensual design elements. The lamp in the lower left-hand corner, designed and built by Maggie Orth (2007), for example, is controlled by a fuzzy hand-tufted capacitive sensor that is also part of the shade's visual design.

Electrodes, for sensing heart rate and other physiological signals, can similarly be constructed from different conductive materials. A number of researchers have used conductive threads and fabrics to embed electrodes into garments (Pacelli et al., 2006a). The company Numetrex (acquired by Adidas in 2008) developed a line of commercial sportswear that used fabric-based electrodes for monitoring heart rate (Numetrex, 2007). Figure 21.7 shows an image of one of these products.

Expanding our material range beyond conductors broadens the possibilities. Resistive

Figure 21.6   Top row from left to right: switches: a wooden knob, a stroke sensor and a tilt sensor (© Hannah Perner-Wilson). Bottom row: capacitive touch sensors on fabric (© Maggie Orth) and paper (© Leah Buechley).

Figure 21.7   Left: a garment with textile-based heart rate monitoring electrodes (© Adidas). Right: a garment with screen-printed bend sensors (© Rita Paradiso).

sensors can be built from materials and composites, the resistance of which changes in response to temperature, mechanical deformation, chemical exposure, light exposure or some other stimulus (Fraden, 2010). For example, a number of materials have resistive properties that change in response to compression. These include elastomers and foams constructed from rubber or plastic compounds and conductive powders (Ruschau et al., 1992; Knite et al., 2004), yarns spun from metal and textile fibers, and plastic films like Velostat (Fraden, 2010; Perner-Wilson et al., 2011). Figures 21.7 and 21.8 show examples of sensors made from these materials: bend sensors made by screenprinting resistive elastomers on to fabric (developed by Paradiso, de Rossi, and Pacelli et al. Paradiso and de Rossi, 2006; Pacelli et al., 2006b) and, from my research lab, a bend sensor made from Velostat and a stretch sensor knit from a yarn, the resistance of which changes when it is compressed (Perner-Wilson et al., 2011; also see Paradiso et al., 2005 and Yoshikai et al., 2009 for a discussion of knit stretch sensors).

To explore a range of paper-based sensors, my student Jie Qi and I built a pop-up book from conductive ink and a range of thin, flexible materials including Velostat and resistive elastomers (Qi and Buechley, 2010). We used the project to explore how mechanical (pop-up) movement could be coupled to and augmented by electrical actuation. We were also interested in exploring the aesthetics of paper-based electronics and the book serves as another example of how electronics constructed from unorthodox materials can expand the aesthetic as well as the technical landscape of interaction design. Figure 21.9 shows examples of pages that employ handmade bend and pressure sensors.

Rosenberg and Perlin explored more sophisticated ultra thin sensors in their UnMousePad project, developing multitouch surfaces from arrays of resistive pressure sensors (Rosenberg and Perlin, 2009). These sensors can be used to produce detailed pressure maps of the compression forces applied to a surface. Their contribution included developing the sensors and developing methods for rapidly interpolating data acquired from them. Similar capacitive-based sensors that turn surfaces into high-resolution pressure and proximity sensors are described in Rekimoto (2002) and Sergio et al. (2002).

A class of chemical sensors, called chemiresistors, also fall into the resistive

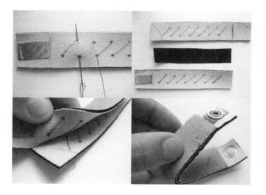

Figure 21.8   Left: a Velostat-based bend sensor. Right: a knit stretch sensor (Hannah © Perner-Wilson).

sensor category. These sensors are constructed by mixing a polymer that swells in the presence of a particular chemical (as it absorbs it) with a conductive powder (Dong et al., 2004). As the material changes shape, the overlaps between embedded conductive particles change, thus changing the material's resistance. These sensors have been employed in a number of 'electronic noses' – devices that are able to detect the presence of different air-born chemicals (Ryan et al., 2004).

Materials that can be used to construct different kinds of inputs on unusual substrates include piezoelectric and fiber optic materials. Piezoelectric crystals produce electrical energy when they are compressed (Steinem and Janshoff, 2007; Fraden, 2010). They can be formed into films, fibers, plates and tubes (Tressler et al., 1998) and are widely deployed in a range of industries. For example, they are used as microphones in guitar pick-ups and as actuating mechanisms in inkjet printers. From the perspective of an interaction designer, off-the-shelf discs and films can be used to build paper-thin vibration, pressure and bend sensors. Piezo sensors can be attached to tabletops to detect when and where objects are hitting their surfaces (Ishii et al., 1999); they can be woven into textiles and used as bend sensors to detect body movements (Edmison et al., 2002) and they can be embedded in a wristwatch to detect the delicate variations in sound of different hand gestures (Deyle et al., 2007).

Optical fibers have been used to build temperature, acceleration and acoustic sensors (Giallorenzi et al., 1982). They can also be used to build simple bend sensors that, unlike most sensors based on the resistive properties of a material, provide consistent readings across a range of temperatures. Engineers and designers have used them to monitor mechanical stress in buildings (Kuang et al., 2002) and bending in fabrics. For instance, Dunne et al. used a custom-made optical bend sensor in a posture-sensing garment (Dunne et al., 2007).

It is also worth mentioning imaging-based inputs – where non-electronic materials are used as interfaces and information is collected via cameras. One noteworthy example in this category involves the use of infrared (IR) ink – which is invisible to the human eye, but detectable with cameras and/or specialized lighting equipment. For example, Rosner and Ryokai used yarns patterned with IR inks to embed hidden messages in hand-knit garments and accessories (Rosner and Ryokai, 2008). In another provocative image-based application, Ishii et al. used a bed of sand to sculpt surfaces that were then captured by video cameras. Infrared light from LEDs mounted underneath the bed was filtered through the sand and picked up by the cameras. This data was then used to construct digital models of the original surfaces (Ishii et al., 2004). This example highlights again

Figure 21.9   Pages from an electronically augmented pop-up book showcasing pressure (left) and bend sensors. Video can be found at: www.youtube.com/watch?v=Al-6wMlaVTc (© Leah Buechley).

how material-based sensors can enable new sensual experiences of technology.

Before I leave the topic of inputs, it is worth mentioning recent research by Kuznetsov et al. on 'natural sensors' (Kuznetsov et al., 2011). Through interviews with zookeepers, biologists and gardeners they investigated how people from different fields track patterns in the natural world. Methods they uncovered included counting the number of insects on a leaf and feeling, by hand, the viscosity of water in a fish tank. This study reminds us that there are numerous ways to extract information from our environment and there are a number of material realms where electronic sensing has challenges to overcome before it can become widely used.

MATERIAL OUTPUTS

In the same way that many materials undergo electrical changes – their resistance changes or they generate an electrical current for example – when they are exposed to stimuli like light, heat or compression, many materials respond to electrical stimulation by changing in some way – by glowing, heating up or moving, for instance. For years researchers (and science fiction authors) have envisioned and investigated 'programmable matter' (Goldstein et al., 2005; Aksak et al., 2005;

Crichton, 2008) – materials that could take on any shape at our command. While this vision is not yet fully realized, substances that change shape (Lagoudas, 2010), color (Siegel et al., 2009), viscosity (Brown et al., 2010) and luminosity (Sugimoto et al., 2004) in electrically controllable ways are bringing us dramatically closer to the programmable matter goal.

To begin with a well-known category of outputs, let me first look at lighting and displays. Traditional light-emitting diodes (LEDs) can be attached to and embedded in a range of different substrates and, though they are hard electronic components, several designers have employed them in remarkable material projects. Figure 21.10 shows two examples where LEDs have turned familiar objects – a dress and a building – into programmable displays.

Organic LEDs (OLEDs), meanwhile, can be used to construct truly flexible displays (Sugimoto et al., 2004). Several companies, including Samsung, LG, and Sony have developed prototypes of flexible OLED and AMOLED (active-matrix OLED) screens, and commercial releases of devices with flexible screens are expected soon. Figure 21.10 shows a Samsung prototype that was demonstrated at the 2011 consumer electronics show. E-Ink is an alternative low-energy flexible display technology (Chen et al., 2003).

Figure 21.10   From left to right: a ballgown by CuteCircuit (© CuteCircuit), the Ars Electronica building in Linz designed by Treusch architecture (© Leah Buechley) and a Samsung prototype of a flexible display screen (© Samsung).

Another noteworthy display – one that is subtler, slower and more ambient than that of traditional screens – is based on thermochromic materials that change color in response to temperature changes. Thermochromic paints, inks and dyes have been used to make non-emissive displays on paper (Siegel et al., 2009) and fabric (Berzowska, 2005b; Orth, 2007) along with a host of other backings. Figure 21.11 shows two examples of dynamic textiles designed by Wakita and Shibutani that employ liquid crystal pigments (Wakita and Shibutani, 2006; Midori, 2011).

Moving into non-visual realms, piezoelectric materials can be used as actuators as well as sensors. They can be employed as ultra thin speakers and mechanical actuators (Kawakita et al., 1997; APC International, Ltd, 2002; Burch and Pawluk, 2009). In actuator mode, the crystals change shape when stimulated by an electrical current. Electromagnetic speakers, which consist of a conductive coil and a magnet, can also be constructed on a wide range of materials (Buechley and Perner-Wilson, 2012). In a playful project, Leclerc and Berzowska created a tuxedo jacket called Accouphène with an array of embroidered textile coils that would emit sound when a wearer brushed a magnet across his chest (Seymour, 2008). This jacket was used to dramatic effect in fashion shows and performances.

Berzowska and her team also designed a striking collection of shape-changing garments called Skorpions from the shape memory alloy Nitinol, one of which is shown in Figure 21.12 (Berzowska and Mainstone, 2008). Nitinol, like thermochromic ink, is a material that is actuated by temperature. In a typical deployment, the metal takes on one

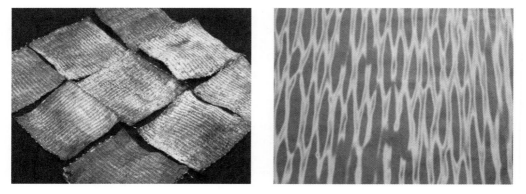

Figure 21.11   Liquid crystal textiles by Wakita and Shibutani (© Akira Wakita).

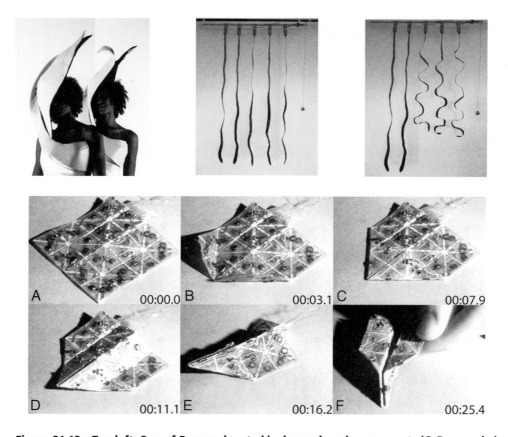

Figure 21.12   Top left: One of Berzowska et al.'s shape-changing garments (© Berzowska). Top right: One of Qi et al.'s animated paper sculptures (© Leah Buechley). Bottom: a programmable dynamic surface by Hawkes et al. (© Robert Wood).

shape at room temperature. When it is heated to an activation temperature, it transitions into another predetermined shape. A designer

sets the predetermined shape that the metal 'remembers' by heating it up to a high temperature, deforming it into the desired shape and then letting it cool at a specified speed (Lagoudas, 2010). Sheets or strands of Nitinol can be used to create structures that soundlessly morph from one shape to another with an eerily life-like motion. The other advantage of shape memory materials is their strength relative to their weight and size. The materials can be hidden in delicate structures that would be impossible to actuate with traditional motors and solenoids.

One example of structures along these lines is the collection of dynamic paper sculptures that Jie Qi is designing in my lab, one of which is shown in Figure 21.12 (Qi and Buechley, 2012). In this piece, called Animated Vines, long thin strips of paper respond to a viewer's proximity by curling, twisting and creaking into new shapes. Another lovely example is the self-assembling surface developed by Hawkes et al., time lapse photos of which are also shown in Figure 21.12 (Hawkes et al., 2010). Here, the same programmable origami sheet can be programmed to take on other forms, including a 'paper' boat and the airplane shown in Figure 21.12.

Another variety of shape-changing/viscosity-changing material is based on the jamming of granular material (Brown et al., 2010). In these systems, a flexible container of a granular material, like sand, remains malleable and soft until all of the air is removed from the container, at which point it stiffens into a rigid solid. Devices based on jammed materials make excellent grippers – one is shown in Figure 21.13 – and can also be used to build dynamic, transformable structures, among them reconfigurable furniture and molds (Oxman and Keating, n.d.).

Electroactive polymers (EAPs) are shape-shifting materials with tremendous promise (Chidsey and Murray, 1986; Shahinpoor et al., 1998; Bar-Cohen, 2004). These soft, malleable substances change shape in response to electrical stimulus; they can be engineered to take on a variety of forms and patterns of movement and are termed 'artificial muscles' for their similarity in elasticity and strength to muscle tissue. Recent projects constructed

from EAPs include a robotic jellyfish robot by Yeom and Oh (2009) and a walking robot developed by researchers at SRI International (n.d.).

CHALLENGES AND OPPORTUNITIES

New materials are compelling for countless reasons: they enable the design of new kinds of technologies, they provide new ways of working with electronics and computation and they can help move technology out of plastic boxes and into the rest of the world. However, employing unorthodox materials in interaction design research can be frustrating.

New materials that are described excitedly by the popular press and documented in detail in materials science journals are often difficult, even impossible, to obtain. Most have been developed in the laboratory, but are not commercially available. Others are produced for niche markets by obscure companies that can be difficult to find. Once found, companies are often reluctant to sell materials in small quantities or reluctant to sell materials to researchers, concerned that academics will attempt to reverse engineer

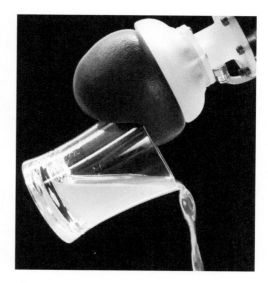

Figure 21.13   A robotic gripping arm based on jamming picks up a glass of juice (© Eric Brown).

their products. Newly developed materials are also usually quite expensive.

Once appropriate materials are sourced and obtained, they are often challenging to work with. Use of a material can require specialized equipment not commonly found in computer science labs or design studios – a chemical hood, a loom, a microscope or vacuum chamber, for example. Moreover, creative exploitation of a material can require deep familiarity with its underlying chemical or physical properties – knowledge that can be time-consuming and difficult to acquire.

Once a researcher has developed feasible ways to work with a particular material she now faces all of the challenges confronted by a more traditional researcher. She needs to design and build a novel and compelling application and prove its viability and originality through user studies and technical experiments. The diversity of expertise that is required and the amount of time it takes to fully realize a project can present significant obstacles to research in material computing.

Yet, in large part because of this unique set of challenges, the domain provides especially fertile ground for research. The practical, expressive and educational potential of a plethora of new materials has yet to be thoroughly explored. There are countless opportunities to build important bridges between materials science, electrical engineering, computer science and design.

As we move further into the twenty-first century, we will have an ever-increasing power to understand, control and animate the material world that surrounds us. This chapter has provided a whirlwind survey of some of the dynamic materials that will define this future. I hope it has provided a glimpse into the rich aesthetic, functional and intellectual landscapes that await us.

REFERENCES

Aksak, B. et al. 2005. *Claytronics*. New york: ACM Press, 299. Available at: http://dl.acm.org.libproxy.mit.edu/citation.cfm?id=1098918.1098964andcoll=DLanddl=ACMandCFID=55897023andCFTOKEN=14339803 (accessed October 11, 2011).

APC International, Ltd 2002. *Piezoelectric Ceramics: Principles and Applications*. Mackeyville, PA: APC International, Ltd.

Bar-Cohen, Y. 2004. *Electroactive Polymer (EAP) Actuators as Artificial Muscles: Reality, Potential, and Challenges, (2nd edn)*. Belling ham, WA: SPIE Publications.

Baughman, R.H., Zakhidov, A.A. and de Heer, W.A. 2002. Carbon nanotubes – the route toward applications. *Science*, 297(5582): 787–792.

Berzowska, J. 2005a. Electronic textiles: wearable computers, reactive fashion, and soft computation. *Textile*, 3(1): 2–19.

Berzowska, J., 2005b. *Memory Rich Clothing*. New York: ACM Press, 32. Available at: http://dl.acm.org.libproxy.mit.edu/citation.cfm?id=1056231anddl=ACMandcoll=DL (accessed October 11, 2011).

Berzowska, J. and Mainstone, D. 2008. *SKORPIONS*. Newyork: ACM Press, 1. Available at: http://dl.acm.org.libproxy.mit.edu/citation.cfm?id=1401032.1401091andcoll=DLanddl=ACMandCFID=55897023andCFTOKEN=14339803 (accessed October 11, 2011).

Brown, E. et al. 2010. Universal robotic gripper based on the jamming of granular material. *Proceedings of the National Academy of Sciences*. Available at: www.pnas.org/content/early/2010/10/18/1003250107.abstract (accessed October 10, 2011).

Buechley, L. et al. 2010. Living wall: programmable wallpaper for interactive spaces. *Proceedings of the International Conference on Multimedia*. (MM '10), ACM, 1401–1402.

Buechley, L., Hendrix, S. and Eisenberg, M. 2009. Paints, paper, and programs: first steps toward the computational sketchbook. *Proceedings of the 3rd International Conference on Tangible and Embedded Interaction*, ACM, 9–12. Available at: http://portal.acm.org/citation.cfm?id=1517664.1517670andcoll=Portalanddl=GUIDEandCFID=67113537andCFTOKEN=17523356 (accessed November 8, 2009).

Buechley, L. and Hill, B.M. 2010. LilyPad in the wild: how hardware's long tail is supporting new engineering and design communities. *Proceedings of the 8th Conference on Designing Interactive Systems*, ACM, 199–207. Available at: http://portal.acm.org.libproxy.mit.edu/citation.cfm?id=1858171.1858206andcoll=Portalanddl=GUIDEandCFID=67113537andCFTOKEN=17523356 (accessed October 12, 2010).

Buechley, L. and Perner-Wilson, H. 2012. Crafting technology: reimagining the processes, materials, and cultures of electronics. *Transactions on Computer–Human. Interaction*, 19(3): article no. 21.

Burch, D.S. and Pawluk, D.T.V. 2009. A cheap, portable haptic device for a method to relay 2-D texture-enriched graphical information to individuals who are visually impaired. *Proceedings of the 11th International SIGACCESS Conference on Computers and Accessibility*, ACM 215–216. Available at: http://dl.acm.org/citation.cfm?id=1639642.1639682andcoll=DLanddl=GUIDEandCFID=46749412andCFTOKEN=44269290 (accessed October 11, 2011).

Chen, Y. et al. 2003. Electronic paper: flexible active-matrix electronic ink display. *Nature*, 423(6936): 136.

Chidsey, C. and Murray, R. 1986. Electroactive polymers and macromolecular electronics. *Science*, 231(4733): 25–31.

Crichton, M., 2008. *Prey*. London: Harper.

Deyle, T. et al. 2007. Hambone: a bio-acoustic gesture interface. *2007 11th IEEE International Symposium on Wearable Computers*, IEEE, 3–10.

Dong, X.M. et al. 2004. Electrical resistance response of carbon black filled amorphous polymer composite sensors to organic vapors at low vapor concentrations. *Carbon*, 42(12–13): 2551–2559.

Dunne, L.E. et al. 2007. A system for wearable monitoring of seated posture in computer users. *Proceedings of the International Workshop on Wearable and Implantable Body Sensor Networks (BSN)*, 203–207.

Edmison, J. et al. 2002. Using piezoelectric materials for wearable electronic textiles. *Proceedings of the International Symposium on Wearable Computers (ISWC)*, IEEE, 41–48. Available at: http://ieeexplore.ieee.org/iel5/8353/26316/01167217.pdf?arnumber=1167217.

Fraden, J. 2010. *Handbook of Modern Sensors: Physics, Designs, and Applications* (4th edn). New York: Springer.

Fraser, D.B. 1973. Sputtered films for display devices. *Proceedings of the IEEE*, 61(7): 1013–1018.

Giallorenzi, T.G. et al. 1982. Optical fiber sensor technology. *IEEE Transactions on Microwave Theory and Techniques*, 30(4): 472–511.

Goldstein, S.C., Campbell, J.D. and Mowry, T.C. 2005. Programmable matter. *Computer*, 38(6): 99–101.

Hawkes, E. et al. 2010. Programmable matter by folding. *Proceedings of the National Academy of Sciences*. Available at: www.pnas.org/content/early/2010/06/24/0914069107.abstract (accessed October 11, 2011).

Hu, L. et al. 2009. Highly conductive paper for energy-storage devices. *Proceedings of the National Academy of Sciences*. Available at: www.pnas.org/content/early/2009/12/04/0908858106.abstract (accessed September 17, 2011).

Inzelt, G. 2010. *Conducting Polymers: A New Era in Electrochemistry*. New York: Springer.

Ishii, H. et al. 1999. PingPongPlus: design of an athletic-tangible interface for computer-supported cooperative play. *Proceedings of the Conference on Human Factors in Computing Systems: The CHI is the Limit*. (SIGCHI '99), ACM, 394–401. Available at: http://doi.acm.org/10.1145/302979.303115 (accessed June 21, 2012).

Ishii, H. et al. 2004. Bringing clay and sand into digital design – continuous tangible user interfaces. *BT Technology Journal*, 22: 287–299.

Ishii, H. and Ullmer, B. 1997. Tangible bits: towards seemless interfaces between people, bits and atoms. *Proceedings of the Conference on Human Factors in Computing Systems (SIGCHI '97)*, ACM, 234–241.

Johnson, S. and Thomas, A.P. 2010. Squishy circuits. Extended Abstracts on Human Factors in Computing Systems (CHIEA '10) ACM 4099. Available at: http://dl.acm.org/citation.cfm?id=1753846.1754109andcoll=DLanddl=ACMandCFID=51149544andCFTOKEN=71387729 (accessed October 3, 2011).

Kawakita, S. et al. 1997. Multi-layered piezoelectric bimorph actuator. *Proceedings of the 1997 International Symposium on Micromechatronics and Human Science,* . IEEE, 73–78.

Knite, M. et al. 2004. Polyisoprene-carbon black nanocomposites as tensile strain and pressure sensor materials. *Sensors and Actuators A: Physical*, 110(1–3): 142–149.

Kuang, K.S.C., Cantwell, W.J. and Scully, P.J. 2002. An evaluation of a novel plastic optical fibre sensor for axial strain and bend measurements. *Measurement Science and Technology*, 13: 1523–1534.

Kuznetsov, S. et al. 2011. Nurturing natural sensors. *Proceedings of the Conference on Ubiquitous Computing (Ubicomp)*, ACM 227–236. Available at: http://dl.acm.org/citation.cfm?id=2030144anddl=ACMandcoll=DLandCFID=46749412andCFTOKEN=44269290 (accessed October 11, 2011).

Lagoudas, D.C. 2010. *Shape Memory Alloys: Modeling and Engineering Applications*. New York: Springer.

Lam, C.-W. et al. 2006. A review of carbon nanotube toxicity and assessment of potential occupational and environmental health risks. *Critical Reviews in Toxicology*, 36(3): 189–217.

Lifton, J., Broxton, M. and Paradiso, J.A. 2005. Experiences and directions Pushpin computing. *Fourth International Symposium on Information Processing in Sensor Networks (IPSN '05)*, IEEE, 416–421.

Midori, S. 2011. Midori Shibutani Design. Available at: http://msdstudio.net/index.html (accessed October 11, 2011).

Numetrex 2007. Numetrex. Available at: www.numetrex.com.

Orth, M. 2007. International fashion machines. Available at: www.ifmachines.com.

Oxman, N. and Keating, S., Morphable Structures (n.d.) *Mediated Matter Group*. Available at: www.media.mit.edu/research/groups/mediated-matter (accessed October 12, 2011).

Pacelli, M. et al. 2006a. Sensing fabrics for monitoring physiological and biomechanical variables: e-textile solutions. *Proceedings of the 3rd IEEE/EMBS International Summer School and Symposium on Medical Devices and Biosensors,* IEEE, 1–4.

Pacelli, M. et al. 2006b. Sensing fabrics for monitoring physiological and biomechanical variables: e-textile solutions. *Proceedings of the 3rd IEEE/EMBS International Summer School and Symposium on Medical Devices and Biosensors*, IEEE.

Paradise, M. and Goswami, T. 2007. Carbon nanotubes: production and industrial applications. *Materials and Design*, 28(5): 1477–1489.

Paradiso, R. and de Rossi, D. 2006. Advances in textile technologies for unobtrusive monitoring of vital parameters and movements. *28th Annual International Conference of the IEEE Engineering in Medicine and Biology Society,* IEEE, 392–395.

Paradiso, R., Loriga, G. and Taccini, N. 2005. A wearable health care system based on knitted integrated sensors. *IEEE Transactions on Information Technology in Biomedicine*, 9(3): 337–344.

Perner-Wilson, H., Buechley, L. and Satomi, M., 2011. Handcrafting textile interfaces from a kit-of-no-parts. *Proceedings of the Fifth International Conference on Tangible, Embedded, and Embodied Interaction. (TEI '11),* ACM, 61–68.

Post, R. et al. 2000. E-broidery: design and fabrication of textile-based computing. *IBM Systems Journal*, 39(3–4): 840–860.

Qi, J. and Buechley, L. 2010. Electronic popables: exploring paper-based computing through an interactive pop-up book. *Proceedings of the Fourth International Conference on Tangible, Embedded, and Embodied Interaction.* (TEI '10), ACM, 121–128. Available at: http://portal.acm.org.libproxy.mit.edu/citation.cfm?id=1709886.1709909andcoll=Portalanddl=GUIDEandCFID=67113537andCFTOKEN=17523356 (accessed August 26, 2010).

Qi, J. and Buechley, L. 2012. Animating paper using shape memory alloys. *Proceedings of the Conference on Human Factors in Computing Systems (SIGCHI '12),* ACM.

Rekimoto, J. 2002. SmartSkin. *Proceedings of the Conference on Human Factors in Computing Systems (SIGCHI '02).* ACM 113–120. Available at: http://dl.acm.org.libproxy.mit.edu/citation.cfm?id=503397andCFID=55897023andCFTOKEN=14339803 (accessed October 11, 2011).

Rosenberg, I. and Perlin, K. 2009. The UnMousePad: an interpolating multi-touch force-sensing input pad. *ACM SIGGRAPH 2009 Papers*. ACM, 1–9. Available at: http://portal.acm.org/citation.cfm?id=1576246.1531371andcoll=Portalanddl=GUIDEandCFID=67113537andCFTOKEN=17523356 (accessed March 15, 2010).

Rosner, D.K. and Ryokai, K. 2008. Spyn: augmenting knitting to support storytelling and reflection. *Proceedings of the 10th International Conference on Ubiquitous Computing,* ACM, 340–349. Available at: http://portal.acm.org/citation.cfm?id=1409635.1409682andcoll=Portalanddl=GUIDEandCFID=67113537andCFTOKEN=17523356 (accessed November 8, 2009).

Ruschau, G.R., Yoshikawa, S. and Newnham, R.E. 1992. Resistivities of conductive composites. *Journal of Applied Physics*, 72: 953.

Russo, A. et al. 2011. Pen-on-paper flexible electronics. *Advanced Materials*, 23(30): 3426–3430.

Ryan, M.A. et al. 2004. Polymer-carbon black composite sensors in an electronic nose for air-quality monitoring. *MRS Bulletin/Materials Research Society*, 29(10): 714–719.

Sergio, M. et al. 2002. A textile-based capacitive pressure sensor. *Proceedings of IEEE Sensors,* IEEE, 2: 1625–1630.

Seymour, S. 2008. *Fashionable Technology: The Intersection of Design, Fashion, Science, and Technology* New York: Springer.

Shahinpoor, M. et al. 1998. Ionic polymer-metal composites (IPMCs) as biomimetic sensors, actuators and artificial muscles – a review. *Smart Materials and Structures*, 7: R15–R30.

Shim, B.S. et al. 2011. Smart electronic yarns and wearable fabrics for human biomonitoring made by carbon nanotube coating with polyelectrolytes. *Nano Lett.*, 8(12): 4151–4157.

Siegel, A.C. et al. 2009. Thin, lightweight, foldable thermochromic displays on paper. *Lab on a Chip*, 9(19): 2775.

SRI International (n.d.) Artificial muscle and biomimetic robots. *SRI: Robots*. Available at: www.sri.com/robotics/epam.html (accessed October 12, 2011).

Steinem, C. and Janshoff, A. 2007. *Piezoelectric Sensors.* New York: Springer.

Sugimoto, A. et al. 2004. Flexible OLED displays using plastic substrates. *IEEE Journal of Selected Topics in Quantum Electronics*, 10(1): 107–114.

Thomas, A.P. 2011. Squishy circuits. *Thomas Lab*. Available at: http://courseweb.stthomas.edu/apthomas/Squishy Circuits (accessed October 11, 2011).

Tressler, J.F., Alkoy, S. and Newnham, R.E. 1998. Piezoelectric sensors and sensor materials. *Journal of Electroceramics*, 2(4): v257–272.

Villar, N. et al., 2006. The VoodooIO gaming kit: a real-time adaptable gaming controller. *Proceedings of the International Conference on Advances in Computer Entertainment Technology*. (SIGCHI '06), ACM, 82. Available at: http://portal.acm.org/citation.cfm?id=1178823.1178919andcoll=Portalanddl=GUIDEandCFID=67113537andCFTOKEN=17523356 (accessed November 8, 2009).

Villar, N. and Gellersen, H. 2007. A malleable control structure for softwired user interfaces. *Proceedings of the 1st International Conference on Tangible and Embedded Interaction* (TEI '07), ACM 49. Available at: http://dl.acm.org/citation.cfm?id=1226980 (accessed October 10, 2011).

Wakita, A. and Shibutani, M. 2006. Mosaic textile: wearable ambient display with non-emissive color-changing modules. *Proceedings of the International Conference on Advances in Computer Entertainment Technology* (ACE '06). ACM, article no. 48.

Weiser, M. 1999. The computer for the 21st century. *SIGMOBILE Mobile Computing and Communications Review* (Special Issue dedicated to Mark Weiser) ACM, 3(3): 3–11.

Yeom, S.-W. and Oh, I.-K. 2009. A biomimetic jellyfish robot based on ionic polymer metal composite actuators. *Smart Materials and Structures*, 18: 085002.

Yoshikai, T. et al. 2009. Development of soft stretchable knit sensor for humanoids' whole-body tactile sensibility. *9th IEEE-RAS International Conference on Humanoid Robots, (Humanoids 2009)*, IEEE, 624–631.

22

Haptic Interfaces

Eve Hoggan

INTRODUCTION

The study of haptic interfaces is an ever-expanding research area focusing on human touch and interaction with the environment through touch. The term haptic can be defined as 'sensory and/or motor activity based in the skin, muscles, joints and tendons' (ISO, 2011 244: 1). The creation of haptic interfaces depends on an in-depth knowledge of the human body and its ability to sense touch and kinesthetic sensations. Our sense of touch is crucial in our everyday interactions with the world. For instance, when looking for an object in your pocket or bag without the use of vision, the haptic sense becomes the primary modality of use. Despite the fact that the haptic modality is an extremely rich bi-directional information channel, it is relatively under-utilized in current interactive systems in comparison to the audio and visual modalities. For example, in terms of interface design, a large number of systems have moved away from mechanical buttons, switches and other widgets and have adopted touchscreen widgets instead. Unfortunately, during this transition, one important feature was lost: the widgets cannot provide the haptic response that physical objects do when touched or clicked. By adding haptic feedback to user interfaces, we can recreate the physical sensation of pressing a button, holding a ball or even create completely new touch sensations.

Within the definition of the term 'haptic', there are several sub-categories as shown in Figure 22.1

The human sense of touch can be divided into two separate channels. *Kinesthetic* perception involves positions, velocities, forces and constraints sensed through the muscles and tendons. Force feedback devices make use of the kinesthetic senses by presenting computer-controlled forces to create the illusion of contact with a rigid surface (Burdea, 1996) *Cutaneous* perception involves the stimulation of the skin surface through direct contact. Cutaneous stimulation can be further separated into the sensations of pressure, stretch, vibration and temperature.

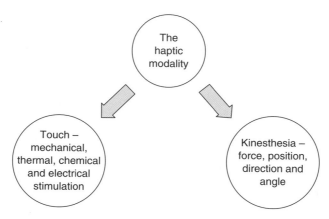

Figure 22.1   Definitions of terminology (ISO, 2011 244).

The concept of cutaneous information transmission has been around for decades. In 1960, Geldard wrote, 'for some kinds of messages the skin offers a valuable supplement to ears and eyes' (1960: 1583). The skin offers a large display space, which can be used to display information. As the skin is often less engaged in other tasks than the eyes or ears, it is always ready to receive information (van Veen and van Erp, 2000). The notion of using the kinesthetic properties of our sense of touch in interactive systems has also been documented in the literature for many decades. One of the very first working examples was the 'Handyman' by Mosher described in Stone's history of haptic feedback (Stone, 2000). The Handyman consisted of two electrohydraulic arms working as a form of exoskeleton.

Recent technological advances have created opportunities to enhance interactive systems with the haptic modality. Haptic feedback is already commonplace in many games controllers and mobile phone ringtones, but this interaction modality has potential value in many other different application areas, including mobile and desktop touchscreen interfaces, medical training simulators, rehabilitation systems, accessible interfaces, robotics and in performative interaction.

The remainder of this chapter will go into detail on many of the aspects of haptic interfaces

that have an influence on research. The next section presents a brief outline of psychophysics, or, the study of human haptic perception capabilities. In the following section, the various different types of haptic actuators currently available will be described, followed by an overview of the haptic interface design process. The final section reviews the most recent novel research in the field, with a focus on a variety of application areas, ranging from mobile touchscreen solutions and multi-touch haptic issues, to the use of haptic creatures for affective touch. The chapter will conclude with a summary of the key research challenges in the field of haptic interface design.

A BRIEF INTRODUCTION TO HAPTIC PERCEPTION

Before designing haptic interfaces it is necessary to gain an understanding of the capabilities of humans to process haptic stimuli. This section begins by providing an overview of the sense of touch and then goes on to present results from the literature regarding perception of the different haptic parameters, drawing conclusions about the implications of these results for the design of haptic interfaces.

There are several key features of our sense of touch that make it ideal for use with interactive systems. For example, our haptic sense is bi-directional because we can perceive and actuate via touch (Klatzky and Lederman, 2002). In terms of interface design, this means that touch can be used as an input and output tool. Haptic perception can also be active or passive (Gibson, 1966). Active haptic perception involves intentional exploration of an object but perception of these objects can also be passive. That is, the object of interest is moved across or against our skin. For example, the vibrotactile actuators inside a mobile phone provide mechanical passive feedback whenever the phone rings.

Our sense of touch has the potential to be one of the most useful input and output tools for digital interaction. The skin can have an area of 1.8 m², a density of 1250 kg/m³ and a weight of 5 kg (Sherrick and Cholewiak, 1986) which means that haptic interfaces have a large area to interact with. The skin is classified as either glabrous (i.e. non-hairy) skin, which is only found on plantar and palmar surfaces, or hairy skin, which is found on the rest of the body (Cholewiak and Craig, 1984), If we take fingertips as an example, there are four types of mechano-receptive fibres that have been identified in glabrous skin: Meissner corpuscle, Merkel cell, Pacinian corpuscle and Ruffini ending. According to Kandel and Jessell (1991) Meissner's corpuscles and Merkel's cells respond to touch, Pacinian corpuscles respond to vibration and Ruffini's endings respond to rapid indentation of the skin. This means that our fingertips can be used to sense a wide range of haptic stimuli, from textures to 2D forms, simple vibrations and pressure. The specific design parameters for these sensations are outlined in the next section.

Cutaneous Perception

The dimensions or attributes of our sense of touch with respect to cutaneous stimuli are detailed below. These are important for haptic interface designers as they help to guide the creation of different sensations. For example, it may be necessary for a haptic interface to include some very noticeable, intense sensations, and knowledge of the attributes associated with our sense of touch can help designers to choose the most appropriate feedback. If we take a navigation aid for visually impaired people as an example, such as the UltraCane system (www.ultracane.com), any haptic information about obstacles in the user's path is extremely important and should be highly distinguishable. The related work on psychophysics detailed below indicates that using a 250Hz square wave presented to the fingertip would be an appropriate obstacle alert.

Frequency

The frequency perception of the skin ranges from 10Hz to 400Hz (Summers et al., 1994) with maximum sensitivity and finer spatial discrimination at around 250Hz (Craig and Sherrick, 1982). Measures for discrimination thresholds of frequency are problematic, as perception of vibratory pitch is dependent not just on frequency but also on the amplitude of stimulation. Geldard (1957) found that subjects reported a change in pitch when frequency was fixed, but amplitude of stimulation was changed. Thus, this interaction between frequency and amplitude should be taken into account when designing stimuli.

Duration

Duration is another important dimension found in the tactile modality. Gescheider (reported in Terhardt, 1974) measured the time difference between two tactile 'clicks' on the fingertip necessary for them to be perceived as two separate sensations and found that the minimum threshold reported was 10 ms. Gunther et al. (2002) suggests that stimuli lasting less than 0.1 seconds may be perceived as taps or jabs, whereas longer-lasting stimuli may be perceived as smoothly flowing tactile phrases.

Rhythm

Rhythm is very important and useful in the design of tactile feedback. Rhythm has been investigated as the main parameter in tacton (tactile icon) design, with recognition rates of over 90 per cent achieved when three different rhythms are used (Brown et al., 2006) The literature suggests that musical principles should be applied to the design of temporal patterns, while tempo (speed) can be used as a distinguishing factor (Blattner et al., 1989).

Location

Body locations that have been identified as most sensitive to pressure and stimulus discrimination include the finger, hand, arm, thigh and torso (Cholewiak and Collins, 1995) In tacton research (Brown et al., 2006) three locations on the forearm have been used. Other research has also shown successful spatial location discrimination on the abdomen, arm and back (Chen et al., 2008; Cholewiak et al., 2004; Tan and Pentland, 1997)

Intensity

Intensity detection in our sense of touch is somewhat limited, with an intensity range of approximately 55dB above the detection threshold. Any vibrations above this threshold feel unpleasant or even painful (Verrillo and Gescheider, 1992). That being said, intensity changes over time have been used successfully in several tactile displays (MacLean and Enriquez, 2003)

Texture

Our sense of touch is extremely good at detecting different textures. In musical composition studies, it has been suggested that waveform can be correlated to the 'texture' of tactile stimuli (Gunther et al., 2002). Enriquez and Mac Lean (2008) achieved recognition rates of 73 per cent for nine haptic icon designs varying in waveform. Hoggan and Brewster (2007) found that square waves could be used to create very rough sensations while sine waves are much smoother.

Kinesthetic Perception

Unlike cutaneous perception, it is very difficult to investigate the kinesthetic sense because it can be problematic to design a study capable of separating the cutaneous sense from the kinesthetic sense. Our kinesthetic sense allows us to detect how fast and in which direction our limbs are moving. Furthermore, we can detect the static position of our limbs once they have stopped moving (also known as proprioception). Kinesthetic interfaces can provide a spatial frame of reference, allow gesture interaction, and also provide force feedback from virtual objects.

The kinesthetic part of our sense of touch is made up of four receptor types located in our muscles, tendons and joints. As outlined in detail in Hale and Stanney's 2004 paper, the Golgi and Ruffini endings are found in or near joints while Golgi tendon organs are, as the name suggests, found in tendons and muscle spindles. Sensory information from the kinesthetic receptors combines with available information from the motor and cognitive systems to produce perceived limb position and movement. The kinesthetic sense bandwidth ranges from 20 to 30 Hz, with proximal joint rotations being more sensitive than those of distal joints. A person often perceives that a movement has occurred before sensing its direction. Thus, a passive kinesthetic stimulus must contain enough movement to indicate direction.

HAPTIC ACTUATOR TECHNOLOGY

One of the biggest factors to limit the progression of haptic interfaces into mainstream technology is the hardware. However, the number and quality of different types of actuator available is increasing all the time. This section will give a brief outline of the general types of actuator available at the moment, ranging from those found in typical mobile devices to more specialized hardware such as artificial muscles.

Linear and Rotational Actuators

Vibrotactile actuators can provide sustained feedback, allowing many different textures

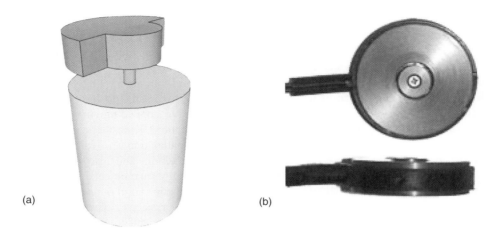

(a)   (b)

Figure 22.2   (a) Illustration of a standard eccentric rotating mass actuator; (b) Engineering Acoustics Inc (EAI) C2 Tactor – a vibrotactile actuator.

and intensities to be presented. Most vibro-tactile devices stimulate the skin by using electromagnetic actuation to drive a mass in either a linear or rotational manner. The majority of mobile phones contain an eccentric rotating mass actuator (ERM) as shown in Figure 22.2(a). The EAI C2 Tactor (www.eaiinfo.com) is shown in Figure 22.2(b). This linear rotational device is resonant at 250 Hz with a much reduced response at other frequencies (which is another reason for the reduced usefulness of frequency as a parameter for tactile interfaces). The advantage of vibrotactile cues is that they can be extremely strong (so can be felt through clothing) and they can also be distributed over the body or device to give spatial cues (often attached to a user's belt around the waist). For a more detailed review of vibro-tactile devices, see the review by Mortimer et al. (2007).

Piezoelectric Actuators

Piezoelectric actuators, as shown in Figure 22.3, can create short, more localized tactile feedback, by moving touchscreen display modules within the device (Laitinen and Mäenpää, 2006. The piezoelectric actuator can

provide displacement very quickly, but with less kinetic energy compared to traditional vibration motor systems. Koskinen et al. 2008) ran three laboratory-based studies to determine which tactile click (from a set of various different designs) is most pleasant to use in fingertip interaction with a mobile touchscreen device. Using two different types of actuator – piezo or standard ERM vibration motor – the experiments allowed the authors to find the most pleasant tactile feedback as perceived by participants. The results show that feedback from piezoelectric actuators is perceived as more pleasant than feedback from vibrotactile motors.

Electroactive Polymers

Electroactive polymers (EAPs) are a relatively new addition to haptic interface technologies. EAPs change shape when voltage is applied to them. The main advantage of using artificial muscles is that they can stimulate the skin without additional electromechanical conduction. There are two different types: dielectric EAPs and ionic EAPs. Dielectric EAPs use an electrostatic force between two electrodes to actuate the polymers. Even though this type of EAP is operated with high

Figure 22.3   Illustration of piezoelectric actuator.

voltages, it only consumes low levels of power. On the other hand, ionic EAPs use a large amount of power and are actuated by the movement of ions inside the polymer. EAPs can be used as an alternative to piezo-electric actuators when an interface requires a higher amount of displacement and wider frequency range.

Skin Stretch

Piezoelectric actuators were mentioned earlier in this section as a type of tactile actuator, but these actuators can also be used for other types of stimulation. Skin stretch displays rely on the motion of individual piezo points to stretch the user's skin when the user touches a display made up of several piezoelectric actuators (Luk et al., 2006). This method allows haptic interfaces to create the illusion of shapes and textures by deforming the skin in different patterns.

Previous research has indicated that mechanoreceptors in the skin respond quickly and accurately to skin strain changes (Bark et al., 2008) and several researchers have successfully created haptic interfaces using this technology (Gleeson et al., 2009).

Force Feedback Devices

Force feedback devices serve two main purposes: to measure the positions and contact forces of the user's hand (and/or other body parts) and to display contact forces and positions to the user (Tan et al., 1994). There are

many different commercially available force feedback devices. These can be categorized according to the degrees of freedom (DOF) provided by each device.

- 1 DOF: steering wheels
- 2 DOF: joysticks, pens and mice
- 3 DOF: high-precision instruments (e.g. PHANToM [Sensable Technologies])
- 6 DOF: extremely high-precision and task-specific devices (e.g. PHANTOM Premium 6 DOF).

Electrovibration

Electrovibration is created using electro-static friction between a surface and a user's skin (Bau et al., 2010). Passing an electrical charge into the insulated electrode creates a small attractive force when a user's skin comes into contact with it. Electrostatic friction can be created to simulate the sensation of localized vibration and friction, thus allowing feedback designers to create many different textures on a smooth surface. Unlike other vibration actuators, electrovibration is relatively silent. In the commercial domain, Senseg has integrated electrovibration feedback with mobile devices using active surfaces called *tixels*, the tactile pixel (Linjama and Mäkinen, 2009). Bau et al. (2010) also created a working example of electrovibration called TeslaTouch. The authors conducted several user studies to investigate perception of electrovibration and potential applications of the TeslaTouch. They found that, unlike mechanical vibration, electro-vibration could support anchored gestures and multi-touch feedback.

DESIGNING HAPTIC INTERFACES

This section will describe the different ways in which haptic interaction can be used in human–computer interfaces, with an outline of the advantages and disadvantages of designing with the haptic modality.

Advantages of Haptic Feedback

There are many reasons why it is advantageous to use touch at the user interface. First and most commonly, haptic feedback can be included in interfaces to reduce the amount of visual information needed on the display. This can help to prevent users from becoming overwhelmed by large amounts of visual feedback and can also help them to manage situational sensory impairments during mobile interaction. Mobile users cannot devote all of their visual attention to the device – they must look where they are going. In this case, the user's visual sense is situationally impaired and visual information may be missed because the user is not looking at the device. Even more importantly, haptic feedback can not only help with situational impairments but also enhance the usability of interfaces for people with visual impairments. Presenting information in a haptic form can allow visually impaired people to use some of the facilities available to sighted users.

Disadvantages of Haptic Feedback

There are three main drawbacks to the use of haptic feedback. The haptic modality is fairly low resolution compared to other modalities such as vision. When using tactile amplitude, for example, only a very few different values can be unambiguously presented (Geldard, 1960). This issue leads on to the next issue of presenting absolute data. This is not advisable with haptic feedback. Most existing systems present information in a relative way – for example, users can feel the difference between two haptic durations and identify which version is longer, but absolute values are difficult. Last, there are many intrasensory dependencies that must be taken into account when creating distinguishable haptic stimuli. Changing one attribute of a haptic cue may affect the others. For example, changing the frequency of tactile stimuli may affect the perceived amplitude and vice versa (see 'A brief introduction to haptic perception' above).

When Should Haptic Interaction Be Used?

The ergonomics ISO on tactile and haptic interactions (2011) states that haptic interaction can be applied to the following application areas: accessibility, desktop interactions, mobile interactions, telerobotics, medical training virtual reality-based tools, gaming and in the arts. Although there is no official set of design guidelines for haptic interfaces, this sub-section outlines the main uses for haptic feedback with respect to a large collection of related research.

Haptic feedback can be divided into two categories: active and passive. As outlined by Gibson (1962), active touch is when the impression on the skin is brought about by the user, whereas passive touch is when the impression on the skin is brought about by an outside agency. In other words, a standard voice call vibrotactile alert from a mobile device could be considered to be passive, as it is not triggered by the user but by the device. However, if the user types on a mobile keyboard and receives vibrotactile feedback for each keypress, this could be considered active feedback.

In general, tactile stimuli are most effectively used in active and passive feedback for alerts and structured information. Kinesthetic stimuli are most useful when considering active feedback for virtual object manipulation. Last, both types of stimuli can be used in collaboration with other modalities, such as audio, to create multimodal or crossmodal interfaces. Some examples of these cases are outlined below.

Active Feedback

A good example of the application of active haptic feedback can be seen in text entry research for touchscreen displays. As mentioned in the introduction, touchscreen widgets cannot provide the haptic response that physical objects do when touched or clicked. Nashel and Razzaque (2003) added tactile cues simulating real buttons to virtual buttons displayed on mobile devices with touchscreens. The experiments conducted found that all participants were able to differentiate between vibration (finger over a button) and no vibration (finger not over any button). Hoggan et al. (2008) also studied the use of tactile feedback for text entry in mobile environments. The experiment compared devices with a physical keyboard, a standard touchscreen and a touchscreen with tactile feedback added in both static and mobile environments. Different vibrations were presented whenever the user touched a touchscreen button or slipped over the edge of a button. The results showed that the addition of tactile feedback to the touchscreen significantly improved performance and brought it close to the text entry on a real physical keyboard. These promising results from existing research indicate that including active haptic feedback in touchscreen interaction can increase user performance and mimic real-world physical objects such as buttons.

Passive Feedback

Tactons and haptic icons can be used as a form of passive feedback. Tactons (Brown et al., 2006) are structured vibrotactile messages that can be used to communicate information non-visually. They are the tactile equivalent of visual icons and can be used for communication in situations where vision is overloaded, restricted or unavailable. Tactons are created by manipulating the parameters or dimensions of cutaneous perception to encode information.

Another approach to developing tactile or haptic icons involves identifying the basic elements, called haptic phonemes, and using these to create different haptic icons. With this method, Enriquez, MacLean and Chita (Enriquez 2006) created a set of nine haptic icons that varied in terms of waveform and frequency. They then trained participants to associate each haptic icon with an arbitrary concept and found that participants learned these associations after about 25 minutes of training and achieved higher identification rates with stimuli that varied in frequency (81 Per cent correct), compared to those that varied in waveform (73 Per cent correct).

Rovers and van Essen (2005) also mention the use of icons with haptic feedback. They state that the message can be designed as a real-world signal, such as a heartbeat, or it can be based on an abstract design. An abstract design requires the use of a set of common rules, for example, two pulses is equal to 'on'. This is the same approach used in tacton design.

Multimodal and Crossmodal Feedback

Haptic interfaces are rarely uni-modal. Being predominantly dependent on a single sense such as touch is unnatural because, in the real world, we receive information from several modalities, as when we feel, hear and see a button being pushed. Blattner and Dannenberg (1992: xviii) discuss some of the advantages of using multimodal interaction: 'In our interaction with the world around us, we use many senses. Through each sense, we interpret the external world using representations and organizations to accommodate that use. The senses enhance each other in various ways, adding synergies or further informational dimensions'. Multimodal and crossmodal interfaces can make use of the haptic modality alongside other modalities such as audio, visual and even olfactory. Crossmodal interaction is a subset of multimodal interaction where the different senses are used to receive the same data (Hoggan, 2010). So far, crossmodal interaction has mainly been investigated with the cutaneous category of the haptic modality. Crossmodal use of the different senses allows the characteristics of one sensory modality to be transformed into stimuli for another sensory modality. Multimodal interaction, however, may also use the different senses to receive different information.

Many more examples of multimodal and crossmodal interaction can be seen in the existing state of the art in haptic interfaces described in the next section.

CURRENT HAPTIC INTERFACE RESEARCH

The remainder of this chapter will review the most recent novel research in the field.

Mobile and Touchscreen Interaction

Given the widespread commercially available touchscreens, it should come as no surprise that touchscreen interaction has become a growing area of research. In terms of touchscreen mobile devices, there is a challenge in designing appropriate interfaces given that there is a large variety of information that needs to be displayed visually on extremely small screens. This places a high demand on the visual sense and explains the extent to which users can often spend more time focused on the screen than on the environment or task in hand. These interfaces also lack the natural touch sensations that users traditionally experience when interacting with physical objects such as buttons and sliders. There has been an increasing amount of research into the addition of the haptic modality to touchscreen output (Fukumoto and Sugimura, 2001; Kaaresoja et al., 2006; Luk et al., 2006; Hoggan et al., 2008; Lee and Zhai, 2009.

Some of the first researchers to investigate the use of tactile feedback on touchscreen devices were Poupyrev and his colleagues (2003). The TouchEngine is a piezoelectric actuator that was embedded in a PDA. In this case, Poupyrev et al. used the tactile feedback as an ambient background channel of information. The authors investigated several applications using touch as the ambient, background channel for mobile communication and conducted a formal user study into the use of tactile feedback with tilting devices.

Participants were required to scroll through a text list using gestures. The results of the study showed that, on average, participants could complete the tasks 22 per cent faster when provided with tactile feedback.

Haptic feedback has also been added to larger surfaces. For example, MudPad (Jansen et al., 2010) is able to give localized active haptic feedback by using magnetorheological fluid and an array of electromagnets to actuate the fluid over a tabletop display (see Figure 22.4). The fluid's viscosity can be controlled by a magnetic field. When it is exposed to a magnetic field it becomes stiff. Multiple electromagnets can be controlled individually to create a tactile sensation at that specific location only. There are several feedback parameters that can be manipulated to create different haptic sensations – for example, duty cycle, frequency and uniformity.

Interpersonal Communication

There are many communication technologies that allow people to stay in touch with remote relatives and friends – for example phones, e-mail and instant messaging. Although there have been several developments in this field, these devices cannot recreate the same feeling of physical connection as meeting someone in person. The use of different communication modalities could overcome these issues. There has been previous research into using alternative modalities such as haptics for interpersonal communication (Brave and Dahley, 1997); Rovers and van Essen, 2005).

One of the first research projects in this area, by Chang et al. (2002) developed ComTouch, a sensory augmentation tool. ComTouch is a device that augments remote voice communication with touch by converting hand pressure into vibration intensity between users in real time. It is a vibrotactile sleeve using small commercial acoustic speakers to transmit vibrations that can be fitted over the back of a mobile phone. A study was conducted using ComTouch to investigate the possible uses of the tactile

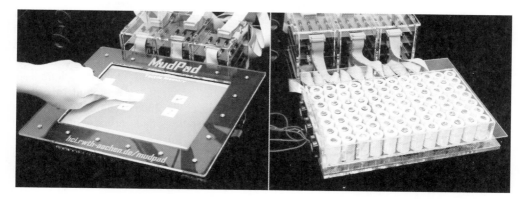

Figure 22.4   MudPad interface and electromagnet array (from Jansen et al., 2010).

channel when used in conjunction with audio and to test the mapping between pressure and vibration. The results showed that users developed an encoding system similar to that of Morse code, as well as three original uses: emphasis, mimicry and turn-taking.

Park et al. (2012) also investigated the use of haptic feedback in interpersonal communication by creating a system called CheekTouch (Figure 22.5). The CheekTouch prototype is based on a standard iPod Touch and makes use of nine vibrotactile actuators positioned in a grid on the front of the device. Whenever a user touches the back of the iPod, the touch location is sent to the other user's device where the corresponding vibrotactile actuator in the matching grid location is activated. A lab study with couples over a period of five days showed that the additional bandwidth provided by the haptic interaction allowed users to communicate in lots of different ways: persuading, conveying

status, delivering information, emphasizing emotion/words, calling for attention and being playful. However, given that a large area on the back of the device was touch sensitive, the users stated that they often accidentally triggered a vibration.

Communicating Emotion

The sense of touch can be very private, subtle and emotional. Traditional methods of face-to-face communication rely heavily on the sense of touch (Smith and MacLean, 2007). In face-to-face interaction, touch is often used to express diverse, private and subtle nonverbal cues. For example, handshakes, holding and squeezing hands, kissing or pats on the back are all common occurrences in everyday communication. Some of these touch interactions occur solely with loved ones, while others are used with everyone. Despite the extent to which touch is used in everyday life to communicate emotion, current interfaces do not fully exploit this. Yohanan and MacLean (2011) created the Haptic Creature to investigate affective haptic displays to support companionship. The Haptic Creature (Figure 22.6) is a robot that imitates a small pet and allows interaction through touch. Whenever the Haptic Creature is touched, it responds by displaying its emotional state through changing the stiffness of its ears, its breathing and through vibrotactile feedback.

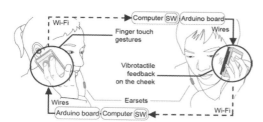

Figure 22.5   Communication through CheekTouch (from Park et al., 2012).

Figure 22.6 The Haptic Creature prototype robot (photo by Martin Dee, used with permission from Steve Yohanan).

The initial studies with the Haptic Creature were conducted in a Wizard of Oz style where participants could touch and interact with the creature. The participants were asked to interact with the creature in various different ways – for example, gently pet – and the creature reacted by rendering an emotional response, which was to be identified by the participants. Overall, the study showed that emotion can be successfully communicated through the haptic modality and, in turn, the user's emotions are often affected, too, much as they would be with a household pet.

Users with Visual Impairments

One of the most important uses for touch is in sensory substitution interfaces for people with visual disabilities. Sensory substitution systems take environmental data that would normally be processed by one sensory system and translate this data into stimuli for another sensory system (Lenay et al., 1997). Sensory substitution using the sense of touch has been used by blind people for many years in the form of Braille. In Braille, the mapping between characters and the tactile equivalents are based on a numerical model where each Braille character consists of a 3-row by 2-column cell, with combinations of raised dots allowing 64 individual patterns. Numerous empirical evaluations have been conducted and the results suggest that Braille code is more effective than using embossed letters in terms of reading speed and accuracy (Sharmin et al., 2005). The main disadvantage of Braille is that very few blind people can read it. For example, it has been estimated that only 2 per cent of British blind people can read Braille (Bruce and McKennell, 1991).

Haptic feedback has also been incorporated into walking canes for people with visual impairments. The commercially available UltraCane uses ultrasound range finders to detect objects in a user's environment and encodes the location information and distance to objects into vibrotactile feedback, which is presented through actuators on the handle of the cane.

Wearable Computing

As mentioned earlier, the skin can be considered as a very large input area so one advantage of using haptic feedback is that actuators can be placed all over the body of the user. For example, Lee and Starner (2010) developed Buzzwear, a tactile display for the wrist that provides alerts for mobile users. The experiments with Buzzwear show that users can identify 24 different tactile patterns on the wrist with up to 99 per cent accuracy after 40 minutes of training. The patterns were designed by manipulating the intensity, starting point, direction and temporal parameters of the haptic feedback. The results indicated that intensity was the hardest parameter to identify on the wrist while temporal changes were much easier.

Taking wearable haptic feedback a step further, Holz et al. (2012) have successfully implanted vibrotactile actuators underneath the skin in an implanted interface. The feedback from the actuators was strong enough to be detected underneath the skin and the authors suggest that implanted interfaces are beneficial as the interface is available at all times.

Navigation

When trying to navigate in various different environments – for example, when piloting a plane, whilst driving a car, walking in the dark or even when navigating through a graphical user interface – our visual sense can often be overloaded or situationally impaired. Haptic interfaces can be used to aid navigation in these situations.

One of the most common implementations of haptic navigation systems involves the use of a haptic belt (Tsukada and Yasumura, 2004). Van Erp and van Veen (2001) created a tactile belt with eight vibrotactile actuators positioned around the user's waist for use in pedestrian navigation. Different directions were mapped to the locations of each actuator and the distance to each target was encoded in different vibrotactile rhythms. A pilot user study showed that users could accurately distinguish directions but distance was less effective.

Haptic feedback can also be used for other forms of navigation such as target acquisition. Oron-Gilad et al. (2007) presented vibrotactile cues to the hand to help guide the user towards a target, in this case, objects in a computer game. The game players received tactile feedback whenever their hand deviated away from the target. The results of a user study showed that the haptic feedback on the hand when combined with visual feedback was extremely effective in helping users to acquire a target.

Art

The haptic modality has also been successfully used in a variety of artistic domains such as music and interactive art installations. In terms of music, Gunther et al. (2002) investigated tactile composition from a crossmodal standpoint where tactile feedback was combined with traditional audio. The authors created a system that enables the composition and perception of musically structured spatio-temporal patterns of vibration on the surface of the body.

Thirteen vibrotactile actuators were placed on the body with three on each limb and one on the lower back. An initial test of the system was conducted in a performance context, which found that the body locations were suitable for presentation of tactile music and music can be composed for the sense of touch.

Haptic interfaces have also been exhibited as a form of interactive art. At the SIGGRAPH 2010 Art Gallery there was an exhibition called 'TouchPoint: Haptic Exchange Between Digits'. The work exhibited ranged from audio and tactile sculptures that respond to touch, to robots that mimic snakes by writhing, wriggling, twisting and squeezing in response to how they are held and touched (Stedman, 2010). One exhibit was 'The Lightness of Your Touch' by Henry Kaufman (Kaufman, 2010), which was an interactive infrared multitouch display of a large torso (see Figure 22.7). When the torso is touched, the skin moves and a hand imprint remains on the display once the hand has been removed.

Figure 22.7 'The Lightness of Your Touch' by Henry Kaufman (© Henry Kaufman).

Although the output is visual, it responds to the sense of touch and encourages people to think about the warmth and other sensations that remain after physical touch.

NEW AREAS OF HAPTIC INTERACTION RESEARCH

Alongside the areas of the haptic modality discussed above, temperature and pressure are also part of our sense of touch. However, until recently, they have rarely been present in interfaces.

Temperature

There has been very little research into the use of temperature for feedback. One reason is that there is no standard hardware; everything must be custom built. It is also a low-bandwidth sense, but could accompany other forms of feedback to create richer mobile displays. It does have some useful properties for interaction: heat often means danger, so temperature could be a way of presenting warnings. It is also an important way to assess the type and material of objects in the real world. In the children's game, 'hotter' and 'colder' are used to give directions to a target, so temperature could be used for guidance in a mobile application.

There has been some work in HCI on thermal interfaces, particularly focusing on heat display. Ottensmeyer and Salisbury (1997) included temperature feedback as part of a PHANToM haptic device to create more realistic haptic experiences. They proposed a range of uses of temperature, including medical simulations. It has also been used in ambient computing – for example, Lovelet (Fujita and Nishimoto, 2004) allowed one user to hold a device and send heat information to another using a Peltier device.

Wettach et al. (2007) developed a display that could give directions based on heat. An informal test showed that people could navigate with it. However, they relied on specific temperature values to convey

information and Stevens suggests this might not be the most effective way as the body is not a good thermometer (Stevens, 1991). Wilson et al. (2011) conducted two studies into how well users could detect hot and cold stimuli presented to the fingertips, the palm, the dorsal surface of the forearm and the dorsal surface of the upper arm. Evaluations were carried out in static and mobile settings. Results showed that the palm is most sensitive, cold is more perceivable and comfortable than warm and that stronger and faster-changing stimuli are more detectable but less comfortable.

Pressure

The most common use of pressure in interaction is in graphics tablets where an increase in pressure may increase line thickness, simulating some of the aspects of drawing with a real pen. This can be very expressive, allowing graphic artists much subtlety in their drawings. A similar application is for electronic music, with pressure-sensitive keyboards allowing rich control of sound. Srinivasan and Chen (1993) studied pressure using the index finger. Participants had to control the amount of force applied to a force sensor under a range of different conditions (including an anesthetized fingertip to examine the effect of removing tactile feedback). They suggest that pressure-based interfaces need to have a force resolution of at least 0.01N in order to make full use of human haptic capabilities. Mizobuchi et al. (2005) suggest that ranges of 0–3N are comfortable and controllable and users can reliably apply around 5–6 levels of pressure (Ramos et al., 2004; Mizobuchi et al., 2005).

Ramos et al. have done some of the key work in HCI on pressure input on graphics tablets. They looked at how pressure might be used in applications such as video editing and proposed a set of 'pressure widgets' (Ramos et al., 2004) for tasks such as zooming and selection based on pressure. Irani and colleagues (Cechanowicz et al., 2007; Shi et al., 2008) looked at adding pressure to a mouse for desktop interactions and at how multiple pressure sensors could be used. Their results showed that users were slower when they had to press harder; they also showed that a click-selection technique was faster than a dwell, although dwell was the most accurate. More recently, Wilson et al. (2010) investigated pressure as an input technique on mobile touchscreen devices. The results showed that touchscreen users could distinguish and apply up to 10 pressure levels with high levels of accuracy when navigating through a standard menu.

THE FUTURE OF HAPTIC INTERFACES

Research into the use of haptic interaction has shown its benefits in a wide range of different applications, from systems for visually impaired people to mobile computing. Four areas are likely to be important in its future growth.

The first is in combining the haptic modality with others (vision, sound, etc.) to create multimodal or crossmodal displays that make the most of all the senses available to users. For instance, it is well known that certain combinations of multimodal stimuli can create sensory illusions. In particular, one of the most famous areas of study is the size/weight illusion (Ross and Gregory, 1970; Murray et al., 1999; van Mensvoort, 2002;) where, when lifting two objects of different volume but equal weight, people judge the smaller object to be heavier. This is an area ripe for further investigation and there are many interesting interaction problems that can be tackled when multiple senses are used together. Moreover, there is an extreme lack of research focusing on intra-modal haptic combinations. For example, using temperature and pressure together in haptic interfaces could increase the amount of information transmitted through the haptic modality or increase the realism of the stimuli. In the real world we are exposed to many different intra-modal sensations. For example, when we hold a loved one's hand, we can feel both the pressure and the warmth of their Person's hand – imagine if your phone could mimic this.

The second area in which haptic feedback could play a large role is with multitouch and tabletop interfaces. Many large touchscreen computers use direct finger-based multitouch input and can support users at any position around the display. This configuration means that users should be able to use the table without restriction no matter where they are positioned around it. In these cases, feedback will be required for separate fingers and also separate users. Including haptic feedback in these interfaces could have advantages over visual feedback alone.

The third important area of research is the development of three-dimensional haptic interfaces. Touch as an input and output technique currently requires direct contact, either with tactile actuators or touch-sensitive surfaces. This limitation of current tactile technology means touch-based interaction is typically limited to the physical space directly in front of or around an interactive display. For large interactive displays, this means only a limited number of users may interact and, in crowded public settings, users must contend for space at the display. For smaller mobile displays, users must carry this device with them at all times in order to access any interactive content. However, recent advancements in technology such as ultrasonic and capacitive haptic feedback mean that distant haptic interaction may be possible.

Last and perhaps most importantly, new types of tactile technologies are becoming available all the time and it will be necessary to investigate whether the results achieved in existing research can be recreated or even improved using different types of actuators. As the technology itself becomes cheaper and more readily available, it will be possible to ensure that haptic interfaces become a part of everyday life, with mobile interaction, electronic Braille applications for visually impaired people, pedestrian navigation systems and all sizes of multitouch devices featuring sophisticated multidimensional haptic feedback.

REFERENCES

Bark, K., Wheeler, J.W., Premakumar, S. & Cutkosky, M.R. (2008) 'Comparison of Skin Stretch and Vibrotactile Stimulation for Feedback of Proprioceptive Information'. Paper presented at Symposium on Haptic interfaces for virtual environment and teleoperator systems, Washington, DC.

Bau, O., Poupyrev, I., Israr, A. & Harrison, C. (2010) 'Teslatouch: Electrovibration for Touch Surfaces'. Paper presented at 23nd Annual ACM Symposium on User Interface Software and Technology, New York.

Blattner, M.M. & Dannenberg, R.B. (eds.) 1992. *Multimedia Interface Design*. NewYork: ACM Press.

Blattner, M.M., Sumikawa, D.A. & Greenberg, R.M. (1989) 'Earcons and Icons: Their Structure and Common Design Principles', *Human Computer Interaction*, 4(1): 11–44.

Brave, S. & Dahley, A. (1997) 'InTouch: A Medium for Haptic Interpersonal Communication'. *Extended Abstracts on Human Factors in Computing Systems* (CHI EA '97), ACM, 363–364

Brown, L.M., Brewster, S.A. & Purchase, H.C. (2006) 'Multidimensional Tactons for Non-Visual Information Presentation in Mobile Devices'. Proceedings of the 8th Conference on Human-Computer Interaction with Mobile Devices and Services, (Mobile HCI '06), ACM, 231–238.

Bruce, I.A. & McKennell, A. (1991) 'Blind and Partially Sighted Adults in Britain: The RNIB Survey'. London: HMSO.

Burdea, G. (1996) *Force and Touch Feedback for Virtual Reality*. NewYork: John Wiley.

Cechanowicz, J., Irani, P. & Subramanian, S. (2007) 'Augmenting the Mouse with Pressure Sensitive Input'. *Proceedings of the Conference on Human Factors in Computing Systems* (SIGHI '07), ACM, 1385–1394.

Chang, A., O'Modhrain, S., Jacob, R., Gunther, E. & Ishii, H. (2002) 'ComTouch: Design of a Vibrotactile Communication Device'. *Proceedings of the 4th Conference on Designing interactive Systems*, (DIS '02), ACM, 312–320.

Chen, H., Santos, J., Graves, M., Kim, K. & Tan, H.Z. (2008) 'Tactor Localization at the Wrist', *Haptics: Perception, Devices and Scenarios*: 209–218.

Cholewiak, R.W., Brill, J.C. & Schwab, A. (2004) 'Vibrotactile Localization on the Abdomen: Effects of Place and Space', *Perception and Psychophysics*, 66: 970–987.

Cholewiak, R.W. & Collins, A.A. (1995) 'Vibrotactile Pattern Discrimination and Communality at Several Body Sites', *Perception and Psychophysics*, 57(5): 724–737.

Cholewiak, R.W. & Craig, J. C. (1984) 'Vibrotactile Pattern Recognition and Discrimination at Several Body Sites', *Perception and Psychophysics*, 35: 503–514.

Craig, J.C. & Sherrick, C.E. 1982. Dynamic Tactile Displays. In: ScHIFF, W. & FoULKE, E. (eds), *Tactual Perception: A Sourcebook*. Cambridge: Cambridge University Press.

Enriquez, M. & MacLean, K. (2008) 'The Role of Choice in Longitudinal Recall of Meaningful Tactile Signals'. Paper presented at IEEE Symposium on Haptic Interfaces for Virtual Environments and Teleoperator Systems, 49–56.

Enriquez, M., MacLean, K. & Chita, C. (2006) 'Haptic Phonemes: Basic Building Blocks of Haptic Communication', *Proceedings of the 8th International Conference on Multimodal Interfaces* (ICMI '06), ACM, 302–309.

Fujita, H. & Nishimoto, K. (2004) 'Lovelet: A Heart-Warming Communication Tool for Intimate People by Constantly Conveying Situation Data'. *Extended Abstracts on Human Factors in Computing Systems* (CHI '04), ACM, 1553.

Fukumoto, M. & Sugimura, T. (2001) 'Active Click: Tactile Feedback for Touch Panels'. *Extended abstracts, Human Factors in Computing Systems* Conference (CHI '01), ACM, 121–122.

Geldard, F.A. (1957) 'Adventures in Tactile Literacy', *The American Psychologist*, 12: 115–124.

Gibson, J. (1962) 'Observations on Active Touch', *Psychological Review*, 69(6): 477–491.

Gibson, J.J. (1966) *The Senses Considered as Perceptual Systems*. Boston,MA: Houghton Mifflin.

Gleeson, B.T., Horschel, S.K. & Provancher, W.R. (2009) 'Communication of Direction through Lateral Skin Stretch at the Fingertip'. Paper presented at World Haptics, Salt Lake City.

Gunther, E., Davenport, G. & O'Modhrain, S. (2002) 'Cutaneous Grooves: Composing for the Sense of Touch'. Paper presented at New Interfaces for Musical Expression, Dublin, Ireland.

Hale, K.S. & Stanney, K.M. (2004) 'Deriving Haptic Design Guidelines from Human Physiological, Psychophysical, and Neurological Foundations', *Computer Graphics and Applications*, IEEE, 24: 33–39.

Hoggan, E. 2010. 'Crossmodal Audio and Tactile Interaction with Mobile Touchscreens'. PhD thesis Glasgow. University of Glasgow,

Hoggan, E. & Brewster, S.A. (2007) 'New Parameters for Tacton Design'. Paper presented at CHI 2007, San Jose, CA.

Hoggan, E., Brewster, S.A. & Johnston, J. (2008) 'Investigating the Effectiveness of Tactile Feedback for Mobile Touchscreens'. Paper presented at CHI' 08, Florence, Italy.

Holz, C., Grossman, T., Fitzmaurice, G. & Agur, A. (2012) 'Implanted User Interfaces'. *Proceedings of the Annual Conference on Human Factors in Computing Systems*, (CHI '12), ACM.

ISO (2011) Ergonomics of Human-Computer Interaction — Part 910: Framework for Tactile and Haptic Interaction. Ref. No. ISO 9241-910:2011.

Kaaresoja, T., Brown, L.M. & Linjama, J. (2006) 'Snap-Crackle-Pop: Tactile Feedback for Mobile Touch Screens'. Paper presented at Eurohaptics '06, Paris, France.

Kandel, E.R. & Jessell, T. M. 1991. Touch. In: KaNDEL, E.R., ScHWARTZ, J.H. & JeSSELL, T.M. (eds), *Principles of Neural Science*. New York: Oxford University Press.

Kaufman, H. (2010) 'The Lightness of Your Touch', *Leonardo – SIGGRAPH 2010 Art Papers and TouchPoint Art Gallery (2010)*, 43(4): 398–399.

Klatzky, R.L. & Lederman, S.J. (2002) 'Touch', *Experimental Psychology*, 4(1): 147–176.

Laitinen, P. & Mäenpää, J. (2006) 'Enabling Mobile Haptic Design: Piezoelectric Actuator Technology Properties in Handheld Devices'. Paper presented at HAVE '06, Canada.

Lee, S. & Starner, T. (2010) 'Buzzwear: Alert Perception in Wearable Tactile Displays on the Wrist'. *Proceedings of the 28th International Conference on Human factors in Computing Systems*, Atlanta, GA.

Lee, S. & Zhai, S. (2009) 'The Performance of Touch Screen Soft Buttons'.*Proceedings of the Conference on Human Factors in Computing Science (CHI '09)*, ACM, 309–318.

Lenay, C., Canu, S. & Villon, P. (1997) 'Technology and Perception: The Contribution of Sensory Substitution Systems'. 2nd International Conference on Cognitive Technology (CT '09), 44.

Linjama, J. & MäKinen, V. (2009) 'E-Sense Screen: Novel Haptic Display with Capacitive Electro-Sensory Interface'. Paper presented at HAID '09, 4th Workshop for Haptic and Audio Interaction Design, Dresden, Germany.

Luk, J., Pasquero, J., Little, S., Maclean, K., Levesque, V. & Hayward, V. (2006) 'A Role for Haptics in Mobile Interaction: Initial Design Using a Handheld Tactile Display Prototype '. Proceedins of the Conference on Human Factors in Computing Systems (CHI '06), ACM, 171–180.

MacLean, K. & Enriquez, M. (2003) 'Perceptual Design of Haptic Icons'. Paper presented at Eurohaptics, Dublin, Ireland.

Mizobuchi, S., Terasaki, S., Keski-Jaskari, T., Nousiainen, J., Ryynanen, M. & Silfverberg, M. (2005) 'Making an Impression: Force-Controlled Pen Input for Handheld Devices'. *Proceedings of Extended Abstracts on Human Factors in Computing Systems* (CHI EA '05), ACM, 1661–1664.

Mortimer, B., Zets, G. & Cholewiak, R.W. (2007) 'Vibrotactile Transduction and Transducers', *The Journal of the Acoustic Society of America*, 121: 2970–2977.

Murray, D.J., Ellis, R.R., Bandomir, C.A. & Ross, H.E. (1999) 'Charpentier (1891) on the Size-Weight Illusion', *Perception and Psychophysics*, 61(8): 1681–1685.

Nashel, A. & Razzaque, S. (2003) 'Tactile Virtual Buttons for Mobile Devices'. *Extended Abstracts on Human Factors in Computing Systems* (CHI EA '03), ACM, 854–855.

Oron-Gilad, T., Downs, J.L., Gilson, R.D. & Hancock, P.A. (2007) 'Vibrotactile Guidance Cues for Target Acquisition', *IEEE Transactions on Systems, Man, and Cybernetics, Part C: Applications and Reviews*, 37(5): 993–1004.

Ottensmeyer, M.P. & Salisbury, J.K. (1997) 'Hot and Cold Running VR Adding Thermal Stimuli to Haptic Experience'. *Proceedings of the Second PHANToM Users Group Workshop, A1 Lab Technical Report* 1617, 34–37

Park, Y.-W., Bae, S.-H. & Nam, T.-J. (2012) 'How Do Couples Use CheekTouch over Phone Calls?'. *Proceedings of the Conference on Human Factors in Computing Systems*, (CHI '12), ACM.

Poupyrev, I. & Maruyama, S. (2003) 'Tactile Interfaces for Small Touch Screens'. Paper presented at 16th Annual ACM Symposium on User Interface Software and Technology, Vancouver, Canada.

Ramos, G., Boulos, M. & Balakrishnan, R. (2004) 'Pressure Widgets'. *Proceedings of the Conference on Human Factors in Computing Science (CHI '04)*, ACM.

Ross, H.E. & Gregory, R.L. (1970) 'Weight Illusions and Weight Discrimination – a Revised Hypothesis', *Quarterly Journal of Experimental Psychology*, 22(2): 318–328.

Sharmin, S., Evreinov, G. & Raisamo, R. (2005) 'Non-Visual Feedback Cues for Pen Computing'. Paper presented at Eurohaptics, Pisa, Italy.

Sherrick, C.E. & Cholewiak, R.W. 1986. 'Cutaneous Sensitivity'. In: Boff, K., Kaufman, L. & Thomas, J.L. (eds), *Handbook of Perception and Human Performance*. New York: Wiley

Shi, K., Irani, P., Gustafson, S. & Subramanian, S. (2008) 'PressureFish: A Method to Improve Control of

Discrete Pressure-Based Input'. *Proceedings of the Conference on Human Factors in Computing Systems* (CHI '08), ACM, 1295–1298.

Smith, J. & Maclean, K. (2007) 'Communicating Emotion through a Haptic Link: Design Space and Methodology', *International Journal of Human–Computer Studies,* 65(4): 376–387.

Srinivasan, M. & Chen, J. (1993) 'Human Performance in Controlling Normal Forces of Contact with Rigid Objects', *Winter Annual Meeting of the American Society of Mechanical Engineers,* 49: 119–125.

Stedman, N. (2010) 'ADB', *Leonardo – SIGGRAPH 2010 Art Papers and TouchPoint Art Gallery (2010),* 43(4): 414–415.

Stevens, J. (1991). 'Thermal Sensitivity'. In: Heller, M. & Schiff, W. (eds), *The Psychology of Touch.* Hillsdale, NJ: Lawrence Erlbaum.

Stone, R. (2000) 'Haptic Feedback: A Potted History, from Telepresence to Virtual Reality'. Paper presented at the Workshop on Haptic Human-Computer Interaction, Glasgow, UK.

Summers, I.R., Dixon, P.R. & Cooper, P.G. (1994) 'Vibrotactile and Electrotactile Perception of Time-Varying Pulse Trains', *The Journal of the Acoustical Society of America,* 95(3): 1548–1558.

Tan, H.Z. & Pentland, A. (1997) 'Tactual Displays for Wearable Computing'. *Proceedings of the 1st IEEE International Symposium on Wearable Computers,* ACM, 84.

Tan, H.Z., Srinivasan, M., Eberman, B. & Cheng, B. (1994) 'Human Factors for the Design of Force-Reflecting Haptic Interfaces'. Paper presented at 3rd International Symposium on Haptic Interfaces for Virtual Environment and Teleoperator Systems, Chicago, IL.

Terhardt, E. (1974) 'On the Perception of Periodic Sound Fluctuations (Roughness)', *Acustica,* 30: 201–213.

Tsukada, K. & Yasumura, M. (2004). 'Activebelt: Belt-Type Wearable Tactile Display for Directional Navigation'. In: Davies, N., Mynatt, E. & Siio, I. (eds), *Ubicomp 2004: Ubiquitous Computing.* Berlin/Heidelberg: Springer.

Van Erp, J.B. & Van Veen, H.A. H.C. (2001) 'Vibro-Tactile Information Presentation in Automobiles'. Paper presented at EuroHaptics 2001.

Van Mensvoort, I.M. (2002) 'What You See Is What You Feel: Exploiting the Dominance of the Visual over the Haptic Domain to Simulate Force-Feedback with Cursor Displacements'. Paper presented at 4th Conference on Designing Interactive Systems: Processes, Practices, Methods, and Techniques, London, UK.

Van Veen, H.a. H.C. & Van Erp, J.B. (2000) 'Tactile Information Presentation in the Cockpit'. Paper presented at First International Workshop on Haptic Human-Computer Interaction, Glasgow, UK.

Verrillo, R.T. & Gescheider, G.A. (1992) 'Perception Via the Sense of Touch', In: I.R. Summers (ed.), *Tactile Aids for the Hearing Impaired:* 1–36.

Wettach, R., Behrens, C., Danielsson, A. & Ness, T. (2007) 'A Thermal Information Display for Mobile Applications'. Paper presented at MobileHCI '07, Singapore.

Wilson, G., Halvey, M., Brewster, S.A. & Hughes, S. (2011) 'Some Like It Hot: Thermal Feedback for Mobile Devices', paper presented at CHI'11, Vancouver, Canada.

Wilson, G., Stewart, C. & Brewster, S. A. (2010) 'Pressure-Based Menu Selection for Mobile Devices'. Paper presented at MobileHCI '10, Lisbon, Portugal.

Yohanan, S. & Maclean, K.E. (2011) 'Design and Assessment of the Haptic Creature's Affect Display'. Paper presented at ACM/IEEE International Conference on Human-Robot Interaction (HRI '11), Lausanne, Switzerland.

23

Contrasting Lab-based and In-the-wild Studies for Evaluating Multi-user Technologies

Yvonne Rogers, Nicola Yuill and Paul Marshall

INTRODUCTION

Multi-user technologies, such as Circle Twelve's DiamondTouch, Microsoft's Surface, SMART Table and multi-touch tablets, offer new opportunities for supporting collaborative activities outside lab settings. The entertainment, hospitality and retail sectors were initially targeted: applications were created to enable tourists to find out where to eat in a city; customers and sales agents to develop together a personalized surfboard; and restaurant goers to order a drink at their table – all at the touch of a fingertip. More recently, large interactive displays have been designed for cultural and educational settings, in museums, schools and shopping malls, to enable groups to play collaborative games, plan tours together, or learn about the galaxy.

Despite much publicity and interest in multi-user displays in a range of settings there has been little research that demonstrates their hoped-for benefits. Much of the research has been in lab-based settings.

However, it is unclear to what extent findings from these more controlled settings can transfer to real-world uncontrolled settings. Do collaborative behaviours and interactions exhibited in lab studies – such as enhanced problem-solving, equitable participation, new forms of collaborative creativity, and better learning – also occur and persist in the real world?

This chapter addresses these questions by comparing the findings from lab-based experiments with emerging in-the-wild studies of the situated use and appropriation of multi-touch interfaces. The next section discusses the pros and cons of lab versus in-the-wild studies. The following section reviews lab-based and in-the-wild studies that have investigated group interactions for large wall displays and tabletops, after which the costs and benefits of the two kinds of studies for researchers and developers are summarized. The final section draws out the implications of adopting the two approaches and how each can inform the further development of the other.

PROS AND CONS OF LAB AND IN-THE-WILD STUDIES

Usability of technology is typically studied in the lab but, increasingly, in the field. These two settings are typically associated with different methodological approaches in HCI. Lab studies often measure systematic and overt recordings of behaviour, such as number and kind of errors, or time on task. A set of rules applies to participants in the lab: they have an explanation of what the study is about, can opt out and will probably receive some sort of compensation, the duration is restricted and the experimenter is on hand to explain and to control variables systematically. In contrast, field studies are more closely associated with a naturalistic and holistic approach. Rather than manipulating the situation and guiding participants, the field observer is at the mercy of the vagaries of the world outside the lab. The social context of the participants is at the fore in field studies, which may take place in airports, shopping malls, homes and workplaces, with participants having a history and a future, rather than being part of a one-off artificial group in the lab. This holistic set-up can require a more flexible design, where the observer might tweak the set-up in response to an observation, and theories may emerge from the data. These differences in approach can also be reflected in methods of analysis: a lab study may more often involve quantitative statistical analysis of responses, errors or time, while a field study, being less controllable, might involve qualitative analysis, case study and an ethnographic approach to analysing behaviour. Experimental methods can be used in the field and qualitative analysis in the lab, but the properties of the two different environments lend themselves to different approaches. Below we explore the pros and cons of these approaches.

Behaviour in the lab can be in stark contrast to how the technology is likely to be used in the real world – for example, where people may use only a small number of functions or be interrupted in the middle of doing something. Unsurprisingly, then, recent studies of technology use *in situ* reveal people using them differently in the wild (Rogers, 2011). Accounts are emerging from in-the-wild studies of how new technologies are being appropriated in different contexts, uncovering *unanticipated* use (e.g. Brown et al., 2011; Peltonen et al., 2008; Van der Linden et al., 2011) – rather than demonstrating *anticipated* performance. For example, Van der Linden et al. (2011) gained new insights – not evident from their lab-based studies – about designing haptic technology to improve children's violin-playing at school. Their *in situ* studies of the Music-Jacket system showed how real-time vibro-tactile feedback was most effective when matched to tasks selected by their teachers to be at the right level of difficulty – rather than by the researchers.

The diversity of uses (and non-use) being discovered suggests that people are inventive and creative but also sometimes frustrated or confused, in ways that would be difficult to predict from lab-based findings and theories (Marshall et al., 2011). Alternative findings are emerging from lightweight, one-off use contexts, such as museums and shopping malls, where the expectation is for brief walk-up-and-use explorations. A challenge of moving into the wild, however, is how best to evaluate technologies used in people's everyday lives. Researchers have to give up control; the lab 'scaffolding' is largely absent and people (not necessarily 'participants') are left to their own devices. This makes it difficult to anticipate what is going to happen and to be present when something interesting does happen. The researcher relies on making inferences from their observations of what appears to be happening and what might be causing it. Moreover, there is an emerging expectation that an in-the-wild study should be sufficiently lengthy to show behaviour patterns over time. Researchers implicitly assume this should be weeks, months or even longer, but this can make studies costly, unwieldy and difficult to conduct (Kjeldskov et al., 2004), especially if they involve equipment that needs to be maintained or many visitors. Ironically, the length of lab experiments is not questioned and, conversely, experiments may be strictly limited to an hour to avoid fatigue, with the focus instead on number of participants (Rogers, 2011).

Another effect of removing experimenter control is that people do not neatly congregate into groups, sit at desks, look at computer screens, etc., in the ways that are simply taken for granted in lab studies (Marshall et al., 2011). In a typical lab study participants are brought in and shown their place by a researcher and given instructions on what to do. This guidance is largely absent in-the-wild, with significant effects for any intended subsequent interactions or collaborations. Also, people in public settings vary in what they notice and try out when encountering technology in situ. The full range of functionality or possible set of interactions may, therefore, be rarely engaged, making it difficult to see whether they are useful, usable or capable of supporting the intended interactions.

A common approach to running lab experiments is to collect a diversity of data using a mix of methods. Typically, this includes performance measures (e.g. task completion, accuracy), logging (e.g. software measuring the frequency and position of where users touch the tabletop), surveys (e.g. preference rating scales, interviews) and video data (e.g. recording gestures and group talk). Whilst specific hypotheses may only need one kind of data (e.g. that display size will affect speed of task completion), there is often an implicit assumption that 'more is more'. Having collected an assortment of data, researchers can mine, triangulate and correlate it in interesting ways to show in detail how people interact and collaborate at a display. The downside of amassing so much data, however, is the time-consuming nature of video transcription and analysis. There is no accepted paradigm as to which analytic framework to use. Moreover, there has been a lack of suitable experimental paradigms for

comparing how different interactive surfaces affect group working (Tan et al., 2008).

A further problem lies in controlling independent and dependent variables. For example, the physical actions involved in touching a tabletop (using fingertips) or gesturing at a wall display are quite different from those involved in manipulating tangibles (handling objects) or using a PC (with mouse and keyboard). Group composition also varies on numerous dimensions (e.g. dominance, familiarity) that can be difficult to control. Instead of trying to investigate the effects of one variable while trying to control all others, an alternative approach (Rogers et al., 2009) is to adopt a less formalized, but still experimental methodology that aims to explore simultaneously the effects of a range of variables that affect collaboration. Rogers et al. (2009) suggested using a higher-level conceptual framework to characterize ways that multi-user interfaces can *invite* groups to participate, which, in turn, will affect how they collaborate. The 'shared information spaces' framework enables comparison of interfaces that vary in how they invite participation. Other conceptual frameworks (e.g. Benford et al., 2005; Yuill and Rogers, 2012) serve similar roles, where the aim is to guide researchers on experimental design, scope of investigation and appropriate analytic frameworks.

In sum, the different contexts of use make it difficult to generalize from lab findings to use in everyday settings. Appreciating their respective value means adopting a different mindset. In particular, it is important to understand how the context – be it social, experimental or physical – primes and subsequently shapes the findings of a study. This includes examining temporal–spatial aspects – for example, the physical environment and how it is used at different times of the day or week – or social-cultural aspects – such as how the technology fits into a particular culture, genre, etc. – as well as taking into account researcher–participants effects – such as trying to do one's best in an experiment to please the experimenter. As Brown et al. (2011) notes, participants have certain expectations about what investigators are looking for in an experiment (known as demand characteristics) that can affect the way they behave. Alternatively, if participants disagree with the perceived hypothesis, they may take care to act in a way that disproves it. Given this proclivity, Johnson et al. (2012) ask, how can researchers account for and manage their influence on participants in-the-wild when so many contextual factors are at play? It is important to understand the effects that different methodological approaches can have on user behaviour and participant performance when comparing and analysing the same kinds of interfaces that are used in them. One simply cannot assume that an effect found in one readily transfers to another. In the next two sections, we compare lab and in-the-wild studies, first for large wall displays and then for tabletop surfaces.

LARGE WALL DISPLAYS

Lab Studies

Much early work on multi-user displays was concerned with how to enable users to interact with digital content on large wall displays. Pioneering research by Guimbretière et al. (2001) investigated how 'natural' gestures and handwriting could be used to create such content. The goal was to support fluid interactions – to sketch, create and manipulate images and text for brainstorming or problem-solving – with minimal disruption from the mechanics of the interface. Having to remember or repeat a gesture because the system does not recognize it requires attentional switching that can disrupt activity. Lab studies could show the usability of the pen and gesturing, but were less successful at demonstrating whether fluid interactions were obtained, or indeed, attention switching was a genuine problem.

The advent of high-resolution wall displays, up to 6 metres on the diagonal with a large matrix of high-quality displays (e.g. 4 x 9 30-inch tiles) enable visualization of very

large datasets, numerous images stitched together and complex simulations. Their sheer size and resolution enables new forms of large-scale gesture-based navigation. People can step back to get an overview of the displayed data and forwards again to see details, such as text. However, this means that parts of the display are out of sight or reach when close up. An assortment of input devices has been experimented with in lab studies to allow users to point in mid-air at displays, including 3D mice, wands and touchless interaction. User performance has been measured in terms of time and errors when comparing gesture-based and handheld input devices for specific tasks, such as zooming and panning (Nancel et al., 2011). Findings are mixed, but gestures have been found to be more error and fatigue prone.

Hence, lab studies have highlighted that a challenge for gesture-based interaction with wall displays is selecting gestures that are easy to perform, map on to the underlying operations in a meaningful way and are distinctive enough from each other for the system to recognize. Gestures should be designed to mimic the iconic aspect of the operation being represented, such as a sweeping movement for dragging (Grandhi et al., 2011).

Many lab-based studies have been concerned with how best to interact with wall displays, using several performance measures. In contrast, studies of large displays in-the-wild have focused more on investigating social issues, such as level of engagement and extent of collaboration. Measures have been more qualitative, gleaned from interviews, ethnographic observations and surveys.

In-the-Wild Studies

Early research explored how communities and the general public would use new kinds of large wall displays situated in public spaces, such as hallways, cafés and malls. A goal was to encourage participation and a sense of community in organizations, city centres and schools. However, it soon became apparent that people did not use them as intended because of self-consciousness in these public settings. Early studies of situated displays found that people felt uncomfortable knowing that their actions and their effects were highly visible to strangers or colleagues, affecting their willingness to participate in a group activity (Brignull and Rogers, 2003). For example, Churchill et al. (2003) found that people needed constant encouragement and demonstration to interact with a shared whiteboard system, Plasma Poster. Agamanolis (2003: 329) also noted how, 'half the battle in designing an interactive situated or public display is designing how the display will invite that interaction'.

A seminal finding arising from an early in-the-wild study of public displays was the 'honey pot' effect (Brignull and Rogers, 2003). As the number of people immediately near a shared display increases, a sociable 'buzz' results. By standing in this space and showing an interest (e.g. visibly looking at the screen), people give a tacit signal to others that they are willing to take part. Bystanders need to cross the threshold from peripheral to focal awareness activities (e.g. from chatting to someone on the other side of the room to moving within view to have a better look). Their understanding of what the display is, and what it has to offer, has to entice them forward.

Familiar PC input methods, such as keyboards, mice and personal phones, have been used to help overcome passer-by reluctance to interact with a public display the use of which is not obvious. The Dynamo system (Izadi et al., 2003) designed the exchange and sharing of digital media on a large shared display to be a lightweight and straightforward activity. Multi-user interaction was supported through wireless keyboards and mice or laptops that were placed freely around a space. A study of the Dynamo system in a school common room showed that many students displayed and exchanged photos, video and music, which they had created themselves or brought in from home, entertaining audiences, posting notices for others,

playing together on the surface, and engaging side-by-side in group discussions and interactions (Brignull et al., 2004). The resulting social atmosphere provided opportunities for them to engage with others who they wouldn't normally talk to. However, a researcher was always at hand, helping the students as and when requested. The common room was also 'owned' by the students who felt comfortable in their own space.

In more 'general' public settings, such as a street or a mall, persuasion or even coercion by a facilitator may be required to encourage participation. For example, O'Hara et al. (2008) found that passers-by required a compère cajoling them into joining a collaborative game played on a large public display in a busy UK shopping mall. The Red Nose Game started with blobs being splattered on the screen, which the passers-by had to push together by using their bodies. These were tracked by an embedded live camera feed; when the image of a player touched a blob it enabled that person to push it around the screen towards the other blobs. O'Hara et al. (2008) suggested that people did not want to participate in case they made fools of themselves in front of others. However, once in the game, people 'joined forces' and worked closely together as groups, developing effective strategies to move their blobs together, such as linking arms and sweeping the blobs together across the screen.

This kind of street-based group action in unfamiliar groups is novel where people won't have any previous experience with a pre-digital version. This may accentuate the difficulties of getting over the threshold to participate. Other street-based public displays have been designed without a need to engage with strangers in a group activity and more as an individual interaction that can still invite others to join in. For example, Peltonen et al. (2008) describe CityWall, a large vertical multi-touch display installed in a city street to enable photo browsing. They detail several phenomena: the influence of users in drawing attention to the display, performative actions to communicate intentions or to engage others in playful activity and patterns of shared use – primarily parallel activity by both strangers and acquaintances, but also working together and conflict resolution where the activity of one user interfered with that of another.

Magical Mirrors (Michelis and Müller, 2011) was deployed in Berlin to enable passers-by to see their own mirror image on four 'mirrors' and interact with virtual objects through gestures and body movements. The four adjoining displays each showed a mirror image of the environment in front of them that reacted to their bodily gestures. An *in-situ* study analysed the behaviour of over 600 passers-by, revealing a number of patterns. These included glancing at a display when passing it, moving arms to cause effects, then approaching the other displays and centring oneself in front of them. Based on the observations, the 'audience funnel' framework was proposed, which suggests how people move between phases of engagement (Brignull and Rogers, 2003). Being able to view the consequence of one's own interactive behaviour was found to be an important factor in determining the extent to which the users went on to explore the other displays (Michelis and Müller, 2011).

However, perhaps most striking is Huang's (2007) observation of public displays. She studied how people used and reacted to 31 large displays in public settings by examining *existing* (i.e. non-research, non-prototype) large ambient information displays that showed non-critical information in a variety of public settings. People rarely did more than glance at situated displays, if at all, and tended only to do so after something else nearby grabbed their attention. The findings suggest that content or salience of information depicted on a public display will not necessarily be eye-catching; location and context will determine whether people will even glance. This challenges the generalizability of lab findings on how people approach and use novel displays – especially if it is all that is physically in the lab space. For example, while Beyer et al. (2011) were able to

show differences between how people moved around a flat and a cylindrical screen, it seems unlikely that these differences would transfer directly to a real-world setting – such as Huang examined. Admittedly, Beyer et al. (2011) discuss the limitations of their study in terms of how other factors may influence the way people move around displays, such as being situated in streets, and the presence of buildings and other pedestrians. Their design implication – that cylindrical displays should be situated in busy areas – may be based on empirical data of what people do in a lab, but perhaps any kind of display, be it cylindrical, flat, small or large, may be subject to display blindness. Maybe passers-by who do stop and glance at it will be part of a small group, rather than individuals, and approach it in quite different ways. As shown in the examples above, an in-the-wild study may reveal quite different behaviours.

Most in-the-wild studies of large wall displays have been conducted in public spaces. However, a few have been conducted in work settings, to determine how they integrate with the existing work set-up. For example, Huang et al. (2007) examined the challenges of deploying interactive displays in the NASA Ames research labs as part of the Mars Exploration Rover mission. They found that the software applications customized specifically for the wall display, and not available on any of the other technological systems, were under-used. They suggest several reasons, including time constraints, overheads involved in learning competent use, technical complexity, organizational priorities and other external pressures.

TABLETOPS

Multi-touch tabletops may support flexible and collaborative ways of interacting (Rogers and Rodden, 2003). The horizontal surfaces are designed to support dynamic fingertip actions, such as swiping, flicking, pinching and tapping that have been mapped on to specific kinds of operations – for example, zooming in and out of maps, moving photos, selecting letters from a virtual keyboard when writing and scrolling through lists. Two hands can also be used together to stretch and move objects, as if stretching an elastic band or scooping objects together.

Similar to wall displays, a core research question has been to determine how easy and seamless it is to shift from watching what is happening at a tabletop to knowing where and how to interact around it. Can groups comfortably access, create, interact with and move digital content in an equitable and free-flowing manner? Lab and in-the-wild studies have investigated a number of factors, including how obvious it is to the participants what to do in such settings, how comfortable people feel and how to take turns in collaborative tasks.

Lab Studies

In contrast to the findings for large wall displays, people appear more willing to interact with multi-touch tabletops in the presence of others (Rogers and Rodden, 2003). This may be because these kinds of shared horizontal surfaces are able to lure people without them feeling intimidated or embarrassed about the consequences of their actions. For example, small groups were more comfortable working together around a tabletop compared with sitting in front of a PC or lined up before a vertical display (Rogers and Lindley, 2004). The familiar and lightweight action of touching a surface may also make it easier for people to take part in a social/public setting. Lab studies have shown how groups of people, new to tabletops, find them easier and more enjoyable to use when sharing and assembling sets of digital images for a variety of collaborative tasks (Ryall et al., 2004; Scott et al., 2003; Shen et al., 2002).

Many lab studies investigating tabletop interfaces have been concerned with isolating specific effects, such as form factors of the interface (table size, orientation, type of input, allocation of personal/shared space), the pros and cons of novel inputs (such as

gestures, digital pens); and the role of group processes (e.g. size, make-up, shyness, dominance, age). Findings suggest overall that co-located groups find it more comfortable to work together on such tables compared with other interfaces adapted from single-user technologies (Rogers et al., 2009). However, the studies have tended to be piecemeal, focusing on specific features. This is not surprising given the many degrees of freedom and possible factors than can influence group interactions.

One strand of this research agenda has compared the efficacy of different interaction techniques at the tabletop interface, such as how best to select, zoom, open, move, shrink and expand images and objects. The focus has been on developing techniques best suited to finger-tip interaction when looking down on to a display – a quite different experience from using the canonical GUI and WIMP styles at a vertical PC display. Early experiments examined different ways of moving images around on a tabletop. For example, Forlines et al. (2005) investigated the efficacy of using different interaction mechanisms for document movement tasks. They proposed a number of hypotheses about the speed and accuracy of document positioning, using three different techniques for sliding them across the interface. The results from their controlled experiment showed participants performing best using one technique (document-centric) and that, in general, the smaller the document, the faster it could be repositioned. Despite this, many participants rated the document-centric technique as their least preferred method. This discrepancy between performance and preference raises important questions about the quality of the user experience, especially as it is unlikely to be revealed by in-the-wild studies since people will not be asked to perform a set of tasks and then rate the different methods used.

Others have explored how two or more variables interact with each other. For example, Ryall et al. (2004) looked at how size of group and tabletop surface affected group interactions. They compared the performance of groups of two, three and four participants constructing poems by moving digital word tiles around two tabletops of different sizes. Tabletop size had no effect on task completion but larger groups were significantly faster at assembling the poems than the smaller groups, apparently because it was harder for smaller groups to search for the words. However, the smaller groups appeared to work better at the larger table. The performance data by themselves do not reveal why size matters. However, the findings led the researchers to conjecture why the behaviour occurred, by drawing from a social psychology theory about the diffusion of responsibility (Darley and Latane, 1968). They suggest that perhaps table size affects the participant's perceptions of who is responsible for which tiles. There appears to be an unspoken consensus that depending on the size of group and table, the surface will be carved up in the minds of the participants in terms of 'their area' and 'everyone's area'. The smaller the table the more space is 'everyone's area' since everyone can reach it and thus it is no specific person's responsibility.

The importance of shared versus own space has led to many further experiments on territoriality, with researchers implicitly or explicitly dividing the tabletop into regions that will influence such behaviours. For example, Scott et al. (2004) observed three different types of tabletop interaction areas: personal, group and storage. These were assumed to help people organize tasks while collaborating with others at the table. In one study, participants used the edge zones nearest them for keeping items they used frequently. Using the closest space as a personal work area seems quite natural when groups work both individually and in parallel (e.g. Rogers et al., 2006).

Others have investigated factors that affect people's willingness and comfort in participating in collaborative tasks on tabletops. Birnholtz et al. (2007) found that type of input used can affect group behaviour, especially influencing how individuals can negotiate. Morris et al. (2006) investigated

whether gestures requiring explicit coordination by the group members increased engagement. Enforcing collaboration in this manner discouraged social loafing, where some people work less in a group than when working independently.

Lab-based tabletop research typically tests specific hypotheses about given social phenomena (e.g. Harris et al., 2009; Marshall et al., 2007; Piper et al., 2006). In a series of lab studies comparing multi- with single-touch, Marshall et al. (2008) found that changing the ways groups can interact at an interface affects the way people contribute physically to the task. More equitable physical interactions resulted from the multi-touch condition, but this did not affect verbal contributions. Dominant people continued to talk most while quiet ones remained quiet. The studies demonstrate how technology cannot force people to speak more or less but it can encourage change in physical interaction at the tabletop. In another study with 7–10-year-olds, comparing multi- and single-touch, Harris et al. (2009) found that children talked more about turn-taking and less about the task when the interface was constrained to only allow one user to be able to interact at a given time. Piper et al. (2006) compared a human moderator enforcing the rules of a turn-taking game with the software enforcing the rules, with the aim of facilitating social skills development in a comfortable and motivating way amongst adolescents with behavioural difficulties. They found that enforcing rules in this manner was more effective than when a human facilitator tried to enforce them. The students found computer-based constraints to be more comfortable and consistent than a human moderator.

This lab-based tabletop research has revealed differences for various conditions and when compared with other technologies. It has also led to the development of various indices of co-located collaboration, including awareness of collaborators' action, density of interaction, social presence and equity of participation. These indicators have provided both quantitative and qualitative ways of characterizing and operationalizing aspects of group processes around interactive displays. Several analytic frameworks have also been outlined that articulate the subtle mechanisms at play. For example, the collaborative learning mechanisms framework analysed the way verbal interactions and physical actions are coupled during collaboration around a tabletop (Fleck et al., 2009). Such frameworks can uncover the nuances of how groups collaborate, providing insights into how to promote certain behaviours while constraining others.

In-the-Wild Studies

Far fewer studies of interactive surfaces have been conducted in real-world contexts. However, with the increasing availability of affordable and portable multi-touch interfaces, coupled with a growing interest in research in-the-wild (Rogers, 2011), more studies are appearing. They are often conducted to see how people come to understand and appropriate technologies for their own situated purposes. One research direction has been 'one-off' encounters – for example, how groups approach walk-up-and-use interactive tabletops in public places. A main focus of these studies is to observe and record what people do *in situ* and how this changes over suitable periods of time, varying from weeks to months. A longer period of time is often considered desirable – to see patterns of habits emerge – but this will be constrained by budget and researcher time.

One of the earliest studies of multi-user surfaces following up lab research in-the-wild was of a novel interactive travel planner – designed for travel agents and customers to build round the world trips (Rodden et al., 2003; Rogers and Halloran, 2007). An initial lab study of a single tabletop display revealed users' physical problems with moving objects and documents around in order to free space to look at others. It also revealed some of the 'social difficulties' of moving objects in a shared space with others looking on. It was assumed that if these happened in the lab,

they would be only heightened in the wild. A revised system used three interlinked semi-horizontal displays. Different features of the travel planner were then 'locked' to specific spatial positions, so that their relative locations persisted. This meant people knew where to look for something. It also enabled both customers and agents to easily refer to the same object, rather than one collaborator having to work out what the other has done in the display or what they were looking at.

The planner was deployed as part of an international trade show in London. The travel company chose to 'dress' it up as an eye-catching showpiece. They embellished it with a large koala bear, piles of travel brochures, posters and a vase of flowers. The effect was to draw all manner of people to the stand – single businesspeople, families, young and old people, couples and groups of friends. Over the four days, about 100 different groupings interacted with the travel planner alongside a trained sales assistant. Sometimes it was left to run unattended and experimented with by staff and members of the public.

When shown the system, the agents quickly saw its potential for their work, while visitors were drawn to the attractive stand, first watching others using it and then having a go themselves. Most agent–customer groups spent 5–10 minutes working together, although some interactions lasted up to 30 minutes. Many of the interactions resulted in complex itineraries being created and, in some cases, actual bookings. Conversation flowed freely and the agents often responded to suggestions and requests put forward by the customers.

Based on these initial positive findings, the travel planner was then placed in the company's flagship store. However, the agents were less enthusiastic about using it in this setting than at the trade show because of the pressures related to their commission-based salary. As a consequence, the system was used much less in the store during regular hours, being largely championed by only one particular agent.

The travel planner illustrates what happens when people are expected to encounter a completely novel technology once and for a short period of time. Having an agent at hand encouraged customers to interact with it. When there is no one at hand, however, quite different experiences can result. O'Hara (2010) described some of the problems that can arise when people sit down at a tabletop system in a café. Sometimes they don't realize it is even an interactive tabletop, highlighting some of the issues that occur when moving between interactive and non-interactive use: the interactivity could draw attention to otherwise innocuous gestures such as tapping on the surface, causing social discomfort.

Tabletops have also been placed in museums. Hornecker (2008) describes a museum multi-touch system that asked users questions about natural history. She noted how it often failed to encourage social interactions. Subtle usability issues also impacted the experience, making it confusing for the user to know what to do next or what had caused an image to get larger or move in a certain way.

Similar usability issues were found by Marshall et al. (2011), who placed a tabletop with a group-planning application in a tourist information centre. Several visitors used 'wrong' fingertip tapping gestures, seemingly drawing from their well-honed repertoire of double-click mouse interactions. The first touch for most was a tap or double tap, suggesting that they treated the tabletop surface like a mouse or phone. However, this did not always result in a successful response. After a few more taps many simply left the tabletop without getting any further. Others attempted 'standard' multi-touch interactions like pinch-zooming. Interviews confirmed that some owned iPhones or other multi-touch devices while others had seen demos of the Microsoft Surface online. Hence, previous experience with multi-touch and other interfaces can influence people's initial finger-tip gestures, which will determine whether they continue or walk away. In this public walk-up-and-use scenario, people often didn't give the interface a second chance if they weren't immediately successful.

Another striking difference from earlier lab studies was the rarity of the envisioned scenario of a 'party of four' planning together and discussing more equitably what they should do. Instead, families, friends and couples were observed often splitting up upon entering the tourist information centre and independently foraging for various information resources around the walls. When they did approach the tabletop, they did so in a staggered buffet style of interaction rather than in a dining-table style (all coming together simultaneously). This observation showed how groups do not form in the ways often taken for granted in the lab, suggesting that the concept of 'multi-user' needs rethinking in this context.

Other tabletops deployed *in situ*, have been in homes and schools where people develop a certain ownership. Hence, a key research question is to determine how they are integrated into that setting and how they are used by a group – family, class – over time. Kirk et al. (2010) deployed the Family Archive multi-touch device in three homes for a month. The system supported scanning and archiving sentimental artefacts and memorabilia. They describe how it disrupted existing family roles and practices and was typically used asynchronously. Cao et al. (2010) developed TellTable, a narrative construction tool, on a Microsoft Surface and installed it in a school library for two weeks, with children able to use it during breaks and some lessons. They describe how the tabletop fitted into the existing school culture, access being controlled through a booking system by the librarian, and its central role in the development of genres, practices of planning and an emerging culture of story-telling reputation. Stories created on TellTable became part of the school's social structure: stories were appropriated and developed by others away from the table, authors became known for their stories and children became conscious of developing a reputation for particularly popular stories. Embedding the table in the school, therefore, contributed to its social fabric.

Interactive and social contexts also influenced the choice of gestures used by visitors to the Vancouver Aquarium in their interaction with a large interactive tabletop (Hinrichs and Carpendale, 2011). In particular, an in-the-wild study showed how gestures are not executed in isolation but linked into integrated sequences where previous gestures shape the formation of subsequent gestures. In contrast to lab-based studies (e.g. Wobbrock et al., 2008; Wu et al., 2006) that have focused on one-to-one mappings for sets of gestures, Hinrichs and Carpendale's findings suggest it may be more useful in such public contexts to support many-to-one mappings between gestures and their actions to enable fluid transitions between gestures and social interactions.

SUMMARIZING THE DIFFERENCES

While a lab-based study enables preliminary usability and user experience aspects to be evaluated, such as a display's legibility, users' ability to perform operations at it and strategies for error-recovery, an in-the-wild study is more likely to reveal the unexpected, which may not fit with the designers' expectations or anticipated use, but can be most illuminating about technology-augmented group behaviour. The differences between lab and in-the-wild studies include the following.

1. **Motivations for participating**: a brief paid experiment or a long period integrating a novel technology into their lives.
2. **Predictability**: controlled lab studies can isolate the effects of particular factors on performance, such as size or orientation of interface. However, in so doing, they can fail to capture many of the complexities of the situations in which the applications will ultimately be placed. In-the-wild studies can better explore how people come to understand, use and appropriate technologies on their own terms and for their own situated purposes. However, it is difficult to ascertain what causes people to behave in certain ways around the displays.

3. **Challenge of the unexpected**: a lab study investigates factors as planned by the experimenter, but in-the-wild studies can reveal unexpected effects of the context of use not anticipated in a lab study.
4. **Time period**: in-the-wild studies can take weeks and months compared to an hour or so in the lab (typically one hour) enabling patterns of behaviour to emerge.
5. **Control over the situation**: rests mainly with the researcher in the lab, but in-the-wild studies largely pass the agenda to the participants.
6. **Control of variables**: it is difficult in the wild to separate independent and dependent variables into a neat experimental design and difficult to collect data in a controlled and predictable way. However, lab studies can tempt a researcher into collecting too much data, which can obstruct an holistic view of the technology in context.

DISCUSSION AND CONCLUSION

Lab studies of multi-touch surfaces are invaluable for providing insights into behavioural phenomena and for testing theories about technology-augmented collaboration – especially for activities not previously feasible (e.g. simultaneous interaction at an interface). However, the findings do not transfer straightforwardly into real-world settings and hence can only partially inform the design of new collaborative applications. In contrast, in-the-wild studies provide a 'contextual backdrop' against which to reflect upon the design of a new technology, sensitizing researchers as to how it *would* (rather than should) be used in practice. They also provide a real-world grounding for proposing new functions and features, for example as suggested by Hinrichs and Carpendale (2011), who proposed designing multi-touch gesture sets as integrated interaction sequences based on their in-the-wild study of gestures – which are different from the suggestions promoted by the lab-based studies that investigated the efficacy of one-to-one mappings between actions and gestures.

An alternative strategy is to attempt the best of both worlds by bolting one on to the other. For example, one approach that has been promoted is the 'living lab', which attempts to simulate a particular environment, such as the home, but which is instrumented to sense all manner of human behaviours (e.g. Intille et al., 2006; Kidd et al., 1999). This enables more of the real-world context to be simulated whilst also maintaining the ability to manipulate variables and measure behaviour. However, compromises have to be made; participants have to agree to a contract with the experimenter and hence a suspension of disbelief, where they know it is not a real house, shop or school, but has been designed to create such a context. In some ways it is a bit like dressing up for a while and acting. While this can be useful for eliciting user feedback at various stages of the design process, it is inevitably artificial and so unlikely to reflect all that will happen in the real world – especially as people are brought in for a time-period, say to pretend to live or cook in the make-shift home, just as in their own house. Anyone who has stayed in a serviced apartment or someone else's home knows how different it is – even if it may be the same in functional ways.

Another idea is to make the wild more like a lab, by augmenting a real-world office, classroom or kitchen with cameras, etc., to record events of interest and see what emerges. An example from the 1980s was Xerox's Media Spaces, technologies designed to support real-time collaboration among people based in different locations (e.g. Gaver et al., 1992). In-house systems, such as RAVE, enabled researchers in the US and Europe to work together in this manner, enabling new patterns of use to emerge in their everyday work settings. The daily experiences of using the RAVE environment led to reflection about the organizational use of such technology-rich environments, including issues of peripheral awareness, communication and privacy concerns in media spaces.

Another kind of augmented environment is RoomQuake (Moher, 2006), where a real classroom was augmented with embedded computers and other instrumentation to enable groups of children in the classroom to locate the epicentre and magnitude of a series of simulated earthquakes. The researchers designed the set-up with the conceit that the phenomena were occurring directly in the room, as if the room were a scaled-down version of a large geographic region. The seismic events occurred randomly throughout a period of several weeks; the students stopped what they were doing, such as learning French, and, using the simulated instrumentation, determined the epicentre and magnitude. The set-up resulted in high levels of engagement and learning, but these might have been partially attributed to the seismic event taking place during a boring math lesson, for example.

Such hybrid approaches have their benefits but may be subject to the Hawthorn effect, where being observed using novel technologies, in itself is what can change behavioural responses. Rather than meshing them together in this way, we can consider how in-the-wild studies could begin to inform the design of lab studies. Whereas lab studies have permitted a controlled investigation of specific variables using precise measuring tools, in-the-wild studies are enabling the unanticipated to emerge, in ways that might then inform further lab studies. Moreover, in-the-wild studies can raise new concerns that may be tacit or overlooked in lab studies. For example, Johnson et al. (2012) discuss how a quite different framing and orientation are required; where researchers reflect upon their accountability when running a study, considering how to manage their influence on their participants (sic). Moreover, they argue that instead of trying to minimize their effect on participants by distancing themselves, researchers should try to understand their impact by being in the 'thick' of in-the-wild studies. This can guide the researcher to know what questions to ask in interviews or what to look for.

In sum, lab approaches can highlight variables of interest, but may miss important factors that only come to light in in-the-wild settings. Careful construction and analysis of in-the-wild studies can, as we have seen, lead to frameworks for analysing factors in context – for example, of shared information spaces that can feed back into the design of lab studies.

ACKNOWLEDGEMENTS

The work for this chapter was supported by a grant from the EPSRC for the ShareIT project (EP/F017324/1). We thank all members of the ShareIT project for their contributions.

REFERENCES

Agamanolis, S. (2003) Designing displays for human connectedness. In O'Hara, K., Perry, M., Churchill, E. and Russell, D. (eds), *Public and Situated Displays*. Dordrecht: Kluwer.

Benford, S., Schnädelbach, H., Koleva, B., Anastasi, R., Greenhalgh, C., Rodden. T., Green, J., Ghali, A., Pridmore, T.P., Gaver, B., Boucher, A., Walker, B., Pennington, S., Schmidt, A., Gellersen, H. and Steed, A. (2005) Expected, sensed, and desired: A framework for designing sensing-based interaction. *Transactions on Computer–Human Interaction* (TOCHI), ACM, 12(1): 3–30.

Beyer, G., Alt, F., Müller, J., Schmidt, A., Isakovic. K., Klose, S., Schiewe, M. and Haulsen, I. (2011) Audience behavior around large interactive cylindrical screens. *Proceedings of the Conference on Human Factors in Computing Systems* (CHI '11), ACM, 1021–1030.

Birnholtz, J.P., Grossman, T., Mak, C. and Balakrishnan, R. (2007) An exploratory study of input configuration and group process in a negotiation task using a large display. *Proceedings of the Conference on Human Factors in Computing Systems* (CHI '07), ACM, 91–100.

Brignull, H. and Rogers, Y. (2003) Enticing people to interact with large public displays in public spaces. *Proceedings of INTERACT 2003*, 17–24.

Brignull, H., Izadi, S., Fitzpatrick, G., Rogers, Y. and Rodden, T. (2004) The introduction of a shared interactive surface into a communal space. *Proceedings of the Conference on Computer Supported Cooperative Work* (*CSCW '04*), ACM, 49–58.

Brown, B., Reeves, S. and Sherwood, S. (2011) Into the wild: Challenges and opportunities for field trial methods. *Proceedings of the Conference on Human Factors in Computing Systems* (CHI '11), ACM, 1657–1666.

Cao, X., Lindley, S., Helmes, J. and Sellen, A. (2010) Telling the whole story: Anticipation, inspiration and reputation in a field deployment of TellTable. *Proceedings of the Conference on Computer Supported Cooperative Work* (CSCW '10), ACM, 251–260.

Churchill, E.F., Nelson, L. and Denoue, L. (2003) Designing digital bulletin boards for social networking. In O' Hara, K., Perry, M., Churchill, E. and Russell, D. (eds), *Public and Situated Displays*. Dordrecht: Kluwer.

Darley, J.M. and Latane, B. (1968) Bystander intervention in emergencies: Diffusion of responsibility. *Journal of Personality and Social Psychology*, 8(4): 377–383.

Fleck, R., Rogers, Y., Yuill, N., Marshall, P., Carr, A., Rick, J. and Bonnett, V. (2009) Unpacking collaboration around the tabletop: Implications for collaborative learning. *Proceedings of the International Conference on Interactive Tabletops and Surfaces* (ITS '09), ACM, 189–196.

Forlines, C., Shen, C., Vernier, F. and Wu, M. (2005) Under my finger: Human factors in pushing and rotating documents across the table. In *Proceedings of INTERACT 2005*, 994–997.

Gaver, W.W., Moran, T., MacLean, A., Lovstrand, L., Dourish, P., Carter, K.A. and Buxton, W. (1992) Realizing a video environment: EuroPARC's RAVE system. *Proceedings of the Conference on Human Factors in Computing Systems* (CHI '92), ACM, 27–35.

Grandhi, S.A., Joue, G. and Mittelberg, I. (2011) Understanding naturalness and intuitiveness in gesture production: Insights for touchless gestural interfaces. *Proceedings of the Conference on Human Factors in Computing Systems* (CHI '11), ACM, 821–824.

Guimbretière, F., Stone, M.C. and Winograd, T. (2001) Fluid interaction with high-resolution wall-size displays. *Proceedings of the 14th Annual Symposium on User Interface Software and Technology*, (UIST '01), ACM, 21–30.

Harris, A., Rick, J., Bonnett, V., Yuill, N., Fleck, R., Marshall, P. and Rogers, Y. (2009) Around the table: Are multiple-touch surfaces better than single-touch for children's collaborative interactions? *Proceedings of the 9th International Conference on Computer Supported Collaborative Learning*, (CSCL '09), ACM, 335–344.

Hornecker, E. (2008), 'I don't understand it either, but it is cool', Visitor interactions with a multi-touch table in a museum. *3rd IEEE International Workshop on Horizontal Interactive Human Systems (Tabletop 2008)*, 113–120.

Hornecker, E., Marshall, P., Dalton, N. and Rogers, Y. (2008) Collaboration and interference: Awareness with mice or touch input. *Proceedings of the Conference on Computer Supported Cooperative Work* (CSCW '08), ACM, 167–176.

Huang, E.M. (2007) When does the public look at public displays? *Companion Proceedings of the Conference on Ubiquitous Computing (UbiComp 2007)*.

Huang, E.M., Mynatt, E.D. and Trimble, J.P. (2007) When design just isn't enough: The unanticipated challenges of the real world for large collaborative displays. *Personal and Ubiquitous Computing: Special Issue on Ubiquitous Computing in the Real World*, 11(7): 537–547.

Hinrichs, U. and Carpendale, S. (2011) Gestures in the wild: Studying multi-touch gesture sequences on interactive tabletop exhibits. *Proceedings of the Conference on Computer Supported Cooperative Work* (CH '11), ACM, 3023–3032.

Intille, S., Larson, K., Tapia, E., Beaudin, J., Kaushik, P., Nawyn, J., Rockinson, R. (2006) Using a live-in laboratory for ubiquitous computing research. *Pervasive'06*, LNCS, Heidelberg: Springer. 349–365.

Izadi, S. Brignull, H., Rodden, T., Rogers, Y. and Underwood, M. (2003) Dynamo: A public interactive surface supporting the cooperative sharing and exchange of media. *Proceedings of the 16th Annual Symposium on User Interface Software and Technology* (UIST '03), ACM, 159–168.

Jacucci, G., Morrison, A., Richard, T.G., Kleimola, J., Peltonen, P., Parisi, L. and Laitinen, T. (2010) Worlds of information: Designing for engagement at a public multi-touch display. *Proceedings of the conference on Human Factors in Computing Systems* (SUGCHI '10), ACM, 2267–2276.

Johnson, R., Rogers, Y., van der Linden, J. and Bianchi-Berthouze, N. (2012) Being in the thick of in-the-wild studies: The challenges and insights of researcher participation. *Proceedings of the conference on Human Factors in Computing Systems* (SIGCHI '12), ACM, 1135–1144.

Kidd, C., Orr, R., Abowd, G., Atkeson, C., Essa, I., MacIntyre, B., Mynatt, E. and Starner, T. (1999) The Aware Home: A living laboratory for ubiquitous computing research. *Proceedings of the Second International Workshop on Cooperative Buildings, CoBuild '99*. London: Springer-Verlag. 191–198.

Kirk, D.S., Izadi, S., Sellen, A., Taylor, S., Banks, R. and Hilliges, O. (2010) Opening up the family archive.

Proceedings of the Conference on Computer Supported Cooperative Work , (CSCW '10), ACM, 261–270.

Kjeldskov, J., Skov, M., Als, B. and Høegh, R. (2004) Is it worth the hassle?: Exploring the added value of evaluating the usability of context-aware mobile systems in the field. In Brewster, S., Dunlop, M.D. (eds), *MobileHCI 2004*, LNCS 3160. Heidelberg: Springer. 61–73.

Marshall, P., Fleck, R., Harris, A., Rick, J., Hornecker, E., Rogers, Y., Yuill, N. and Dalton, N.S. (2009) Fighting for control: Children's embodied interactions when using physical and digital representations. *Proceedings of the conference on Human Factors in Computing Systems* (CHI '09), ACM, 2149–2152.

Marshall, P., Hornecker, E., Morris, R., Rogers, Y. and Dalton, N. (2008) When the fingers do the talking: A study of group participation for different kinds of shareable surfaces. *Proceedings of IEEE Tabletops and Interactive Surfaces' 08*, 37–44.

Marshall, P., Morris, R., Rogers, Y., Kreitmeyer, S. and Davies, M. (2011) Rethinking 'multi-user': An in-the-wild study of how groups approach a walk-up-and-use tabletop interface. *Proceedings of the conference on Human Factors in Computing Systems*. (CHI '11), ACM, 3033–3042.

Michelis, D. and Müller, J. (2011) The audience funnel: Observations of gesture-based interaction with multiple large displays in a city center. *International Journal of Human–Computer Interaction*, 27(6): 562–579.

Moher, T. (2006) Embedded phenomena: Supporting science learning with classroom-sized distributed simulations. *Proceedings of the Conference on Human Factors in Computer Systems* (CHI '06), ACM, 691–700.

Morris, M.R., Piper, A.M., Cassanego, A., Huang, A., Paepcke, A. and Winograd, T. (2006) Mediating group dynamics through tabletop interface design. *IEEE Computer Graphics and Applications*, Sept/Oct 2006, 65–73.

Nancel, M., Wagner, J., Pietriga, E., Chapuis, O. and Mackay, W. (2011) Mid-air pan-and-zoom on wall-sized displays. *Proceedings of the Conference on Human Factors in Computer Systems* (CHI '11), ACM, 177–186.

O'Hara, K. (2010) Interactivity and non-interactivity on tabletops. *Proceedings of the Conference on Human Factors in Computer Systems* (CHI '10), ACM, 2611–2614.

O'Hara, K., Glancy, M. and Robertshaw. S. (2008) Understanding collective play in an urban screen game. *Proceedings of the Conference on Computer*

Supported Cooperative Work (CSCW '08), ACM, 67–76.

Peltonen, P., Kurvinen, E., Salovaara, A., Jacucci, G., Ilmonen, T., Evans, J., Oulasvirta, A. and Saarikko, P. (2008) It's mine, don't touch!: Interactions at a large multi-touch display in a city centre. *Proceedings of the Conference on Human Factors in Computer Systems,* (CHI '08), ACM, 1285–1294.

Perry, M., O'Hara, K., Beckett, S. and Subramanian, S. (2010) WaveWindow: Supporting performative gestural interaction in a public setting. *Proceedings of the International Conference on Interactive Tabletops and Surfaces* (ITS '10), ACM, 109–112.

Piper, A.M., O'Brien, E., Morris, M.R. and Winograd, T. (2006) SIDES: A cooperative tabletop computer game for social skills development. *Proceedings of the Conference on Computer Supported Cooperative Work* (CSCW '06), ACM, 1–10.

Rodden, T., Rogers, Y., Halloran, J. and Taylor, I. (2003) Designing novel interactional workspaces to support face-to-face consultations. *Proceedings of the Conference on Human Factors in Computer Systems,* (CHI '03), ACM, 57–64.

Rogers, Y. (2011) Interaction design gone wild: Striving for wild theory. *Interactions*, 18(4): 58–62.

Rogers, Y. and Halloran, J. (2007) Supporting collaboration for choosing holidays. Online case study. Available at: www.id-book.com/casestudy_n_3.php

Rogers, Y. and Lindley, S. (2004) Collaborating around vertical and horizontal displays: Which way is best? *Interacting with Computers*, 16, 1133–1152.

Rogers, Y. and Rodden, T. (2003) Configuring spaces and surfaces to support collaborative interactions. In O'Hara, K., Perry, M., Churchill, E. and Russell, D. (eds), *Public and Situated Displays*. Dordrecht: Kluwer. 45–79.

Rogers, Y., Connelly, K., Tedesco, L., Hazlewood, W., Kurtz, A., Hall, B., Hursey, J. and Toscos, T. (2007) Why it's worth the hassle: The value of in-situ studies when designing UbiComp. J. Krumm et al. (eds), *UbiComp 2007*, LNCS 4717, Berlin, Heidelberg: Springer-Verlag. 336–353.

Rogers, Y., Lim, Y. and Hazlewood, W. (2006) Extending tabletops to support flexible collaborative interactions. *Proceedings of the First International Workshop on Horizontal Interactive Human-Computer Systems* (Tabletop '06), IEEE, 71–78.

Rogers, Y., Lim, Y., Hazlewood, W. and Marshall, P. (2009) Equal opportunities: Do shareable interfaces promote

more group participation than single users displays? *Human-Computer Interaction*, 24(2): 79–116.

Ryall, K., Forlines, C., Shen, C. and Morris, M. (2004) Exploring the effects of group size and table size on interactions with tabletop shared-display groupware. *Proceedings of the Conference of Computer Supported Cooperative Work* (CSCW '04), ACM, 284–293.

Scott, S.D., Carpendale, M.S.T. and Inkpen, K.M. (2004) Territoriality in collaborative tabletop workspaces. *Proceedings of the Conference of Computer Supported Cooperative Work* (CSCW '04), ACM, 294–303.

Scott, S.D., Grant, K. and Mandryk, R. (2003) System guidelines for co-located, collaborative work on a tabletop display. *Proceedings of the Eighth European Conference on Computer Supported Cooperative Work* (ECSCW '03), 159–178.

Shen, C., Lesh, N., Vernier, F., Forlines, C. and Frost, J. (2002) Sharing and building digital group histories. *Proceedings of the Conference of Computer Supported Cooperative Work* (CSCW '02), ACM, 324–333.

Tan, D.S., Gergle, D., Mandryk, R., Inkpen, K., Kellar, M. Hawkey, K. and Czerwinski, M. (2008) Using job-shop scheduling tasks for evaluating collocated communication. *Personal and Ubiquitous Computing*, 12(3): 255–267.

Van der Linden, J., Johnson, R., Bird, J., Rogers, Y. and Schoonderwaldt, E. (2011) Buzzing to play: Lessons from an in the wild study of real-time vibrotactile feedback. *Proceedings of the Conference on Human Factors in Computing Systems* (SUGCHI '11), ACM, 533–542.

Wobbrock, J.O., Morris, M.R. and Wilson, A.D. (2008) User defined gestures for surface computing. *Proceedings of the Conference on Human Factors in Computing Systems* (CHI '08), ACM, 1083–1092.

Wu, M., Shen, C., Ryall, K., Forlines, C. and Balakrishnan, R. (2006) Gesture registration, relaxation, and reuse for multi-point direct-touch surfaces. *Proceedings of the First International Workshop on Horizontal Interactive Human-Computer Systems* (Tabletop '06), IEEE, 185–192.

Yuill, N. and Rogers, Y. (2012) Mechanisms for collaboration: A design and evaluation framework for multi-user interfaces. *Transactions on Computer–Human Interaction* (TOCHI), ACM, 19(1): 1–25.

24

Ubiquitous Virtual Reality Environments

Yoosoo Oh and Woontack Woo

INTRODUCTION

Ubiquitous virtual reality (ubiquitous VR) has been researched for the last five years to maximize synergies between virtual reality and ubiquitous computing technologies. Virtual reality (VR) is a term that applies to computer-simulated environments which can simulate a physical presence in places in real space, as well as in imaginary spaces (Steuer, 1992). The concept of virtual reality was popularized in mass media by movies such as *Johnny Mnemonic* (1995) and *Avatar* (2009). Movies such as these contain VR themes where sentient programs interact with humans through VR equipment. Ubiquitous computing has created a paradigm shift in computing from a technology-orientated experience to a user-centered experience. Technologies in ubiquitous computing are invisible but present everywhere in our daily lives, whether we are aware of it or not. Weiser called this 'calm technology' (Weiser, 1991). In ubiquitous computing, we can have access to computers placed in our offices, walls, clothing, cars,

planes, organs, etc., without our attention being drawn to the technologies. When driving a car, for instance, our attention is centered on the road or our passenger, but not on the operation of the engine. Some movies illustrate the world of ubiquitous computing through, for example, the gestural interfaces seen in *Minority Report* (2002).

Ubiquitous VR can be defined as realizing VR on ubiquitous computing environments, i.e., making VR pervasive in our daily lives (Kim et al., 2006b; Lee et al., 2009a). Ubiquitous VR has some fundamental properties, such as reality, context (i.e. information to characterize the situation of a person, a place, a device, or a time) and human activity (Lee et al., 2009a). Ubiquitous VR supports human social connections within the context of dual (real and virtual) reality spaces. Hence, a ubiquitous VR environment can be described as a future smart environment for realizing the technologies of ubiquitous computing and VR (Lee et al., 2009a, 2009b).

Ubiquitous VR technology creates and manages digital content (e.g. text, image, sound and video; Kim et al., 2009, 2011). The created content stimulates a user's senses by seamlessly registering virtual contents to a physical device through context (Oh et al., 2011). Virtual content refers to digital content when a user represents virtual space information as metadata in order to use the information. The contents of ubiquitous VR environments involve information about virtual and real spaces that are linked through context. For example, a virtual character in a smart room, behaves adaptively according to situations in real space (Kim et al., 2011). This virtual character can move from virtual spaces to real spaces and vice versa. The virtual character can directly control a physical device and intelligently respond to it. Content in ubiquitous VR environments is created, maintained, shared, evolved and consumed with user participation. Based on such a content life cycle, the content forms a sustainable content ecosystem. There are several research challenges in maintaining a content ecosystem. The challenges are related to content creation, sharing and distribution among users, consumption and destruction.

With the growth of Internet services developed within the Web 2.0 paradigm, service ecosystems are generated and maintained through user participation (O'Reilly, 2005). Internet blogs manage user records and information sharing through social relationships with other users. As an example, in Second Life, users perform various activities through social relationships in virtual spaces (Second Life, 2012). In media-sharing sites, such as YouTube or Facebook, users upload their pictures or movie clips and share them with socially related persons (YouTube, 2012; Facebook, 2012). LinkedIn recommends related information to users in order to facilitate social relationships between users by utilizing member profiles and settings (LinkedIn, 2012). Such online sites play an important role in improving users' social relationships by sharing information and knowledge with other persons. Twitter is a Web-based social networking service that provides a short message service (SMS; Twitter, 2012). Twitter provides an environment in which users write messages and share them with other users in real time. Recently, various Web services have begun to provide new services by connecting with heterogeneous services, such as maps, job openings, rentals and pictures. In these open Web service environments, users freely participate in social networks and create and share content. Such network environments can maintain their content ecosystems. Such a sustainable content ecosystem can also be realized in a ubiquitous VR environment by using a mix of real and virtual spaces.

This chapter explores different concepts surrounding the topic of ubiquitous VR systems that support superimposing aspects of virtual reality on to physical spaces. With the advent of new computing technologies, this chapter offers a definition of ubiquitous VR, outlining its key features and describes its application in different scenarios. Beyond this, ubiquitous VR is described using several concepts from various research fields.

FUTURE COMPUTING ENVIRONMENTS

Ubiquitous VR has been developed to create seamless connections between real and virtual spaces. Ubiquitous VR produces intelligent spaces, combining real and virtual spaces to increase human quality of life. By combining real and virtual spaces using technology, the placements of entities (e.g. sensors, devices, places, users, actuators, content, services, etc.) are intertwined. This chapter illustrates a ubiquitous VR environment that is capable of simulating intelligent spaces by enabling the addition of new virtual entities to existing real entities and vice versa.

Conceptualization

Ubiquitous VR historically includes several concepts from various research fields. In

1991, Mark Weiser discussed ubiquitous computing and computing resources that are embedded into our daily lives. In the 1990s, wearable computing had been researched actively in the form of mobile augmented reality at MIT and CMU. In 1997, Azuma presented a survey on augmented reality (AR) that allowed users to see real spaces combined with virtual devices. Using AR, virtual entities are augmented on to a real space as if the virtual one originally existed in the real space. Steve Mann described his wearable computer invention as a tool to describe a reality that can remove certain undesired devices (1997). With this historical background, researchers can think about how these computing paradigms will change in the future. In 2005, ubiquitous VR was described as 'a new paradigm combining virtual reality with ubiquitous computing. This can provide users with various applications according to the context of users or environments' (Lee et al., 2009a). Here, 'context' is a key factor in combining virtual reality and ubiquitous computing. To this end, Lee et al. (2009a) specified key characteristics of ubiquitous VR based on reality, context and human activity.

Ubiquitous Computing

Future computing environments will promote interaction between humans and computers by sinking many kinds of computing resources into our living spaces. Over the last decade, computing technology has steadily developed. Ubiquitous computing, called the fourth revolution, is being reflected in our daily lives (Weiser, 1991). Ubiquitous computing environments include smart homes, smart offices, smart cars, etc. A smart home is a representative example of ubiquitous computing environments, as it is easily implemented due to its small scale. Significant research related to smart homes has been undertaken. This includes the Adaptive House (University of Colorado; Mozer, 1999), AwareHome (GATECH; Dey et al., 1999), EasyLiving (Microsoft; Shafer, 2000) and House_n Project (MIT;

Intille, 2002). Most of the related research focuses on the infrastructure's interface with a users and how it provides services that correspond to the user's intention. Such related research also supports convenient user-centered interfaces and intelligent control of the ubiquitous computing environment.

Context Awareness

Ubiquitous computing environments require context awareness and intelligent services. Intelligent services are especially important for future computing environments. Also, it is necessary to further develop context-aware technology and find out how contextual information can be applied to the design of various sensors and applications in intelligent services (Dey, 2001). Researchers have focused on context-aware systems in numerous computing research studies (Schilit et al., 1994; Oh et al., 2010). Many systems have been developed using various sensors, which act under different contexts. The system communicating with sensors can produce different results (e.g. personalized service) according to the context. For example, the media control system in a smart space provides services that recommend media contents to a user according to the user's profile; this system also controls the space based on various context inputs from the embedded sensors. As the first commercial context-aware applications, location-aware applications (e.g. vehicle navigation systems) and location-based services demonstrate the usefulness of context-aware systems. In particular, the development of life-cycle support for context-aware systems (e.g. simulating, installing, debugging and maintenance, adding new entities, removing old entities and upgrading system components) are issues of interest.

Various types of software architecture have been developed in recent years to support context-aware systems. Examples from different domains include the Context-Toolkit (Salber et al., 1999), TEA context acquisition architecture (Schmidt et al., 1999), JCAF middleware (Bardram, 2005), and UCAM (Oh et al., 2010). As indicated in a recent work, architecture improvements in algorithms for contexts obtained from sensors can simplify the early developmental phases of context-aware systems for large-scale computing environments by handling a large amount of sensory data (Oh et al., 2010). Therefore, a system must provide a process for context inputs from various sensors.

Virtual Reality

Until now, virtual reality (VR) has striven to create a computer-generated virtual space that enables a user to feel realism through interaction that stimulates the five senses. Advances in computer graphics (CG), multimedia, parallel/distributed computing and high-speed networking technologies make it especially possible to construct virtual spaces (Lui, 2001). Participants in virtual spaces can collaborate with each other by exploiting audio, video and 3D graphics, even though they are located around the space (Singhal and Zyda, 1999). Nevertheless, VR is still far from bringing users into a real physical space.

At the same time, ubiquitous computing in our daily lives allows access to computing resources and services anywhere and at any time (Weiser, 1991). Furthermore, context-aware services can be supplied to users based on information obtained from distributed but invisible computing resources (Shin et al., 2005). Context awareness offers intelligent services to users by utilizing contextual information on the spaces and users (Schilit et al., 1994; Dey, 2001). These features of ubiquitous computing can be employed to realize VR spaces.

Dual Reality

Ubiquitous VR environments can be regarded as dual reality environments. There are several dual reality-related research activities. The Metaverse Roadmap project defines the fusion environment of virtualization-enhanced real spaces and the mirrored space

as the metaverse. This project predicts computing environments that utilize augmented reality, Lifelogging, Mirror Worlds and Virtual Worlds by 2017 (Metaverse, 2007). Additionally, the International Organization for Standardization (ISO) is preparing MPEG-V (MPEG for Virtual Worlds), which will be a standard to connect real and virtual spaces (MPEG-V, 2009). The Cross-Reality project suggests a dual-reality environment that provides bidirectional context in real and virtual spaces (Lifton, 2009). Multimedia content includes simple text, audio and visual information. Recently, however, the next generation content being developed includes intelligence, immersiveness and mobility. For instance, an e-book reader like Amazon's Kindle is an example of content consumption that can read e-books and other intangible digital information. The ISO defines RoSE (Representation of Sensory Effect) as an intractable content standard (MPEG-V, 2009). In addition, Web 2.0-orientated Wiki or Twitter users participate in content production, sharing and reuse. Web-based social network services (SNS) maintain service ecosystems by participating in content production and consumption. Micro-blogs enable users to share their blog contents on their mobile phones (Gaonkar et al., 2008). Mobile users leave personal or public messages from their location in a real space by using WiFi, accelerometers, cameras and GPS services embedded into smartphones. These messages make links with Internet blogs and create social interaction between mobile users in virtual spaces. Through this blog example, social tagging messages in real and virtual spaces can be shared and distributed in a dual reality environment. If this tagging activity continues, the content of a ubiquitous VR environment can be sustainably maintained (Soroker et al., 2008).

WHAT IS UBIQUITOUS VR?

Ubiquitous VR extends virtual reality capabilities into a physical space, rather than confining virtual reality to a simulated space. In other words, ubiquitous VR combines virtual space and physical space seamlessly, instead of only focusing on the generation of an ideal virtual space. Furthermore, the contents of ubiquitous VR can be interacted with in real space, as well as virtual space, so that they can be systemically associated with services in a real space. Consequently, ubiquitous VR is defined as 'a concept of creating ubiquitous VR environments, which make VR pervasive in our daily lives and ubiquitous by allowing VR to meet a new infrastructure, i.e. 'ubiquitous computing' (Kim et al., 2006a). To realize ubiquitous VR, the following three features must be addressed.

Ubiquitous VR Features

Ubiquitous VR enables collaborators to share necessary resources (devices) and contents to carry out tasks in both real and virtual environments. The contents of ubiquitous VR stimulate the senses and perceptions of users. The devices are smart devices, which are prevalent in real environments and include both wireless and wired devices. In ubiquitous VR environments, users can collaborate not only within a virtual environment by using conventional VR user interfaces but also within a real environment. They can collaborate with each other by sharing realistic content in both real and virtual environments. Thus, users can share their goal with each other through ubiquitous VR environments.

Users are provided with personalized services with the help of wearable devices that manage personal information, anywhere and at any time, in ubiquitous VR environments. Personal information is managed through wearable devices that reflect the users' desires, needs and preferences regarding services. Furthermore, the intimacy of a user interface must be maintained so that users can utilize wearable devices conveniently. Ideally, the user interface is transparent to users so that they can concentrate on their tasks without the necessity of being conscious of the user interface.

A user interacts with devices or realistic content through a transparent user interface. For this purpose, context-aware augmentation techniques make it possible to project intelligent and realistic content into a real space based on contextual information of spaces and users. It is important that intelligent and realistic content is integrated seamlessly into a real space to provide a seamless experience which preserves the users' senses. Augmented realistic contents are required to be especially responsive, in order to react intelligently to users.

In future computing environments, many computing resources will be pervasive and invisible to users. In addition, they will have various hardware needs according to their purposes. Thus, requirements for high-performance computing resources to implement VR systems can be satisfied by using computing resources that are pervasive in ubiquitous VR environments. Heterogeneity in future computing environments may cause new issues in areas such as access, modification, transmission and maintenance. Furthermore, networks must be real-time, concurrent, persistent, consistent and secure to effectively manage distributed information, devices and computational power. This is especially important because sharing of data, devices and computational power is essential for ubiquitous VR environments to carry out their tasks efficiently.

In general, personal information needs to be managed effectively in order to deal with a user's desires, needs and preferences. First of all, data management must be achieved to collect and process user-related information from diverse sources. This is also required to learn the user's behavior patterns and infer user preferences, so that learning results can be reflected in the dynamic updates of user preferences. In addition, application developers need to provide users with flexible ways to control when, to whom and at what level of detail personal information can be disclosed.

Contents need to be responsive so that they may respond intelligently to users. In other words, the content should understand context in real space, as well as in virtual space. Moreover, the contents should also figure out user propensities and preferences. In conventional VR systems, artificial intelligence and artificial life techniques are employed to create intelligent characters or natural interactions in spite of the high computational complexity required. Equally, new frameworks or algorithms need to be supported for intelligence in a distributed environment, since centralized servers are discouraged in ubiquitous computing environments.

Personalized contents are augmented and provided to users in ubiquitous VR environments. It is worthwhile to note that only personalized information is provided to the user, even though many users are interested in the same device. In order to present personalized content to each user, it is essential to be aware of the context for each user, as well as the space. However, augmented content should be integrated seamlessly with the real space. A user may retrieve necessary information through wearable devices anywhere and at any time.

Content Ecosystem in Ubiquitous VR Environments

The concept of 'contents life cycle' includes resources, contents production, growth, evolution, reproduction and destruction in ubiquitous VR environments. Based on this life-cycle, a ubiquitous VR system produces high-quality content by using interaction from end-users and digital contents. In the content ecosystem, resources include the original forms of digital content (e.g. video, images, sound and text). At this step, people create preliminary resources related to their interests. All preliminary resources are retrievable by context (e.g. user/device ID, keywords and tags). At the production step, multiple preliminary contents are produced and registered to specific devices in real or virtual spaces. Incorrect information is discarded by a system in ubiquitous VR environments. In the next step, content is shared among users and extended to real and virtual spaces. For content to grow, a system must react to a user's actions according to intelligence of content based on context-awareness. Through sustainable growth, content is self-contained and changes into new types of content that are adaptable to a user's situation. If some content is not accessible by users, then the content will be destroyed. The content's destruction means that content and the devices are separated. Also, low-quality, incorrect or old content can be destroyed. On the other hand, if the content survives, then the content is changed by production, growth and evolution. To this end, the content in ubiquitous VR environments can be seamless, proactive and adaptive.

An ecosystem is a system defined by the interactions of a community of organisms with each other and with their environment. In a natural ecosystem, the natural environment is one in which interdependent organisms, such as plants and animals, and physical factors of the environment, such as rocks and soil, interact (Van Dyne, 1969). Likewise, the interactions between the organisms in computing resources (e.g. devices, systems) that include users can be defined as non-natural ecosystems; some examples include digital, mobile and content ecosystems.

Many researchers describe digital ecosystems (Villalba et al., 2009; Hadzic and Chang, 2010) as ecosystems formed by computing resources (e.g. devices, services, data), in order to acquire new perspectives on information systems. Some researchers focus on mobile ecosystems (Xia et al., 2010) and, more rarely, on content ecosystems. Mobile ecosystems are digital ecosystems in which the mobility of users drives interactions and evolutions. Typically, a mobile device informs a service about its position, orientation or movement and the service processes this information to suggest or provide appropriate content. For example, a smartphone may use its embedded GPS to send its position to a navigation service that calculates and displays a path to the user's destination, indicating restaurants on the way. Content ecosystems are digital ecosystems that preserve the order of contents that are created, maintained, modified, evolved or destroyed by means of users' participation. Typically, the devices provide stable standard hooks (e.g. API and codecs) and are considered more or less equivalent from a systems perspective; rather than organisms, they become parts of the environment (which does not preclude improvements).

Both mobile and content ecosystems can be extended, but are basically disjointed; their features can promote services in ubiquitous VR but are difficult to combine in their current form. Consequently, it is important to define and realize a new type of ecosystem for content, based on mobile and content ecosystems – ideally with minimal fundamental changes to effectively and efficiently reuse existing works. Content should be 'appropriate for a ubiquitous computing-enabled space, making VR pervasive into our daily lives' (Oh et al., 2011: 378). The content highlights the importance of users: they are continuously immersed in a mix of real and virtual spaces, their contexts and activities influence the relevance of content and services and they may both produce and consume content.

In a content ecosystem, content should be precisely matched to users' characteristics (e.g. abilities, preferences, viewpoints), the physical world (e.g. location, luminance, noise), events, social situations, and to goals of producers and consumers. A sustainable participation of users as mobile producers and consumers implies simple content creation, flexible distribution, easy retrieval and seamless interaction.

The concept of ecosystems has been applied to the domains of advertisement, energy and health. Hadzic and Chang (2010) have also suggested digital business ecosystems, digital government ecosystems, digital law ecosystems and digital industry ecosystems, although they did not clarify any potential components of these ecosystems. In

the domain of advertisements, Villalba et al. (2009) discussed an advertising ecosystem. For example, in MyAds (Di Ferdinando et al., 2009), information and advertising agents could move through screens, looking for people with specific user profiles, and then try to control a selected screen with their own content (Villalba et al., 2009). In 'display niches', information is fed to user agents, who relay the information to advertiser agents; then, display-bound monitoring agents selectively spread information and guide advertiser agents, forming a feedback loop – ecosystem laws define how user agents make user profiles available in the displays (Villalba et al., 2009). In cloud-based ad printing services (Yang et al., 2010), Web content is automatically created with automatic contextual ads and coupon insertion when the Web content is printed.

In an energy Web system (Carreras et al., 2010), autonomous agents control the production and usage of distributed renewable energy sources to optimize energy use and allow for rapid innovation. It is envisioned based on five layers: distribution, control, prosumer, community and incentive (from more technical to more social). A digital health ecosystem has been proposed to focus on electronic health records, which, notably, involves hospitals, health services, general practitioners, pharmacies, health systems, and health information resources (Hadzic and Chang, 2010). According to Estrin and Sim (2010), ecosystems may increase the efficiency of, and innovation in, mHealth (e.g. disease prevention and management) by providing flexibility.

In the domain of P2P applications, the BitTorrent ecosystem is currently one of the most successful open Internet applications. Zhang et al. (2011) crawled popular torrent-discovery sites and studied in depth the ecosystem's tracker, peers, user behaviors and content landscapes. In the domain of digital museums, information and knowledge of museum artifacts have been digitally collected to construct the ecosystem of the virtual museum (Eklund et al., 2009).

Additionally, virtual museums encourage sustainable user participation by containing a wiki component, where the ecosystem enables the creation of user-defined perspectives with semantic themes.

All of these ecosystems could be merged into a universal digital ecosystem (Hadzic and Chang, 2010), in the same sense that our natural ecosystems can be merged into an Earth ecosystem. Conversely, many digital ecosystems could be split into lower-level ecosystems; for example, a digital ecosystem may be split into a digital dental ecosystem, a digital physiotherapy ecosystem, a digital surgery ecosystem, and so on (Hadzic and Chang, 2010).

Ubiquitous VR environments can be built based on fundamental concepts of ubiquitous VR (Lee et al., 2009a) and support links between real and virtual spaces with additional information (e.g. immersive five-sense augmentations of content). Moreover, ubiquitous VR environments enable bi-directional interaction between contents on the fly in linked dual space. In such environments, content is utilized and shared, as shown in Figure 24.1. In this environment, virtual content is seamlessly registered to interested real devices. The content is augmented as the linking information between real and virtual spaces. Figure 24.1 shows a concept of

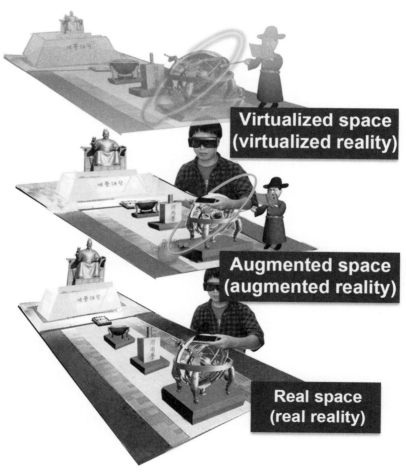

Figure 24.1 **Concept diagram of content ecosystem in ubiquitous VR environments.**

a reality continuum among three spaces for ubiquitous VR.

The following application scenario, as shown in Figure 24.2, could explain a concept of ubiquitous VR environments. In Sejong Street, in Seoul, Republic of Korea, a user wearing an HMD can experience educational AR simulation by looking at real/physical historical assets (e.g. armillary sphere, rain gauge, altitude dial, etc.) when the real and virtual spaces are linked together. Controlled augmented devices (e.g. a virtual clock connected to the armillary sphere) can use a mobile device to simulate virtual movement of constellations or virtual weather changes for education through immersive five-sense augmentation.

In addition, users in real space can also make and share information (real-time personal broadcasting, social media, etc.) with users in various spaces (virtual or remote space) on the fly through bidirectional interaction that links dual spaces. Moreover, for more effective educational simulations, intelligent AR contents and characters can be augmented and provide additional explanation for the simulation. The scenario could be applied to other time/space-transcended work (remote immersive conference), next-generation experimental education (science, math, English, art, history, etc.), immersive simulations (climate, ecology, environment, military, urban planning, architecture, disaster, medical, etc.), video-based information surveys (traffic, security, etc.) or immersive entertainment.

APPLICATION AND CONCLUSION

In a smart space, the placements of the entities (e.g. sensors, devices, places, users, actuators, content, services, etc.) are intertwined. Ubiquitous VR simulators (Oh et al., 2009) are a concrete example of ubiquitous VR environments, because the simulation operates by

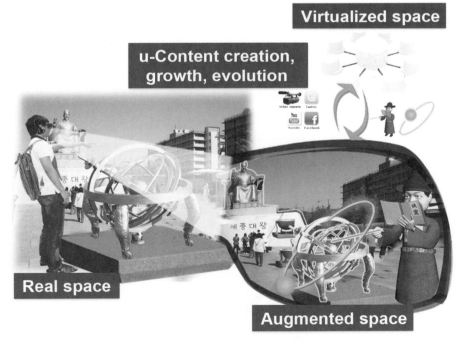

Figure 24.2   Application scenario for next-generation experimental education in ubiquitous VR environments (u-Content means 'content in ubiquitous VR environments').

adapting every feature of ubiquitous computing and VR. This simulator bi-directionally controls real and virtual spaces and manipulates both devices in both spaces.

Simulations of ubiquitous computing environments have investigated potential scenarios, in terms of sensors, services and environments (Reynolds et al., 2006). To simultaneously test entities in a real space typically requires a great deal of time and effort. For instance, changing existing entities or adding new entities incurs financial costs, due to installation, experimentation and maintenance of the entities. It is also expensive to test the entities in a real space, which commonly involves the testing of various sensors, actuators and a huge number of contexts (Nishikawa et al., 2006). Moreover, exhaustively deploying all usable entities is difficult whenever a given framework or infrastructure is being tested. Thus, adding or extending entities in a real smart home environment is difficult because of its predefined structure and high financial cost (Armac and Retkowitz, 2007).

Equally, simulating entities in a virtual space can reduce financial costs, although it is hard to maintain real system operation, such as real-time debugging, reactions and system behavior (Nishikawa et al., 2006; Reynolds et al., 2006). Thus, though simulating entities in both real and virtual spaces can be complicated, each case has its advantages. Simulations in a real space can be more realistic and accurate, whereas simulations in a virtual space are relatively more cost-effective; accordingly, there are a number of benefits to testing in both. To this end, a ubiquitous VR environment (Lee et al., 2008) enables computing in both real and virtual spaces, such that the simulations can be used in the development of specialized applications. The benefits of combining a virtual simulator with a real ubiquitous VR system include the advantages of both spaces, such as real-time measurement and realistic and systematic testing using physical devices, as well as time-saving and cost-effective evaluation using simulated devices.

As a representative example of a ubiquitous VR environment, there are some simulators that only consider newly designed entities. Those works include UbiREAL (Nishikawa et al., 2006), eHomeSimulator (Armac and Retkowitz, 2007), TATUS (O'Neill et al., 2005), UBIWISE (Barton and Vijayarahgavan, 2002), CASS (Park et al., 2007), CAST (Kim et al., 2006) and C@SA (De Carolis et al., 2005). Mixing both entities is important to completely link real and virtual environments. A ubiquitous VR simulator seamlessly connects entities in real and virtual environments by exploiting context, based on the concept of ubiquitous VR. Using this simulator, a smart space can be simulated and new virtual entities, such as sensors, actuators and services, can be integrated into existing real entities.

Based on surveyed previous research, we have created a list of feasible criteria for a ubiquitous VR environment. The list modifies and adds criteria to UbiREAL's list (Nishikawa et al., 2006).

- Networked entities in a ubiquitous VR environment should be easily deployed.
- Realistic and rich context should be generated in both real and virtual spaces.
- Virtual entities should be intuitively visualized, and the visualization should show how entities work based on context.
- The flow of context should be clearly represented between real and virtual spaces.
- Bidirectional control of entities should be supported. Physical entity manipulation through virtual simulation and virtual entity control through real operation should also be supported.
- Existing established infrastructure should be used with virtual entities for cost-effectiveness.
- To seamlessly link real and virtual environments by context, a context-aware process (e.g. middleware, framework) should be utilized in a ubiquitous VR environment.

A simulator operates consistently and is maintained in real time. A simulator also allows a developer to easily check context flow using 3D visualizations and rapidly debug each entity when problems occur. Also, simulators are cost-effective, as they use existing real devices and simply deploy the entities. A ubiquitous VR simulator represents a dual reality; augmented reality technologies can be applied to a simulator and enable a simulator to overlay entities on real or virtual spaces using a mobile phone.

In this chapter, we explored different concepts in the area of ubiquitous VR, which are systems that support superimposing aspects of VR on to the physical world. With the advent of new computing technologies, ubiquitous VR has appeared for seamless connection between the real and virtual worlds. Ubiquitous VR produces intelligent spaces that are the convergence of real and virtual spaces to increase human quality of life. By combining real and virtual spaces with technology, the advantages of virtual spaces move to real spaces, and the advantages of real spaces move to virtual spaces. This chapter has illustrated ubiquitous VR as realizing mirrored VR in ubiquitous computing-enabled physical spaces, such that the dual (both real and virtual) realities are tightly coupled and synchronized through context. This chapter presented ubiquitous VR with concrete examples. It described a ubiquitous VR environment that is capable of simulating a smart space by enabling the addition of new virtual entities to existing real entities and vice versa. Moreover, this chapter has explained the usefulness of ubiquitous VR environments: (1) real-time management of physical and virtual devices; (2) personalized content augmentation; (3) realistic and systematic testing using both physical and virtual spaces; and (4) time saving and cost-effective evaluation.

REFERENCES

Armac, I. and Retkowitz, D. (2007) 'Simulation of Smart Environments', Proceedings of IEEE ICPS2007, IEEE, 322–331.

Azuma, R.T. (1997) 'A Survey of Augmented Reality', Presence: Teleoperators and Virtual Environments 6(4): 355–385.

Bardram, J. (2005) 'The Java Context Awareness Framework (JCAF): A Service Infrastructure and Programming Framework for Context-Aware Applications', Proceedings of the 3rd International Conference on Pervasive Computing (Pervasive '05), ACM, 98–115.

Barton, J.J. and Vijayarahgavan, V. (2002) 'Ubiwise: A Ubiquitous Wireless Infrastructure Simulation Environment', Technical Report HPL2002-303. HP Labs, HP.

Carreras, I., Miorandi, D., Saint-Paul, R. and Chlamtac, I. (2010) 'Bottom-Up Design Patterns and the Energy Web', Transactions on Systems, Man, and Cybernetics: A IEEE, 40(4): 815–824.

De Carolis, B., Cozzolongo, G., Pizzutilo, S. and Plantamura, V.L. (2005) 'Agent-Based Home Simulation and Control', Proceedings of ISMIS2005, LNCS3488, 404–412.

Dey, A.K. (2001) 'Understanding and Using Context', Personal and Ubiquitous Computing, 5(1): 4–7.

Dey, A.K., Salber, D. and Abowd, G.D. (1999) 'A Context-Based Infrastructure for Smart Environments', GVU Technical Report; GIT-GVU-99-39, Georgia Institute of Technology.

Di Ferdinando, A., Rosi, A., Lent, R., Manzalini, A. and Zambonelli, F. (2009) 'MyAds: A System for Adaptive Pervasive Advertisements', Pervasive and Mobile Computing 5(5): 385–401.

Eklund, P., Wray, T., Goodall, P., Bunt, B., Lawson, A., Christidis, L., Daniels, V. and Van Olffen, M. (2009) 'Designing the Digital Ecosystem of the Virtual Museum of the Pacific', Proceedings of 3rd International Conference on Digital Ecosystems and Technologies, 805–811.

Estrin, D. and Sim, I. (2010) 'Open mHealth Architecture: An Engine for Health Care Innovation', Science 330(6005): 759–760.

Facebook (2012) www.facebook.com

Gaonkar, S., Li, J., Choudhury, R.R., Cox, L. and Schmidt, A. (2008) 'Micro-Blog: Sharing and Querying Content through Mobile Phones and Social Participation', Proceedings of the 6th International Conference on Mobile Systems, Applications, and Services (MobiSys '08), 174–186.

Hadzic, M. and Chang, E. (2010) 'Application of Digital Ecosystem Design Methodology Within the Health Domain', IEEE Transactions on Systems, Man, and Cybernetics-Part A: Systems and Humans 40(4): 779–788.

Intille, S.S. (2002) 'Designing a Home of the Future', IEEE Pervasive Computing 1(2): 80–86.

Kim, I., Park, H., Lee, Y., Lee, S., Lee, H. and Noh, B. (2006) 'Design and Implementation of Context-Awareness Simulation Toolkit for Context Learning', Proceedings of IEEE SUTC2006, IEEE, 96–103.

Kim, K., Han, J., Kang, C. and Woo, W. (2011) 'Visualization and Management of u-Contents for Ubiquitous VR', *Proceedings of the 14th International Conference on Human-Computer Interaction* (HCII '11), LNCS6774, ACM, 352–361.

Kim, K., Oh, S., Han, J. and Woo, W. (2009) 'u-Contents: Description and Representation of Contents in Ubiquitous VR', *Proceedings of the International Workshop on Ubiquitous VR* (ISUVR 2009), 9–12.

Kim, S., Lee, Y. and Woo, W. (2006a) 'How to Realize Ubiquitous VR?', *Proceedings of Pervasive: TSI Workshop*, 493–504.

Kim, S., Suh, Y., Lee, Y. and Woo, W. (2006b) 'Toward Ubiquitous VR: When VR Meets ubiComp', *Proceedings of ISUVR2006*, 1–4.

Lee, Y., Oh, S., Shin, C. and Woo, W. (2008) 'Recent Trends in Ubiquitous Virtual Reality', *Proceedings of the International Workshop on Ubiquitons Virtual Reality* (ISUVR '08) IEEE, 33–36.

Lee, Y., Oh, S., Shin, C. and Woo, W. (2009a) 'Ubiquitous Virtual Reality and Its Key Dimension', *Proceedings of the International Workshop on Ubiquitous Virtual Reality* (ISUVR '09), IEEE, 5–8.

Lee, Y., Shin, C. and Woo, W. (2009b) 'Context-Aware Cognitive Agent Architecture for Ambient User Interfaces', *Proceedings of HCI International*, LNCS5612, 456–463.

Lifton, J., Laibowitz, M., Harry, D., Gong, N.W., Mittal, M. and Paradiso, J.A. (2009) 'Metaphor and Manifestation: Cross-Reality with Ubiquitous Sensor/ Actuator Networks', *Pervasive Computing* 8(3): 24–33, IEEE.

LinkedIn (2012) www.linkedin.com

Lui, J.C.S. (2001) 'Constructing Communication Subgraphs and Deriving an Optimal Synchronization Interval for Distributed Virtual Environment Systems', *Transactions on Knowledge and Data Engineering* 13(5): 778–792, IEEE.

Mann, S. (1997) 'Wearable Computing: A First Step Toward Personal Imaging', *Computer* 30(2): 25, 32. doi: 10.1109/2.566147.

Metaverse (2007) 'Metaverse Roadmap Overview,' Available at: www.metaverseroadmap.org/overview

Mozer, M.C. (1999) 'An Intelligent Environment must be Adaptive', *IEEE Intelligent Systems and their Applications* 14(2): 11–13.

MPEG-V (2009) http://mpeg.chiariglione.org/standards/mpeg-v

Nishikawa, H., Yamamoto, S., Tamai, M., Nishigaki, K., Kitani, T., Shibata, N., Yasumoto, K. and Ito, M. (2006) 'UbiREAL: Realistic Smartspace Simulator for Systematic Testing', *Proceedings of UbiComp2006*, LNCS4206, ACM 459–476.

O'Neill, E., Klepal, M., Lewis, D., O'Donnell, T., O'Sullivan, D. and Pesch, D. (2005) 'A Testbed for Evaluating Human Interaction with Ubiquitous Computing Environments', *Proceedings of TRIDENTCOM2005*, IEEE, 60–69.

O'Reilly, T. (2005) 'What is Web 2.0: Design Patterns and Business Models for the Next Generation of Software'. Available at: http://oreilly.com/web2/archive/what-is-web-20.html

Oh, Y. and Woo, W. (2007) 'How to Build a Context-Aware Architecture for Ubiquitous VR', *Proceedings of ISUVR2007*(CEUR-WS), 032–033.

Oh, Y., Duval, S., Kim, S., Yoon, H. and Woo, W. (2011) 'Foundations of a New Digital Ecosystem for u-Content: Needs, Definition, and Design', *Proceedings of the 14th International Conference on Human-Computer Interaction* (HCII '11), LNCS6774, 377–386.

Oh, Y., Han, J. and Woo, W. (2010) 'A Context Management Architecture for Large-Scale Smart Environments', *Communications Magazine* 48(3): 118–126, IEEE.

Oh, Y., Kang, C. and Woo, W. (2009) 'U-VR Simulator Linking Real and Virtual Environments Based on Context-Awareness', *Proceedings of International Workshop on Ubiquitous Virtual Reality*, 052–055.

Park, J., Moon, M., Hwang, S. and Yeon, K. (2007) 'CASS: A Context-Aware Simulation System for Smart Home', *Proceedings of IEEE SERA2007*, IEEE 461–467.

Reynolds, V., Cahill, V. and Senart, A. (2006) 'Requirements for an Ubiquitous Computing Simulation and Emulation Environment', *Proceedings InterSense*, ACM, 138: 1.

Salber, D., Dey, A., Abowd, G. (1999) 'The Context Toolkit: Aiding the Development of Context-Aware Applications', *Proceedings of the Conference on Human Factors and Computing Systems* (CHI '99), ACM 434–441.

Schilit, B., Adams, N. and Want, R. (1994) 'Context-Aware Computing Applications', *Proceedings of the First Workshop on Mobile Computing Systems and Applications*, (WMCSA '94), IEEE, 85–90.

Schmidt, A., Aidoo, K., Takaluoma, A., Tuomela, U., Laerhoven, K. and Velde, W. (1999) 'Advanced Interaction in Context', *Proceedings of the 1st International Symposium on Handheld and Ubiquitous Computing* (HUC '99), 89–101.

Second Life (2012) http://secondlife.com

Shafer, S., Brumitt, B. and Meyers, B. (2000) 'The EasyLiving Intelligent Environment System', *CHI Workshop on Research Directions in Situated Computing*.

Shin, C., Oh, Y. and Woo, W. (2005) 'History-Based Conflict Management for Multi-Users and

Multi-Services', *Proceedings of the Workshop on Context Modeling and Decision Support* (Context '05 Workshop).

Singhal, S. and Zyda, M. (1999) *Networked Virtual Environments: Design and Implementation*. Reading, MA: Addison-Wesley Professional.

Soroker, D., Paik, Y.S., Moon, Y.S., McFaddin, S., Narayanaswami, C., Jang, H.K., Coffman, D., Lee, M.C., Lee, J.K. and Park, J.W. (2008) 'User-Driven Visual Mashups in Interactive Public Spaces', *Proceedings of the 5th Annual International Conference on Mobile and Ubiquitous Systems: Computing, Networking, and Services*, (Mobiquitons '08), ICST, article no. 43.

Steuer, J. (1992) 'Defining Virtual Reality: Dimensions Determining Telepresence', *Journal of Communication* 42(4): 73–93.

Twitter (2012) www.twitter.com

Van Dyne, G.L. (1969) *The Ecosystem Concept in Natural Resource Management*. New York: Academic Press.

Villalba, C., Mamei, M. and Zambonelli, F. (2009) 'A Self-Organizing Architecture for Pervasive Ecosystems', *1st International Workshop on Self-Organizing Architectures* (SOAR), LNCS 6090, 275–300.

Weiser, M. (1991) 'The Computer for the 21st Century', *Scientific American* 265(3): 94–104.

Xia, R., Rost, M. and Holmquist, L.E. (2010) 'Business Models in the Mobile Ecosystem', *Ninth International Conference on Mobile Business*, Ninth Global Mobility Roundtable, 1–8.

Yang, S., Jin, J., Parag, J. and Liu, S. (2010) 'Contextual Advertising for Web Article Printing', *The 10th Symposium on Document Engineering* (DocEng '10), ACM, 195–198.

YouTube (2012) www.youtube.com

Zhang, C., Dunghel, P., Wu, D. and Ross, K.W. (2011) 'Unraveling the BitTorrent Ecosystem', *Transactions on Parallel and Distributed Systems* 22(7): 1164–1177, IEEE.

Location-based Environments and Technologies

Ty Hollett and Kevin M. Leander

INTRODUCTION

Location-based services (LBS) and location-based environments (LBE) exist at the intersection of three separate technologies: information and communication technology (ICT), geographic information systems (GIS) with spatial databases and the Internet (Steiniger et al., 2006). Location-based technologies – such as mobile phones, mobile networks and global positioning systems (GPS) – can be employed in order to deliver information that is relevant to one's location or to use one's location as an information resource. Users can, then, interact with these location-based environments (LBE) for social (e.g. checking in to a restaurant), educational (e.g. exploring a science-based augmented reality world) or informational (e.g. checking the traffic on Google Maps) purposes. With continually evolving advances in mobile technology, users do not merely use their device in a specific location; rather, location, as Gordon and de Souza e Silva argue, sets the 'conditions for interaction and provides the

context from which information is interpreted and used' (Gordon and de Souza e Silva, 2011: 11).

In a fiction piece from *Wired Magazine*, Bruce Sterling (2007) writes from the perspective of a 'hyperlocal' blogger – an 'early-adapter webceleb in the hyperlocal biz' – in the year 2017. He breaks down the definition of hyperlocal for his readers: 'Hyper, as in linked and local, as in location'. All of the databases of this 'new Web', he explains, are 'stuffed with geographic coordinates. Real positions. Real distances'. In this hyperlocal future, corporations have 'used their skills and capital to weld the virtual world firmly to the actual world'. 'Hyperlocality', he writes, 'is transforming our lives at every scale: bodyware, roomware, streetware, cityware, nationware, and global ware'. In 2017, bodyware, for instance, is the stuff carried in pockets and bags that broadcasts its location to nearby devices; roomware maintains rooms, the sensors that control temperature and remotes that control locks; streetware is his 'mobile's navigator, plus social tags, ad filters, and all those black-and-white barcode blotches painted on walls like graffiti'.

Today, we witness the early predecessors of Sterling's hyperlocal future: smartphone applications such as FourSquare, SCVNGR and Facebook Places have earned popularity by allowing users to 'check-in' to their current location, share information, stories and even participate in location-specific activities; food trucks broadcast their location via Twitter and consumers seek them out; applications such as Yelp and Urbanspoon determine nearby restaurants based on a user's location; photos uploaded to Flickr and Google Photos can be geotagged and mapped by location; computer chips in key fobs can even communicate with billboards for personalized advertisements (Scholz, 2011). In a more nuanced view of location, popular applications for smartphones use the accelerometer, for instance, to wake a user as the phone senses a certain degree of movement in the morning; IP addresses are tracked and provide location-based advertisements; and power consumption can be monitored and altered by mobile devices. Location-based services (LBS) as a whole are forecast to increase their overall profits from $515 million in 2007 to $13.3 billion in 2013 – making them the fastest-growing Web technology sector (ABI Research, 2009 as cited in Gordon and de Souza e Silva, 2011: 9).

This chapter will draw on examples from research using location-based technologies and environments. We will consider at length the general problem of tracking and understanding users, focusing first on how researchers have collected and analyzed aggregate data on users and locations, then turning to approaches to collecting and analyzing data on individual users. Of course, these larger categories of aggregate and individual are not tightly bound, yet they offer a means of organizing the different research goals, technologies employed and perspectives on the relationship between locations, human activity and digital tools and spaces. We then shift to a more in-depth consideration

of sensing technologies as a way to suggest how technologies are opening up opportunities to embed rich environmental and human data within location data, even offering real-time feedback loops. A discussion of sensors provides a variety of ways to use location-based technologies as well as a variety of research questions to pursue. While a number of research fields and contexts have drawn on location-aware technologies and have designed location-aware environments, we focus in the following section on two exemplary contexts: health research and research on education and game-based learning. Finally, we will discuss the issues and challenges in conducting research with location-based technologies and within location-based environments.

The research surveyed here ranges from studies using early location-based technologies and environments to more contemporary work and technologies. Its goal is to probe some of the over-arching questions revolving around research with location-based technologies and within location-based environments: how have such technologies pursued issues surrounding individual and group mobility? How have researchers used them to learn more about the movement within space, changes to space or ways to improve space? What can such technologies tell us about human behavior? How do such technologies interface with the human body for the collection of data? By seeking to answer such questions, this chapter aims to better understand the gradual shift from, as Sterling's blogger suggests, the local to the hyperlocal.

TRACKING AND UNDERSTANDING USERS

Aggregate Users and Locations

Research focused on aggregate users and locations tends to be interested in questions of large-scale patterns of use and distribution, such as the distribution of online activity in/with physical space or the ways in which virtual and physical worlds overlay and intersect with one another. This area also includes research that has taken new directions through new technologies and, theoretically, the 'spatial turn' (Soja, 1996) to forge new forms of analysis.

GIS and spatial humanities

Scholars in the humanities have begun to use geospatial tools for their research. A *New York Times* article (Cohen, 2011) highlighted the research of those working in the 'spatial humanities' who are using GIS 'to re-examine real and fictional places like the villages around Salem, Massachusetts., at the time of the witch trials; the Dust Bowl region devastated during the Great Depression; and the Eastcheap taverns where Shakespeare's Falstaff and Prince Hal 'caroused'. Knowles (2008) has used GIS to recreate Robert E. Lee's vantage point during the battle of Gettysburg; Cunfer's (2008) investigation of the causes of the Dust Bowl used GIS to drastically shift the time/space scale from 'an intense two-county case study to a broad two-hundred-county region at a coarser resolution, but one which allows for systematic analysis and a broad context' (2008: 118). Ray's mapping (2002) of the accusations made during the Salem Witch Trials emphasized its epidemic-like qualities. As evidenced here, GIS technologies provide a way to investigate time and space from new vantage points and scales as well as ways to visualize data from previously unexplored perspectives.

WiFi

A number of early location-based studies explored behavior and spatial and computational dynamics afforded by wireless Internet (WiFi) and wireless local area networks (WLAN). Tang and Baker (2000) sought to better understand the ways in which users on a college campus took advantage of wireless networks by examining overall user behavior, network traffic and load characteristics. They found users were primarily generating Web traffic by using synchronous chat applications. Henderson and colleagues (2004) expanded upon this work to trace the kinds of computer-mediated actions taking place over WiFi, such as peer-to-peer sharing, the streaming of multimedia and voice over (VoIP) traffic. They found that, despite the increase of mobile devices – PDAs and mobile VoIP clients – users were not nearly as mobile as predicted, remaining close to home nearly 98 per cent of the time. Sevtsuk and colleagues (2008) probed the intersection between work and mobility on a college campus. They then mapped the spatial patterns of WiFi usage in real time, provided the visualization and allowed the university community to act on that knowledge.

Geo-referencing Digital Activity

Location-based technologies can also be used to track 'digital footprints' (Girardin et al., 2009). By exploring geo-referenced photos shared publicly by individuals, as well as records generated by mobile phone users placing calls and sending text messages, Girardin and colleagues (2009: 128) explored the 'attractiveness of the urban space over time'. By summing aggregate data of mobile calls, texts and AT&T satellite traffic within the proximity of a specific location, they were able to measure overall cellular network activity. This data was then coupled with information gathered from the public photo-sharing website Flickr – specifically, photos anchored to their location by latitudinal and longitudinal coordinates. The mapping of this data allowed Girardin and colleagues to 'detect the main areas of photographic activities in New York because the accumulation of georeferenced photos over a period of time reveal[ed] the boundaries of areas of interest in a neighborhood' (2009: 131). In the end, the study allowed the team to trace the development of the attractiveness of major points of interest by following the density of digital footprints. When combined with the flows of visitors at popular locations, the data supplemented traditional measures, such as surveys and manual counts, to quantify the impact of a public space or event.

Similarly, Currid and Williams (2010) mapped 'the geography of buzz'. They used Getty Images to geo-code over 6000 events and 300,000 photos taken in Los Angeles and New York City. In other words, they provided a digital footprint to images through their geo-codes. All events were affiliated with the art, fashion, music, design and film industries – creators of social 'buzz'. Then, they conducted a macro-analysis of geographical patterns through GIS and spatial statistics. Through such an analysis, they observed 'particular nodes within the city' (2017: 3) at which cultural events often occurred, drawing connections between specific places and goods, and achieving a better understanding of the specific places that attract the media.

Michael Batty posits that this kind of work maps 'the pulse of the city' (2010: 576). Tracing 'buzz' or digital footprints, like Flickr photos, though, does not allow information to be processed in real time. Geo-located messages from Twitter (Tweets) and FourSquare 'check-ins' have begun to be mined as data and can provide real-time figures. Fabian Neuhaus (2010) detailed the ways in which new 'landscapes' were formed through the topographic visualization of such data by geographic location – plains, for instance, in areas of little usage and hills and peaks in areas of high usage. The digital landscape changed throughout the day: from the morning, to lunch, rush hour and so on. Such an understanding of the pulse of the city, Batty argues, will 'provide us with new views of urban structure and pattern that could well demonstrate to us that cities are much less stable structures than we have previously perceived' (2010: 576).

Individual Users and Locations

While location data on individual activity – whether virtual or physical – is fascinating data, on its own such data is relatively thin and telegraphic to interpret. Researchers therefore have found an increasing array of ways to augment the collection of geo-referenced data with ethnographic data, with participatory data collection and other approaches. Following this, we give a sense of the augmentation of location data by moving stepwise through a range of technologies or ensembles of technologies that have been used in order to augment location data and embed it in other means of understanding social and cultural lives. We survey, in sequence, research on individual users and locations with mobility kits, GPS, participatory GIS and mobile phones.

Mobility Kits

Studies of mobility have sought to better understand the portable objects people carry with them and the ways in which such objects have allowed them to interface with their local environment. Nippert-Eng (1996) investigated notions of privacy through the contents of participants' wallets. Strickland's 'Portable EFFECTS' (2001) project prompted participants to consider the ways in which they were 'nomadic designers' through the objects they carried with them on a daily basis. Other studies have focused on the urban routines established through mobility and the 'mobility kits' (Ito et al. 2009; Mainwaring et al., 2005) that fostered them.

Roth and colleagues (2010) took a major component of a Londoner's mobility kit – the Oyster card – to analyze patterns of urban movement throughout London. The Oyster card is used as an electronic ticket to access public transport services in Greater London. In their study, Roth and colleagues collected data from the Transport of London documenting 11.2 million trips from 2.03 million Oyster cards. Their analysis led to the conclusion that intra-urban movement was primarily organized around specific activity centers. Such a study provided an example of inferring individual movement and mobility patterns through the aggregation of location-based data generated by physical artifacts.

The work on mobility kits acts as a precursor to current location-based research conducted through pervasive, ubiquitous computing devices, like mobile phones. In their study, Mainwaring and colleagues detail the ways in which the mobile kit signaled the participants' interaction with the urban environment in three cities: Los Angeles, London and Tokyo. Through the kit, along with the way the users networked with personal contacts and sought their own private spaces, or 'cocoon', and the researchers mapped the travel patterns of the participants through their 'collecting places' (Mainwaring et al., 2005: 281), in the form of business, loyalty, transit and ID cards.

Okabe et al. (2005) analysis extended the discussion of mobility kits through their further emphasis on location within data collection. The participants in each city, however, captured their own personal diaries through different media: Los Angeles through an audio diary; London through paper-and-pencil; and Tokyo through a GPS-enabled mobile blogging system (moblogging). Participants using the moblogging device could choose to document their personal diary by taking pictures, recording a movie or writing a text message. When doing so, they also provided contextual information, such as whom they were with and what they were doing. The device logged the GPS coordinates, time and date after each entry. Ito and Okabe's study provides a strong example of the analysis of location, both through movement through space via physical artifact and the digital tracking of data entries.

GPS

Evolving technologies have led to similar tracking studies conducted through the use of GPS. Van der Spek and colleagues (2009) explored the possibilities of using GPS as sensor technology in measuring the activities of people. Their study observed the patterns of tourists in historic cities in order to improve the visitor experience. They developed tools to measure the overall impact of city improvements – such as city beautification, additional street furniture and lighting – on visitors. Street interviews coupled with the tracking technologies allowed the team to develop maps of 'great public spaces' (Van der Spek et al., 2009: 3044).

GPS has been employed in order to understand the connection between travel, activity and location. In an early study, Wolf and colleagues (2001) sought to eliminate the travel survey through the use of GPS data loggers. Their research sought to use GPS point data and GIS to derive trip purpose. Kochan and colleagues (2010) developed PARROTS (PDA system Activity Registration and Recoding of Travel Scheduling) to collect activity and travel data as well as location data through GPS during trips. PARROTS automatically recorded location data and allowed users to document the activities they performed at each location. When response rates with PARROTS were compared with those of a paper and pencil survey, Kochan and colleagues found fewer dropouts during the survey period. They also reported a greater number of executed trips, higher-quality data and the convenience of immediate access to electronic data (Kochan et al., 2010: 103).

GPS capabilities extend beyond tracking, or sensing, human activity – toward predicting it as well. Krumm and Horvitz (2004) implemented 'Predestination', a method developed to predict where a driver is going as a trip progresses. Ashbrook and Starner (2003) collected GPS data from a single user and then 'developed algorithms to extract places and locations from that data' (2003: 283). Such a method, they concluded, 'may be able to find locations that are semantically meaningful to the user' (2003: 283). Liao and colleagues (2007) detailed a system that could create a probabilistic model of a user's travel patterns through GPS data. Using GPS data, they sought to infer a user's mode of transportation and predicted when the user would shift to a new mode. Their research also predicted future movements and inferred when a user had deviated from his or her typical routine by error. GPS, as such, has primarily been used as a means to understand human movement through, and within, specific spaces.

Participatory GIS and Urban Tomography

Similar research has provided users with data-collecting tools – or employs the users' own devices – to collect location-based data. Walker and colleagues (2009) used a mixed-methods approach by combining qualitative methods and GIS in order to learn more about the regular journey that students took to school. To do so, they provided young teenagers with a mobile phone. A previously installed application, GeoBlog, allowed participants to write text, take pictures and annotate their image. The phone was wirelessly connected to a Bluetooth-enabled GPS device in order to track participants' journeys. Through their research, Walker and colleagues found it was really the qualitative component – 'the talk' – that truly established the narrative surrounding the student's journey and the interesting things captivating them during it. Such an approach allowed learners from all levels to 'engage in the research process in different ways and their involvement provided a richness to the data set and to our understanding of the school journey as contingent and complex' (Walker et al., 2009: 120).

Urban tomography (Krieger et al., 2010) imbues individual users with agency as data-collectors. With current smartphone technology, urban tomography 'makes use of multiple media records of city life to provide a multi-aspectival view of urban activity' (2010: 22). In other words, users document city life by providing a multitude of 'slices' (2010: 22) of activity. Urban tomography depends on the most recent affordances of smartphones: high-quality video documentation, time/location tagging, 3G wireless and uploading capabilities. Once videos are uploaded to a server, they can then be displayed on a webpage. Kreiger and colleagues also note that urban tomography captures the flows of people within a given location. Each of these examples demonstrates the individual, participatory nature of data collection with location-based technologies.

Mobile Phones and Networks

Mobile phones and networks have been leveraged to research human behavior and the daily locations that structure individuals' lives. Ratti and colleagues (2006: 737) used the 'pervasiveness of cell phones to capture extensive urban dynamics'. They sought to make sense of the 'unlimited flow of data from the cell phone infrastructure' in order to better visualize the flowing nature of urban life. Through their research, they questioned the ways in which pedestrians moved through the city; the ways people self-organized in response to disturbances; and the 'critical points of urban infrastructure' (2006: 728).

In their reality mining study, Eagle and Pentland (2006) exploited the use of short- and long-range networks of mobile phones. By tracing the connections that mobile phones made to these networks, Eagle and Pentland then used their data to reveal the structures of behavior that guided actions by individuals and organizations. Using an information entropy metric, they quantified the amount of 'predictable structure in an individual's life' (2006: 258). They analyzed the large-scale dynamics of human behavior by continually logging a user's activity, location, and proximity to other people (2006: 263). Such a method of data collection could provide, they argue, a wide range of possibilities – ranging from the ability to predict the locations of colleagues at specific times to aiding research in computational epidemiology as it seeks to 'build more accurate models of airborne pathogen dissemination' (2006: 263).

In later studies, the reality mining dataset (2006) was used to identify repeating structures in human behavior (Eagle and Pentland, 2009). Eagle and Pentland called these characteristic behaviors of people *eigenbehaviors*. With such a 'behavioral caricature'

(2009: 1058) in place, predictions could be made about the subsequent behavior of an individual. Eagle and Pentland then demonstrated the ways *eigenbehaviors* could be applied to whole groups as they moved through a 'behavior space' (2009: 1058) as well as 'to extract the underlying structure in the daily patterns of human behavior, infer group affiliations, and predict subsequent user behavior' (2009: 1059).

In his discussion of 'human sensing' – the use of aggregate data from mobile phones to track human activity – Shoval (2007) points to Ahas and Mark (2005) as the first to undergo such a method of data collection. In their study, Ahas and Mark introduced the social positioning method (SPM) in order to study the flows of mobile phone users in time and space through the analysis of the location coordinates of a mobile phone and the social identification of its user. In each of these studies, the primary impetus for using mobile communication was to create lightweight, low-cost systems that could track the mobile patterns of users through their own mobile phones.

ENHANCED LOCATION AND HUMAN DATA THROUGH SENSING

Sensors automatically detect, measure and observe events – motion, pressure, temperature, sound and pollutants, for instance – within close proximity. The spectrum of sensors stretches from, at one end, opportunistic sensors to, at the other end, participatory, human-driven sensors (Lane et al., 2008). Current applications use sensor-based technology in order to collect, process, share and visualize sensed information (Parker et al., 2006). An early opportunistic sensing application, Jetsam (Paulos and Jenkins, 2005), for instance, sought to visualize urban public trash through an augmented trash can in order to illustrate to city dwellers the ways in which they participated in their 'newly emerging digital city landscape' (2005: 350). Such work preceded the rise of mobile

devices. The ubiquity of mobile devices, and their evolving sensing capabilities, have made them ideal for people-centric sensing systems – where humans are the focal point of sensing activity – in urban and social settings (Campbell et al., 2006; Eisenman et al., 2006). Mobile devices, then, provide new opportunities for participants involved in data collection to interact with their local environments in ways that are 'uniquely relevant to the interests of individuals, groups, and communities as they seek to understand the social and physical processes of the world around them' (Reddy et al., 2011).

SENSEable City

MIT's 'SENSEable city lab' (Martino et al., 2010) has employed a variety of data collection methods, including pervasive systems, like mobile phone networks and crowdsourced materials, as well as participatory and opportunistic sensing. Specifically, the lab sought a better understanding of the flow of bodies through urban spaces and how such systems could lead to more efficient and sustainable living situations. Through the collection of information by 'smaller urban actors' with 'localization tags' – like GPS, WiFi spots and Bluetooth – the researchers at the SENSEable Lab then created visualizations of 'dynamic localized flows' (Martino et al., 2010). Such visualizations, they argued, effectively depicted human behavior once overlaid on top of traditional maps.

The Copenhagen wheel study (Outram et al., 2010), for instance, used environmental sensors to detect carbon monoxide, nitrogen dioxide, temperature, noise and humidity as the bicyclist rode through the city. The Copenhagen wheel, including sensors located in a hub-controller, was linked to a smartphone on the handlebars of the bike. While the Copenhagen wheel offered health benefits for the user – and encouraged frequent rides – Outram and colleagues noted other potential benefits of such a device, ranging from helping determine routes to measurements of heat, noise and pollution fluctuation within the city. Key to the project, though, was the visualization created and overlaid on top of traditional maps. Such a visualization translated the data into a form that could lead to change in the mobility patterns and energy consumption once viewed by the participant.

Other studies used visualizations of data to explore the relationships between place, behavior and interaction in real time. They questioned the ways in which such real-time information altered the mobility choices made by people. Real Time Copenhagen (Martino et al., 2010) and WikiCity Rome (Calabrese et al., 2008) displayed the 'pulse of the city' (Martino et al., 2010: 8) or the real-time mapping of urban flows during an evening. The visualization, broadcast for passers-by on a large screen in an accessible area of the city, detailed the movement of people through the city based upon the intensity of mobile phone use. Viewers could alter their routes based on the available information. By doing so, they factored into the ever-shifting nature of their surrounding urban environment.

The SENSEable City Lab also noted other projects with a similar ethos. The Amsterdam Real Time project by the Waag Society and Esther Polak (2002) examined the movements of citizens through GPS and created a map of the city based solely on their mobility. The Real Time Rome Project (Rojas et al., 2008) analyzed the 'emotional landscape' (2008: 4) of Rome through the aggregation of cell phone logs during the 2006 World Cup Final between Italy and France, as well as a Madonna concert. Nold's 'Bio mapping' (2007) used galvanic skin response to measure people's psychological reactions to different areas within a city. This work, in many ways, helps posit urban areas as flowing, fluctuating entities, altered by human movement and affective intensities.

Sound Sensors

Sound sensors have also been deployed for a variety of research purposes (Peltonen et al., 2002; Choudhury, 2003). SoundSense (Lu et al., 2009) used the microphone of the Apple iPhone to distinguish both general sounds, like music and voices, and sounds specific to individual users. It differentiated itself from previous studies through the real-time classification of sounds as opposed to offline analysis and classification. Through its Audio Daily Diary capability, SoundSense used opportunistic sensing to produce a log of acoustic events throughout the user's daily activities. SoundSense also acted as a music detector. In this application, when it detected music, the device prompted users to take a picture of their current location. That image, then, was uploaded to a Web portal where other users could view the stream of images connected to music and location. Users then witnessed the locations of music and could choose whether or not they wanted to go to that event based on the provided visuals. This emphasis on sound highlights foundational work on opportunistic sensing connected to popular mobile devices.

Sensing Proximity through CenceMe

CenceMe (Campbell et al., 2006; Miluzzo et al., 2008), used sensor-enabled mobile phones to infer the actions of users and then shared those actions with their selected social networks, like Facebook or Twitter. It collected sound samples through the microphone and motion through the accelerometer; it scanned Bluetooth and MAC (Media Access Control) addresses in the user's vicinity; it also took GPS readings, as well as 'random' pictures when a keypad was pressed or a call was made. CenceMe classifiers used specific algorithms to categorize audio, motion and images. The phone classifier retrieved audio bits, both voice and background noise; the activity classifier captured data from the accelerometer and classified it as, for instance, sitting, standing, walking, or running. Further, social context was inferred based on sensed conversations with a group of other CenceMe users, and could also recognize when the user was alone. With training, CenceMe could classify the user as attending a party based on ambient sounds. The accelerator could also determine if the user was dancing at that party.

Musolesi and colleagues (2008) sought to extend the capabilities of CenceMe by uniting it with the virtual world of Second Life. In Second Life, users interact with one another through digital representations of themselves called avatars. When creating an avatar, though, as Musolesi and colleagues noted, users could nearly mirror their physical appearance, but they could not translate their real-world actions into the virtual realm. Musolesi and colleagues sought to bridge real-world actions with virtual world representations through the use of mobile phones. Through the activity inference capabilities of CenceMe, physical actions in the real world were mapped into the virtual world. For example, when a user was running in the real world, his avatar was running in Second Life.

EXEMPLARY RESEARCH CONTEXTS

Perhaps it comes as no surprise that we might select as exemplary research areas for new directions with location-aware technologies, work in health and environmental research on the one hand and education and games for learning on the other. Both areas involve broadly distributed systems and the use of technologies to change human behaviors and human learning. Of course, as of writing, these technologiea are becoming more ubiquitous in daily life: 'bodyware' like the Fitbit and Nike's Fuelband are increasingly common. Moreover, it is also noteworthy that these broad areas are undergoing 'spatial turns' (Soja, 1996) in their own right, including the intersection of physical and virtual spaces. Both areas, for instance, are becoming increasingly interested in studying the individual outside of the laboratory and in the context of everyday, mobile life. How does one live a healthy life across space and time? How does one learn across space and time or how might such learning contexts be designed? Such expanded visions of living on the move, and life as distributed, inform new directions in health, environmental and education research.

Health and Environmental Studies

The nexus of location-based technologies and health information exemplifies data collection at the individual level. Such forms of data collection employ location-based technologies, ranging from the use of mobile networks tracking and sensing technologies. Anderson and colleagues (2007) extended the tracking of mobility to include the tracking of exercise habits through mobility. With their implementation of the application Shakra, they sought to design a system that would motivate adults who were not achieving their daily levels of physical activity by tracking and categorizing daily activity. Such activity was inferred through fluctuations in GSM signal strength.

Other studies have used commercial pedometers in order to motivate users towards a more active lifestyle. Lin and colleagues (2006) created 'Fish'n'Steps', a program that used the step-count gathered by a pedometer to determine the growth of a virtual fish displayed on a monitor. Users could then compete with one another as they witnessed fitness growth through the visual of their respective fish's growth. Similar programs, such as 'Houston' (Consolvo et al., 2006) and 'Chick Clique,' (Toscos et al., 2006) emphasized the social component of fitness as they used data collected by a pedometer to monitor users' step-counts and to share that data with other participants via mobile phone.

Expanding on studies concerning the health of bodies, others have used location-based technologies to gather data to benefit the environment. Chamberlain and colleagues' 'Professor Tanda' (2007) collected both environmental and location-based data to first determine users' environmental footprints and then to persuade the to change their or her travel habits. Similarly, Froelich and colleagues (2010) explored eco-feedback technology, which bridges an environmental literacy gap by automatically sensing a user's activities, like driving and showering, and providing the user with appropriate information through a computerized interface.

PEIR (Mun et al., 2009) – the personal environmental impact report – included a variety of location-based sensing capabilities – GPS location, activity classification and other contextual data – to provide personalized estimates of environmental impact and exposure for users (2009: 55). PIER focused on 'mobility-related impacts and exposures, using only the commodity sensors built into everyday smartphones' (2009: 56). Specifically, PIER provided its user with two types of her environmental impact – carbon and sensitive site – and two types of her environmental exposure – smog and fast food. The former impact stemmed from the use of transportation that left a carbon footprint and the release of airborne particulate matter emissions near sites like schools and hospitals; the latter related to the user's proximity to airborne particulate matter emissions as well as her proximity to fast-food restaurants. PIER demonstrated the 'emerging class of adaptive, human-in-the-loop sensing systems that combine the distributed processing of the web with the personal reach of mobile technology to engage people in exploring the previously unobservable relationships of their actions to the world around them' (Mun et al., 2009: 67).

Education and Games for Learning

Location-based technologies have been used to create interactive environments in (and out of) classrooms, especially through augmented reality environments. Roschelle and Pea (2002) provided an overview of the affordances of wireless devices in learning venues, including a description of the ways in which they could augment physical space through such technologies as sensors and wearable badges. In 'Savannah: mobile gaming and learning', Facer and colleagues

(2004) explored how using mobile technologies could create an engaging learning experience. Specifically, students were given handheld devices that – depending on their physical location – provided information related to the Savannah (e.g. 'A lion is fifty feet in front of you, what do you choose to do?'). This allowed students 'to "see", "hear" and "smell" the world of the Savannah as they navigate the real space outdoors' (Facer et al., 2004: 400).

Klopfer and Squire (2008) designed 'Environmental Detectives', 'a multiplayer, handheld augmented reality simulation game designed to support learning' in science (2008: 5). As an augmented reality game, students used personal digital assistants that would augment the physical world by providing a virtual layer of data accessed through the device: the data on the handheld was connected to the physical location. The game itself simulated the investigation of a toxic spill and encouraged players to conduct their own desktop research and fieldwork as well as taking part in investigations constrained by time and budgets (2008: 13).

Squire and Jan (2007) created the 'Mad City Mystery' – an augmented reality game that challenged students to seek the cause of the death of a citizen. Using their devices, students traveled to various locations to interact with citizens and gain new information about the death. In the end, students had to use their evidence, as well as scientific enquiry, reasoning and logical argument skills, to decide the cause of death. Built-in opportunities for reflection as well as multilayered challenges encouraged further critical thinking. Similarly, Dunleavy and colleagues' 'Alien Contact!' (2009) established an augmented world for students to explore as they interviewed virtual characters, collected digital items and solved a variety of math, language, arts and science puzzles. Such an environment allowed the researchers to investigate the ways in which teachers and students described their learning and teaching experiences in an augmented reality space.

ISSUES AND CHALLENGES

Privacy

Of all the sources of skepticism around research involving location-based technologies, privacy stands as the dominant issue (Beresford, 2005; Dobson, 2000). Spurred to action by the recent controversy surrounding Apple and location tracking by iPhones, US representative of Minnesota Al Franken (2011) recently drafted the Location Privacy Protection Act of 2011. Fearing just what may come from such overt awareness of location, Dobson and Fisher (2003) coined the term geoslavery and asserted that it was 'a new form of slavery characterized by location control' and that it 'now looms as a real, immediate, and global threat' (2003: 47). In their discussion of the social positioning method, Ahas and Mark (2005) acknowledged privacy issues, but also noted that such fears of surveillance often correspond with new technologies: mobile positioning is simply more readily perceivable, due to its seemingly panoptic nature.

Recently, Gasson and colleagues (2011) wrote of the 'privacy implications of behavioral profiles drawn from GPS enabled mobile phones'. In their short case study tracking four people, they argued that 'personal, and in some cases, sensitive information [could] be revealed' (Gasson et al., 2011: 255). By tracking four people, they were able to make inferences – but not assertions – about residence and place of work, gender, social status, family life, routine. More sensitive information, like religion, sexual orientation, and health were more inconclusive. Months, and possibly years, are needed to come to conclusions about some personal and sensitive information.

Even more recently, location-based applications on smartphones have led to concerns surrounding privacy. As mentioned, applications such as FourSquare and Facebook allow users to 'check in' to specific locations. To do so, a user actively logs into the application and notes his or her current location to a social network, which, to some critics, invites potential security risks to a user's now unprotected home. It is not only location-based apps, though, that make use of data connected to one's location. Other apps include the option for location data to be accessed. Instagram, for instance, a popular picture-taking application, can geo-tag images based on one's location. At the time of writing, a new generation of apps now interface with that data in order to provide location information for social purposes. Sonar and Banjo, as examples, integrate with Facebook, Twitter, Instagram and FourSquare to show friends, or friends-of-friends, who is nearby.

The notion of geo-fencing has some users concerned about privacy matters as well (Avalos, 2012). Essentially, geo-fencing allows companies to target consumers via their mobile devices as those users become proximate to their store. Best Buy, McDonalds and Victoria's Secret have the capability, for instance, to interface with the popular Pandora Radio app in order to deliver advertisements to a user depending on his or her proximity to their outlets.

Privacy has been a major topic of discussion in relation to many forms of new media and technologies. danah boyd (2011) has written and spoken extensively on teen privacy in 'networked publics', or 'publics that are restructured by networked technologies' (2011: 7). Teens, she argues, are managing privacy on their own accord. boyd reports that teens perform social steganography – the act of hiding in plain sight – in public messages posted on social networks. As a whole, boyd argues:

> Privacy is in a state of flux not because the values surrounding it have radically changed, but because the infrastructure through which people engage with each other has. Networked technologies introduce new challenges, particularly in environments that are public by default. Privacy cannot be assumed, especially when powerful individuals or entities are interested in leveraging newfound opportunities for access to undermine social norms. (2011: 26)

Technical challenges and limitations

Dunleavy and colleagues (Dunleavy et al., 2009) documented significant problems they faced in the implementation of their augmented reality game, 'Alien Contact'. They encountered hardware and software issues, primarily in the form of GPS error. They reported GPS failure rates of 15–30 per cent. These failures were likely a result of software instability and incorrect set-up of the handheld configuration by the research team (p. 16). Students reported being annoyed by the technical glitches, especially when it impacted other team members. Such a glitch then had a ripple effect on team members as it negatively impacted group management, learning, cohesiveness and collaboration.

When conducting experiments resulting in data collected through Bluetooth - enabled mobile phones, Eagle and Pentland (2006) detailed the issues they confronted. By using Bluetooth, which can penetrate through some types of walls and has up to a 10-meter range, people who were not physically proximate were logged as such. Further, since the device scanned for proximity every five minutes, it may have missed a person if he or she were proximate for only a short period of time. Moreover, human-induced errors occurred when the participant left his phone in a different location, allowed the battery to run out or turned off the phone completely.

Van der Spek and colleagues (2009) noted difficulties in data processing when using GPS. They noted the battery life of GPS-enabled devices as an issue confronted in data collection and the need for its preservation. They also described difficulties of gaining adequate GPS data in an urban environment, primarily from lost signals that often occur when entering buildings.

CONCLUSION

Location-based technologies and environments have played a role in ushering forth new ways for us to experience local spaces. Here, we have surveyed research that has employed location-based technologies and environments to research notions surrounding mobility, human behavior and space; digital footprints, geo-tagging and visualizations; participatory sensing systems, augmented reality, and more. As technologies continue to provide us with a further global reach, they have also enabled us to better understand our own local community and environment. How do we plan for and improve it? How do we participate in it and take part in its upkeep? How is the community continually reconfigured by new technologies? How do we work, play and learn in it? Further, as boyd notes, how does the private continue to take shape? The (networked) public? Research with new technologies will allow us to continually revisit such questions, all the while embedding us further within our local spaces.

REFERENCES

ABI Research (2009). Applications, platforms, positioning technology, handset evolution, and business model integration. Available at: http://goo.gl/gp55P

Ahas, R., & Mark, Ü. (2005). Location based services – new challenges for planning and public administration? *Futures*, *37*(6), 547–561.

Anderson, I., et al. (2007). Shakra: tracking and sharing daily activity levels with unaugmented mobile phones. *Mobile Networks and Applications*, *12*(2), 185–199.

Ashbrook, D., & Starner, T. (2003). Using GPS to learn significant locations and predict movement across multiple users. *Personal and Ubiquitous Computing*, *7*(5), 275–286.

Avalos, G. (2012) Location-based 'geofencing apps' raise privacy concerns. *The News Tribune*, Tacoma, WA,. January 8.

Batty, M. (2010). The pulse of the city. *Environment and Planning B: Planning and Design*, *37*(4), 575–577.

Beresford, A.R. (2005). Location privacy in ubiquitous computing. *Pervasive Computing, IEEE*, 2(1), 46–55.

boyd, d. (2011). Social steganography: privacy in networked publics. Available at: http://www.danah.org/papers/2011/Steganography-ICAVersion.pdf

Calabrese, F., Kloeckl, K., & Ratti, C. (2008). WikiCity: real-time location-sensitive tools for the city. In M. Foth (ed.), *Handbook of Research on Urban Informatics: The Practice and Promise of the Real-Time City*. Hershey, PA: Information Science Reference, IGI Global.

Campbell, A.T., Eisenman, S.B., Lane, N.D., Miluzzo, E., & Peterson, R.A. (2006). People-centric urban sensing. *Proceedings of the 2nd Annual International Workshop on Wireless Internet* (WINCOM '06), ACM, article no. 18.

Chamberlain, A., et al. (2007). Professor Tanda: greener gaming & pervasive play. *Proceedings of the 2007 Conference on Designing for User eXperiences* (DUX '07), ACM, article no. 26.

Choudhury, T.K. (2003). Sensing and modeling human networks. PhD thesis. MIT, Cambridge, MA.

Cohen, P. (2011). Digital maps are giving scholars the historic lay of the land. *New York Times,* 26 July.

Consolvo, S., Everitt, K., Smith, I., & Landay, J.A. (2006). Design requirements for technologies that encourage physical activity. *Proceedings of the (SIGCHI '06), Conference on Human Factors in Computing Systems.* ACM, 457–466.

Cunfer, G. (2008). Scaling the Dust Bowl. In Anne Kelly Knowles (ed.), *Placing History: How Maps, Spatial Data, and GIS are Changing Historical Scholarship.* Redlands, CA: ESRI Press. pp. 95–121.

Currid, E., & Williams, S. (2010). The geography of buzz: art, culture and the social milieu in Los Angeles and New York. *Journal of Economic Geography*, 10(3), 423.

Dobson, J.E. (2000). What are the ethical limits of GIS? *Geo World*, 13(5).

Dobson, J. E., & Fisher, P. F. (2003). Geoslavery. *Technology and Society Magazine, IEEE*, 22(1), 47–52.

Dunleavy, M., Dede, C., & Mitchell, R. (2009). Affordances and limitations of immersive participatory augmented reality simulations for teaching and learning. *Journal of Science Education and Technology*, 18(1), 7–22.

Eagle, N., & Pentland, A. (2006). Reality mining: sensing complex social systems. *Personal and Ubiquitous Computing*, 10(4), 255–268.

Eagle, N., & Pentland, A.S. (2009). Eigenbehaviors: identifying structure in routine. *Behavioral Ecology and Sociobiology*, 63(7), 1057–1066.

Eisenman, S.B., Lane, N.D., Miluzzo, E., Peterson, R.A., Ahn, G.S., & Campbell, A.T. (2006). Metrosense project: People-centric sensing at scale. *Proceedings from First Workshop on World-Sensor-Web* (WSW '06).

Facer, K., Joiner, R., Stanton, D., Reid, J., Hull, R., & Kirk, D. (2004). Savannah: mobile gaming and learning? *Journal of Computer Assisted Learning*, 20(6), 399–409.

Franken, A. (2011). The Location Privacy Protection Act of 2011, S. 1223, Bill Summary. Retrieved from http://goo.gl/QvnTW

Froehlich, J., Dillahunt, T., Klasnja, P., Mankoff, J., Consolvo, S., Harrison, B., & Landay, J.A. (2009). UbiGreen: investigating a mobile tool for tracking and supporting green transportation habits. *Proceedings of the 27th International Conference on Human Factors in Computing Systems.* (SIGCHI '09), ACM, 1043–1052.

Froehlich, J., Findlater, L., & Landay, J. (2010). The design of eco-feedback technology. *Proceedings of the 28th International Conference on Human Factors in Computing Systems.* (SIGCHI '10), ACM, 1999–2008.

Gasson, M.N., Kosta, E., Royer, D., Meints, M., & Warwick, K. (2011). Normality mining: privacy implications of behavioral profiles drawn from GPS enabled mobile phones. *Systems, Man, and Cybernetics, Part C: Applications and Reviews, Transactions on Systems, Man*, and Cybematics, C, IEEE, 41(2), 251–261.

Girardin, F., Vaccari, A., Gerber, A., & Ratti, C. (2009). Quantifying urban attractiveness from the distribution and density of digital footprints. *Journal of Spatial Data Infrastructure Research*, 4, 175–200.

Gordon, E., & de Souza e Silva, A.S. (2011). *Net Locality: Why Location Matters in a Networked World.* Chichester, West Sussex: Wiley-Blackwell.

Gordon, E., & Manosevitch, E. (2011). Augmented deliberation: merging physical and virtual interaction to engage communities in urban planning. *New Media & Society*, 13(1), 75.

Henderson, T., Kotz, D., & Abyzov, I. (2004). The changing usage of a mature campus-wide wireless network. *Proceedings of the 10th Annual International Conference on Mobile Computing and Networking* (MobiCOM'04), ACM, 187–201.

Ito, M., Okabe, D., & Anderson, K. (2009). Portable objects in three global cities: the personalization of urban places. In R. Ling and W.S. Campbell (eds), *The Reconstruction of Space and Time: Mobile Communication Practices.* pp. 67–87. New Brunswick, NJ: Transaction Publishers.

Jain, S. (2002). Urban errands: the means of mobility. *Journal of Consumer Culture*, 2(3), 385–404.

Klopfer, E., & Squire, K. (2008). Environmental detectives: the development of an augmented reality platform for environmental simulations. *Educational Technology Research and Development*, 56(2), 203–228.

Knowles, A.K., & Hillier, A. (2008). *Placing History: How Maps, Spatial Data, and GIS are Changing Historical Scholarship.* Redlands, CA: ESRI Press.

Kochan, B., Bellemans, T., Janssens, D., Wets, G., & Timmermans, H.J.P. (2010). Quality assessment of location data obtained by the GPS-enabled PARROTS survey tool. *Journal of Location Based Services*, *4*(2), 93–104.

Krieger, M.H., Ra, M.R., Paek, J., Govindan, R., & Evans-Cowley, J. (2010). Urban tomography. *Journal of Urban Technology*, *17*(2), 21–36.

Krumm, J., & Horvitz, E. (2004). Locadio: Inferring motion and location from wi-fi signal strengths. *Proceedings from First Annual International Conference on Mobile and Ubiquitous Systems: Networking and Services* (Mobiquitous '04), 4–13.

Krumm, J., & Horvitz, E. (2006). Predestination: inferring destinations from partial trajectories. *UbiComp 2006: Ubiquitous Computing*, 243–260.

Lane, N.D., Eisenman, S.B., Musolesi, M., Miluzzo, E., & Campbell, A.T. (2008). Urban sensing systems: opportunistic or participatory? *Proceedings of the 9th Workshop on Mobile Computing Systems and Applications*.

Liao, L., Patterson, D.J., Fox, D., & Kautz, H. (2007). Learning and inferring transportation routines. *Artificial Intelligence*, *171*(5–6), 311–331.

Lin, J., Mamykina, L., Lindtner, S., Delajoux, G., & Strub, H. (2006). Fish'n'Steps: encouraging physical activity with an interactive computer game. *UbiComp 2006: Ubiquitous Computing*, 261–278.

Lu, H., Pan, W., Lane, N. D., Choudhury, T., & Campbell, A.T. (2009). SoundSense: scalable sound sensing for people-centric applications on mobile phones. *Proceedings of the 7th International Conference on Mobile Systems, Applications, and Services* (MobiSys '09), ACM, 165–178.

Mainwaring, S.D., Anderson, K., & Chang, M.F. (2005). Living for the global city: mobile kits, urban interfaces, and ubicomp. *UbiComp 2005: Ubiquitous Computing*, 269–286.

Martino, M., Britter, R., Outram, C., Zacharias, C., Biderman, A., & Ratti, C. (2010). Senseable city. Available at: http://goo.gl/zRiv0.

Miluzzo, E., et al. (2008). Sensing meets mobile social networks: the design, implementation and evaluation of the Cenceme application. *Proceedings of the 6th Conference on Embedded Network Sensor Systems* (SenSys '08), ACM, 337–350.

Mun, M., et al. (2009). PEIR: the personal environmental impact report as a platform for participatory sensing systems research. *Proceedings of the 7th International Conference on Mobile Systems, Applications, and Services* (MobiSys '09), ACM, 55–68.

Musolesi, M., Miluzzo, E., Lane, N.D., Eisenman, S.B., Choudhury, T., & Campbell, A. T. (2008). The Second Life of a sensor: integrating real-world experience in virtual worlds using mobile phones. *Proceedings from 5th Workshop on Embedded Networked Sensors* (HotEmNets '08).

Neuhaus, F. (2010). New city landscapes: interactive tweetography maps. Available at: http://goo.gl/lkks

Nippert-Eng, C. (1996). Calendars and keys: The classification of "home" and "work". *Sociological Forum 11*(3), 563–582.

Nold, C. (2007). Bio Mapping Emotion Mapping. Available at: www.biomapping.net

Okabe, D., Anderson, K., Ito, M., & Mainwaring, S. D. (2005). Location-based moblogging as method: New views into the use and practices of personal, social and mobile technologies. *Hungarian Academy of Science Conference: Seeing, Understanding, Learning in the Mobile Age*. Retrieved from http://www.itofisher.com/mito/moblogmethod.pdf

Outram, C., Ratti, C., & Biderman, A. (2010). The Copenhagen Wheel: an innovative electric bicycle system that harnesses the power of real-time information and crowd sourcing. *Proceedings of EVER Monaco International Exhibition & Conference on Ecologic Vehicles & Renewable Energies*.

Parker, A., et al. (2006). Network system challenges in selective sharing and verification for personal, social, and urban-scale sensing applications. *Irvine Is Burning*, 37.

Paulos, E., & Jenkins, T. (2005). Urban probes: encountering our emerging urban atmospheres. *Proceedings of the Conference on Human Factors in Computing Systems* (SIGCHI '05), ACM, 341–350.

Peltonen, V., Tuomi, J., Klapuri, A., Huopaniemi, J., & Sorsa, T. (2002). Computational auditory scene recognition. *Proceedings from IEEE International Conference on Acoustics, Speech, and Signal Processing, 1993*. ICASSP-93.

Polak, E. (2002). Amsterdam real time. Available at: www.waag.org/project/realtime

Ray, B. (2002). Salem witch trials document archive. Available at: www2.iath.virginia.edu/salem/home.html

Ratti, C., Williams, S., Frenchman, D., & Pulselli, R. M. (2006). Mobile landscapes: using location data from cell phones for urban analysis. *Environment and Planning B: Planning and Design*, *33*(5), 727–748.

Reddy, S., Estrin, D., & Srivastava, M. (2011). Network services for mobile participatory sensing. In D. Raychaudhuri

& M. Gerla (eds), *Emerging Wireless Technologies and the Future Mobile Internet*. New York: Cambridge University Press. pp. 154–177.

Rojas, F., Calabrese, F., Krishnan, S., & Ratti, C. (2008). Real Time Rome. *Networks and Communication Studies*, 20(3), 247–258.

Roschelle, J., & Pea, R. (2002). A walk on the WILD side: how wireless handhelds may change CSCL. *Proceedings of the Conference on Computer Support for Collaborative Learning: Foundations for a CSCL Community*. (CSCC '02), ISLS, 51–60

Roth, C., Kang, S.M., Batty, M., & Barthelemy, M. (2010). Commuting in a polycentric city. *Technical report CNRS*. Available at: http://www.sistemasdeingenieria.cl/Urbanics/Batty_Full.pdf

Scholz, T. (2011). Your mobility for sale. In M. Shepard (ed.), *Sentient City: Ubiquitous Computing, Architecture and the Future of Urban Space*. Cambridge, MA: MIT Press.

Sevtsuk, A., Huang, S., Calabrese, F., & Ratti, C. (2008). Mapping the MIT campus in real time using WiFi. In M. Foth (ed.), *Handbook of Research on Urban Informatics: The Practice and Promise of the Real-Time City*. Hershey, PA: Information Science Reference, IGI Global. pp. 326–338.

Shoval, N. (2007). Sensing human society. *Environment and Planning B: Planning and Design*, 34(2), 191–195.

Soja, E.W. (1996). *Thirdspace: Expanding the geographical imagination*. Oxford: Blackwell.

Squire, K.D., & Jan, M. (2007). Mad City Mystery: developing scientific argumentation skills with a place-based augmented reality game on handheld computers. *Journal of Science Education and Technology*, 16(1), 5–29.

Steiniger, S., Neun, M., & Edwardes, A. (2006). Foundations of location based services. Available at: http://goo.gl/ZcKTd

Sterling, B. (2007). Dispatches from the hyperlocal future. *Wired* 15(7). Retrieved from http://www.wired.com/techbiz/it/magazine/15-07/local

Strickland, R. (2001). Portable effects: a survey of nomadic design practice. *O Monografias*, 1, 82–117.

Tang, D., & Baker, M. (2000). Analysis of a local-area wireless network. *Proceedings of the 6th Annual International Conference on Mobile Computing and Networking* (MobiCOM '00), ACM, 1–10.

Toscos, T., Faber, A., An, S., & Gandhi, M.P. (2006). Chick clique: persuasive technology to motivate teenage girls to exercise. *Extended Abstracts on Human Factors in Computing Systems*. (CHI EA '06), ACM,1873–1878.

Van der Spek, S., Van Schaick, J., De Bois, P., & De Haan, R. (2009). Sensing human activity: GPS Tracking. *Sensors*, 9(1.4), 3033–3055.

Walker, M., Whyatt, D., Pooley, C., Davies, G., Coulton, P., & Bamford, W. (2009). Talk, technologies and teenagers: understanding the school journey using a mixed-methods approach. *Children's Geographies*, 7(2), 107–122.

Wolf, J., Guensler, R., & Bachman, W. (2001). Elimination of the travel diary: experiment to derive trip purpose from global positioning system travel data. *Transportation Research Record: Journal of the Transportation Research Board*, 1768(1), 125–134.

26

Mobile Learning in the Majority World: A Critique of the GSMA's Position

Niall Winters

INTRODUCTION

The interest in mobile learning in the majority world[1] is understandable. The figures alone show why: Asia and Africa are the world's two largest mobile phone markets. Asia Pacific has over 3 billion mobile connections (GSMA-Kearney, 2011a) and Africa has 620 million (GSMA-Kearney, 2011b). With such high numbers and with continued growth predicted in these regions, a sensible question to ask is how can mobile technologies support learning and education? However, depending on which literature one reads, and the community one comes from – academia, industry or development – very different answers to this question emerge.

In this chapter, we investigate the position taken by industry by focusing on two recent reports published by the Global System for Mobile Communications (originally Groupe Spécial Mobile) Association (GSMA):

- *Transforming Learning Through mEducation* (McKinsey-GSMA, 2012)

- *mLearning: A Platform for Educational Opportunities at the Base of the Pyramid* (GSMA, 2010).

The GSMA was chosen because it is the representative body of the mobile telecommunications industry worldwide. In its own words, it has 'taken a leadership role on behalf of mobile network operators (MNOs) and vendors, in understanding how mLearning is being used today and its potential for the future' (GSMA, 2010: 5) and its 2010 report focuses directly on 'the current landscape of mLearning in the developing world to assess the ways in which mobile devices are being used as an intervention in learning and to consider the future of this powerful tool' (GSMA, 2010: 5). The 2012 report with McKinsey is broader in scope, focusing on its understanding of mobile technology in education worldwide.

These particular GSMA documents were chosen primarily for three reasons. First, they encapsulate the state of the art envisaged by the GSMA and McKinsey and describe what they view as key mobile learning projects.

They are important in understanding how the mobile telecommunications industry believes it can support mobile learning. Second, mobile learning for development raises unique research challenges that very often require high levels of interdisciplinary knowledge to solve. Therefore, academia needs to better understand the perspective of industry in order to support any potential collaboration and build bridges across the domains of expertise. This position is echoed by the GSMA:

> MNOs and vendors have the networks and hardware to deliver content, and experience developing sustainable and commercial enterprises around this. However, pedagogues, researchers, content providers and governments all have vital roles to play, ensuring relevance of learning materials and best practices in delivering educational resources via mobile devices. (GSMA, 2010: 30)

However, interdisciplinary engagement comes with its own challenges and implied assumptions. These must be well understood and conceptualized in order to provide a strong underpinning for any joint public–private endeavour. While there is not space in this chapter to deal with the complexities of interdisciplinary practices in depth (for a discussion of the issues within a technology-enhanced learning (TEL) context, see Winters and Mor 2008) one very important issue for mobile learning in the majority world is the *nature* of the public–private relationship. This brings us to the third reason for picking the two documents, namely that they are a concrete manifestation of what Kenway et al. (1994) describe as a 'markets/education/technology triad [that] has the capacity significantly to alter the ways in which education is produced, conducted and consumed' (1994: 322).

Simply put, such triads break down the traditional boundaries between the public and private, 'informed mainly by market values'. In this way, analysis of mobile learning in the majority world provides an insight into the way technology-enhanced learning is approached and framed by key players in the private sector and how they view the role of public organizations.

The exploration of the private sector's (i.e. the GSMA and McKinsey) position on these issues within the two documents is framed by two key questions.

- What is the justification for mobile learning?
- How is mobile learning positioned conceptually?

Investigation of the justification and conceptualization of mobile learning is important for two reasons. First, it provides an increased understanding of the GSMA's educational rationale for engaging in mobile learning. It allows for judgement of the seriousness placed by the GSMA on educational approaches and theories and the extent to which they draw on the mobile learning research undertaken by educational researchers over the past ten years or so. Second, the private sector – including international telecoms organizations – fund and support many TEL projects in the majority world and these projects are viewed as 'setting the standard' for others to follow. However, to do so often requires a large resource commitment by, for example, Ministries of Education, and therefore any work that seeks a better insight into the justification and conceptualization issues as understood by the private sector is worthwhile for critiquing the need for investment in mobile learning.

WHAT IS THE JUSTIFICATION FOR MOBILE LEARNING?

The justification for mobile learning in the two documents is more complex than it seems at first reading. Our analysis provides three main underlying themes: (1) the relationship to formal education; (2) the nature of innovation; and (3) developing a market.

The Relationship to Formal Education

Formal education provides the key underpinning for the justification of mobile learning in the documents, but in a somewhat unexpected manner. There is a dichotomy between how it is viewed currently and the role it can play in the future. These two views provide very different justifying roles. In brief, the current state of the formal education system is shown to be weak. In contrast, as we will see later, some of the pedagogical ideas discussed for the future of mobile learning will rely on a strong formal educational system with a central role for teachers as designers of learning.

The weakness of the formal education system is highlighted by the GSMA as a key rationale for the implementation of mobile learning, drawing on work by international agencies such as UNESCO. To take just two examples of the GSMA's interpretation of this work:

> Adhesion to the national or local curriculum is especially a problem in developing countries where often the educational resources are outdated due to a lack of funding and availability. (GSMA, 2010: 9)

> Many schools do not provide the management support, resources or tools that teachers need to carry out their job. (GSMA, 2010: 10)

The high numbers of students not in formal education is also highlighted. For example, the McKinsey-GSMA report states:

> Enabling and facilitating access to education is a key challenge in education today. For example, almost 70 million children ages 6 to 12 are not enrolled in schools, 60 million of them in developing countries. Access to education remains a critical problem for reasons ranging from insufficient school coverage and low household incomes to limitations in the quality of locally available materials. (McKinsey-GSMA, 2012: 9)

Thus, one argument being made for mobile learning is that it can overcome weaknesses in formal education. In particular, this can be achieved by using mobile networks to *provide access to educational content for marginalized learners*. In contrast to the formal education system, the geographical reach of these networks is emphasized. Chris Locke, Managing Director of the GSMA Development Fund puts it as follows:

> [W]e feel that mobile has a unique role to play in reaching those who are outside of the scope of traditional schooling, and yet who will benefit immensely from access to simple educational programmes. (GSMA, 2010: 2)

Moreover, clear gaps in the formal education system provide an opportunity for mobile learning to benefit marginalized learners through the provision of learning opportunities:

> It is often debated that mLearning should only be seen as a final resort for learning or teaching, however, for many people it is a way to incorporate education into their lives when they may have previously been denied the opportunity, therefore becoming an enhancement to their livelihood. (GSMA, 2010: 13)

By directly addressing the 'access issue', mobile learning is viewed as having a potentially transformative affect on the educational system and on the learning opportunities provided to those most in need. While highlighting the issue of access to educational opportunities is laudable, the way in which it is presented is problematic as tensions and trade-offs are raised, the consequences of which are not clear currently.

First, access is not well conceptualized. Put simply, access is about much more than the provision of content. To frame it in this way misses much of the complexity of what education and learning are about and how they are to be supported. Even if we were to focus on content development, this requires the skills of teachers, which in turn relies on a strong education system, including teacher training and classroom experience. Therefore, framing mobile learning as an adjunct to the formal system does not seem to be sustainable. A more beneficial approach would involve exploration of the intricacies of working within the formal system, not outside of it. This is true particularly where learning designs are more complex than those focused on information dissemination. Indeed, there is much education research from which to draw in conceptualizing access in relation to formal education. To take just one example, since 2006 the DFID-funded Consortium for Research on Education Access, Transitions and Equity (CREATE) has worked to understand the different dimensions of access, with a focus on meaningful learning, sustained access over time and equitable access provision. In line with Millennium Development Goal 3, 'Achieve Universal Primary Education', they do not decouple content from formal schooling. Instead, access is conceptualized to include the following five aspects:

- secure enrolment and regular attendance
- progression through grades at appropriate ages
- meaningful learning that has utility
- reasonable chances of transition to lower secondary grades, especially where these are within the basic education cycle
- more rather than fewer equitable opportunities to learn for children from poorer households, especially girls, with less variation in quality between schools. (Lewin, 2007: 21)

More work is needed to investigate the ways in which mobile learning can help address the above issues through appropriate implementation strategies: the idea that access to content via a mobile phone is an end in and of itself is too limiting. The complexity of this implementation challenge when working within existing structures is highlighted by a recent UNESCO review of mobile learning, which found that:

> In spite of the potential of mobile learning to help achieve UPE [Universal Primary Education], research for this review found little evidence of the use of mobile phones to expand access to formal primary schooling for children who are not in school. (UNESCO, 2012: 21)

This serves to illustrate the difficulties in addressing the needs of marginalized learners appropriately. However, working within existing educational structures is not the only challenge. If mobile learning is to work across the formal and informal sectors, the complexities of the lives of those in marginalized communities must be given very focused consideration. Again, an example will help demonstrate the everyday difficulties faced by

the most marginalized. A recent cross-country analysis from Kenya, Ghana and Mozambique by the Institute of Education and Action Aid on the Stop Violence Against Girls in School (SVAGS) project found that:

> Girls in the three project districts experience multiple forms of violence, with 86% of girls in the project area in Kenya, 82% in Ghana, and 66% in Mozambique reporting some form of violence in the past 12 months. (Parkes and Heslop, 2011: 11)

They also note the impact of poverty on completion rates for girls:

> While increasing numbers of younger girls are enrolling in the project schools, in the later years of primary school girls' enrolment drops, most markedly in the Kenyan schools where the number of girls in the last class of primary school in 2009 was almost ten times lower than in the first year. Poverty intersects with gendered inequalities in creating barriers to schooling for girls, with girls missing out on schooling because of household chores and childcare, farm work, inability to pay school fees, early pregnancy and marriage. In the schools themselves, particularly in the project areas in Kenya and Ghana, there is a shortage of well-qualified women in teaching and management positions, and gendered attitudes favouring boys, gendered division of labour, and poor conditions and resources hinder girls' capacity to enjoy, achieve and thrive in school. (Parkes and Heslop, 2011: 11)

Thus, the idea that access can be addressed through content delivery alone seems somewhat idealistic. Far more research is needed into how mobile learning activities can be designed to support marginalized learners and how the context of their lives affects their learning opportunities. The depth of this challenge should not be underestimated.

The Nature of Innovation

The second key underpinning justification for mobile learning is its innovative nature. Primarily, the reports focus on technical innovation. Mobile devices and networks are viewed as the conduit through which access can be delivered. For example, the McKinsey-GSMA report states:

> Any portable device, such as a tablet, laptop or mobile phone, that provides access to educational content through mobile connectivity (2G, 3G, or 4G complemented by mobile-based Wi-Fi) can be a tool for mEducation. (McKinsey-GSMA, 2012: 4)

In particular, this 'innovative' use of mobile technology is seen as having great potential because it has worked on other sectors, with the conduit idea again clearly evident:

> We have seen over the past years how mobile is playing an increasing role in addressing development issues – such as providing access to banking, to health information, to agricultural services reaching rural farmers. The scale and ubiquity of mobile networks means they are often the only infrastructure in remote and rural areas, and the mobile industry has shown its innovative approaches to solving these needs using mobile technology. (GSMA, 2010: 2)

This success provides one justification for exploring opportunities in education, with the McKinsey-GSMA report noting:

> Mobile operators can seize this exciting opportunity and shape the market if they understand how new technologies and initiatives will impact education around the world – and if they can develop smart strategies and implement them quickly. (McKinsey-GSMA, 2012: 4)

It is interesting to note the focus on understanding how technologies can 'impact' on education. We saw one instantiation of this effort to 'shape the market' earlier with regard to the provision of content, but also noted the broader issues that were not addressed. As such, the provision of access required little technical or pedagogical innovation. It is no surprise then that it was viewed as not needing to be linked in any way to the formal education system: provision of content is simple, hence can be done outside of the formal education system. Here, we add to this critique by noting the provision of content on a mobile phone is neither pedagogically nor technically innovative. From a conceptual perspective, focusing on provision positions mobile learning in a place where research was more than ten years ago. A key critique that moved the

field forward came from Roschelle (2003: 9), who made what still remains today, a compelling argument for 'unlocking' the potential of mobile devices for learning. His work resulted in part from his concern that technology in education is often blinded by complex views of technology without placing sufficient emphasis on social practices. Instead, he makes a strong argument for the identification of the 'simple things that technology does extremely and uniquely well' – that is, its affordances and the need to 'understand the social practices by which those new affordances become powerful educational interventions'. However, while mobile phones are very good at providing access to content, in line with Roschelle's first point, it is less clear (as we have seen) how this can become a powerful educational intervention, particularly outside of the support offered by formal schooling. However, it is reasonable to ask what other learning activities (beyond access) are promoted – that is, what is the nature of pedagogical innovation in the documents? Pedagogical innovation is premised primarily around personalization.

Personalization

The potential for mobile learning to support personalization is discussed in different forms:

> mLearning provides a personal way of accessing educational content with the ability to build an extensive learning community. Activities can be tailored to meet the individual user needs. (GSMA, 2010: 12)

This simple version of personalization can be interpreted as meaning that educational material is always available to the learner. However, to provide someone with a personal way to learn where activities 'can be tailored to meet the individual user needs' implies the need to draw on the personalization literature within TEL. This is even more true when personalization is linked to adaptivity:

> It personalizes education solutions for individual learners, helping educators customize the

teaching process, using software and interactive media that adapt levels of difficulty to individual students' understanding and pace. (McKinsey-GSMA, 2012: 4)

This is a far more complex issue, requiring design and technical expertise that enables educational software to be tailored in realtime to a learner's needs. The resources required to design, develop and implement such an approach are vast and would certainly mean an investment in education beyond what is achievable currently for most Ministries of Education in the majority world.

Personalization is built upon to promote mobile learning as a means to customize context and support learner collaboration:

> Educators can now assess students' understanding using wireless assessments on handheld devices. These provide real-time updates on individual student progress, allowing educators to track class progress and tailor instruction for students requiring remedial support. (McKinsey-GSMA, 2012: 15)

> With mobile technology, students can source or create their own content, share it with peers, share different learning paths and evolve better answers through collaboration. (McKinsey-GSMA, 2012: 8)

Entire communities of TEL researchers work in the technically innovative areas of artificial intelligence in education (AIED) and computer supported collaborative learning (CSCL), each field having its own respective conferences and journals. While it is indeed true that mobile learning can have a role to play in each of these areas, the level of expertise required to design, develop and implement such interventions is large. Again, this requires a strong formal education system to support the development of pedagogically and technically innovative interventions. This is a massive challenge for the majority world (and, I would add, the UK, too) and more research is needed to explore ways to better support the development of innovative mobile learning approaches (including personalization) by teachers working as learning designers (Laurillard, 2012). Drawing

on Roschelle (2003) here, setting the agenda for mobile learning in the majority world should be less about a retrospective rationalization of its use to address current problems (e.g. providing access to content) and more about looking at social practices, the affordances of mobile devices and ecosystems to determine how we can address current weaknesses in a fundamental manner.

Mobile Learning as a Market Opportunity

The third main justification for mobile learning is as a market opportunity, with the McKinsey-GSMA report stating that 'mEducation is poised to become a USD 70 billion market by 2020' (2012: 4), leading them to classify more than a hundred commercial mobile learning offerings into seven product types. While not all are suitable for mobile learning in the majority world, the GSMA does frame marginalized communities' lack of access to education as a serious business opportunity:

> [W]ith 98% of the world's illiterate or semi-literate population residing in developing countries, where access to schools and resource materials is at a minimum, such regions present the greatest areas of need. These markets therefore represent the greatest opportunities for mLearning programmes and products. (GSMA, 2010: 5)

However, in order to achieve such targets the GSMA notes that 'a sustainable and robust business case' (GSMA, 2010: 5) remains to be developed. While from a corporate point of view the discussion of business opportunities is appropriate, from an education perspective, it raises questions as to whether the ideals of access for marginalized communities can be addressed while seeking a strong business case. As noted by the GSMA '[c]ontent and the provision of it costs money and it is not yet clear who should pay – governments, local authorities, the consumer or other' (GSMA, 2010: 24). This tension is a serious issue of concern when the potential

beneficiaries of mobile learning are determined by their ability to pay:

> There is greater and more immediate value in
> vocational forms of mLearning where the end
> user is paying for the service. Health education,
> language lessons and general life skills are seen by
> mobile customers as valuable and worth paying
> for. (GSMA, 2010: 30)

This should raise serious concerns about the
agenda of mobile operators if, as educators,
we are concerned about building long-term
and sustainable access to educational opportunities. What about subjects that do not fit
with the above agenda, communities who do
not fit the business model or subjects that are
seen by a small minority of 'mobile customers as valuable'? The GSMA's stated position
of providing access to marginalized communities takes on a very different motivation
when the 'value' of mobile learning is a
market-driven one or when the subjects
deemed worthy of support are those who fit
neatly within a corporate social responsibility agenda. In addition, if market values is
the overarching concern, then this will do
little to increase access at secondary level as
'[a]ccess to secondary schooling is very
strongly household-income-related in all
poor countries' (Lewin, 2007: 8). Indeed, this
is recognized by the GSMA in discussions
regarding who will pay for their services:

> The global goal set by the UN Millennium
> Development Goals is for universal primary
> education for all, however who pays for education
> beyond this point is widely debated and varies
> from one country or region to the next. (GSMA,
> 2010: 24)

Such an agenda is a good exemplar of
Kenway et al.'s triad and their concern with
postmodern forms of education:

> Modern forms can be identified as those which
> are primarily state funded, identified also as
> public, institutionalized, formal, largely print-
> based, mass-orientated, steered and serviced
> largely by education professionals and informed
> mainly by educational values. In contrast,
> postmodern forms can be categorized as those
> being produced and consumed outside the

institutions of the state; identified as private,
market funded, national and international,
de-institutionalized, de-territorialized, informal,
largely image-based, niche-orientated, steered
and serviced mainly by commercial, cultural and
technology professionals and informed mainly by
market values. (1994: 327)

Equally the GSMA report also promotes the
notion of taking a hands-off approach and
'conducting "business as usual", implementing and growing their network coverage,
they are increasing the opportunities for
those in rural and ultra regions to "get connected" and use mLearning services, thus
increasing customer base and revenues'
(GSMA, 2010: 30). This could be interpreted
as letting educators, practitioners and other
interested parties develop mobile learning
interventions, who would then determine
their own connectivity needs.

HOW IS MOBILE LEARNING POSITIONED CONCEPTUALLY?

Overall, the reports do not engage in any
significant way with recent mobile learning
literature on how mobile learning is conceptualized. Mobile learning is considered to be
learning that happens 'anywhere, anytime',
in line with very early research work in the
area (e.g. Quinn, 2000). It is not surprising,
then, that mobile learning is conceptualized
in a techno-centric manner:

> We define mEducation as technology-enabled
> learning solutions available to learners anytime,
> anywhere. Any portable device, such as a tablet,
> laptop or mobile phone, that provides access to
> educational content through mobile connectivity
> (2G, 3G, or 4G complemented by mobile-based
> Wi-Fi) can be a tool for mEducation. (McKinsey-
> GSMA, 2012: 4)

> mLearning is the ability to access educational
> resources, tools and materials at anytime from
> anywhere, using a mobile device. (GSMA, 2010: 6)

Many case studies in the GSMA report are in
line with this position. To take just one example,
as part of the Millennium Villages project,
Ericsson has developed 'mLearning modules'

that community health workers can download to their mobile phones. While there is no doubt that having access to this content is welcome, the benefits of having such content on a low-end mobile phone (rather than a book, for example) are not immediately clear. No particular affordances of mobile technologies are being leveraged in this case in a way that enables learning to happen in new ways. However, some cases proved more successful because they addressed a real need: Janala from the BBC World Service is one example. This service provides three-minute English language lessons and associated text-based content to Bangladeshis on their mobile phones.

Mobile learning research began with an initial focus on mobile technologies and their ability to support learning 'any time, any place, any where'. From there, it has moved to a more complex view involving the learner, the technology and the context. Theories of mobile learning have been developed (see e.g. Sharples et al., 2007) building on a socio-cultural understanding of learning. Such theories emphasize the importance of context in mobile learning:

> The common denominator is context: physical, technological, conceptual, social and temporal contexts for learning. … Context, then, is a central construct of mobile learning. It is continually created by people in interaction with other people, with their surroundings and with everyday tools. (Kukulska-Hulme et al., 2009, 21)

Frohberg, Göth and Schwabe (2009) built on the idea of using context as a classification criterion for mobile learning projects. However, their view of context is limited to 'where the learning takes place' (2009: 313) but this nevertheless provides a useful starting point for determining innovation in mobile learning. Their classification, from least to most complex, is as follows:

- **Independent context**: The mobile learning activity has no relationship to the learner's environment (e.g. doing drill-and-practice quizzes on a phone).
- **Formalizing context**: The mobile learning activity is dependent on a formal learning setting (e.g. classroom response systems).

- **Physical context**: The mobile learning activity is dependent on the location (e.g. museum guides).
- **Socializing context**: The mobile learning activity is dependent on the social setting (e.g. a community of learners supporting each other through peer learning).

This discussion and exemplification of mobile learning in the document can be characterized by the first two classifications. As noted by Frohberg, Göth and Schwabe (2009), such forms of mobile learning have very little pedagogical innovation. This undermines the rationale for their development. Why should mobile learning interventions that offer very little more than conventional approaches be funded? The Janala case study works because it leverages the audio affordances of mobile phones needed to support language learning.

While the reports do not take a contextual view of mobile learning, some of the case studies presented in the GSMA (2010) document do. One such case is the Yoza/M4Lit project funded by the Shuttleworth Foundation to support reading by South African youth. The stories were developed in a participatory manner, and the social aspect of learning is emphasized with reader interaction supported via links with the MXit social network. Referring again to Roschelle's research here, mobile learning applications become 'powerful educational interventions' through leveraging the key things that mobile phones do well, while supporting existing social practices.

However, overall it can be claimed that where the reports discuss more innovative pedagogical approaches that could be underpinned by the latest mobile learning research, they again take a techno-centric stance:

> Portable device form factors are rapidly evolving. Increased availability and penetration of smart portable devices with advanced functionalities, such as accelerometers that sense motion, will lower costs and open a world of new possibilities for mEducation solutions. (McKinsey-GSMA, 2012: 4)

There is no effort to understand how mobile learning design can incorporate such technical innovation. This is perhaps not surprising, given the complex nature of this challenge.

DISCUSSION

Mobile learning in the majority world is a nascent area of research. There are clear differences in approaches between the corporate and academic communities that need to be addressed. The purpose of this chapter has been to highlight some of these issues in support of productive future dialogue and collaboration. Three key issues emerge from the analysis presented so far: transformation, scale and sustainability.

Transformation

Mobile learning is viewed in the reports as having a transformation potential:

> Mobile technology's power to transform education is difficult to overstate, given the importance and impact of learning that takes place outside a traditional classroom environment. (McKinsey-GSMA, 2012: 4)

This focus on transformation outside of formal schooling can be linked to two contrasting approaches in the literature. Within the mobile learning community, there is general support for this stance, tempered by the need to support teachers' professional development in the use of mobile learning in the classroom:

> It is important to consider the perspective of teachers (at all education levels) and the opportunities they have for professional development in this area of technology use. At European and individual state level, there appears to be little teacher development or training activity addressing mobile learning. (Kukulska-Hulme et al., 2009: 25)

However, more serious concerns are raised by those researchers working in education and development, in particular with respect to ensuring equitable access to educational opportunities as exemplified by the access discussion in this chapter. In general, too, the GSMA frames transformation as occurring because of technical innovation and availability of mobile phones and new technologies. In contrast, many education researchers prefer to view transformation change as a bottom-up process driven by understandings of existing structures, barriers and social circumstances. Each approach has its own merits and both are needed if mobile learning is to be pedagogically successful in the majority world.

Scale

One determinant of the power of mobile learning for the GSMA is the question of scale, through the geographical reach of mobile phone networks. As such, many projects exemplify this approach through mobile learning applications that have the potential to be used by millions of people and are easily replicable by mobile phone network operators in different countries. Where scale is a concern for educational researchers (although, for some it is not), the approach of the GSMA contrasts with their position. Particularly, for educational researchers working in development, issues of scale are directly related to the formal education system. Scale should be achieved by working to support the formal educational systems of countries in a way that makes learning meaningful to students. Equity is a key driver here, as exemplified by the debates regarding who – in the public and private sectors – places what values on which aspects of education.

Sustainability

The sustainability of mobile learning projects is a key motivation for both the private and academic sectors. However, their approach to the problem differs. The private sector focuses on the need to make mobile learning interventions sustainable by making them cost-effective. The well-made argument is that if learners place enough value on mobile learning, they will pay for it, just like any other service. This will make the intervention viable in the long term.

In general, the education community approaches sustainability from a different angle. It works from the premise that, if an intervention directly addresses the needs of learners in a profound manner, is designed to work well within the context of use and engages learners in meaningful forms of participation, then it will be sustainable. Thus, mobile learning interventions are often designed in a participatory manner (Kensing and Blomberg, 1998) that emphasizes the role of the learners in the design and development of the tools they will use.

CONCLUSION

Mobile learning in the majority world is a nascent area of research and so can be developed along many parallel avenues, informed by corporate needs, development agendas or educational research, amongst other issues. However, none of these is sufficient. This chapter has argued that mobile learning in the majority world needs to draw more on the large body of mobile learning research available, but only where appropriate. It needs to be open to working with the corporate sector, while not necessarily being led by its understanding of what learning and education should be. Perhaps most importantly though, it needs to work from a development perspective to ensure that interventions are sustainable and directed towards improving the lives of marginalized learners.

NOTE

1. The term 'majority world' was coined by Bangladeshi photographer Shahidul Alam, to refer to what had been called the 'Global South' or 'developing world'. Majority world 'defines the community in terms of what it is, rather than what it lacks' (see: www.appropedia.org/Majority_world).

REFERENCES

Frohberg, D., Göth, C. and Schwabe, G. (2009) Mobile learning projects: a critical analysis of the state of the art, *Journal of Computer Assisted Learning*, 25: 307–331.

GSMA-Kearney (2011a) Asia Pacific Mobile Observatory 2011. London: GSMA.

GSMA-Kearney (2011b) African Mobile Observatory 2011. London: GSMA.

GSMA (2010) mLearning: A Platform for Educational Opportunities at the Base of the Pyramid, London: GSMA .

Kensing, F. and Blomberg, J. (1998) Participatory design: issues and concerns, *Computer Supported Cooperative Work*, 7: 167–185.

Kenway, J., Bigum, C., Fitzclarence, L., Collier, J. & Tregenza, K. (1994) New education in new times, *Journal of Education Policy*, 9 (4): 317–333.

Kukulska-Hulme, A., Sharples, M., Milrad, M., Arnedillo-Sánchez, I. and Vavoula, G. (2009) Innovation in mobile learning: a European perspective, *International Journal of Mobile and Blended Learning*, 1(1): 13–35.

Laurillard, D. (2012) *Teaching as a Design Science: Building Pedagogical Patterns for Learning and Technology*. Abingdon, Oxon: Routledge.

Lewin, K.M. (2007) Improving access, equity and transitions in education: creating a research agenda, *CREATE Pathways to Access, Research Monograph, No. 1*.

McKinsey-GSMA (2012) Transforming Learning Through mEducation. Mumbai: McKinsey & Company.

Parkes, J. and Heslop, J. (2011) Stop violence against girls in school: a cross-country analysis of baseline research from Kenya, Ghana and Mozambique. London: ActionAid. Available at: www.actionaid.org/publications/cross-country-analysis-baseline-research-kenya-ghana-and-mozambique

Quinn, C. (2000) mLearning: mobile, wireless, in-your-pocket learning. *LineZine*. Available at: www.linezine.com/2.1/features/cqmmwiyp.htm

Roschelle, J. (2003) Unlocking the learning value of wireless mobile devices. *Journal of Computer Assisted Learning*, 19(3): 260–272.

Sharples, M., Taylor, J. and Vavoula, G. (2007) A theory of learning for the mobile age. In R. Andrews and C. Haythornthwaite (eds), *The Sage Handbook of E-learning Research*. London: Sage.

UNESCO (2012) Turning on mobile learning in Africa and the Middle East: illustrative initiatives and policy implications. UNESCO Working Paper Series on Mobile Learning. Available at: www.unesco.org/new/en/unesco/themes/icts/m4ed/mobile-learning-resources/unescomobilelearningseries

Winters, N. and Mor, Y. (2008) IDR: a participatory methodology for interdisciplinary design in technology enhanced learning. *Computers & Education*, 50(2): 579-600.

27

Online and Internet-based Technologies: Gaming

Catherine Beavis

INTRODUCTION

There are high expectations of the educational value of online games and immersive, virtual worlds. Digital games, if used for educational purposes, it is argued, have the capacity to transform learning, engage disenchanted students, address major differences in experience and orientation in out-of-school and in-school contexts, teach twenty-first-century literacies and prepare learners to become tech savvy, critical and agential participants – 'knowledge workers' (Gee et al., 1996) – in a technologically saturated world. As governments and other institutions seek to respond to rapid and massive technological change, the use of games-based learning via online and mobile Internet-based technologies is seen as providing much potential for innovative, effective and accessible contemporary teaching and learning. While there is considerable enthusiasm for the use of games to support learning, however, research focusing on these areas is fragmented and provides a more mixed picture of their use.

Games studies as an interdisciplinary field is informed by quite different research traditions and trajectories, each with its own epistemological framework and assumptions. In this chapter, the focus is on games and learning as they relate to formal educational contexts, particularly schools. The chapter begins with an overview of some of the key issues and assumptions related to the use of online games for educational purposes. This is followed by a discussion of four major disciplinary orientations towards games: perspectives and approaches drawn from humanities and social science; explorations of literacy and media production from media and cultural studies; the study of games themselves within the formally designated, eponymous area of games studies; and technical and educational approaches based in the fields of e-learning and instructional design. This section concludes with an extended discussion of the use of Second Life, which canvasses the main approaches to its use, and goes on to provide a detailed account of the use of Second Life to foster argumentative knowledge construction in Singapore secondary schools (Jamaludin et al., 2009). This is followed by an account of ongoing tensions between research traditions and a discussion of issues for further development and insights for the future. The chapter concludes with a summary of key points and suggestions for further reading in the games and learning area.

KEY IDEAS/DEBATES

Games to Teach, Serious Games

The enthusiasm for online games to support learning derives from many sources, and is fuelled by a range of observations, hopes and agendas, some of which have a stronger basis in research than others. Chief amongst these are views about the nature of games and their affordances and the ways in which these might be used to teach (Facer, 2002; Kirriemuir and McFarlane, 2004; Schaffer et al., 2005); the ways in which games are seen to embody good learning principles (Gee, 2003) and function effectively as 'learning machines'; the problematic notion of the 'digital native'; and the need for schools, universities and other educational contexts to overtly recognize and connect with the technologically mediated nature of much of the contemporary world. At a less visible level, assumptions about the nature of learning, the nature of learners and the nature of knowledge shape the design and content of games used to support learning, while, conversely, beliefs and values about learning, learners and knowledge may be constrained to what the technology most readily makes possible. Assumptions relate also to the role of context and intervention and to where, as well as how, learning takes place. Sociocultural perspectives on games and learning, for example, emphasize the socially situated nature of play, the importance of discussion around the game and the role of reflection in learning, including that prompted or fielded by the teacher (e.g. Squire, 2002). Perspectives

drawing on new media and cultural studies traditions, new literacies and games studies, see situation as central to game play (e.g. Stevens et al., 2008). Some instructional design-based perspectives, however, focus on developing games that utilize learning taxonomies such as Bloom's, but seek to deliver self-contained learning material and sequences directly to the player regardless of the teacher's intervention or expertise, and irrespective of the context in which they are played.

However, despite increasing interest in the possibilities offered by digital games to support learning in a variety of educational contexts and areas, there is as yet no settled agreement on how to describe the field. The term 'serious games' and the ideas associated with the serious games movement gained widespread currency in the early years of the twenty-first century, through programmes such as the Wilson Center 'serious games initiative' established in 2002, which aimed 'to help usher in a new series of policy education, exploration, and management tools utilizing state-of-the-art computer game designs, technologies, and development skills' (Serious Games Initiative, n.d.). Serious games have been defined as games 'designed with the intention of improving some specific aspect of learning' (Derryberry, 2007) and as 'games which aim at providing an engaging, self-reinforcing context in which to motivate and educate the players' (Kankaanranta and Neittaanmäki, 2010: v). Serious games utilize a variety of genres and platforms and are used in fields including education, business, health, tourism, the armed forces and more.

Games and Learners: Which Platforms, Which Games?

Games developed to support learning draw on games design principles that reflect their provenance, while approaches to using commercial or 'educational' games in educational contexts, similarly, reflect the disciplinary field from which they arise, and the perspectives and priorities of the teachers and researchers concerned.

Using games as the focus of a wider ecology of learning presents some difficulties. Games designed specifically for educational purposes constitute one section only of a movement that also includes the use of commercial off-the-shelf games (COTS) and free-to-download network games and the use of game-like environments such as virtual worlds as contexts in which to teach. However, defining serious games (or any games) according to the structures and design of games does not signal the active engagement of players with the game and the role of teachers and others in creating an environment for learning and reflection through the use of games. Terms such as 'games-based learning' and 'learning through play' are variously used to more broadly to represent different aspects of games, learning, teaching and play and their intersections

Games and Play

Both the term and concept, however, continue to raise issues. Leaving aside the question of which games or which games platforms are entailed, and the role of players, teachers and contexts definitional questions concern what constitutes a game and what constitutes (game) play. For many scholars and game players, the term 'serious games' is either tautologous or an oxymoron. Games are quintessentially 'playful' and therefore not serious, even though challenging demands and design within the game, on the one hand, and sustained and serious commitment from the player, on the other, are essential to successful game play (Gee, 2003). Difficult games take players seriously, while a game that is too easy is not likely to succeed commercially. Games provide 'hard fun' (Papert, 2002) – 'the kind of fun you have when you work on something difficult, something that you care about, and finally master it' (Schaffer, 2006: 21). Good games demand 'serious play' (de Castell and Jensen, 2003). Oppositions between work and

play, and perspectives associating fun with what is easy, and a lack of pleasure with what is hard, are problematized or refused.

What is at issue from this perspective are definitions of play, links between play and learning, whether 'hard play' can be fun and, more contentiously, whether appropriating out-of-school pleasures and engagement for educational purposes in the form of serious games or games-based learning can still be 'fun' and 'play'. Drawing on the work of Sutton-Smith (2001), Huizinga (1950), Caillois ([1961]1979) and others, play theory in games studies attends to several characteristics of play: play as freedom – freedom to fail, freedom to experiment, freedom to fashion identies, freedom of effort and freedom of interpretation (Klopfer et al., 2009: 4); play as 'not real' – that is, outside the parameters of the normal and everyday; the functions and outcomes of play; distinctions between game forms that have winners, such as competitive sports, and game forms that do not, as in children's improvised dramatic or narrative play; play as private, interior and individualistic, and as collaborative rule-governed behaviour.

A central but contested tenet of games studies and game play is Huizinga's notion of the 'magic circle' of play: the bounded space or site within which game play occurs – the playground, the card table, the theatre and so on. The image encapsulates a number of dimensions of games and game play: the imaginative entry to the world presented by the game, including rules, goals and parameters; the tacit agreement of players to abide by these rules; the situated nature of game play; and the active co-operation between players and the game in the co-construction of game play. A more contentious feature signalled by the image, however, is the separation of the game from everyday life – a view of 'space' as bounded and separate , and one significantly at odds with understandings of space in virtual contexts, and the fluid cohabitation and transversal of on and offline worlds that players undertake. Further, the structuralist assumptions underpinning the notion undervalue the active role of players, contexts and practices that surround the game.

Which Platforms, Which Games? When is a Game Not a Game?

The field of games-based learning encompasses a variety of platforms and games, including commercially available or network games; 'serious games' developed for specifically educational purposes on platforms ranging from desktop computers through to iPads and mobile phones; and commercially available or purpose-built virtual, often persistent, worlds. While earlier literature and research into the potential of games to promote learning (e.g. Gee, 2003; McFarlane et al., 2002) primarily referred to computer or video games, these terms have become too limited to describe games and game play that include but extend beyond computer-based play. The term 'digital games' encompasses a wider range of both games and platforms, which includes but is not limited to this form. 'Digital games' include computer, video and console games, Internet games including massively multiplayer online role play games (MMORPG), and games utilizing mobile technologies such as mobile phones, iPhones or iPads – sometimes referred to as wireless, mobile ubiquitous technologies in education (WMUTE) games. Other digital forms that are not, strictly speaking, games but are often encompassed under the games-based learning umbrella include social networking sites such as Facebook, and virtual worlds, such as Second Life, which provide contexts for learning in a 'playful' environment. 'Sandbox' games such as The Sims, which provide tools for play but no 'game' in the sense of a driving narrative or set of problems to solve, are also used, while some serious games, such as Revolution or Legends of Alkhimia, adapt games engines designed for MMORPG – in this instance, Neverwinter Nights and Diablo 2 respectively. In addition, in the context of games and learning, the term 'games' is used to encompass a range of Internet-enabled or stand-alone simulation contexts and games.

Kinds of Learning and Kinds of Games

The use of games-based learning for acquiring disciplinary/school curricula knowledge includes the acquisition and practice of specific skills and understandings, as for example in some maths-based 'skill and drill' games; and the development of detailed understandings of curriculum content areas, either through 'serious' games designed specifically for that purpose, for example Twisted Physics, Tafelkids, Legends of Alkhimia, or through the use of commercial games, as for example Civilization III. It also includes the use of virtual worlds whether commercial (e.g. Second Life) or specifically designed for educational purposes (e.g. Whyville) as sites for interaction, teaching, learning and design, together with open-ended simulation games or sandbox environments such as The Sims. Game-making software, for example Mission Maker, Game Maker or Kahootz, is used to teach ICT proficiency and provide resources and opportunities for creativity and design, enabling students to be makers as well as consumers. Simulation 'games' or scenarios are used to teach complex processes in 'real-world' situations, as for example in medicine and disability studies; and the study of games as cultural artefacts may feature in curriculum areas such as English, design, art and media studies.

Underlying the use of games to teach in specific disciplinary or existing curriculum areas lies a set of debates about formulations of disciplinary knowledge and the usefulness or appropriateness of these formulations in contemporary times. Game design and usage in some areas, in particular those arising from e-learning and instructional design positions, and in simulation-based usage, but also commonly in research and practice drawing on social science and humanities perspectives, are posited on a belief in the capacity of games to enable learners to more effectively understand iterations of knowledge that remain largely unchanged from those prescribed in a traditional, conventional curriculum. A contrary view, however, coming particularly from games studies perspectives, but also implicit in much writing and research from humanities and social sciences positions, argues not only that the affordances of games alter the nature of knowledge, requiring and enabling a different form of both learning and knowledge (see, for example, de Castell's (2011) notion of 'ludic epistemology', or Bogost's (2007) 'procedural rhetoric') but also the kinds of disciplinary knowledge required in the present time are significantly different from traditional constructions. In this view of games and learning, the present time is constructed as radically unstable and uncertain, schools are regarded as failing and young people (learners) are viewed as 'digital natives', at home in the digital world but not in the out-of-date, print-based world presented to them at school.

PERSPECTIVES AND APPROACHES FROM THE HUMANITIES AND SOCIAL SCIENCES

Perspectives drawn from the humanities and social sciences build on constructivist approaches to learning and see learning as contextual, socially situated and relational. Much of the impetus for observations and research in this area arises from observations of the kinds of engagement that games players, particularly but not exclusively young people, have in their out-of-school lives. Observations here cover a variety of foci – the social and collaborative nature of much game play, the kinds of learning players seem to be engaged in and the learning practices in evidence, such as the shared nature of problem solving, the generation and use of distributed knowledge, ethical and scientific reasoning, and distributed situated cognition

(Egenfeldt-Neilsen, 2005; Simkins and Steinkuehler, 2008).

An important focus of much literature in this tradition, particularly associated with Gee (2007) and others, considers 'good games' as exemplars of ways to promote 'good learning' (by contrast to supposed general practice in schools). Within the organization, in structure and design of 'good' games – games that are 'long, complex and difficult' (Gee, 2005: 5), games like Deus Ex, The Elder Scrolls III: Morrowind, or Rise of Nations – key learning principles, it is argued, can be seen at work. Games take players seriously; players are immediately inducted into the games world (rather than subjected to 'dummy runs') and treated as members of the game world narrative and community. Games reward risk taking and trial and error. In games it is safe to fail because you can always try again; games become progressively harder and more challenging so that players become increasingly knowledgeable, expert and skilled. Unlike school pedagogy and the curriculum, games are based on 'performance before competence' (Gee, 2007) and are driven by choices and problem solving rather than content. As Pelletier (2009) notes, an odd consequence of such positions is that content is seen almost as abstraction, merely a means to an end, with the focus, rather, on the learning processes entailed.

Many scholars in the humanities and social sciences field do not argue that games should be introduced into schools, but, rather, schools and education authorities should learn about learning from watching young people play games and observing the literacy and learning practices they engage in and the ways in which games are structured to bring about the high levels of commitment, problem solving and collaboration that they require. For others, however, observations about games, game-play and games players are the impetus to designing and using games in schools, whether purpose built or off the shelf. These writers identify what are said to be twenty-first-century skills that are exemplified by game play. They look to find ways to build the capacity to develop these skills into games designed for use in schools. Such skills include 'problem solving, analytic thinking, systems thinking, credibility and judgment of information, technology fluencies, the ethics of fair play, collaboration in cross-functional teams and accessing knowledge networks' (Klopfer et al., 2009: 33).

These observations were the basis for much of the early research into the use of commercially available, off-the-shelf games (COTS) to support teaching in the content areas. UK group TEEM BECTA, for example, tracked the use of a range of games in eight primary and four secondary classrooms in formal curriculum settings, using a range of games including Sim City, Age of Empires, Roller Coaster Tycoon and Championship Manager, with games trialled 'on more than one occasion' over the summer term, 2001. Games were chosen by the research team that were seen to have specific curriculum relevance, in relation to content and learning styles of skills. Games were given to pairs of teachers in different schools and teachers were asked to evaluate them in relation to one of the Key Stages, with a class they taught, if possible as part of mainstream teaching. Teachers were given a framework/ questionnaire to give to parents. Focus group interviews with teachers were conducted, and 800 students were surveyed, with over 700 completed responses. The findings of this research were that while problem-solving skills, working collaboratively, the development of mathematical skills and the like – many of the skills, practices and dispositions the curriculum sought to teach – were much in evidence, teachers struggled to find ways to make the content 'fit' within the formal structures of the curriculum (McFarlane et al., 2002). This was one of the first studies grounded in the actual experience of teachers, students and schools, albeit for a very limited period, and one of the few that looked across schools and across a range of teachers, curriculum areas, children and games.

Squire (2004), in a more extended study of the use of one game, Civilization 111 in history teaching, similarly found that the use of games deepened (successful) students' conceptual understandings of history and geography, that 'engagement was a complex process of appropriation and resistance whereby the purposes of game play were negotiated among students' identities, classroom goals and the affordances of Civilization 111', but that the game 'presented challenges for teachers 'in integrating such a complex game within classroom settings' (Squire, 2004: 4). Using case study methodology, Squire worked with practising teachers to develop two 'design experiments' 'to explore what happens to classroom culture and learning when a complex computer game such as Civilization 111 is used as a tool for learning' (2004: 96). With the teachers, Squire designed a game-based curriculum for teaching history, using Civilization 111, then as a participant-observer researched the material as it was taught in three settings – an interdisciplinary world history course for US high school freshmen, a group of students from the same group in a week-long summer camp, and an after school computer club for middle school students (2004: 96). As Squire notes, this was 'not a controlled experimental study, nor … a direct comparison of how learning occurs in each setting. Rather, the purpose of this study is to understand how learning occurs in each context and to generate "petite generalizations"' (2004: 97). Drawing on Vygotskyan and sociocultural views of learning, Squire's account provides a sustained and textured view of classroom teaching and learning. Squire resists taking instructional design as his starting point, arguing rather that 'it is far more useful to build such an emerging pedagogy based on empirical studies of the software in actual use than to design instructional programs and theories a priori' (Squire, 2004: 124–125).

Research here also explores the nature of the connections generated amongst groups of players. Some advocates of the use of games to support learning argue that game play, in and of itself, generates communities of practice amongst players and, hence, extrapolate from that to assume such communities of practice will be reproduced in classroom uses (Oblinger, 2006). This conflation bedevils much of the gamification literature and advocacy for game-based learning. As such, it is part of a broader pattern of boosterism that assumes an unproblematic transfer of games and game play from out-of-school to in school, a naïve perspective on learning, and glosses compulsory compliance with willing participation, ignoring questions of identity, relationships, context, community and the like (Sodestrom et al., 2006), as well as issues of performance and performativity (Chee 2011). Most perspectives drawn from humanities and social sciences traditions tend to be more sceptical. Calling on Wenger's accounts of communities of practice, they point out the absence of key features and warn against romanticism and the dangers of assuming unproblematic transfer into games-based learning in classroom contexts.

A point also worth noting is the ways in which some online 'communities' work actively to judge and exclude would-be members who do not conform or demonstrate the requisite level of skill in play or adherence to group discourses, values and norms. The reproduction in online contexts of power relationships, toxic social practices and social structures evident elsewhere and the effective emergence of online dictatorships in some 'communities' also preclude assumptions that the communities established around games are always-already, automatically open and benign. The utopian view of online interactions as democratizing and participatory needs to be juxtaposed against perspectives that highlight 'thick' distinctions regarding access and opportunity (Burbules and Callister, 2000) and a second-level digital divide associated not with access but with use (Organisation for Economic Co-operation and Development/Centre for Educational Research and Innovation [OECD/CERI], 2009).

EXPLORATIONS OF LITERACY AND MEDIA PRODUCTION DRAWN FROM MEDIA AND CULTURAL STUDIES

Research into video games and game play drawing on cultural studies traditions centres on the place of games in everyday life: the uses to which they are put; their role in projects of identity, friendships and community; representation and the role of games and game play; games paratexts and games mastery and the establishment of 'gaming capital' (Consalvo, 2007) amongst players.

How to conceptualize games as emergent cultural forms shapes the ways in which they are understood and analysed. Within literacy research, new literacies and multiliteracies frameworks suggest that digital games might be seen as textual cultural forms that epitomize multimodal literacies 'in the wild' (Beavis, 2013). Games require a high level of multimodal literacy from players in order for them to be able to play, including the capacity to identify and attend to a wide range of textual elements and their interrelationships, simultaneously. Linguistically based research here also explores the negotiation of relationships and representations through intersections between game-play expertise, values and linguistic/multimodal form. As with other literate and textual forms, experience, interactions and relationships are seen as textually mediated, whether this be through the representation of self to others through game play or the interpretation of others through their representations and game play; reading, playing and creating the game all happen through textual action. Game play is similarly seen as purposeful and socially situated and as calling on and activating 'new' literacies such as the capacity to critically evaluate multimodal forms and the effects of the juxtaposition of multiple elements.

Notions of discourse in this context, particularly with respect to online games, have been the focus of some of the most interesting research into game-based literacy practices, relationships and identity. Considerable attention has been paid to the establishment and role of discourse in multiplayer games and related websites and, in the context of MMORPG particularly, to discourse and cognition, discourse across print and digital literacies, and to discourse, values and identity (Steinkuehler, 2006, 2007, 2008; Thomas, 2007).

Research arising within media and cultural studies traditions considers distinctions between 'old' and 'new' media. Digital games are seen as quintessentially products of the 'new' media, and features of new media distinguish them from 'old media' forms and shape conceptual understandings and analytic approaches to be employed (Dovey and Kennedy 2006). Consistent with media studies approaches that emphasize audiences, texts and institutions (Buckingham, 2000), media studies approaches to games draw on such notions as the circuit of culture (du Gay et al., 1997). The circuit of culture framework requires that, in studying a cultural text or artefact, five aspects receive attention: representation, identity, production, consumption and regulation. For du Gay et al. (1997: 14), 'meaning is constructed – given, produced – through cultural practices – it is not simply found in things'. With respect to games and learning, media studies approaches consider games as specific cultural forms, often drawing on semiotic and linguistic frameworks (e.g. Burn, 2009) to strengthen understandings of semiotic elements and their contribution to form. Research and scholarship here emphasizes production elements, including game making and games in schools, and analysing games as cultural phenomena in their own right (Buckingham and Burn, 2007; Carr et al., 2006).

GAMES STUDIES

Questions about whether games should be understood in terms of play (ludology) or narrative (narratology) were fiercely argued for a number of years (Frasca, 2003), with hostility towards the importation of frameworks developed for other cultural forms. Rather, games needed to be recognized as cultural phenomena in their own right. As more disciplinary traditions have contributed to the mix that is games studies, more flexibility has emerged. Games need to be recognized as specific interactive forms of entertainment and as part of contemporary culture.

The field of games studies embraces a wide range of areas and is quintessentially multidisciplinary, but also lays claim to being a specific disciplinary field in its own right. Of particular relevance to this chapter are concerns with the nature and affordances of games and what they are most suited for in relation to games-based learning and serious games: an insistence on understanding games as action, recognizing the role of the machine, game design and game algorithms in game play; and the contrast between recreational, leisure-based play and attempts to harness or domesticate games for educational purposes. Games utilize the logic of 'procedural rhetoric' (Bogost, 2007), whereby learning takes place through the actual playing of/creating the game in a manner analogous to drama-based experiential learning.

A related set of questions address the interaction between online and offline worlds, the investment of self in game play and issues of representation, performativity and identity, and contrasts between the formal structures and expectations of schooling and those supposedly embedded in games. Attention to the role of context in game play, to game play as socially situated and purposeful, the place of games in players' lives, digital culture, consumption and the broader global economy within which game play occurs, also variously shape the ways in which the use of games for educational purposes is viewed from the games studies tradition, raising corresponding questions for research.

Critiques made by games studies scholars of attempts to design games for educational purposes concern such matters as assumptions about the importation of leisure time expectations, orientations, pleasures and commitment into formal learning contexts; ignoring social contexts and purposes for play; the trivialization and/or misappropriation of games and games' design capacities; the substitution of mechanized and lock-step progress through desired content areas for what games can really achieve and the more free-ranging and anarchic possibilities of leisure time play and the attempt to domesticate play through 'gamification' in what is regarded as naked commercialization and a crass and unwarranted marketing endeavour (Bogost, 2011). Other critiques concern the assumption that online games in and of themselves foster development of critical skills and characteristics, the dangers of technological determinism and the epistemological limitations of behaviourist conceptions of learners, learning and knowledge that may be embedded in such games.

E-LEARNING AND INSTRUCTIONAL DESIGN

Most serious games are developed according to the principles and practice of instructional design (ID). The capacity of games to engage learners, present complex worlds and teach skills and conceptual understandings make them an ideal subject for ID. However, games designed for educational purposes rarely have the complexity or array of dazzling aesthetics and high production values evident in large-scale commercially successful games. Drawing on cognitive and behaviourist psychology, ID is organized around a staged process that identifies needs, defines desired outcomes and creates structures and strategies that will enable learners to achieve observable outcomes in measureable ways. Common models include the Analyse, Design, Develop, Implement, Evaluate (ADDIE) model, rapid prototyping – a variant of the ADDIE model – and Dick, Carey and Carey's 'systems approach' model (2005).

Within ID parameters, serious games are seen as 'learning tools' and are centrally concerned with outcomes, measurement and

change: 'what sets serious games apart from the rest is the focus on specific and intentional learning outcomes to achieve serious, measurable, sustained changes in performance and behavior' (Derryberry, 2007: 4). Unlike cultural and media studies approaches to games, which tend to focus on the investment of self and socially situated engagement, in this tradition motivation is seen as fundamental to game play, with games elements providing the stimuli necessary for the achievement of desired learning outcomes. The design elements of games that help make this happen include narrative context, rules, goals, rewards, multisensory cues and interactivity (Dodlinger, 2007), from a set of games attributes that includes backstory and storyline, game mechanics, rules, immersive graphical environment, interactivity, challenge/competition, risks and consequences (Derryberry, 2007: 4).

A central tenet of the ID approach to games is that it is possible to maintain the entertainment aspects of games while simultaneously repurposing them for explicitly didactic purposes. Meaningful play is seen as approximating learning, and serious games based on the above principles are seen as ideal delivery vehicles for twenty-first-century learning and for addressing the needs of digital natives – 'today's worker or learner' (Derryberry, 2007: 11) – unlike schools that are anchored in print-based, nineteenth-century modes of operation rather than reflecting the twenty-first-century world. 'Digital natives' are said to prefer:

- receiving information quickly from multiple multimedia sources
- parallel processing and multitasking
- processing pictures, sounds and video before text
- random access to hyperlinked multimedia information
- interacting/networking simultaneously with many others
- learning 'just in time'
- instant gratification and instant rewards
- learning that is relevant, instantly useful and fun (Derryberry, 2007: 11).

It is a short step from listing these characteristics, already problematic in their essentializing and technological determinism, to seeing all students as 'digital natives' and, in a somewhat circuitous argument, seeing games as 'a natural environment' for contemporary learners, one moreover that 'fosters development of critical skills and characteristics' (Dewberry 2007: 11).

Critiques of ID approaches from within and outside the field focus on a number of issues. Iuppa (2010) points to tensions between instructional designers and games designers in creating serious games, where the absence of ID principles can make serious games less effective, but where games designers at the same time may regard ID as competing with and detracting from games design. Iuppa also points out that while ID properly includes both analysis and design, pressures of time and money may result in the analysis stage being short-circuited or minimized, to the detriment of the game. Iuppa similarly points to the need to be specific in recognizing the areas in which serious games are able to effect change, and the need to reflect these in the development of games design. Analysis needs to address both the needs of the learners and the needs of the task. Iuppa outlines three areas to be included in the analysis of the learning context: environmental, motivational and skills/knowledge, with this last properly the focus of the game. Where blockers to change are located in environmental and motivational areas, there is little a game can do to bring about change. Iuppa cites the example of the Alaskan firefighters for whom he designed a game to promote greater efficiency in combatting fires. While best practice was clearly not to send sick firefighters to fight fires, in the 'real world' a sick worker who stayed home rather than fighting fires, however ineffectively, would be penalized financially. Nothing the game did was therefore likely to change real-world practice in this regard.

VIRTUAL WORLDS: SECOND LIFE

Virtual worlds are not games in and of themselves, but, rather, engaging and immersive environments in which to play. Virtual worlds offer the opportunity for geographically dispersed groups of participants to 'meet' together in the online world in real time, to socialize and accomplish tasks ranging from the design and creation of avatars, buildings and furnishings through to viewing and discussing artworks, putting on plays or presenting or attending lectures. As teaching sites, they provide a bridge between formal and traditional modes of teaching and learning and the digital experiences that characterize large parts of young people's lives.

In educational contexts, perhaps the best-known and most frequently used of virtual worlds is Second Life, with many universities buying their own islands to use as sites for teaching and research, extending pedagogy and location into the virtual sphere. The appeal of virtual worlds include their immersive, game-like environment, the use of avatars to represent the 'player' within the space of the virtual world and the shared environment in which activities can take place. Educational uses of virtual worlds include activities in which students are actively involved in the design and building of online spaces, attendance at virtual lectures, the development and display of artworks and performance – both classic and of the student's own design – virtual field trips and the use of alternative learning spaces to meet for diverse purposes. Of particular value is the capacity of virtual worlds to provide for the co-presence of groups of students in the one space, regardless of their 'real-life' proximity or time zones, sociality and the experience of presence and embodiment through their visual and interactive affordances.

What virtual worlds actually 'are' however, and the ways they are perceived, varies. In a study exploring the value of Second Life as a space within which to teach, Carr, Oliver and Burn (2010: 20) argue that as Second Life can be used in different ways and as definitions reflect use:

there is scope to define it in various ways (as collective text, programme, social networking platform, tool, public space, etc.). This suggests that definitions will be provisional, and reflect the perspective of the user (or the disciplinary perspective of the researcher). The preferred definition of *Second Life* will inform the research questions that are devised, and the methodologies that are adopted. All of which will impact on analysis and findings. None of this is 'bad' of itself, but failing to recognise this circularity could be detrimental.

Drawing on Second Life's capacity to provide engaging online contexts for discussion through virtual presence in the form of avatars and as meeting points for students from different locations to engage in common tasks, Jamaludin et al. (2009) conducted a study exploring the teaching of argument amongst a group of 45 final-year secondary school students in Singapore. The students came from two classes, one an arts stream class and one a science stream class. The students were described as being of 'average ability', based on standardized test scores in the subjects they were studying. The compulsory general paper (GP) taken by students as part of their A levels requires not only a good command of English, but also strong argumentative and critical thinking skills and the capacity to produce as well as critique argument. The research hypothesized that:

> faced with the requirement to defend their assertions and to critically evaluate those of their peers, students may be led to produce a more articulated discourse … to elaborate meanings, to clarify views, and to modify or adjust their degrees of commitment towards their assertions. (Jamaludin et al., 2009: 317)

Argument was constructed as 'dialogical and dialectical' (Jamaludin et al., 2009: 318), with the view that participating in a context where multiple voices and positions were heard would contribute to the 'co-construction of critical thinking and writing skills' (2009: 318) in dialogic space. The research sought to investigate 'whether contextualized enactive role play fosters a positive shift in students' epistemic interaction processes,

argumentative moves and social interaction processes' (2009: 319).

The project particularly capitalized on the opportunities provided by Second Life for the creation and representation of character, through players' emotional and psychological commitment to their avatars, particularly in the context of dramatic role play, the centrality of socialization and interaction within Second Life and the power of the immersive environment to foster engagement, agency and ownership. It coupled these qualities and affordances with a structured programme of improvisational role play, to help students develop complex understandings and multiple perspectives on the controversial topic at the heart of the role play – euthanasia. The players' commitment to their characters and the dilemmas they faced were central to the development of these insights and the research set out to use the visual and interactive affordances of Second Life to enable students to develop a strong sense of agency over their avatars' behaviour and emotions, 'enabling opportunities for deeper experiential and situated learning in an immersive, collaborative learning environment' (Jamaludin et al., 2009: 321). In a combination of traditional classroom-based lessons and technologically facilitated discussions using Second Life and a discussion board, the study took ten groups of students through five in-world sessions where students, in their avatar form, were put into role play and required to consider a range of issues related to euthanasia, from ethics through morality and religion, according to the perspectives of the characters they were playing as they progressed from teenagers through middle age to incapacity across the role play sequence and the groups for whom they were advocating. Each enactive role play in Second Life was followed by argumentative discourse using the discussion board, with this session in turn providing input into the next Second Life role play.

The study used a mixed methods approach, which included a range of data-gathering mechanisms, including pre- and post-test essays, log transcripts from the Second Life role plays, with Second Life set up to automatically record these, the argumentation on the structured discussion board and a survey including space for open-ended responses and interviews with participating teachers. The study 'revealed salient differences in the nature of epistemic interactions, the patterns of argumentative moves, and the patterns of social interactions between students from two classes' and showed 'that students valued the embodied experience afforded by the immersive virtual environment' (Jamaludin et al., 2009: 317). Differences were noted in the effectiveness of epistemic and social dimensions in building critical thinking and writing skills for students of higher or lower abilities, recognizing both what was achieved and what not. The study concluded by affirming the value and potential of such sites and experience to provide collaborative, experiential learning environments supportive of the development of students' critical thinking and writing skills.

This research moved beyond approaches where simple presence in the virtual world is assumed to provide engagement and higher-order thinking. It benefited from the mix of on- and offline teaching sites and the careful pedagogical theorization and analysis that informed the research design. The complex model of argument adopted, the use of structured role play and the recognition of the importance of agency, performance and play helped create a teaching context geared to developing high-level argumentative writing skills. This tight conceptualization enabled student participation that was closely aligned with formal curriculum requirements and explicitly provided for staged progress towards that end.

SUMMARY, TENSIONS AND DEBATES: DIRECTIONS FOR FUTURE RESEARCH

The use of digital games and immersive, virtual worlds for teaching and learning is a rapidly developing field. Competing disciplines and theoretical frameworks, purposes and expertise coexist in the search to maximize the possibilities and affordances of games to support learning in a world characterized as global and digital and peopled by learners whose orientations and expectations of learning have been shaped by their participation in it. Games are seen as a way to meet twenty-first-century learning needs, engage otherwise disenfranchised learners, deliver deeper conceptual understandings and prepare learners to participate in a technologically saturated world.

Perhaps the most strident tension in the field of games and games-based learning is that between ID perspectives on the one hand and, on the other, those typical of games and cultural studies views of the nature of games and game playing. Where games and cultural studies scholars emphasize the social and socially situated nature of game play, the investment of self, the rules and freedom of play and immersive commitment to the world of the game, ID perspectives posit a somewhat different view. Against engagement and issues of representation, identity and community, the circuit of culture or of media consumption, production and critique, Instructional Design offers motivation, the achievement of change, a staged sequence of activities to teach specific skills and the measurable outcomes of observable goals. Against views of games and game play that foreground the unregulated and potentially anarchic pleasures of play, global contexts of media convergence, circuits of culture and escalating social and technological change, it offers an ordered vision of learning, learners and knowledge, and the harnessing of games' affordances for teaching purposes in the interests of managed, beneficial change.

A related set of tensions surrounds the epistemological understandings of learners, learning and knowledge and the ways in which these are envisaged, affected and enacted in different traditions of games-based learning and game play. The degree to which learning is seen as co-constructed and contextual varies across traditions, with a set of related debates about agency and what 'interactivity' actually means in relation to learners' real choice within the broader context of the algorithms of the game and their capacity to influence outcomes. In their most limited form, games developed using ID guidelines construct learning as the transmission and acquisition of information, construct learners as 'empty vessels' and narrow what is to be learnt down to discrete if interconnected skills and knowledge rather than a more conceptually deep and flexible view of curriculum and disciplinary areas. Conversely, claims made about informal learning or formal learning utilizing commercially available 'off-the-shelf' games may overestimate the degree of formal learning taking-place that closely matches current iterations of curriculum areas. Concepts such as 'ludic epistemology' (de Castell, 2011) are indicative of the ways in which what counts as knowledge may be reconstructed by games or at least put under review.

A further set of tensions relates to what is seen as the appropriation of the 'pure' world of games for educational purposes and the inevitable distortion that ensues, both of games themselves – their purposes, structures and design – and of the contexts and investments made by players in games and gaming culture and the role of games in everyday life. Finally, there is well-warranted concern that the attractiveness of the concept of using games to support formal learning in a wide range of areas, coupled with the extensive costs of development, may mean games are designed and sold by companies and designers with limited conceptions of learning, pedagogy and curriculum and, that once in place, games will become fixtures in schools and educational contexts in limiting ways and/or draw funding away from other areas. Nonetheless, the potential of digital games and immersive worlds to enhance learning is considerable, with the high hopes held for the field likely to eventuate in further dynamic and productive growth.

FURTHER READING

Humanities and Social Sciences

Shaffer, D., Squire, K., Halverson, R. and Gee, J. (2005) 'Video games and the future of learning', *Phi Delta Kappan*, 87(2): 104–11.

Literacy, Media and Cultural Studies

Buckingham, D. and Burn, A. (2007) 'Game literacy in theory and practice', *Journal of Educational Media and Hypermedia*, 16(3): 323–49.
Consalvo, M. (2007) *Cheating: Gaining Advantage in Videogames*. Cambridge, MA: MIT Press.
Dovey, John and Kennedy, Helen (2006) *Game Cultures: Computer Games as New Media*. Maidenhead: Open University Press.
Gee, J.P. (2003) *What Video Games Can Teach Us About Learning and Literacy*. New York: Palgrave.
Steinkuehler, C. (2007) 'Massively multiplayer online gaming as a constellation of literacy practices', *eLearning*, 4(3): 297–318.

Games Studies

Aarseth, E. (1997) *Cybertext: Perspectives on Ergodic Literature*. Baltimore, MD: Johns Hopkins University Press.
Bogost, I. (2007) *Persuasive Games: The Expressive Power of Videogames*. Cambridge, MA: MIT Press.
Galloway, A. (2006) *Gaming: Essays on Algorithmic Culture*. Minneapolis, MN: University of Minnesota Press.
Taylor, T.L. (2006) *Play Between Worlds*. Cambridge, MA: MIT Press.

Instructional Design

Dodlinger, M.J. (2007) 'Educational video game design: A review of the literature', *Journal of Applied Educational Technology*, 4(1): 21–31.
Iuppa, N. and Borst, T. (2010) *End-to-End Game Development: Creating Independent Serious Games and Simulations from Start to Finish*. Oxford: Elsevier.
Kankaanranta, M.H. and Neittaanmäki, P. (eds) (2010) *Design and Use of Serious Games*. New York: Springer.

Learning in Virtual Worlds

De Freitis, S. (2006) *Learning in Immersive Worlds: A Review of Games-Based Learning*. Bristol: JSC.

Peachey, A., Gillen, J., Livingstone, D. and Smith-Robbins, S. (eds) (2010) *Researching Learning in Virtual World*. London: Springer.

REFERENCES

Beavis, C. (2013) 'Multiliteracies in the wild: Learning from computer games', in Guy Merchant, Julia Gillen, Jackie Marsh and Julia Davies (eds), *Virtual Literacies: Interactive Spaces for Children and Young People*. Abingdon, Oxon: Routledge. pp. 55–74

Bogost, I. (2007) *Persuasive Games: The Expressive Power of Videogames*. Cambridge, MA: MIT Press.

Bogost. I. (2011) 'Gamification is bullshit'. Available at: www.theatlantic.com/technology/archive/2011/08/gamification-is-bullshit/243338 (accessesd October 5, 2011).

Buckingham, D. (2000) *After the Death of Childhood: Growing Up in the Age of Electronic Media*. Cambridge: Polity Press.

Buckingham, D. and Burn, A. (2007) 'Game literacy in theory and practice', *Journal of Educational Media and Hypermedia*, 16(3): 323–49.

Burbules, N. and Callister, T. (2000) *Watch IT: The Risks and Promises of Information Technology for Education*. Boulder, CO: Westview Press.

Burn, A. (2009) *Making New Media: Semiotics, Culture and Literacies*. New York: Peter Lang.

Caillois, R. ([1961] 1979) *Man, Play and Games*. Trans. M. Barash. New York: Schocken Books.

Carr, D., Buckingham, D., Burn, A. and Schott, G. (2006) *Computer Games: Text, Narrative and Play*. Cambridge: Polity Press.

Carr, D., Oliver, M. and Burn, A. (2010) 'Learning, teaching and ambiguity in virtual worlds', in A. Peachey, J. Gillen, D. Livingstone and S. Smith-Robbins (eds), *Researching Learning in Virtual Worlds*. Londin: Springer. pp. 17–30.

Chee, Y.S. (2011) 'Learning as becoming through performance, play and dialogue: A model of game-based learning with the game Legends of Alkhimia', *Digital Culture and Education*, 3(2): 98–122.

Consalvo, M. (2007) *Cheating: Gaining Advantage in Videogames*. Cambridge, MA: MIT Press.

de Castell, S. (2011) 'Ludic epistemology: What game-based learning can teach curriculum studies', *Journal of the Canadian Association for Curriculum Studies*, 8(2): 19–27.

de Castell, S. and Jensen, J. (2003) 'Serious play', *Curriculum Studies*, 35(6): 649–665.

Derryberry, A. (2007) 'Serious Games: Online games for learning'. White paper ©Adobe Systems Incorporated. San Jose, CA. Available at: www.adobe.com/resources/elearning/pdfs/serious_games_wp.pdf

Dick, W., Carey, L. and Carey, J.O. (2005) *The Systematic Design of Instruction* (6th edn). Boston, MA: Allyn & Bacon.

Dodlinger, M.J. (2007) 'Educational video game design: A review of the literature', *Journal of Applied Educational Technology*, 4(1): 21–31.

Dovey, J. and Kennedy, H. (2006) *Game Cultures: Computer Games as New Media*. Maidenhead: Open University Press.

du Gay, P., Hall, S., Janes, L. MacKay, H. and Negus, K. (1997) *Doing Cultural Studies: The Story of the Sony Walkman*. Milton Keynes: Open University Press.

Egenfeldt-Nielsen, S. (2005) 'Beyond edutainment: Exploring the educational potential of computer games'. Unpublished PhD thesis, IT-University of Copenhagen, Copenhagen.

Facer, K. (2002) 'Computer games and learning: Why do we think it's worth talking about computer games and learning in the same breath?' A discussion paper. Futurelab.

Frasca, G. (2003) 'Ludologists love stories too: Notes from a debate that never took place', *GamesStudies*, Available at: www.ludology.org/articles/Frasca_LevelUp2003.pdf (accessesd October 1, 2011).

Gee, J.P. (2003) *What Video Games Can Teach Us About Learning and Literacy*. New York: Palgrave.

Gee, J.P. (2005) 'Learning by design: Good video games as learning machines', *E-Learning and Digital Media*, 2(1): 5–16.

Gee, J.P. (2007) *Good Video Games and Good Learning: Critical Essays on Video Games, Learning and Literacy*. New York: Peter Lang.

Gee, J., Lankshear, C. and Hull, G. (1996) *The New Work Order: Behind the Language of New Capitalism*. St Leonards, NSW: Allan Unwin.

Huizinga, J. (1950) *Home Ludens: A Study of the Play Element in Culture*. Boston, MA: Northeastern University Press.

Iuppa, N. (2010) 'Instructional design vs game design: Part 1: End-to-end game development', Available at: www.youtube.com/watch?v=hyt_oCyaFfQ

Jamaludin, A., Chee, Y.S. and Ho, C.M.L. (2009) 'Fostering argumentative knowledge construction through enactive role play in Second Life', *Computers & Education*, 53: 317–29.

Kankaanranta, M.H. and Neittaanmäki. P. (eds) (2010) *Design and Use of Serious Games*. New York: Springer.

Kirriemuir, J. and McFarlane, A. (2004) *Literature Review in Games and Learning*. Report 8. Futurelab. Available at: http://hal.archives-ouvertes.fr/docs/00/19/04/53/PDF/kirriemuir-j-2004-r8.pdf

Klopfer, E., Osterweil, S. and Salen, K. (2009) 'Moving learning games forward: obstacles, opportunities and openness. *The Education Arcade*. Available at: http://education.mit.edu/papers/MovingLearningGamesForward_EdArcade.pdf (accessed September 30, 2011).

McFarlane, A., Sparrowhawk, A. and Heald, Y. (2002) *Report on the Educational Use of Games: An Exploration by TEEM of the Contribution Which Games Can Make to the Education Process*. Cambridge: TEEM/BECTA.

Oblinger, D. (2006) 'Games and learning', *Educause Quarterly*, 28(3): 5–7.

Organisation for Economic Co-operation and Development/Centre for Educational Research and Innovation (2009) *The New Millennium Learners*. Available at: www.oecd.org/document/10/0,3746, en_21571361_49995565_38358154_1_1_1_1,00. html (accessed May 31, 2012).

Papert, S. (2002) 'Hard fun', *Bangor Daily News*, Maine. Available at: www.papert.org/articles/HardFun.html

Pelletier, C. (2009) 'Games and learning: What's the connection?', *International Journal of Learning and Media*, 1(1): 83–101.

Raybourne, E.M. (2008) *Simulation Experience Design Method for Serious Games*. Albuquerque, NM: Sandia National Laboratories. Available at: www. cs.tut.fi/ihte/CHI08_workshop/papers/Raybourn_UXEM_CHI08_06April08.pdf

Robertson, J. (2008) 'Conceptualizing boys (and) video gaming: Communities of practice?'. PhD dissertation, University of Auckland. Available at: https://researchspace.auckland.ac.nz/bitstream/handle/2292/3300/01front.pdf?sequence=10 (accessed September 30, 2011).

Schaffer, D.W. (2006) *How Computer Games Help Children Learn*. New York: Palgrave Macmillan.

Shaffer, D., Squire, K., Halverson, R. and Gee, J. (2005) 'Video games and the future of learning', *Phi Delta Kappan*, 87(2): 104–11.

Serious Games Initiative (n.d.) Available at: www. seriousgames.org/about2.html

Simkins, D. and Steinkuehler, C. (2008) 'Critical ethical reasoning & role play', *Games & Culture*, 3: 333–55.

Sodestrom, T., Hamilton, D., Dahlgren, E. and Hult, A. (2006) Premises, promises: Connection, community and communion in online education. *Discourse: Studies in the Cultural Politics of Education*, 27(4): 533–549.

Steinkuehler, C.A. (2006) 'Massively multiplayer online videogaming as participation in a discourse', *Mind, Culture & Activity*, 13(1): 38–52.

Steinkuehler, C. (2007) 'Massively multiplayer online gaming as a constellation of literacy practices', *eLearning*, 4(3): 297–318.

Steinkuehler, C. (2008) 'Cognition and literacy in massively multiplayer online games', in D. Leu, J. Coiro, C. Lankshear and K. Knobel (eds.), *Handbook of Research on New Literacies*. Mahwah, NJ: Erlbaum. pp. 611–34.

Stevens, R., Satwicz, T. and McCarthy, L. (2008) 'In-game, in-room, in-world: Reconnecting videogames to the rest of kids' lives', in K. Salen (ed.), *The Ecology of Games: Reconnecting Youth, Games and Learning*. Cambridge, MA: MIT Press. pp. 41–66.

Squire, K.D. (2002) 'Rethinking the role of games in education', *Game Studies*, 2(1). Available at: http://gamestudies.org/0201/Squire/ (accessed August 31, 2005).

Squire, K. (2004) 'Replaying history: Learning world history through playing Civilization III'. PhD dissertation, Indiana State University, Indiana. Available at: http://website.education.wisc.edu/kdsquire/REPLAYING_HISTORY.doc (accessed September 30, 2011).

Sutton-Smith, B. (2001) *The Ambiguity of Play*. Cambridge, MA: Harvard University Press.

Thomas, A. (2007) *Youth Online*. New York: Peter Lang.

28

Online and Internet-based Technologies: Social Networking

Kirsty Young

INTRODUCTION

This chapter examines the rise of second-generation Internet platforms, which enable social engagement via the Internet and specifically the ways in which online content created through social media provides insights into the lives of individuals and the transformation of communities and society, both locally and globally.

The term Web 2.0 was introduced in 2004 and, although there is currently no one agreed definition, the term encapsulates a move away from static sites, where individuals are bound to a single computer and website, to dynamic sites that provide the infrastructure for publication, human collaboration and social interaction. Web 2.0 applications require no specialized technical knowledge but allow the user to interact directly on the website. Web 2.0 technologies are user-centred and enable uploading of audio, video and photo files and interactive information sharing. Examples of the transition from first-generation Web technology to Web 2.0 include: Britannica online to Wikipedia; personal websites to blogging; page bookmarking to social bookmarking (e.g. del.ic.ious).

Web 2.0 has enabled social media to come to the forefront of Internet activity. Social media is Web-based technologies that transform communication to interactive dialogue. Social media is the blend between social interaction and technology for the co-creation of content knowledge and takes many forms on the Internet, as depicted in Table 28.1.

Social media is distinct from industrial media (i.e. newspapers, magazines, television, film) in a number of ways. First, it is inexpensive for individuals to publish online compared with industrial media, which require significant resources and where publication is limited to a select few companies. Also, forms of industrial media are static and once created exist in that state forevermore while social media is accessible for continued modification, not only by its creator but also often members of the wider community. Finally, industrial media is delayed due to production constraints, whereas social media

Table 28.1   Forms of online social media

Reviews/opinions	Product, service and business ranking, rating and reviews; question and answer forums
Collaboration	Document editing/sharing; social bookmarking; wikis
Communication	Blogging; microblogging; online social networking
Multimedia	Photo sharing; music and audio sharing; presentation sharing

is noted for its immediacy in reaching a global audience.

Social media sites are found to dominate the most popular website charts, with eight distinctive social media sites in the Top 20 global sites based on website traffic (Table 28.2).

The popularity of these sites is undeniable and although sites may come and go (such as the rapid decline of the once dominant site MySpace) the overall impact of social media will be enduring. The content produced through social media provides the greatest source of authentic social data to exist. The publication of content by individuals, the collaboration between people and the immediacy of sharing information and experiences has presented social science researchers with a plethora of data through which they can examine language, gender, generations and cultural, intra- and interpersonal experiences.

This chapter presents an overview of the popular forms of research emerging in relation to social media. Specifically, the chapter is devoted to online social networking (OSN)

Table 28.2   Popular social media sites (as at January 2012)

Ranking	Site	Launched	Description
2nd	Facebook	2004 Harvard 2006 public	800 million active users as at January 2012. Individuals set up a personal profile and add other users as 'friends'. Friends can then communicate online. Features include posting/tagging photos, status updates, posting on the 'wall', private e-mail, creating events, playing games and creating and joining groups.
3rd	YouTube	2005	A video-sharing website. Content is generally uploaded by individuals and includes a variety of genres: video blogs, short movies, movie clips, television clips and music videos. Individuals can comment on and rate clips. A new phenomenon – to go viral – occurs when a clip is viewed by millions of people in a short period of time.
6th	Wikipedia	2001	A free, collaborative online encyclopedia. Articles on Wikipedia have been written collaboratively by volunteers and most are open to being edited by anyone who has access to the site.
8th	Blogspot (blogger)	1999	A dedicated blog-publishing service allowing personal journals to be published that are time-stamped and presented in reverse chronological order. Blogs are generally interactive, allowing visitors to leave comments.
10th	Twitter	2006	A microblogging site where users send text-based updates of up to 140 characters. Users can subscribe to 'follow' other users' tweets. 'Trending topics' are identified when a word, phrase or topic is tagged at a greater rate than others.
16th	LinkedIn	2003	A business-focused social networking site. The aim of the site is to allow registered users to make 'connections' to build a network of professional contacts.
18th	Wordpress	2005	A dedicated blog-publishing service similar to blogspot.

Source: Alexia (2012).

and social networking sites (SNS) with a focus on *friend* sites (such as Facebook, MySpace) rather than those of *followers* (Twitter, Blogspot, Wordpress) or *content sharing* (YouTube, Wikipedia). Six areas of social science research are examined: functions; education; generations; gender and cultures; identity; and social ties/networks. The chapter then moves to describe in detail a study using the social networking site, Facebook. The chapter concludes by identifying both the challenges and potential for social science researchers using SNS as a context and source of research data.

REVIEW OF THE LITERATURE

Research on the effects of OSN and research using SNS as a source of data are in their relative infancy. Given this, the literature review presented here is not a critique of methodology or methods. Instead, it presents an overview of the types of research problems being investigated and the various methods currently being engaged.

Originally online research focused on the anonymity of Internet interaction, which enabled identity experimentation, such as large gaming websites where users take on avatars to participate. It was presumed that the anonymity afforded during online communication encouraged users to take on a persona, a desirable self. The rise of OSN, however, has changed this perception and it is now accepted that to actively engage on a SNS you must present a resemblance of your 'offline' self – if you do not it is impossible to develop the trust and reciprocity necessary to gather online friends. Given this, the research explored in this section reflects emerging insights into authentic people engaging with social media for authentic social activity.

As noted, this review of research using SNS is constructed around six themes: (a) functions; (b) education; (c) generations; (d) gender and cultures; (e) identity; (f) social ties, social networks and social capital.

Functions of Social Networking Site Usage

Content analysis and surveys are predominantly used to explore the how, when, where and why of OSN activity. Large-scale projects have been engaged, such as that by Caverlee and Webb (2008) who analysed 1.9 million MySpace profiles in an effort to understand who is using these networks and how they are being used.

Survey design enables basic demographic data to be collected and such data present a comprehensive picture of who is engaged with OSN and what their OSN experience looks like, including number of friends, time spent on site and the various applications used on site (e.g. Young, 2009; Watkins & Lees, 2010). Survey data has also revealed motivations, perceptions and acceptance of risk (Lusoli & Miltgen, 2009) and the gratifications being met through these sites (Raacke & Bonds-Raacke, 2008). These studies contribute to knowledge about individual use and comparisons across gender, culture and age. Survey data on basic SNS use has also been used to identify those who do not use social networking sites and possible digital inequality (Hargittai, 2007).

Acknowledging the limitations of survey design, researchers are employing additional qualitative methods. In one such example the Australian Communication and Media Authority (2008) used telephone surveys, three-day time use diaries (N = 751) and a short self-complete survey to examine the nature of online engagement and time spent online. Pempek, Yermolayeva and Clavert (2009) similarly enquired as to how and why undergraduate students use these sites through a diary-like measure each day for a week and a follow-up survey.

These types of studies are fundamental to understanding how and why SNS are being used and the characteristics of those who make use of such sites. However, these studies are often limited because they may not assess or explain the consequences of OSN or the possible effects of SNS on individuals and groups across a variety of contexts, including formal education.

Education

The use of SNS in formal education contexts are prevalent in the literature examined across school and tertiary levels, as researchers attempt to understand the potential of bringing students' outside activity into the classroom to facilitate learning.

One issue is the impact of SNS on digital literacy. The potential of SNS to present authentic texts for the promotion of digital literacy education has been recognized (Perkel, 2006; Young, 2011b) but significant difficulty exists in conducting research to examine such hypotheses while many formal education systems reject calls for SNS to be accessed in schools and OSN to become part of the school curriculum. The essential need for digital literacy education is widely recognized, but Web 2.0 technologies, and particularly SNS, present unique challenges and tensions (Crook, Cummings et al., 2008; Livingstone & Brake, 2010; Johnson et al., 2011). A step forward is research that focuses on school students' everyday use of, and learning with, Web 2.0 technologies outside the classrooms.

To facilitate understanding of student learning, there is a call for education researchers to cultivate online lives (and identities) as part of their professional development (Greenhow et al., 2009). Research is beginning to emerge that examines the role of teachers using SNS, such as Mazer, Murphy and Simonds (2007), who examined the effects of computer-mediated teacher self-disclosure on student motivation, affective learning and classroom climate. It must be noted, however, that the merging of educators/researchers' professional and social lives online is ethically problematic.

Research at the tertiary level is significantly more advanced. This level of education acknowledges that students are active users of SNS and so research seeks to uncover the ways in which SNS use might impact learning and university life generally.

Tertiary-level research has concluded that, at this point in time, sites such as Facebook are generally used for social reasons and to exchange logistical or factual information, rather than for formal teaching purposes. Sometimes SNS are used for supplication and moral support with regard to assessment or learning or the promotion of oneself as academically incompetent and/or disengaged (Selwyn, 2009; Madge et al., 2010).

As noted above, formal education can best make use of social networking sites once their purpose in daily life is understood. The following sections examine individual and cultural uses of OSN.

Generations

The OSN activity of tertiary students is often researched given their proximity and availability to the academic community. Outside the education context, adolescent users have been of primary interest to researchers. Older populations are, however, becoming prevalent as they continue to take up OSN in growing numbers.

Adolescent Users

The adolescent years are a unique period and it has traditionally proved difficult for adult researchers to intrude and immerse themselves in the personal and social lives of adolescents. Today, the design of SNS enables researchers to capture and analyse the authentic experiences of adolescents without intrusive research designs that disrupt authentic social activity. This research does, however, require careful consideration of ethical principles of respect, consent and anonymity (Young, 2012).

Ongoing studies enable researchers to ascertain changes in adolescent activity. The PEW American Life Project has conducted extensive large-scale quantitative surveys, and several of these have been focused on adolescent use of OSN. In one such study Lenhart and Madden (2007) report issues of privacy and how teens manage their online identities and personal information in the age of MySpace. More recently, Lenhart, Purcell, Smith and Zickuhr (2010) report data of 800 adolescents aged between 12 and 17 in 2009, claiming that 73 per cent of wired American teens now use social networking websites, which is a significant increase from previous surveys (55 per cent in November 2006 and 65 per cent in February 2008).

Qualitative research has been at the forefront to uncover the consequences of OSN on adolescent development. Such research has revealed that the goal of adolescent SNS use is to look cool and receive peer validation and for intimacy and self-expression (boyd, 2006; boyd & Heer, 2006; Livingstone, 2008). Williams and Merten (2008) conclude that adolescent profiles create an overall picture of adolescent behaviour, through intimate, candid and observable self-disclosure and peer interaction. Important insights can be drawn from these profiles about adolescent self-esteem and well-being, identifying areas for possible adult intervention (Valkenburg et al., 2006; Valkenburg and Peter, 2007).

Whilst adolescents and tertiary students have dominated as research participants in both small-scale qualitative and large-scale survey data, older adults are emerging as important participants as they increase in numbers in their OSN activity.

Older Persons

Large-scale survey design has been used to establish the extent of older persons' OSN use. Telephone survey (landline and mobile phones) conducted in the United States revealed that social networking use among Internet users aged 50 and older has nearly doubled – from 22 per cent (2009) to 42 per cent (2010) and 47 per cent of Internet users aged 50–64 and one in four (26 per cent) of users aged 65 and older now use SNS, and 1 in 10 (11 per cent) 50–64-year-olds and 1 in 20 (5 per cent) aged 65+ use Twitter or another service to share updates about themselves or see updates about others (Madden, 2010).

Generally, positive outcomes have been associated with OSN use by older people. An online survey (N = 222) was used to investigate how the Internet affects seniors' social capital and well-being (Sum et al., 2008), while Hogeboom and colleagues (2010) used telephone and face-to-face data collection methods (N = 2284) and found that Internet use and social networking by over 50s strengthens social networks.

Older persons' experiences with OSN provides insights into issues affecting older users and enables cross-generational analysis. Pfeil, Arjan and Zaphiris (2009) explore age differences through comparisons of user behaviour between two groups (60+ and 13–19 years). They did this using locally developed Web crawlers to collect large sets of data from MySpace's user profile pages. Content analysis was applied to investigate age-related differences and the researchers concluded that older people's online friend networks are smaller than adolescents' networks, but more diverse in age distribution.

OSN sites present sources of data to investigate cross-generational comparisons, similarly the genders can be explored and also various cultures and groups.

Gender and Cultural Studies

Gender

As both males and females are represented on SNS their authentic displays of communication and social activity are valuable sources of data for those interested in gender studies. Researchers, such as Magnuson and Dundes (2008), use these sites to explore gendered identity. In this instance, one researcher drew upon 100 members of her own personal online network (aged 17–29) and engaged an abbreviated content analysis of their profiles. Based on number of references a person made about their partner, it was concluded from this preliminary study 'that women's identity may still be largely determined by the men in their lives, given their greater propensity for including their significant other in their presentation of [online] self' (Magnuson and Dundes, 2008: 241).

Gender-based research also extends to analysis of language contrasts. One such example, Thelwall (2007) examined cursing and gender in MySpace. The program Soc-SciBot was used to download 40,000 profile pages. Each page was parsed in full to extract complete lists of words used and a central vocabulary was created of all words used. Three strengths of swear words were identified and a count was made of frequency of use. The findings revealed males using more moderate-level swearing on MySpace than females in the United States and United Kingdom samples, but that there was no gender difference in strong swearing amongst the United Kingdom sample, reflecting the 'ladette' culture.

Current Affairs

In addition to providing researchers with authentic data to understand generational and gender similarities and differences, analysis of SNS data also provides insights into individual and collective responses to specific cultural events. This is highlighted in a special edition of the peer-reviewed journal *Mass Communication and Society* in late 2010. This issue of the journal focused on the United States presidential election and the various impacts of social media (Johnson & Perlmutter, 2010).

Understanding the potential role and impact of new online social media sources, in contrast to traditional modes of media, were of interest to Kushin and Yamamoto (2010) and Hanson and colleagues (2010). In the first study, Kushin and Yamamoto (2010) used an online survey of college students (N = 407) to contrast political self-efficacy and situational political involvement through traditional Internet sources and social media. It was concluded that for this sample traditional Internet sources had greater influence and the researchers suggest that social media alone would not have profound and lasting effects on young adults' political decision making (Kushin and Yamamoto, 2010: 626).

Through distribution of a hard-copy questionnaire (N = 467) Hanson et al. (2010) sought to understand how SNS, video sharing and blogging relate to users' self-reported levels of political cynicism. The researchers concluded, using media that help people connect with those they know may lead them to find sources they deem to be more trustworthy than the mainstream media. For the participants in the study, the use of SNS predicted a lower level of cynicism.

Content analysis of Facebook pages proved a popular research design during this period. Two examples drawn from the special edition of *Mass Communication and Society* took a similar approach to answer vastly different research questions.

More than 1,000 Facebook pages were created that focused on the two election candidates and Woolley, Limperos and Oliver (2010) applied content analysis to these user-generated political Facebook groups. The researchers set out to explore how group membership differed and how each candidate was portrayed through the positive and negative references to race, religion and age and the use of profanity. They found social media does enable individuals to express their beliefs, and such expressions are often partisan and polarizing.

Fernandes et al. (2010) also applied content analysis to Facebook groups. In this instance, the researchers analysed nine Facebook pages created at seven different universities. The content analysis concluded that university students were using Facebook to facilitate dialogue and civic political involvement.

Creating SNS pages around political candidates/parties is one form of group that has emerged directly as a result of OSN. Another way to analyse how groups function online is by looking at how existing offline groups make use of OSN.

Groups

Alignment with, and participation as a collective within, a defined group has important and meaningful consequences for a significant proportion of the population. Groups are varied and can include consumer groups, support groups for people with illness, fan groups for a celebrity, a team or athlete, products/brands, gaming communities, hobby groups or clubs and so on. Rainie, Purcell and Smith (2011) found that 75 per cent of all American adults are active in some kind of voluntary group or organization. Following from the online 'groups' identified to have emerged from the US elections, OSN activity has also had an impact on the activities of pre-existing offline groups, such as charitable fundraising and for emotional support, as outlined in Figure 28.1.

Rather than interrupting group dynamics with intrusive research methods, investigation of how OSN supports group activity provides a wealth of authentic research data.

Up to this point the review of literature has focused on collectives – students, adolescents, older persons, male/female and groups. The review now moves on to explore the SNS experience of the individual, then the individual experience within social networks.

Identity

As noted previously, OSN is not considered a vehicle for identity experimentation. That is not to say, however, online identity is not constructed and manipulated, but this is done within the confines of representations of an authentic self (Chan, 2006; Dwyer, 2007).

Mixed methods research incorporating quantitative and qualitative methods is evident in research into online identity. For instance, Tufekci (2008) examined social grooming and the presentation of the self from the perspectives of both users and non-users. She commenced with qualitative focus groups (N = 51) and then developed a survey instrument in the light of her findings, which was administered to college students in class in paper and pencil format. Zhao, Grasmuck and Martin (2008) applied content analysis to 63 Facebook accounts and interviewed the 63 Honours students to explore how Facebook identities differ from those constructed in the anonymous online environment. It was concluded that in the Facebook context, identities are implicitly, rather an explicitly,

% among those who are members of a group that achieved each goal in the preceding 12 months

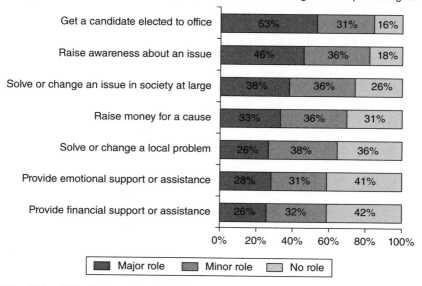

Figure 28.1   Role of the Internet in achieving group goals.

Source: Pew Research Center's Internet & American Life Project, November 23–December 21, 2010, Social Side of the Internet Survey. N = 2,303 adults 18 and older, including 748 reached via mobile phone.

expressed; users 'show rather than tell' and stress group and consumer identities over personally narrated ones.

Experimental designs are also evident in exploratory studies of OSN and identity. In one, experiment participants were given rating forms to rate themselves and their four well-acquainted friends. Eight months later they completed measures that included ideal-self ratings and ratings of how the targets believed they were viewed on the basis of their Facebook profiles. Facebook profiles were identified for 133 participants and saved at that point (to avoid participants manipulating/changing their profiles). Nine undergraduate research assistants independently rated the personality traits of all 133, based solely on an examination of the targets' Facebook profiles. The researchers conclude that OSN are a relevant and valid means of communicating personality (Gosling et al., 2007).

In another experimental design Utz (2010) used screenshots to examine how far extroversion of the target (self-generated information), extroversion of the target's friends (friends-generated information) and number of friends (system-generated information) influence the perceived popularity, communal orientation and social attractiveness of the target. The findings suggest that to make a good impression, it is not enough to carefully construct one's profile; it is also wise to carefully select one's friends. The accumulation of these online friends develops into a social network, which is the final area to be addressed in this literature review.

Social Ties, Social Networks and Social Capital

Social Ties and Social Networks

The term 'friend' has been appropriated by several SNS and the nature of these new online friendships has been a source of research interest. Wang and Wellman (2010) examined changes in adult friendship network sizes from 2002 to 2007 and found that, despite panic about possible decline, due to the Internet, friendships are abundant and grew in the period under study.

Using surveys and interviews, researchers have sought to describe how adults perceive online friendship, exploring the connection between online friends and the users' offline relationships and the strength of online friendships (Fono & Raynes-Goldie, 2006; Subrahmanyam et al., 2008; Lewis & West, 2009; Young, 2009, 2011a).

Traditional social network analysis has been used to investigate online social networks. This has been applied to researchers' own friendship networks (e.g. Hogan, 2008) and adolescent friend networks (e.g. Heer & boyd, 2005). These studies have found that online social networks are beneficial. Positive findings from social network analysis are supported by studies using other research methods. A large-scale survey design examined the role of the Internet in core social network interactions and found Internet use in general and use of SNS such as Facebook in particular are associated with having a more diverse social network (Hampton et al., 2009). Social ties provide many benefits, including companionship, access to information and emotional and material support. Increasing the number of and access to these ties increases access to the benefits, SNS forming a technology that can bring this about (Donath, 2007). This is an increase in social capital.

Social Capital

Yoder and Stutzman (2011) report that a number of studies have identified a robust relationship between the use of social network sites, particularly Facebook, and positive outcomes such as increased social capital. Yoder and Stutzman (2011) identify the elements of the Facebook interface influencing perceived social capital and note that reciprocal wall postings are of particular importance in fostering connectedness and build bridging social capital.

Research suggests, amongst other things, that Facebook enables users to broadcast requests for information and support and may be a useful resource for providing a diverse set of perspectives and information (Ellison et al., 2007, 2011). Longitudinal analysis of Facebook data suggests that Facebook affordances help reduce barriers that individuals with lower self-esteem might experience in forming the kinds of large, heterogeneous networks that are sources of bridging social capital (Steinfield et al., 2008).

Recognizing the limitations of self-reporting (which has been a dominant research approach) Burke, Kraut and Marlow (2011) set out to validate this data using empirical data from Facebook. The researchers investigated the role of directed interaction between pairs (wall posts, comments, and 'likes') and consumption of friends' content (status updates, photos, and friends' conversations with other friends). They found that directed communication is associated with greater feelings of bonding social capital and lower levels of loneliness, but has only a modest relationship with bridging social capital, which is primarily related to overall friend network size.

A research study that examined online identity, social ties and social networks is now presented to demonstrate in detail the application of these concepts to OSN research design, analysis and reporting.

ONLINE IDENTITY, SOCIAL TIES AND NETWORKS: A STUDY OF ADULT FACEBOOK USERS

Research Questions and Design

Online identity, online friendship and social activity are all interrelated and an exploratory qualitative study set out to examine these three areas. A series of eight research questions (grouped under three headings) underpinned the study.

Identity

- What Facebook tools do adults use to construct online identity?

- How do adults' diverse social networks on Facebook affect identity creation?

Online Friendship

- How do adult Facebook users accumulate online friends?
- How do they define these online friends?
- How do they manage large numbers of online friends?
- What is the relationship between online and offline friendships?

Social Ties and Networks

- How do adults engage socially on Facebook?
- What tools do adults use on Facebook to socialize?

To canvass the experiences of SNS users, data was collected using an online survey (N = 752). The results of this survey have been reported in detail in Young (2009).

To investigate issues arising from the survey and explore in more depth the experiences of adult users, individual face-to-face sessions were held with 18 active Facebook users in Sydney, Australia. Participants self-selected from the original survey. The sessions incorporated semi-structured interviews and verbal protocols, where each participant viewed his or her online profile while talking aloud to the researcher about its construction and contents (see Young (2005) for detail of the think-aloud method). The sessions were captured using video/audio/screen capture software. The age range of participants was 21–57 years (6 males and 12 females).

Findings and Discussion

Identity

The Facebook tools adults use to construct their online identity are status updates, joining groups/pages and posting photographs.

At this point there is relatively limited published research data on the use of status updates and the joining of groups/pages to reflect online identity. The study revealed that status updates are a valued means of communicating with one's audience, particularly by projecting humorous or insightful comments. Individuals who do not adhere to unspoken rules about the appropriate tone of status updates and continually post the negative or mundane are viewed critically. Status updates are not made to stand as isolated comments, the author uses the updates to provoke a response from his/her audience and serve as a source of interaction between the parties. This interaction between online 'friends' is then open for analysis by other members of the network, who, although not part of the online dialogue, have access to the 'conversation'.

Participants recognized that joining groups/pages projects an image of themselves and were beginning to change their practices, becoming increasingly more selective about joining groups/pages. In part this is because their membership of a group did not result in action, it was merely a statement of their interest.

Posting photographs serves to connect with the participant's offline social group. In the past photographs were used as a personal memento to be treasured, bringing back memories of important times and special people. Today, photos are posted online in bulk and used to demonstrate to others that the individual is engaged socially and undertaking interesting activities. However, there is a certain level of social acceptability, which, if crossed, can negatively impact the perceptions the audience has of the person. The most unique aspect of posting pictures online is the tagging of these photographs and public comments being made about each photo. This represents the co-constructed nature of identity as a result of online social networking activity. Comments made on a photo can significantly impact how an individual perceives that image, which is broadcast to the wider friend network.

The management of diverse networks on one's Facebook site was a primary issue identified in the data. Participants in this study did not make use of privacy settings to manage the content available to different groups within their online 'friend' collection. Instead, the participants were very conscious of the public nature of their postings and took measures to monitor their own online behaviour. Participants were aware that even though their profile may only be accessible by 'friends', that scope was wide-reaching and all of their activity on Facebook left a permanent trace, which could come under scrutiny at a later point in time.

Identity has always been impacted by social interactions but these interactions traditionally occurred between individuals and groups who interact within specific boundaries (e.g. family members, work colleagues, social groups). This has meant that individuals can maintain (to some extent) different identities in different contexts. Online social networking melds a person's family, friends and colleagues (amongst others) together. Each interaction that occurs publicly on a Facebook page is open for interpretation by all the associated Facebook networks. This gives insights into a person's otherwise separate identities (professional and private). Harter (1999) suggests that false self-behaviour involves the suppression of one's opinions, thoughts and feelings, while true self-behaviour is associated with expressing your thoughts and opinions. In the case of online social networking users might make efforts to suppress their thoughts and opinions to present a desirable image of themselves, but the postings of others can make public these private beliefs/actions. This, in a sense, is the co-construction of one's social identity through a broad range of networks.

It is also noted that Facebook keeps a permanent record of social interactions. Identity has always been subject to change based on social interaction but this can often go unnoticed as it occurs gradually over time. The digital age has enabled a permanent record of one's identity evolution to be captured on one's social networking site. It is now possible to review someone's online profile (which could extend back several years) and see changes in that person's life: appearance, relationships, employment, family and so on.

The online context sees personal identity formation as public and permanent.

Friends

Psychological research asserts that the limit of social relationships an individual can achieve is about 150, with about 5 in the innermost closest circle (Dunbar, 2007). Similarly, Spencer and Pahl (2006) report the number of close friends people have ranged from 5 to 41. The survey and interview data revealed that, although Facebook users can now claim a large number of online friends, the number of close friendships being maintained has not been affected by their use of this online SNS.

Older people experience more difficulty accumulating online friends, predominately because their peers are not using SNS to the same extent as younger generations. Older persons may turn to family and colleagues to form a strong friendship base. Their inclusion as an online friend can be viewed negatively by younger people because it forces them to manage their social identity within the boundaries imposed by family or the workplace. This is not such an issue for older persons who are more likely to have consistent representations of self in these various contexts.

Intergenerational online friends can possibly have two very positive outcomes. First, older persons who engage in online social networking are provided with insights into issues affecting younger generations, which would not usually be available to them. They are also exposed to language and cultural references that usually divide generations. Second, the acceptance of older friends (whether family members or work colleagues) may force younger users to more closely monitor their online postings and this forced self-censorship could benefit them in the longer term to avoid harm to their reputation.

When SNS emerged, the early rhetoric revolved around how knowledgeable adolescents were at mastering these sites with skill and perception unknown to older generations. In accordance with the traditional master/apprentice notions of social–cultural theories

of learning the data of the study presented in this chapter revealed that adolescents could benefit from observing the ways in which adult users manage their online profiles to project an appropriate image of themselves, across diverse social networks. This contradicts popular media reports of adolescents utilizing social media with advanced knowledge; while the adolescents may have the technical savvy, adults have the life experience to recognize the potentially negative consequences of their online social networking activity and so adjust their activity accordingly.

It was apparent from this research that online friends are deeply connected with offline social life and Facebook users are engaging with the site to gain insights into the lives and personalities of their offline friends. What occurs between friends offline influences their interactions online and, conversely, what occurs online can impact upon offline interactions.

Social Ties and Networks

Facebook was used by the participants of this study as an economical way to get in touch and stay in contact with people met casually or to communicate regularly with people with whom they already have well-established relationships. An online social life does not detract from offline relationships – as Gefter (2007) noted, they are simply different manifestations of the same network of friends. The Internet has not replaced traditional forms of communication – rather, it is an extension of the kinds of interactions that people have daily by phone, text message and e-mail, so the line between what is real and what is virtual is beginning to fade. Facebook is a social tool but not an alternative to social activities.

Posting photos and making status updates is not necessarily about promoting the individual – rather, they are used as conversation starters. Photos and status updates are also important tools for strengthening bonds with offline friends, by publicly displaying offline interactions or posting an abstract comment or picture that only has meaning to a person or select group of friends.

One consequence of individuals posting photos, making status updates and posting comments to friends' walls is 'Facestalking'. Facestalking is a new phenomenon to emerge as a result of OSN and involves in-depth perusal of another's online profile to gain insights and understanding of their life, without actually making your presence known. The participants of this study all alluded to engaging in this activity to varying degrees. There is a sense that while Facestalking is inevitable, most people would not admit publically to engaging in the activity.

The combination of survey, interview and think-aloud data was invaluable in giving voice to current SNS users. The survey enabled canvassing of the broad issues affecting individuals, which shaped the interview schedule. Having previously had success using the think-aloud method, the researcher was keen to apply this to an online context. Rather than the researcher viewing profiles and making decisions about their construction, the think-aloud process enabled clarification of the pictures, status updates, wall posting and groups and pages present on an individual's profile.

While this study proved enlightening on many issues affecting OSN users, several research challenges and limitations emerged, particularly in relation to research design and ethics.

ONLINE SOCIAL NETWORKING RESEARCH CHALLENGES

One specific limitation, which must be kept at the forefront of mind, is that although the numbers of people engaging with Web 2.0 technology is enormous, it is only a percentage of the population. We cannot make generalizations about society without taking account of the non-users. Those who post the most online data may reflect a certain type of person. As Keen (2007: 15) provocatively suggests, '… the Internet was the law of

digital Darwinism, the survival of the loudest and most opinionated …'.

Ethical issues present the greatest challenge for researchers of Web 2.0 technologies. The ethical principles of respect, beneficence and justice are held to apply to online research but at present there is no consensus in the research community on the public/private nature of online spaces. There are some Web 2.0 contexts where it is challenging for researchers to determine the applicability of certain ethical principles, particularly around issues of consent and anonymity. There is concern that, in some instances, researchers are accessing data that, although available publicly, were, to all intents and purposes, created by the individual for private purposes. Alternatively there are instances where researchers had access to data but did not initially obtain access for the purpose of research – for instance, being a member of an online group for social reasons, then using the online communication of that group for research purposes or accepting friend requests and using these online friendships for data-collection purposes.

It is somewhat common for interpretive researchers to be members of a group/community and it is this emic perspective that enables depth of data collection and analysis. So it is expected that circumstances will arise where a researcher has access to online groups/communities that are valued for research purposes. When and how should informed consent be obtained from members of this group? Lurking (visiting but not actively participating) is not appropriate. Researchers need to make their presence known. Naturally, if this is done prior to data collection it risks changing group dynamics and participation. If consent is sought after data collection has been completed, this risks alienating those in the research group who feel violated (Young, 2012).

Another issue that arises is obtaining informed consent of all 'friends' within a SNS network. If an individual chooses to participate in a research study, we need to establish the level of consent required from that person's online friendship networks, particularly where a third party may be identifiable as a result of online engagement with the participant.

There are also ethical concerns with SNS research and reporting. The visual nature of much online data requires different levels of consent for reporting purposes. It is essential to obtain appropriate consent for the level of anonymity required for screenshots of a person's online profile that may be used in journals, books and conference presentations. It is important to note that, while a person might use an alias (common on sites such as Twitter, uncommon on sites such as Facebook), the alias may actually be an identifiable trait of the participant if used extensively as part of his or her online life. Consent to use aliases and images of the participant must be obtained.

These are issues for researchers, the research community and human research ethics committees to continue to negotiate. While there are clearly new ethical challenges that emerge from online research, the value of research conducted using Web 2.0 technologies makes it worth experiencing these ethical challenges and even constraints. Web 2.0 and particularly OSN has opened up the private experiences and social communications of individuals, providing insights into human behaviour, friendships and relationships not previously available to researchers. SNS capture data that enables analysis of individuals, genders, generations and cultures, whether a snapshot at a given point of time or longitudinally through the permanency of online data.

ONLINE SOCIAL NETWORKING: RESEARCH POTENTIAL

The literature review and illustrative example of Facebook research presented in this chapter demonstrate that OSN lends itself to research situated in a range of paradigms. For many researchers one of the most exciting opportunities presented by Web 2.0 technologies is having a new environment in which to apply traditional research methodologies and methods.

Positivist researchers can apply large-scale survey design to uncover broad generalizations about OSN populations. This work enables comparisons to be made between genders, generations, groups and cultures. Experimental design can be used to test theories of SNS experience and it is also possible to apply social network analysis to assess the composition of large sets of 'friend' networks.

Interpretive researchers can elicit enormous volumes of qualitative data through traditional research methods (interviews, focus groups, questionnaires). However, of most value to the interpretive researcher is the ability to capture data through non-intrusive means. Analysis of OSN profiles presents authentic social interactions in real time. This data can be captured at a given point in time or longitudinally, with thematic analysis applied to understand the experiences of individuals and groups.

In addition to giving new life to traditional research methods, the data that can be extracted and the enormity of data produced on SNS is unlike anything previously available to researchers. This corpus presents extensive authentic data through which we can examine linguistic and language elements; it provides a global perspective on communication acts.

At a given point in time, a snapshot can be obtained of society – the issues at the forefront of people's minds are represented on SNS. Also of value to researchers, Web 2.0 sites exist at some level of permanency. This is particularly valuable for the analysis of group dynamics and to create a cohesive picture of local and global events, as experienced by the thousands, or even millions, of individuals who express their thoughts and feelings through Web 2.0.

An important area for further research is to examine profiles longitudinally. For example, analysis of a Facebook profile will reflect an individual's social identity at that point. The permanence of one's profile presents a valuable source for analysis of identity transition. The collection of status updates, photographs and membership to groups/pages, combined with the comments made by others on these elements is a chronological portfolio of a person's progression through life, enabling reflection on how they have transformed over time.

The scope of OSN and SNS for social science researchers is enormous. SNS present a unique environment for the study of human behaviour and interactions. OSN activity provides publicly accessibly insights into the activities of individuals and groups at a given point in time and across time. This knowledge can be obtained through the application of traditional social science methodologies in new ways, which presents exciting challenges and opportunities for the social science researcher.

FURTHER READING/LINKS TO MEDIA SOURCES

http://aoir.org/reports/ethics2.pdf
http://digitalyouth.ischool.berkeley.edu/report
http://pewresearch.org/topics/internetandtechnology/
www.nmc.org/horizon-project/horizon-reports

REFERENCES

Alexa. (2012). Alexa Top 500 Global Sites. Retrieved from http://alexa.com/topsites

Australian Bureau of Statistics. (2008). *Australian Social Trends, 2008: Internet Access at Home*. Retrieved from www.abs.gov.au/AUSSTATS/abs@.nsf/Lookup/4102.0Chapter10002008

Australian Communication and Media Authority. (2008). *Telecommunications Today: Report 6*. Canberra: Internet Activity and Content.

boyd, d. (2006). 'Identity Production in a Networked Culture: Why Youth Heart MySpace'. Paper presented at the American Association for the Advancement of Science.

boyd, d., & Heer, J. (2006). 'Profiles as Conversation: Networked Identity Performance on Friendster', *Proceedings of the 39th Hawaii International Conference on System Sciences*. New York: IEEE.

Burke, M., Kraut, R., & Marlow, C. (2011). Social Capital on Facebook: Differentiating Uses and Users.

Conference on Human Factors in Computing Systems. (ACM CHI 2011) ACM, 571–580.

Caverlee, J., & Webb, S. (2008). A Large-Scale Study of MySpace: Observations and Implications for Online Social Networks. *Proceedings of the 2nd International Conference on Weblogs and Social Media, 36–44*.

Chan, A. (2006). *Social Interaction Design Case Study: MySpace*. Gravity7.

Crook, C., Cummings, J., Fisher, T., Garber, R., Harrison, C., Lewin, C., et al. (2008). *Web 2.0 Technologies for Learning: The Current Landscape – Opportunities Challenges and Tensions*. Becta (former government agency).

Crook, C., Fisher, T., Harrison, C., Logan, K., Luckin, R., Oliver, M., et al. (2008). Web 2.0 Technologies for Learning: The Current Landscape – Opportunities, Challenges and Tensions. Available at: http://partners.becta.org.uk/upload-dir/downloads/page_documents/research/web2_technologies_learning.pdf

Donath, J. (2007). Signals in Social Supernets. *Journal of Computer-Mediated Communication, 13*(1).

Dunbar, R. (2007). *Oxford Handbook of Evolutionary Psychology*. Oxford: Oxford University Press.

Dwyer, C. (2007). Digital Relationships in the 'MySpace' Generation: Results from a Qualitative Survey. *Proceedings of the 40th Hawaii International Conference on System Sciences*, Hawaii.

Ellison, N.B., Steinfield, C., & Cliff, L. (2011). Connection Strategies: Social Capital Implications of Facebook-enabled Communication Practice. *New Media & Society* 13(6): 873–892.

Ellison, N.B., Steinfeld, C., & Lampe, C. (2007). The Benefits of Facebook 'Friends': Social capital and College Students' Use of Online Social Network Sites. *Journal of Computer-Mediated Communication, 12*(4). Available at: http://jcmc.indiana.edu/vol12/issue4/ellison.html

Fernandes, J., Giurcanu, M., Bowers, K.W., & Neely, J.C. (2010). The Writing on the Wall: A Content Analysis of College Students' Facebook Groups for the 2008 Presidential Election. *Mass Communication and Society,* 13(5), 653–675.

Fono, D., & Raynes-Goldie, K. (2006). Hyperfriends and Beyond: Friendship and Social Norms on LiveJournal. Association of Internet Researchers Conference, New York.

Gefter, A. (2007). The Difference Between 'Real' and Online is no Longer Clear-Cut. In S. Engdahl (ed.), *Online Social Networking*. Detroit: Greenhaven Press.

Gosling, S.D., Gaddis, S., & Vazire, S. (2007). Personality Impressions Based on Facebook Profiles. International Conference on Weblogs and Social Media.

Greenhow, C., Robelia, B., & Hughes, J. E. (2009). Web 2.0 and Classroom Research: What Path Should We Take Now? *Educational Researcher,* 38(4), 246–259.

Hampton, K.N., Sessions, L.F., Her, E.J., & Rainie, L. (2009). *Social Isolation and New Technology: How the Internet and Mobile Phones Impact American's Social Networks.* Washington, DC: PEW Internet & American Life Project.

Hanson, G., Haridakis, P.M., Cunningham, A.W., Sharma, R., & Ponder, J.D. (2010). The 2008 Presidential Campaign: Political Cynicism in the Age of Facebook, MySpace and YouTube. *Mass Communication and Society,* 13(5), 584–607.

Hargittai, E. (2007). Whose Space?: Differences Among Users and Non-Users of Social Network Sites. *Journal of Computer-Mediated Communication,* 13(1).

Harter, S (1999) The Construction of Self: A Developmental Perspective. New York, NY: The Guildford Press.

Heer, J., & boyd, d. (2005). Vizster: Visualizing Online Social Networks. IEEE Symposium on Information Visualization.

Hogan, B. (2008). A Comparison of On and Offline Networks through the Facebook API. *Proceedings from European Science Foundation for the Quantitative Methods in the Social Sciences,* Amsterdam, The Netherlands.

Hogeboom, D.L., McDermott, R.J., Perrin, K.M., Osman, H., & Bell-Elison, B.A. (2010). Internet Use and Social Networking Among Middle-aged and Older Adults. *Educational Gerontology,* 36(2), 93–111.

Johnson, L., Adams, S., & Haywood, K. (2011). *The NMC Horizon Report: 2011 K-12 Edition.* Austin, TX: The New Media Consortium.

Johnson, T., & Perlmutter, D.D. (2010). Introduction: The Facebook Election. *Mass Communication and Society, 13*(5), 554–559.

Keen, A. (2007). *The Cult of the Ameteur: How Today's Internet is Killing our Culture.* New York: Doubleday.

Kushin, M.J., & Yamamoto, M. (2010). Did Social Media Really Matter?: College Students' Use of Online Social Media and Political Decision Making in the 2008 Election. *Mass Communication and Society, 13*(5), 608–630.

Lenhart, A., & Madden, M. (2007). *Teens, Privacy & Online Social Networks: How Teens Manage their Online Identities and Personal Information in the Age of MySpace.* Washington, DC: PEW Internet & American Life Project.

Lenhart, A., Purcell, K., Smith, A., & Zickuhr, K. (2010). *Social Media & Mobile Internet Use Among Teens and Young Adults.* Washington, DC: PEW Internet & American Life Project.

Lewis, J., & West, A. (2009). 'Friending': London-based Undergraduates' Experience of Facebook. *New Media Society,* 11(7), 1209–1229.

Livingstone, S. (2008). Taking Risky Opportunities in Youthful Content Creation: Teenagers' Use of Social Networking Sites for Intimacy, Privacy and Self-Expression. *New Media & Society,* 10(3), 393–411.

Livingstone, S., & Brake, D.R. (2010). On the Rapid Rise of Social Networking Sites: New Findings and Policy Implications. *Children & Society,* 24, 75–83.

Lusoli, W., & Miltgen, C. (2009). *Young People and Emerging Digital Services: An Exploratory Survey on Motivations, Perceptions and Acceptance of Risk.* JRC European Commission.

Madden, M. (2010). *Older Adults and Social Media: Social Networking Use Among Those Aged 50 and Older Nearly Doubled in the Past Year.* Washington, DC: PEW Internet & American Life Project.

Madge, C., Meek, J., Wellens, J., & Hooley, T. (2010). Facebook, Social Integration and Informal Learning at University: 'It is More for Socialising and Talking to Friends about Work than Actually Doing Work'. *Learning, Media & Technology,* 34(2), 141–155.

Magnuson, M.J., & Dundes, L. (2008). Gender Differences in 'Social Portraits' Reflected in MySpace Profiles. *CyberPsychology & Behavior,* 11(2), 239–241.

Mazer, J.P., Murphy, R.E., & Simonds, C.J. (2007). I'll see you on 'Facebook': The Effects of Computer-Mediated Teacher Self-Disclosure on Student Motivation, Affective Learning, and Classroom Climate. *Communication Education,* 56(1), 1–17.

Pempek, T., Yermolayeva, Y., & Calvert, S. (2009). College Students' Social Networking Experiences on Facebook. *Journal of Applied Developmental Psychology,* 30, 227–238.

Perkel, D. (2006). Copy and Paste Literacy: Literacy Practices in the Production of a MySpace Profile. Available at: http://people.ischool.berkeley.edu/~dperkel/media/dperkel_literacymyspace.pdf

PEW Internet & American Life Project. (2008). Internet Activities. Available at: www.pewinternet.org/trends/Internet_Activities_7.22.08.htm

Pfeil, U., Arjan, R., & Zaphiris, P. (2009). Age Differences in Online Social Networking: A Study of User Profiles and the Social Capital Divide among Teenagers and Older Users of Myspace. *Computers in Human Behavior,* 25(3), 643–654.

Raacke, J., & Bonds-Raacke, J. (2008). MySpace and Facebook: Applying the Uses and Gratifications Theory to Exploring Friend-Networking Sites. *CyberPsychology & Behavior,* 11(2), 169–174.

Rainie, L., Purcell, K., & Smith, A. (2011). *The Social Side of the Internet*. Washington, DC: PEW Internet & American Life Project.

Selwyn, N. (2009). Faceworking: Exploring Students' Education-Related Use of Facebook. *Learning, Media & Technology*, 34(2), 157–174.

Spencer, L., & Pahl, R. (2006). *Rethinking Friendships. Hidden Solidarities Today*. Princeton, NJ: Princeton University Press.

Steinfield, C., Ellison, N.B., & Lampe, C. (2008). Social Capital, Self-Esteem, and Use of Online Social Network Sites: A Longitudinal Analysis. *Journal of Applied Developmental Psychology*, 29: 434–445.

Subrahmanyam, K., Reich, S.M., Waechter, N., & Espinoza, G. (2008). Online and Offline Social Networks: Use of Social Networking Sites by Emerging Adults. *Journal of Applied Developmental Psychology*, 29, 420–433.

Sum, S., Mathews, M.R., Pourghasem, M., & Hughes, I. (2008). Internet Technology and Social Capital: How the Internet Affects Seniors' Social Capital and Wellbeing. *Journal of Computer-Mediated Communication*, 14(1), 202–220.

Thelwall, M. (2007). Fk yea I swear: Cursing and gender in Myspace. Available at: http://www.scit.wlv.ac.uk

Tufekci, Z. (2008). Grooming, Gossip, Facebook and MySpace: What Can We Learn About These Sites from Those Who Won't Assimilate? *Information, Communication & Society*, 11(4), 544–564.

Utz, S. (2010). Show Me Your Friends and I Will Tell You What Type of Person You Are: How One's Profile, Number of Friends, and Type of Friends Influence Impression Formation on Social Network Sites. *Journal of Computer-Mediated Communication*, 15, 314–335.

Valkenburg, P.M., & Peter, J. (2007). Online Communication and Adolescent Well-Being: Testing the Stimulation Versus the Displacement Hypothesis. *Journal of Computer-Mediated Communication*, 12(4).

Valkenburg, P.M., Peter, J., & Schouten, A.P. (2006). Friend Networking Sites and their Relationship to Adolescents' Well-Being and Social Self-Esteem. *CyberPsychology & Behavior*, 9(5), 584–590.

vanHalen, C., & Janssen, J. (2004). The Use of Space in Dialogical Self-Construction: From Dante to Cyberspace. *Identity: An International Journal of Theory and Research*, 4(4), 38–405.

Wang, H., & Wellman, B. (2010). Social Connectivity in America: Changes in Adult Friendship Network Size from 2002 to 2007. *American Behavioral Scientist*, 53(8), 1148–1169.

Watkins, S.C., & Lee, H.E. (2010). *Got Facebook?: Investigating What's Social About Social Media*. Austin, TX: The University of Texas at Austin.

Williams, A.L., & Merten, M.J. (2008). A Review of Online Social Networking Profiles by Adolescents: Implications for Future Research and Interventions. *Adolescence*, 43(170), 253–274.

Woolley, J.K., Limperos, A.M., & Oliver, M.B. (2010). The 2008 Presidential Election, 2.0: A Content Analysis of User-Generated Political Facebook Groups. *Mass Communication and Society*, 13(5), 631–652.

Yoder, C., & Stutzman, F. (2011). Identifying Social Capital in the Facebook Interface. *Proceedings of the Conference on Human Factors in Computing Systems*, ACM, 585–588.

Young, K. (2005). Direct from the Source: The Value of 'Think-Aloud' Data in Understanding Learning. *Journal of Educational Enquiry*, 6(1), 19–33.

Young, K. (2009). Online Social Networking: An Australian Perspective. *International Journal of Emerging Technologies*, 7(1), 39–57.

Young, K. (2011a). Social Ties, Social Networks and the Facebook Experience. *International Journal of Emerging Technologies and Society*, 9(1), 20–34.

Young, K. (2011b). Applying Multimodal Analysis to MySpace: An Instructional Framework to Develop Students' Digital Literacy. In A.V. Stavros (ed.), *Advances in Communications and Media Research* (Vol. 8). Hauppauge, NY: Nova Publications.

Young, K. (2012). Researching Young People's Online Spaces. In K. te Riel & R. Brooks (eds), *Negotiating Ethical Challenges in Youth Research*. New York: Routledge.

Young, K. (2013). Friendship in the Facebook era. *Webology* 10(2).

Zhao, S., Grasmuck, S., & Martin, J. (2008). Identity Construction on Facebook: Digital Empowerment in Anchored Relationships. *Computers in Human Behavior*, 24, 1816–1836.

29

Learner Modelled Environments

Kaśka Porayska-Pomsta and Sara Bernardini

INTRODUCTION

Learner modelled environments (LMEs) are digital environments that are capable of automatically detecting learners' behaviours in relation to a specific knowledge domain, to reason about those behaviours in order to assess learners' performance, skills, socio-emotional and cognitive needs and to act accordingly in a pedagogically appropriate manner. Digital environments that possess such capabilities are typically referred to as intelligent learning environments or, more traditionally, as intelligent tutoring systems (ITSs).

LMEs research is motivated by the need to understand how meaningful interactions between a teacher and a learner develop, what contributes to successful learning and how the pedagogy can and ought to be adapted to the individual learners. Adapting the instruction to the needs of the individuals constitutes an important way in which to support them in their transformation from novices to experts in a specific subject domain.

Theories from diverse disciplines serve as the basis for tackling the above questions. Increasingly, the existing theories can be combined, tested and improved through computational means – more specifically through computational models of learning and communication processes and through the dynamically generated and updated models of individual learners themselves. The advantages that such computational models bring to the learning sciences are their systematic, objective and inspectable nature, as well as the fact that, in contrast to models created by and held in human teachers' heads, they lend themselves to being manipulated, changed and repeatedly tested. Importantly, they can be used to make predictions about the relationships between the specific elements included in the models, the pedagogy selected on the basis of such models and the learning outcomes that may ensue as a consequence of learners being exposed to the specific pedagogy. Such computational models can therefore serve to both support learning of a specific subject domain and to progress research about learning and teaching. In other words, LMEs research is of relevance to anyone interested in understanding how learners learn and especially how successful learning and teaching can be supported through technology.

MODELLING THE LEARNER

A core element of a LME is its *learner model*, traditionally also known as a *student model*. It is a core element, because it is responsible for interpreting the learner's observable behaviours against the background knowledge available and, through this, for providing a system with information necessary for making pedagogical decisions (e.g. whether or not the student is ready to progress to the next problem), as well as for choosing the appropriate feedback (e.g. whether at any specific point the learner would benefit from, say, a hint, positive feedback or a prompt).

Learner modelling (LM) is concerned with *real-time* diagnosis of the learners' knowledge and with diagnosis of their cognitive and affective states during learning. LM is important within computer-based systems intended to promote learning, because it guides the *just-in-time* adaptation of the interaction to the individual learner (Dillenbourg and Self, 1992). Many educators consider individually adapted instruction as ideal means for facilitating effective learning, with the *2-sigma effect* (Bloom, 1986) being often cited as the main evidence of the increased learning outcomes that ensue from such instruction and as one factor that motivates many an investment in technology-enhanced solutions for education (Koedinger and Corbett, 2006). Specifically, in comparing the effect of one-on-one tutoring with the effect of a number of other variables and treatments, Bloom found that the average student who had been tutored in a one-to-one context performed better than 98 per cent of the students who had been taught in the classroom (none of the other treatments produced an effect as large as 2 sigma).

The ability to model others is inherent in all human social behaviour, of which educational interactions are one instance. In general, humans engage implicitly in modelling others on a moment-by-moment basis (e.g. Sperber and Wilson, 1995). Observing and reasoning about others' intentions, beliefs and goals is also crucial to facilitating relevant and effective learning – that is, learning which results in deep understanding of concepts and skills, is long-term and transferrable to new situations. Teachers continually assess learners' needs, abilities and progress, and adapt the difficulty of the problem accordingly. For example, they modify the explanation, give hints, alter the form and strength of praises and remediative feedback according to how they interpret the learner's actions in context. Teachers are *trained* to pursue the explicit goal of helping to advance the individual learner's knowledge and skills.

For example, consider a simple problem of double-column subtraction, where the number 65 needs to be subtracted from 500. Let us suppose that the learner's answer is 565, which is incorrect: the learner's mistake is to add 65 to 500 instead of subtracting from it. In order to provide the learner with appropriate support, apart from spotting the error, the teacher needs to diagnose the root of the error: is it due to a simple slip and the inattention of the learner or is it due to some deeper underlying misconception? In order to be accurate, this diagnosis needs to be based on the teacher's knowledge of the learner, the immediate circumstances in which the error occurred, the history of the previous problems given to the learner, the way the learner solved them (correctly or incorrectly), the instruction that was available to the learner prior to and during problem solving and so on. Such information enables the teacher to decide *what* to say to the learner. Also the information that the teacher infers about the learner's emotional and motivational states contributes to decisions about *how* to communicate the support to the learner – a learner lacking confidence may need a hint and encouragement, whereas a confident but bored learner may benefit from feedback which identifies the mistake and instructs.

Thus, the main purpose of LM is to enable an LME to choose the most appropriate pedagogical actions for the moment – that is, actions which have the best chances of leading to the desired short- and long-term learning outcomes for specific individuals. LM also constitutes a paradigm within which to study learning and provides tools for learning about learning. In other words, the *computational* LM tools embedded in LMEs provide means for representing the covert mental processes (both cognitive and affective) that relate to learning, for specifying, testing and improving theories about cognition and for making predictions about learning outcomes in the domains of interest.

Learner Modelling as Knowledge Representation

LM is a computational approach to representing knowledge about the learner during a learning task. The result (the output) is a *learner model* – a qualitative, approximate and possibly partial representation of knowledge about the learner and about the learner's knowledge and learning. The construction of a learner model depends on the information about the learner's specific behaviours (the input) in relation to the domain knowledge being learned.

Knowledge representation is key to enabling the construction of learner models. It is concerned with two primary questions: (i) *what* knowledge to represent and (ii) *how* to represent it. These two seemingly simple questions lead to further, fundamental questions in education and cognitive science research, including: what is knowledge; what constitutes and influences learning; and how do we know that someone has reached the level of mastery in a domain?

To support learner modelling in a LME, at least four types of knowledge need to be represented: (1) domain knowledge; (2) knowledge about the learner; (3) pedagogic knowledge; and (4) communication knowledge. These types of knowledge are interdependent in that domain knowledge representation is essential to enabling the modelling of the learner's progress and it determines the type of pedagogic knowledge required to facilitate the learner's progress. Representation of the knowledge about the learner is essential for enabling decisions about the appropriate pedagogy and the communication strategies that will best realize the pedagogy. Whilst all four types of knowledge are crucial to a fully fledged LME, with domain knowledge representation of particular importance to enabling the assessment of learner's progress and to choosing optimal support (Sison and Masamichi, 1998; Woolf, 2009), this chapter focuses specifically on LM.

Knowledge about the Learner

Knowledge about the learner is a representation of what the learner knows, including specific skills and misconceptions at any given time during and after a tutoring interaction. It is crucial to enabling a teacher (whether human or computer tutor) to provide appropriate feedback to the learner – that is, feedback which will promote long-term learning. We refer to this type of information as the *learner's knowledge*.

Moreover, it has been long recognized that learning does not happen purely in the learner's head, but also depends on the environment in which it takes place (Sawyer, 2006). While prior knowledge along with the learner's misconceptions provide the necessary information for assessing the progress of the learner in some domain knowledge, it does not always suffice for choosing feedback that supports the learner's motivation. Learning is a process that involves the learner as a person, with emotional predispositions and transient affective states such as anxiety, boredom, frustration or joy, which impact their motivation and ultimately both their learning experiences and their learning outcomes (Lepper et al.,

1993). We refer to the information about the learner's psycho-emotional states as *learner's affect*.

Learner's Knowledge

The term *learner's knowledge* encompasses both the factual knowledge, stored in a person's declarative (factual) memory, and the relevant skills needed to use, apply and update the facts (procedural knowledge). More specifically *learner's knowledge* typically refers to *what* facts the student knows, the *extent* to which the learner is believed to know them and the learner's *skills*, along with the extent to which they are believed (by the tutor) to have been mastered, what knowledge is erroneous (*misconceptions*) and what the learner does not know (*missing knowledge*).

There are a variety of different ways in which a model of learner's knowledge and skills can be conceptualized, represented and inferred. The type of representation and inference used often depends on the theoretical and/or computational perspectives taken by the individual researchers and the research questions they seek. Put simplistically, cognitive scientists are typically concerned with *cognitive fidelity* of the models that they create and they may use LM techniques and tools as a way of verifying the validity of those models (e.g. Anderson et al., 1990). However, ITS researchers are typically concerned with supporting the learner in achieving mastery of a subject matter or a subset thereof, with LM tools providing a way of diagnosing a learner's knowledge to enable optimal feedback by the system in *real time*. Thus, although cognitive fidelity of a LM tool is highly desirable to the ITS researchers, *computational efficiency* is of the essence (Dillenbourg and Self, 1992).

It is important to bear in mind that the theoretical and pedagogical question of how knowledge is or should or can be represented and acquired by people is somewhat, but not altogether, separate from the issue of how the learner's knowledge acquisition can be modelled computationally during a learning task – that is, the specific computational techniques which can be used to infer the model of the learner. This muddled relationship (and the tension) between a theory preferred and a computational approach favoured is a known facet of the research related to building and deploying LMEs. This is because such environments are inextricably cognitive as well as computational (Ohlsson and Mitrovic, 2007). The difficulty of disentangling the cognitive and the computational considerations is also largely responsible for the apparent lack of an integrated, overarching definition of the LM field as a whole and the difficulties with presenting it as a coherent collection of findings, principles and guidelines for the novice researchers who wish to learn about it and to explore it.

Representing Learner's Knowledge

Assuming the existence of a domain knowledge representation, building a LM requires one to choose the appropriate way in which to represent the existing and evolving knowledge of the learner. Although there are a number of differing approaches to representing learner knowledge, many ITSs contain a representation of an *ideal* learner model, which, typically, is a representation of expert knowledge in the subject domain. Specifically, learner's knowledge can be represented as (i) overlaying the ideal learner model – that is, as a subset of what the expert knows (Figure 29.1(a)), (ii) as a differential model, whereby learner's knowledge is represented in terms of two categories: knowledge that has been already presented to the student v. knowledge that has never been introduced (Figure 29.1(b)), or (iii) a perturbation model (Figure 29.1(c)), which combines the standard overlay model with a representation of learners' misconceptions.

The perturbation model and bug libraries (BLs), used most notably in the BUGGY systems (Brown and Burton, 1978; Burton

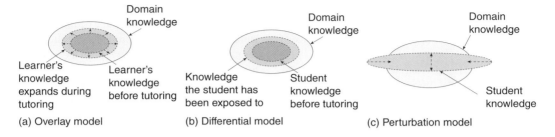

Figure 29.1   Lecture slides by Sara Bernardini (London Knowledge Lab, 2011).

and Brown, 1982), are the most successful of the three types of representations and are still in use (in improved forms) in some LMEs such as the cognitive tutors (Koedinger et al., 1997).

BLs specify common misconceptions and errors in terms of lists of *mal-rules*, reflecting the *typical* erroneous knowledge in a domain. They allow a computer tutor to identify incorrect knowledge through accounting for *errors of commission* – that is, errors which can be observed overtly through the actions of learners – for example a step in a solution or an answer to a question. In general, BLs are useful for: (i) diagnosing step-by-step the state of the learner's knowledge and possible misconceptions; (ii) predicting the reasoning path taken by the learner in a specific problem; (iii) predicting the learner's future errors and answers; and (iv) adapting the feedback accordingly. However, they are not practical for modelling learners' knowledge in ill-defined domains such as architecture, where the value of a solution path may not be known until the solution is reached. They are even less suitable for domains such as social interaction skills, where the appropriateness of learner's actions is relative to the context of a specific situation, because the specification of mal-rules is absolute and discrete rather than, as such domains demand, relative and probabilistic.

Nevertheless, BLs serve to demonstrate the role of LM and the difficulties involved in representing and inferring learner's knowledge. They highlight the relationship between domain knowledge and different types of errors that one may make in relation to it and raise questions as to what are misconceptions, what constitutes a typical misconception in a domain and what learners' behaviours count as manifesting mastery and lack thereof. As such, BLs provide an invaluable tool for thinking about learning, the best ways to support learning, showing that even the simplest and best defined problems such as double-column subtraction can result in a potentially intractable number of possible errors that originate from different sources.

Cognitive Tutors and Cognitive Modelling

Cognitive tutors (CTs) are an example of and improved and hugely successful application of the BLs. CTs, originally developed as a test-bed for the ACT* and then ACT-R theory of cognition (Anderson, 1983), combine a specification of a domain knowledge, using ACT-R, with the cognitive diagnosis tools – specifically *model tracing* and *knowledge tracing*. ACT-R aims to describe human cognitive processes involved in problem solving in procedural domains such as mathematics, in order to inform our understanding of the structure of knowledge, of how knowledge is stored and how it is acquired and enhanced by humans over time when exposed to appropriate triggers. ACT-R makes a distinction between declarative knowledge (e.g. '*When both sides of the equation are divided by the same value, the resulting quantities are equal*') and procedural knowledge (e.g. '*IF the goal is to solve an equation for variable X and the equation is of the form aX = b, THEN divide both sides of the equation by a to isolate X*', (see Corbett et al.,1997) and it envisions that the factual and procedural knowledge components are fundamentally distinct from one another.

Cognitive fidelity is paramount in ACT-R, because the purpose of the modelling in which it engages – the cognitive modelling – is to enhance our understanding of how humans really learn and how the cognitive processes are reflected in the neural activities of the human brain. Thus, it is important to appreciate that the purpose of cognitive modelling is not necessarily to enhance education as such, but, rather, it is to explain how humans acquire and use knowledge; at best it is to mimic the real processes. Ensuring cognitive fidelity of a model is therefore first on the agenda in cognitive modelling – that is, LM within CTs – and provides the means for verifying the models that ensue.

Cognitive diagnosis consists of keeping track of (i) the learner's progress through a solution and (ii) the growth of learner's knowledge over time, respectively. The diagnosis is based on the specification of production rules that apply, given a particular stage in the problem-solving episode. The production rules are annotated for the correctness and specificity of the solutions that they offer. During problem solving a CT keeps track of the solution steps committed by the learner and identifies the production rules in its database that correspond to the learner's solution steps. The annotations associated with each production rule in the database provide the basis for the assessment of the correctness of the learner's step and for decisions about appropriate feedback.

The system employs *model tracing* to infer a solution path for individual learners given a specific goal. CTs assume that a cognitive model represents the *ideal* learner model. Given a description of declarative and procedural knowledge for a specific domain and given the actions of the learner at the interface level (*learner's behaviours*), the system can generate (*trace*) a sequence of possible steps through a problem space alongside the steps that the learner is producing. Since different elements of the interface in a cognitive tutor correspond to a representation of the problem-solving goal and to the relevant information needed to achieve it, the tutor is able to use the information about the learner's behaviours at the interface level to infer the learner's current goal. Furthermore, by applying its production rules to the learner's goal, the tutor is able to generate a set of possible rules which could satisfy that goal (Corbett et al., 1997).

While model tracing is concerned with generating a generic learner model step-by-step, the purpose of *knowledge tracing* – the second diagnosis mechanism employed in CTs – is to assess the growth of the learner's knowledge over time. Knowledge tracing provides the basis for selecting the appropriate problem given the changing mastery level of the learner. Similar to model tracing, it consists of the tutor generating a set of production rules at each level of the curriculum taught and assessing the extent to which the learner is believed to have mastered those rules. This assessment is done through assignment of probabilities to each of the productions generated by the tutor based on the learner's behaviours whenever an opportunity arises for a specific rule to be applied.

Cognitive modelling has been very successful in supporting the learning of certain aspects of well-defined domains – for example, mathematics. It is important to bear in mind, however, that its success in LM environments came out of the benefit of hindsight, in particular from the early research in intelligent tutoring systems, which highlighted the importance of aiming for cognitive fidelity in the representation of expert knowledge, because the less cognitively truthful the model, the greater the chance that the pedagogic support offered might be inadequate. Cognitive fidelity seems especially important in relation to (i) a system's ability to reason about the problem in the same way as humans do and (ii) a system's ability to apply the same mechanisms as humans do when searching for a solution. Early tutoring systems that fail with respect to the first requirement – for example, SOPHIE I (Brown et al., 1973) tend to lack the ability to support – learner's reflection: while they are able to recommend to the learner actions that are optimal in each problem-solving context, they fail to describe the context and the rationale for the solution, leaving the learner to construct these for themselves. Tutoring systems that fail with respect to the second requirement – for example, GUIDON (Clancey, 1983) – also fail to advise the learner on what to do next, because they have no idea about what learning trajectory the learner follows (Corbett et al., 1997).

Constraint-based Modelling

An approach, developed in parallel with the CTs, is constraint-based modelling (CBM) (Ohlsson, 1992). CBM emerged as a response to the issue of the *over-specificity* of the learner knowledge descriptions that were characteristic of the early LMEs such as DEBUGGY and later CTs (e.g. Ohlsson and Mitrovic, 2007). Over-specificity of learner models refers to the observation that, in its general form, LM is intractable (Holt et al., 1994; Self, 1990). Independently of the technique chosen to represent learner's knowledge, any mechanism that aims to infer a complete model of a learner is bound to rely on the specification of hundreds, sometimes thousands, of individual knowledge chunks. In practice, this level of specificity cannot be handled effectively, even when sophisticated authoring tools are employed to construct the appropriate knowledge representations. Indeed, if human tutors are taken as the gold standard for building digital learning environments, then the question arises of whether it is sensible to expect a fully specified learner model at all. Empirical studies show that human tutors do not tend to rely on complete and fully coherent models of learners

when they teach, yet can be very effective (Holt et al., 1994; Leinhardt and Ohlsson, 1990). Ohlsson's CBM approach builds on this observation by introducing a technique that requires only a partial, but nevertheless effective, learner model.

In the CBM approach, the problem of over-specificity is addressed through abstraction, whereby domain knowledge is represented through a set of constraints corresponding to correct solutions for that domain. Each constraint is an ordered pair $<C_r, C_s>$, where C_r is the relevance condition that identifies the class of problem states for which the constraint is relevant and C_s is the satisfaction condition which identifies the class of states in which the constraint is satisfied. Each member of the pair can be thought of as a set of properties of a problem state, with the constraint representing a statement such that, *if the properties C_r hold, then the properties C_s have to hold also*. If the properties C_s do not hold, this means that an error has occurred. Therefore, indirectly each constraint also represents a set of erroneous solutions – that is, the solutions which violate the constraint. A constraint-based learner model consists of the set of constraints that the learner does and does not violate. A violation of a constraint on the part of the learner signifies missing or incomplete knowledge.

Formulating a set of constraints for a domain that are also valid pedagogically is a difficult task, but by addressing the over-specificity problem this task is eased greatly, making CBM a computationally efficient solution to LM. As Ohlsson argues, a set of constraints does not necessarily have to be complete in order to be useful. An incomplete set of constraints may lead to rare errors not being caught, but as long as the most common mistakes are addressed, the system is still able to provide learners with valuable feedback. Importantly, formulating the knowledge in terms of constraints (i.e. only correct knowledge requires to be fully specified) enables reusability of such representations across different populations of learners.

While ensuring computational efficiency, CBM also claims to be a cognitively truthful approach: it is based on a psychological theory of learning that asserts learning occurs primarily when students catch themselves making mistakes. This theory makes a distinction between *generative* and *evaluative* knowledge (Ohlsson, 1996). Generative knowledge produces problem-solving actions and can be expressed by means of a set of *rules*. Evaluative knowledge evaluates the actions' outcomes in terms of their desirability with respect to a specific goal, which can be expressed by a set of *constraints*. Ohlsson conjectures that the acquisition of a new cognitive skill is (at least, partially) based on the transfer of knowledge from the evaluative to the generative component (Ohlsson 1996). Hence, in the context of this theory of learning, a constraint-based modelled environment constitutes an amplified evaluative knowledge base that contains the constraints representing the evaluative expert knowledge. By giving the learner access to the evaluative knowledge, the system can speed up and augment the transfer of information from the evaluative to the generative component, so, ultimately, it can support the learning process.

Learner's Affect

None of the approaches to LM reviewed so far incorporate a mechanism that accounts in any way for the influence that emotions and context of educational interactions have on the experience and effectiveness of learning. Yet, the importance of learners' affect to learning has been long recognized in education. As early as 1908, Yerkes and Dodson discussed the relationship between a learner's emotional arousal and learning performance, observing that moderate arousal is always preferable to the extremes of low or high arousal, which can be detrimental to learning (Yerkes and Dodson, 1908). This observation still holds, with a number of recent studies in neuroscience and psychology providing further evidence that cognition and emotions are profoundly intertwined (Damasio, 1994). Experiencing emotions helps humans make decisions given the

demands of their environment, with the three main cognitive processes involved in learning – attention, memorization and reasoning – being strongly influenced by the specific emotions experienced. Positive emotions enhance attention, facilitate memorization and improve efficiency and efficacy in problem solving and decision making (Isen, 2000). Negative emotions, however, can be detrimental to concentration, may disturb the retrieval of information, make memorization less effective and can induce convergent and sequential reasoning (Lissetti and Schiano, 2000).

An increasing number of LMEs incorporate modules that recognize, interpret and respond to learners' affective states. The goal of such modules is to generate real-time data that informs our understanding of what constitutes the experience of emotion(s) during learning and of the relationship between such experiences and learning outcomes. In such LMEs, the learner model incorporates information about learners' affective states and updates it regularly based on the interaction between the learner and the system.

Modelling learner's affect automatically requires a designer to: (i) specify the emotions of interest in a given domain; (ii) define the emotions in terms of observable behaviours; and (iii) specify a mechanism for detecting and interpreting the behaviours in terms of the emotions identified. These three requirements apply regardless of the theoretical or technical standpoint of the individual designers. Real emotional experiences are far messier than contemporary digital implementations allow for: emotions are transient, co-occurring and overlapping; they are short-lived and often triggered by multiple events (D'Mello and Graesser, 2011) and they depend on medium- and long-term affective predispositions and the goals and beliefs of the people who experience them (Gebhard, 2005). Educational contexts bring about very specific types of affective experiences, bearing heavily and explicitly on the types of goals, attitudes and beliefs that learners may harbour in a given context about a specific domain, the support they are receiving and about themselves as learners.

What Affect to Model?

The past hundred years of research on emotions produced many diverse theories that could potentially serve as the basis for developing systems to detect and interpret learners' affect. An excellent review by Calvo and D'Mello (2010) provides a taxonomy of the most prominent theories of emotion used specifically to inform the automatic detection of emotions of a generic user. The theories considered include those that are used to inform the design of *generative* models – that is, models of how emotions should be produced by artificial agents in real time (either through physical expressions of emotions (Darwin, 2009; Ekman, 1971) or a combination of physical and physiological expressions (James, 1884). Alternatively, the existing theories are used to *predict* (i.e. model) user emotions based on specific triggers in the environment, such as objects, events and other people (e.g. *cognitive appraisal* theories (Scherer, 2005) or the OCC theory (Ortony et al., 1988). Many of the theories, save for the ones developed in the *social constructivist* perspective (Averill, 1980; Stets and Turner, 2008), consider emotions outside of the environment in which they come about or the task at hand. Those that do consider the context as integral to the experience and understanding of emotions are, however, far less amenable to being automated than those that consider emotions in isolation.

Although most designers of LMEs remain agnostic about the theories available, the main considerations taken into account in choosing a particular theory in this context are: (a) how well a theory lends itself to being implemented in a computer system and (b) how well it is able to support the selection of appropriate feedback for a given learner. Similar to modelling learner's knowledge, considerations of computational efficiency and psychological fidelity play a role in what

approaches are adopted. An additional consideration is whether a theory is pedagogically viable. How well a theory lends itself to being implemented may conflict with its applicability to a wide range of LMEs – for example, arguably the nature, diversity and duration of emotional experiences within an educational game environment are different from the affect experienced in exploratory environments, because the nature of tasks and rewards for achieving the tasks therein are different. The focus on educational viability also highlights the question of how well a given theory of affect can account for the affective states users experience during learning – that is, whether the affective states modelled are relevant to learning.

All of the six approaches reviewed by Calvo and D'Mello, individually and sometimes in combination, have been used to inform various implementations of the automatic affect detection systems. Some of them, such as the one adopted by D'Mello et al. (2007), combine physiological and behavioural signal detection, such as skin conductance, eye gaze tracking and posture detection, with linguistic cues analysis, in order to infer motivational states such as boredom, flow, etc. In doing so, such approaches combine different theoretical perspectives – in D'Mello et al.'s case the theories of emotions as expressions, embodiments as well as theories of motivation that tend to be socio-constructivist by nature. In the rest of this chapter we focus on two relatively coherent trends that have been substantially used in the context of modelling learner's affect: the cognitive appraisal and social constructivism.

Cognitive Appraisal v. Social Constructivism

The OCC theory (Ortony et al., 1988) is one of the most popular theories of emotions in the affective computing field, not least because it is computationally viable. It represents 22 emotions categorized into a hierarchy of 6 groups, such as well-being; joy and distress or attraction; love and hate. The categories are based on valenced (i.e. positive/negative) responses of a person to current states of the world and to their causes. OCC accounts for how emotions arise from people's appraisal of the current situation, where any situation consists of objects, events and agents. OCC allows one to specify the intensity of emotions based on the likelihood of an event taking place or of a person's familiarity with an object. The model is very detailed, but its limitation is that it accounts only for prime emotions, leading some researchers to question its relevance to learning (Calvo and D'Mello, 2010).

In contrast with the cognitive appraisal theories, in the social constructivist perspective emotions are seen as socio-cultural constructs that are co-constructed with others through social interactions and (linguistic) communication (Averill, 1980; Glenberg et al., 2005). While, most approaches within affective computing rely on other theories available, arguably, affective computing research that focuses on education is by necessity socio-constructivist. This is because in education, the learning processes, including the affective states that impact a person's learning, are inextricably perceived as outcomes of the interaction between the learner, the teacher and the learner's performance in relation to a learning task at hand. Specifically, there is substantial and maturing research pointing to learners' *motivation* as a form of affective experience of particular relevance to learning (Craig et al., 2004). Motivational states are sometimes understood as 'the determinants of thought and action' (Weiner, 1992: 17) and, unlike the short-term, prime emotions such as joy, distress, love or hate, etc., motivation is a composite of emotions, the long- and short-term goals of a person and their beliefs, as well as their emotions experienced in specific contexts.

The two theoretical perspectives are not incompatible, even if they approach the affect modelling from different directions: the cognitive appraisal implementations use prime emotions to describe learner's affect, whereas the social constructivists focus broadly on motivation, differences in learners' behaviours and the influence of motivation regulation on learner's affect and performance (see Figure 29.2). As Conati and Maclaren (2009) point out, the ultimate goal of LM research is to devise methods and approaches that combine low-level emotion detection and interpretation with higher-level motivational state modelling.

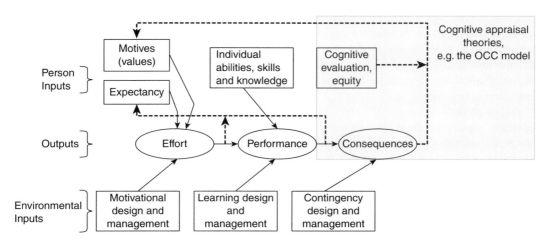

Figure 29.2   Keller's model of motivation, performance and instructional influence (Keller, 1983: 392).

We use Keller's (1983) model (Figure 29.2) to illustrate both the relevance and compatibility of the different approaches used to date. The figure represents Keller's understanding of motivation and motivation regulation as a loop between effort, performance and consequences, whereby the level of effort that the learner dedicates to a task depends on their values and goals. The level of effort dedicated impacts on the quality of the learner's performance and results in the intrinsic and extrinsic outcomes (in the figure: 'consequences'), including the learners' emotional responses and social and material rewards. The consequences are appraised by the learner (through cognitive evaluation) and these appraisals feed back into 'motives', either reinforcing or changing the values of the learner.

Keller's model shows explicitly how motivation depends on the input from the environment and especially how appropriate pedagogical design and management fits in with learners' motivation and emotional regulation (environmental inputs). Therefore, the model is explicitly socio-constructivist in nature in Averill's sense. Second, the model explicitly accounts for learners' cognitive appraisal of the outcomes of their actions, which is where the approaches concerned with the prime emotions, such as the OCC model, fit – these are highlighted by us in the top right part of the figure. While none of the existing implementations include all of the aspects of this (enhanced) model, it can be thought of as representing, possibly with some modifications, the ideal to which the affective LM field aspires. In the following sections we describe two approaches to affective LM as exemplars of the existing implementations that adopt different theoretical entry points.

Modelling the Learner's Affect Based on a Cognitive Appraisal Theory

Many LMEs rely solely on the OCC model, although seldom do the learner models

generated include all 22 emotions specified therein. Instead, the designers tend to select the most relevant emotions that fit the domain or the mode of interaction afforded by their environment. One example is the Prime Climb, two-player game developed to support children in learning number factorization (Conati and Maclaren, 2009). Each player controls a different character and the two players need to cooperate to allow the respective virtual characters to climb a mountain and arrive at the top. The mountain is divided into numbered sectors. Each character can only move to a numbered sector that does not share any factors with the sector occupied by the other character. If a player makes a wrong move, then her/his corresponding character falls down the mountain.

The learner model in Prime Climb includes only the six most relevant emotions to the target domain: *joy/distress*, *admiration/reproach* and *pride/shame*. Prime Climb uses the video game paradigm to exploit the intrinsic motivation that such games have been found to afford (Johnson and Rickel, 2001). Apart from modelling the learner's progress through the number factorization exercises, the environment is also enabled with a diagnostic model responsible for detecting and interpreting children's emotions as they progress through the game. The model also assesses how multiple goals, which the students typically have while playing the game, influence learners' appraisal of the game interaction and, consequently, their emotions. Five high-level student goals have been considered in Prime Climb: *Have Fun*, *Avoid Falling*, *Learn Math*, *Beat Partner* and *Succeed By Myself*.

Because the relations between student personality, goals, game states and emotions are not deterministic, the student model in Prime Climb is implemented through dynamic Bayesian networks to manage uncertainty in real time. The Prime Climb affective model represents the first known probabilistic representation of the OCC theory. The purpose of this model is to facilitate a nuanced selection of feedback and ultimately to enhance both the experience and learning outcomes. In order to maintain learner engagement while providing instructional learning support, Prime Climb incorporates an animated pedagogical agent, which responds to learners' requests for hints or offers hints when it decides that the student needs help. The agent relies on a probabilistic model of the student's knowledge about factorization to decide when to intervene and what hints to provide. The evaluation studies of Prime Climb with users in primary schools offer promising evidence that providing personalized instruction, in this case by means of an animated pedagogical agent, can improve learning over traditional methods (Conati and Zhao, 2004).

Modelling Learner's Affect Based on Interaction Analysis (Socio-Constructivism)

In Porayska-Pomsta et al. (2008) the specific context of a situation along with the interaction between a learner and a teacher are integral to both regulating learners' emotions and to being able to recognize and act on them in pedagogically viable ways. Our starting point is the natural language dialogue that takes place between the teacher and the learner. In line with the socio-constructivist theories of emotions, our premise is that modelling learner's emotions and the delivery of appropriate support requires an understanding of the interaction context in which the emotional states are experienced. We define the educational interaction loop (also known as the *affective loop* – see D'Mello et al., 2007) as involving a teacher: (i) observing the learner's behaviours in context; (ii) modelling the learner's affective states based on such observations; (iii) deciding how to support the learner based on the model; and in turn (iv) influencing the learner's affective states by acting on his/her decisions as to how to provide the support.

We used the language enhanced, user adaptive, interactive e-learning for mathematics (LeAM) system as a context for this research. LeAM consists of a learner model, a tutorial component, an exercise repository, a domain reasoner and natural language dialogue capabilities. We conducted a number of studies, over two years, in order to inform the design of the learner and the natural language dialogue models. In order to ensure ecological validity of the data generated, we restricted the student–teacher communication to a chat interface (Figure 29.3), with no visual or audio channels, to resemble the interface of the final learning environment. Five experienced teachers and 28 learners were asked to adapt their responses to each other by accommodating to this limited channel of communication. The data collected included natural language dialogue logs, verbal protocols and semi-structured interview data from teachers and learners and, crucially, teachers' real-time annotations of the situations in the context of which they were providing feedback to the learners, which included affective states such as the learners' confidence, interest and effort, as well as other contextual information such as the teachers' perception of the universal and relative difficulty of the task given to the learner, the correctness of learners' answers, amount of session time left and the learner's aptitude, knowledge and goals. The data collected provide us with rich information, not only about the context that teachers take into account when diagnosing learners' affect but also with the specific examples of the feedback that the teachers deem appropriate for the individual learners in the *heat of the moment*, given the specific tasks and given their own teaching and communication styles.

The findings from these studies informed the design of the LeAM's learner model. The model, implemented as a Bayesian network, represents causal relationships between a specific combination of contextual factors and enables a link to be made between the learners' affective states that the teachers have inferred from the learners' observable behaviours and the ways in which the teachers act on such inferences (Porayska-Pomsta et al., 2008).

Preview of the feedback to student

Dialogue Window

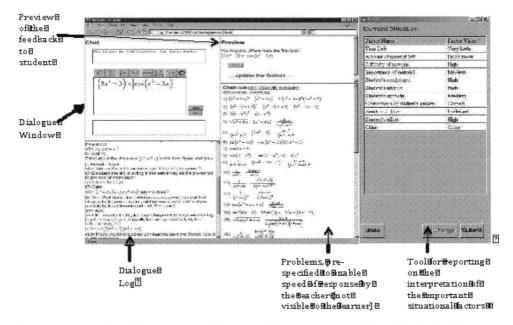

Dialogue Log

Problems, pre-specified to enable speed of response by the teacher (not visible to the learner)

Tool for reporting on the interpretation of the important situational factors

Figure 29.3 Chat interface tools used by Porayska-Pomsta, Mavrikis and Pain (2008) to collect data about how teachers diagnose and respond to student affect in context.

DISCUSSION AND CONCLUSION

This chapter introduced the concept of LMEs as a method through which learning can be both supported and studied. In particular, LM was presented as the core, defining characteristic of any learner-modelled environment and as a necessary prerequisite for a technology-enhanced educational tool capable of adapting the interaction and the pedagogical support to the individual learners' needs in real time.

Automatic LM, although still an emerging discipline, is not brand new and much research is available that illustrates different approaches to it. Early LM tools focused solely on modelling learners' knowledge. As the field matured and new technologies, such as physiological sensors, became available and affordable, the focus extended to learners' affect. However, few researchers in the field approach the task of LM and of building LMEs from the same theoretical, technological or pedagogic perspectives. This means that the field lacks uniformity, which hinders comparisons between the different implementations and a coherent account of both the trends and achievements therein.

The aim of this chapter was to bring the many approaches and points of view, as much as possible, into a coherent description of the LM field, in order to enable a novice reader to understand its main tenets and perspectives. We identified two broad types of uses of learner models: (i) to implement a theory about human learning processes and (ii) to support HCI in educational contexts. Although these research aims are not incompatible, the choice of one over another typically requires the research focus being placed on achieving in such models either cognitive and/or psychological fidelity, or computational efficiency. Throughout the chapter we provided exemplar implementations reflecting these two foci, as well as examples of approaches that, like the constraint-based approach to learner's knowledge modelling or Conati's and Maclaren's approach to affect modelling, attempt to achieve a bit of both. We selected the approaches to LM in both the cognitive and the affective branches that represent the predominant trends in a still maturing field and allow us to highlight the main questions related not only to the design of LMEs but also to education research in general. There are many more notable instances of LMEs than described here. There are examples of LMEs that illustrate the state-of-the-art achievements in LM and many other examples of research that inform the design of such environments. Using the references we provide to support our conclusions throughout the chapter, we hope that the interested reader will be able to trace and pursue those approaches.

In concluding this chapter, we would like to reflect on some of the criticisms that the field sometimes receives and are encapsulated in the question of whether or not LM is at all a feasible endeavour. There are some who point to the intractability of the LM question and to the fact that, thus far, the question has been addressed with little success: the models created are limited in scope and are difficult to port to other domains and different types of learners; they focus only on the cognitive, but not the affective or they model learners with respect to states which are only tangentially relevant to learning. Whilst these criticisms are valid and they point to what is the most likely reason for the lack of complete coherence in the field, we believe that the question of feasibility is misguided. Models are by definition a reduced representation of the real phenomena and this is precisely why they are so useful for research. They lend themselves to being tested, manipulated and changed and through this they inform our understanding of the phenomena studied. In relation to LM, it is easy to forget that as much as they are tools through which believable and educationally beneficial interactions can be facilitated, such models are first and foremost research tools through which we can explore what it means to learn.

BIBLIOGRAPHY

Anderson, J.R. (1983) *The Architecture of Cognition.* Boston, MA: Harvard University Press.

Anderson, J.R., C.F. Boyle, A.T. Corbett and M.W. Lewis (1990) 'Cognitive modelling and intelligent tutoring'. *Artificial Intelligence* 42: 7–49.

Averill, J.R. (1980) 'A constructivist view of emotion'. In R. Plutchik and H. Kellerman (eds), *Emotion: Theory, Research and Experience.* New York: Academic Press.

Bloom, B. (1986) 'The 2 sigma problem: the search for methods of group instruction as effective as one-to-one tutoring'. *Educational Researcher* 13(6): 4–16.

Brown, J.S. and R.R. Burton (1978) 'Diagnostic models for procedural bugs in basic mathematical skills'. *Cognitive Science* 2: 155–192.

Brown, J.S., R. Burton and F. Zdybel (1973) 'A model-driven question-answering system for mixed initiative computer-assisted instruction'. *IEEE Transactions on Systems, Man and Cybernetics,* 248–257.

Burton, R.R. and J.S. Brown (1982) 'An investigation of computer coaching for informal learning activities'. *Intelligent Tutoring Systems,* 157–183.

Calvo, R.A. and S. D'Mello (2010) 'Affect detection: an interdisciplinary review of models, methods, and their approaches'. *IEEE Transactions on Affective Computing,* 1(1): 18–37.

Clancey, W.J. (1983) 'GUIDON'. *Journal of Computer-Based Instruction* 10(1): 8–14.

Conati, C. and H. Maclaren (2009) 'Empirically building and evaluating a probabalistic model of user affect'. *User Modeling and User-Adapted Interaction* 19(3): 267–303.

Conati, C. and X. Zhao (2004) 'Building and evaluating an intelligent pedagogical agent to improve the effectiveness of an educational game'. *Proceedings of International Conference on Intelligent User Interfaces* 6–13.

Corbett, A.T., K.R. Koedinger and J.R. Anderson (1997) 'Intelligent tutoring systems'. In M.G. Helander, T.K. Landauer and P. Prabhu (eds), *Handbook of Human-Computer Interaction.* Amsterdam: Elsevier.

Craig, S.D., A.C. Graesser, J. Sullins and B. Gholson (2004) 'Affect and learning: an exploratory look into the role of affect in learning with autotutor'. *Journal of Educational Media* 29: 241–250.

Damasio, A.R (1994) *Descartes' Error: Emotion, Reason, and the Human Brain.* New York: G.P. Putnam.

Darwin, C. (2009) *The Expression of the Emotions in Man and Animals.* London: Penguin.

Dillenbourg, P. and J. Self (1992) 'A framework for learner modelling'. *Interactive Learning Environments* 2(2): 111–137.

D'Mello, S. and A.C. Graesser (2011) 'The half-life of cognitive-affective states during complex learning'. *Cognition and Emotion* 25(7): 1299–1308.

D'Mello, S.K., R.W. Picard and A.C. Graesser (2007) 'Towards an affect-sensitive autotutor'. *IEEE Intelligent Systems, Special Issue on Intelligent Educational Systems* 22(4): 53–61.

Ekman, P. (1971) *Universals and Cultural Differences in Facial Expressions of Emotion.* Lincoln, NE: University of Nebraska Press.

Gebhard, P. (2005) 'ALMA – a layered model of affect'. *Proceedings of the Fourth International Joint Conference on Autonomous Agents and Multiagent Systems,* 29–36.

Glenberg, A., D. Havas, R. Becker and M. Rinck (2005) 'Grounding language in bodily states: the case for emotion'. In R. Zwaan and D. Pecher (eds), *The Grounding of Cognition: The Role of Perception and Action in Memory, Language and Thinking.* Cambridge: Cambridge University Press. pp. 115–128.

Holt, P., S. Dubs, M. Jones and J.E. Greer (1994) 'Student modeling: the key to individualized knowledge-based instruction'. In J. Greer and G. McCalla (eds), *The State of Student Modelling.* Berlin: Springer-Verlag. pp. 3–35.

Isen, A.M. (2000) 'Positive affect and decision making'. In M. Lewis and J. Haviland-Jones (eds), *Handbook of Emotions.* New York: Guilford Press. pp. 417–435.

James, W. (1884) 'What is an emotion?' *Mind,* 188–205.

Johnson, W.L. and J. Rickel (2001) 'Research in animated pedagogical agents: progress and prospects for training'. *International Conference on Intelligent User Interfaces.*

Keller, J.M. (1983) 'Motivational design of instruction'. In C.M. Reigeluth *Instructional-Design Theories and Models: An Overview of their Current Status.* Hillsdale, NJ: Lawrence Erlbaum pp. 383–434.

Koedinger, K.R. and A.T. Corbett (2006) 'Cognitive tutors: technology bringing learning science to the classroom'. In K. Sawyer (ed.), *The Cambridge Handbook of the Learning Sciences.* New York: Cambridge University Press. pp. 61–78.

Koedinger, K.R., J.R. Anderson, W.H. Hadley and M.A. Mark (1997) 'Cognitive tutors: lessons learned'. *Journal of the Learning Sciences* 4(2): 30–43.

Leinhardt, G., and Ohlsson, S. (1990) 'Tutorials on the structure of tutoring from teachers.' *Journal of Artificial Intelligence in Education* 2: 21-46.

Lepper, M., M. Woolverton, D. Mumme and J. Gurtner (1993) 'Motivational techniques of expert human tutors: lessons for the design of computer-based tutors'. In S. Lajoie and S. Derry (eds), *Computers as Cognitive Tools.* Hillsdale, NJ: Lawrence Erlbaum pp. 75–105.

Lisetti, C. and D. Schiano (2000) 'Automatic facial expression interpretation: where human-computer interaction, artificial intelligence and cognitive science intersect'. *Pragmatics and Cognition* 81(1): 223–239.

Lynch, C., K. Ashley, V. Aleven and N. Pinkwart (2006) 'Defining ill-defined domains: a literature survey'. *Proceedings of the Workshop on Intelligent Tutoring Systems for Ill-Defined Domains at the 8th International Conference on Intelligent Tutoring Systems.* pp. 1–10.

Mitrovic, A. and S. Ohlsson (1999) 'Evaluation of a constraint-based tutor for a database language'. *International Journal of Artificial Intelligence in Education* 10: 238–256.

Mitrovic, A., K.R. Koedinger and B. Martin (2003) 'Comparative analysis of cognitive tutoring and constraint-based modeling'. *9th International Conference on User Modelling.* Lecture Notes in Computer Science, pp. 313–322.

Newell, A. (1990) *Unified Theories of Cognition.* Cambridge, MA: Harvard University Press.

Newell, A. and H.A. Simon (1995) 'GPS, a program that simulates human thought'. In G.F. Luger (ed.), *Computation and Intelligence: Collected Readings.* Cambridge, MA: AAAI and MIT Press. pp. 415–428.

Ohlsson, S. (1992) 'Information-processing explanations of insight and related phenomena'. In M.T. Keane and K.J. Gilhooly (eds), *Advances in the Psychology of Thinking.* New York: Harvester Wheatsheaf.

Ohlsson, S. (1996) 'Learning from performance errors'. *Psychological Review* 103: 241–262.

Ohlsson, S. and A. Mitrovic (2007) 'Fidelity and efficiency of knowledge representations for intelligent tutoring systems'. *Technology, Instruction, Cognition and Learning* 5(2): 101–132.

Ortony, A., Clore, G.L. and Collins, A. (1988) *The Cognitive Structure of Emotions.* Cambridge: Cambridge University Press.

Porayska-Pomsta, K., M. Mavrikis and H. Pain (2008) 'Diagnosing and acting on student affect: the tutor's perspective'. *User Modeling and User-Adapted Interaction,* 125–173.

Sawyer, R.K. (2006) *The Cambridge Handbook of the Learning Sciences.* Cambridge: Cambridge University Press.

Scherer, K.R. (2005) 'What are emotions?: And how can they be measured?' *Social Science Information* 44(4): 695–729.

Self, J.A. (1990) 'Bypassing the intractable problem of student modeling'. In C. Frasson and G. Gauthier (eds), *Intelligent Tutoring Systems: At the Crossroads of Artificial Intelligence and Education*. Norwood, NJ: Ablex Publishing.

Sison, R. and S. Masamichi (1998) 'Student modelling and machine learning'. *International Journal of Artificial Intelligence in Education* 9: 128–158.

Sperber, D. and D. Wilson (1995) *Relevance: Communication and Cognition*. Oxford: Blackwell.

Stets, J. and Turner, J. (2008). 'The Sociology of Emotions'. In M. Lewis, J. Haviland-Jones, and L. Barrett (eds) *Handbook of Emotions* third edition, pp. 32–46. New York, NY: The Guilford Press.

Weiner, B. (1992) *Human Motivation: Metaphors, Theories, and Research*. Thousand Oaks, CA: Sage.

Woolf, B. (2009) *Building Intelligent Interactive Tutors: Student-Centered Strategies for Revolutionizing e-learning*. Burlington, MA: Morgan Kaufmann.

Yerkes, R.M and J.D. Dodson (1908) 'The relation of strength of stimulus to rapidity of habit formation'. *Journal of Comparative Neurology and Psychology* 18: 459–482.

30

The Interplay Between Research and Industry: HCI and Grounded Innovation

Lars Erik Holmquist

Successful industry-orientated research is a matter of balancing many different stakeholders with different and sometimes conflicting goals. While a company's objective is to produce products that can be sold to customers, the researcher's main goal is to produce and document new knowledge. It may be natural to assume that new knowledge can contribute directly to innovative products, but experience shows us that this is in fact rarely the case – innovation is a much more complex pursuit than simply turning knowledge into products. In the case of HCI research, it is not enough to create a great interface; there is a large number of other factors that contribute to success, such as finding the right cost, defining compelling use cases, determining market position and so on. While some of these may fall outside the definition of academic research, it certainly does not hurt to have an awareness of them, and even a basic understanding of the path to successful innovation can help with making research results more relevant in the larger picture.

So what is the role of HCI research in the development of successful products? There is no question that our field has contributed to an awareness of human factors in the development of computer software and interactive artifacts. Many methods that are used in system and Web development can be derived in whole or part from earlier research efforts. However, beyond the established benefits of better usability and standardized evaluation methods, many HCI researchers also have the ambition to introduce new interfaces and interaction techniques. In fact, a sizeable portion of the proceedings of HCI-related academic conferences contain technical inventions that purport to improve interaction with digital systems. Increasingly, we are also seeing entirely new concepts for interactive artifacts and services being presented in an academic context. These new concepts are usually verified by some kind of user evaluation, be it lab studies or tests in more naturalistic settings, but the real measure of a new idea goes beyond formal studies. In fact, a more relevant measure of success for a new interactive artifact should be if it gets taken up and used in the real world – in other words, if it constitutes an *innovation*.

Computer scientist and innovation writer Peter J. Denning has given some helpful definitions of innovation and related terms. He first makes a distinction between innovation and invention: '*invention* means simply the creation of something new', he says. Innovation, however, means 'the adoption of a new practice in a community ... [It] requires attention to other people, what they value and will adopt' (Denning 2004). In other words, an invention does not become an innovation until people actually start to use it. To achieve this, it is not enough that something is new; it is necessary to understand the prospective users and what they want. This obviously resonates well with the general thrust of HCI research, but still requires a measure of qualification.

Currently, much of the development of new interactive artifacts in HCI is bundled under the general term of 'design'. However, what is taken for design in the HCI research community is often very different from what is otherwise considered the practice of design. One reason is that creating new artifacts in a research context is essentially a *design process without a goal*. This goes against the traditional view of design. Nobel laureate Herbert Simon claimed that to design means devising 'courses of action aimed at changing existing situations into preferred ones' – in other words, taking the steps of going from a current state into another, better state (Simon [1969] 1996). This is a highly instrumented view of the design process, suggesting that it is a rational search for a solution and it can be formalized in a set of rules or even a computer program (in the 1960s, Simon was influential in the formative field of artificial intelligence). Although this model can fit quite well with more formalized design processes in large teams with clear goals – for instance, software development – it is less well suited to those instances where the main motivation is to generate entirely new ideas.

However, it is important to understand that even though design can in some sense be 'innovative', design and innovation are *not* the same thing. A designer is typically trained to come up with a solution to a problem given by a client, whether it is the layout of a magazine, the shape of the steering wheel in a new car or the size and colors of the buttons on a smartphone interface. There is room to be innovative within such boundaries, but there will be limits. A smartphone interface designer might be free to invent a new layout for the buttons, but the functions that these buttons trigger will most likely already have been decided by the engineering department that produced the phone's operating system. The software engineering department in turn is bound by a set of specifications from other parties, such as the hardware manufacturer that produces the device and the network carrier that is the ultimate customer.

Innovation, on the other hand, is the process of giving solutions to problems that do not yet exist (or we did not know existed). The innovator works with the materials at hand, but instead of being bound by them, a true innovator will transcend the material and create something unexpected. If we continue the case of the smartphone, there can be inventions at many different levels, from hardware features such as sensors and buttons and software functions such as music players and cameras, to systemic inventions such as different data plans or methods for distributing applications. Such new ideas cannot originate in a design process with a pre-set specification. It is not a coincidence that arguably the most innovative phone of recent years, Apple's iPhone, was created by a company that had no previous experience in designing phones and no existing ties to the other actors such as wireless network operators. This allowed Apple to introduce new ideas in every area of the phone including hardware (such as multi-touch screens), software distribution (the App Store) and relationships with operators (for instance fixed-rate data plans). These innovations go far beyond the design of the user interface (impressive as that is) and would have been very hard to carry through by a company with more established ties and practices in the area. Similarly, another disruptive phone paradigm, the Android operating system, is being developed by Google, a company with no in-house consumer hardware manufacturing. This allowed them to abstract the software layer and create a platform that is now manufactured by many different vendors to a common specification, resulting in savings in development cost and improved software interoperability.

We can thus think of the traditional design process as being bounded by a *specification* (such as the brief from a client) and the *material* (such as the screen of a smartphone). An innovation, on the other hand, has no predefined specification and strives to transcend the limitations of the material. This bounding by specification and material is reflected in most of the design methods used in commercial contexts, which explicitly or implicitly assume there is a specification or an 'ideal state' that will be reached through the design process. However, if we consider most of the methods developed for user-orientated research in academia, there is usually no specification to begin with. Instead, the researcher will employ various methods to make an enquiry in the user setting, such as observational studies, questionnaires, interviews and so on. The researcher then uses the results from this enquiry, alongside his or her knowledge of technology, to produce a new artifact that solves the problems which have become apparent during the process. In essence, while there is no original design specification, the results of the user-orientated enquiry will here replace the design brief given by the client to a professional designer.

If we look at how new ideas make it to market, it turns out that having a great invention is not the only thing required for an innovation to be successful – in fact, it is not even the *most important* thing! This is true even if the invention is the result of a careful examination of user needs, such as those we

see in HCI. In a study of successful entrepreneurs, Peter Drucker found that, contrary to popular belief, entrepreneurs are not risk-takers who simply go out on a limb for an exciting idea (Drucker 2006). Instead, they work very systematically in a careful process with a number of clearly identifiable steps that exist specifically to reduce the risk of failure. Drucker identified five steps taken by successful entrepreneurs:

1. searching for opportunity – from one of seven innovation sources
2. analysis – risks and benefits
3. listening – to the community
4. focus – on a simple articulation
5. leadership – mobilizing people and market for the product.

The full process of creating a successful business from these steps is beyond the scope of this article; I recommend the literature on entrepreneurship as an avenue for that, as well as consulting with successful entrepreneurs and investors. Here, we will focus on the sources of innovation that Drucker identified. This is typically what happens *before* the decision to go forward to start a company or launch a business is made and is the process that takes place in universities, research labs, design studios and start-up factories. Drucker's analysis will help put the typical research activities in perspective.

Drucker said that the first step of an innovation process consists of finding opportunities. The first four sources of innovation that Drucker found are internal to a business and can usually be pursued without worrying about external competition.

- **The unexpected** – An unexpected success, an unexpected failure or an unexpected outside event can be a symptom of a unique opportunity.
- **The incongruity** – A discrepancy between reality and what everyone assumes it to be or between what is and what ought to be can create an innovative opportunity.
- **Innovation based on process need** – When a weak link is evident in a particular process, but people work around it instead of

doing something about it, an opportunity is present to the person or company willing to supply the 'missing link'.

- **Changes in industry or market structure** – The opportunity for an innovative product, service or business approach occurs when the underlying foundation of the industry or market shifts.

The following three sources are dependent on outside factors.

- **Demographics** – Changes in the population's size, age structure, composition, employment, level of education and income can create innovative opportunities.
- **Changes in perception, mood and meaning** – Innovative opportunities can develop when a society's general assumptions, attitudes and beliefs change.
- **New knowledge** – Advances in scientific and non-scientific knowledge can create new products and new markets.

All of these are useful to identify and can be harnessed to come up with potential innovations, but note that 'new knowledge' – the end result of almost all activity in academic and industrial research labs – only appears as number 7 (and last) in Drucker's original list! Drucker called knowledge-based innovation 'temperamental, capricious and hard to manage' and noted that it has by far the longer lead time of all sources (Drucker 2006: 107–108). Given the competition from other, more promising, innovation sources, it seems that there is very little chance that a new scientific result, no matter how exciting, will ever make it into the real world. More generally, of all the inventions produced in science labs and universities, almost none will become actual innovations in Denning's sense.

This is not as depressing as it may sound. First of all, researchers are already aware of this; most of what they do adds to an existing knowledge base without being directly useful in the short term. Second and more important, for those who do want to produce lasting innovations, there is a whole wealth

of other opportunities to pick from that can help propel their results into the mainstream – as long as they are prepared to be flexible in how these new inventions are eventually used! To give a popular example, in 1968 Spence Silver, a researcher at 3M was interested in producing a better and more long-lasting glue for the company's products. Unfortunately, what he came up with turned out to be the opposite – a glue that was so weak, paper treated with it would peel off the surface it had been attached to without leaving any trace. He ransacked his brain for years trying to come up with a use for this new invention. Another 3M employee, Art Fry, finally came up with the idea of applying the glue to small pieces of paper, which could for instance be used as markers in a book without leaving a mark (he got the inspiration when the paper bookmarks in his church choir book scattered to the wind when he stood up to sing). The product, Post-It Notes, was finally launched in 1980 and is now a mainstay in offices and homes around the world. Technically, Silver's invention was a complete failure considering what the goal was, but by carefully considering the result and combining the properties of the invention with other activities that they had not originally considered, the researchers could come up with another use for it. It is quite likely every research lab houses a multitude of such seemingly useless results that could be turned into products.

In addition to those identified by Drucker, Denning (2004) suggested a final source of innovation that resonates well with some recent work in HCI.

- **Marginal practices** – A different field offering a novel solution.

This is a way of harnessing innovation by looking at what other smart people are already doing, outside of one's own immediate surroundings. A practice that may seem irrelevant in your field might in fact offer a solution to a problem. In a similar vein, Erich Von Hippel coined the term *user-driven innovation*, which is based on the idea of *lead users* who are ahead of the curve with a given technology (Von Hippel 1986). By finding somebody who is already an advanced user of a technology, a company can find uses that they did not think of. For instance, in another example, 3M used this method to develop a breakthrough surgical drape product by assembling a team of lead users, which included a veterinary surgeon, a make-up artist, doctors from developing countries and military medics. It is also possible to go out in the field to get inspiration from how a product is already being used by advanced users.

When it comes to digital innovation, the desktop computer is a striking example of how hard it can be to turn even the best inventions into a successful product. It seems that as early as 1974, the Alto workstation developed at Xerox PARC already contained almost all of the vital components of today's computers: a user-friendly interface based on the desktop metaphor, powerful document editing complete with high-quality laser printing and networking capabilities to tie it into a larger web of information. Clearly, the Alto was so way ahead of its time, it would take the rest of the world decades to catch up. Yet, when Xerox introduced its STAR workstation in 1980, which incorporated many of these inventions, it was a commercial failure. It took another company – Apple – to package the ideas into a successful product.

When creating the Macintosh, Apple exploited a number of opportunities for innovation that had eluded Xerox. First, they took advantage of a change in the industry, where computers went from being bought by large corporations to also being sold to hobbyists and small businesses. To do this they had to adapt their product to the new market. This included setting a much lower price point. While Xerox was intent on producing the 'best' desktop computer with the most advanced user interface, Apple worked within the limitations of existing technology to produce the best computers that could be affordably produced.

This meant stripping out some of the more advanced features of the Alto and making a less pretty – but still functional – experience. However, the result was that, unlike the Star, the Macintosh 'felt' more usable, since it responded faster and was optimized for speed rather than appearance.

The Macintosh also had a clear use-case. In contemporary advertisements, Xerox claimed that with the Star you could 'create documents with words and pictures'. However, the corporations that they hoped would buy their machines already had advanced systems, attended by highly skilled workers, which let them do exactly that. This was done in advanced page description languages and editing facilities and was far from the simple 'what you see is what you get' (WYSIWYG) interface of the Star – but that did not really matter, because the end result was the same. However, Apple set its sights on small businesses that could not afford to hire dedicated personnel to create their printed materials. By making a user-friendly computer, they let the same person perform many different tasks with little or no training. For a small company, it was suddenly possible, for an investment of a couple of thousand dollars (a Macintosh and a laser printer), to perform the job that previously would have gone to an external company. Of course, quality suffered somewhat – there are many examples of amateur designers going overboard with the new possibilities in the early days of desktop publishing – but the results still looked infinitely more professional than what had previously been created by hand and, in the long run, were much less expensive than having the work done by an outside contractor.

We have now seen how entrepreneurs can work with a set of opportunities to create successful innovations, but let us try to tease out a few aspects of innovation that have not been covered by previous analysis. First of all, remember the distinction between innovation and *invention*. While a great invention is not enough to produce a successful innovation, it is a necessary component of the process.

However, coming up with new ideas is perhaps the *easiest* thing to produce in an innovation process! For example, a group manager at Apple Research Labs was quoted as saying that as he walked from his office to the cafeteria he could 'easily come up with 20 new ideas for products' (Rogers and Belotti 1997). While there are many methods – brainstorming, bodystorming, bootlegging, etc. – that can help anyone come up with new ideas, ultimately, it comes down to too many factors to control completely. There is no simple way to consistently create new ideas with any kind of guarantee as to their quality or quantity. Great ideas often come from random sources and the talent for picking up such impulses is probably partly a function of personal qualities, such as creativity, perseverance and intelligence, but a large part of the process can also be learnt and improved. While we do not know exactly how ideas work, most great ideas do not strike at random, but are the result of a very conscious and laborious process that can take months or years to produce results.

A successful innovation also has to provide a useful function in the real world. To distinguish those inventions that are simply blue-sky from those which actually have a chance of working, we need to perform *enquiries* in order to understand the world better. Enquiries can include many different domains and methods. For instance, if the goal is to create a solution to a specific problem at a workplace, it is often useful to go and spend some time at the actual site or talk to people involved in the work. However, if it is important to use a particular emerging technology, it is a good idea to understand as much as possible what the technology is capable of. When it comes to user-orientated enquiry, there are many established methods to gather data on what users need (or think they need), such as questionnaires, interviews, focus groups and so on. In HCI research, it has become popular to adapt observation and analysis methods from social science, in particular ethnography and ethnomethodology. Here, in order to identify problems encountered in a workplace or other situation, the researcher observes it from outside, without interfering. The idea is that this will give a truer and less prejudiced picture of the setting, which can then be turned into system requirements.

However, while both invention and enquiry are necessary components, they also both present problems. Even considering brainstorming methods and other methods that include a selection phase to find the best results, with blue-sky idea generation, there is no guarantee that the inventions arising are actually realistic or have any bearing on reality. Enquiry does not guarantee success either. It is entirely possible to spend months or even years of study getting to know the minutiae of a workplace setting, but this does not mean the outcome will be ideas that can actually change or improve the situation. Rather, it is quite possible that the researcher becomes so entrenched in the current situation it is difficult or impossible to look beyond it to new solutions. Thus the risk is ending up with technologies that support existing work practices with little of the additional benefit which could have been had with the introduction of new technologies and ideas.

In reality, most methods that are commonly used in research and product development today are situated in the middle ground, drawing on both invention and enquiry. With more or less success they attempt to combine both idea generation and user studies to produce novel and/or useful systems and products, but different methods and techniques emphasize invention and enquiry to different degrees. Some are firmly rooted in the data gathered from users and strive to design systems that address very specific problems; others take this information as just one of many inputs to the design process. Approaches that try to involve users more creatively in the process, and thus might include a higher level of invention, can, for instance, be found in the area of *participatory design*. Here, the prospective users are involved in the system design from the start, not just as study objects but on equal footing with the designers (Muller and Kuhn, 1993). This has become used as a product development strategy and participatory design projects with a high level of user involvement have produced innovative results in areas ranging from home electronics to waste water treatment. Various techniques can be used to improve collaboration between designers and users. For instance, the users can act out and film videos of different scenarios for technology use. This can be used to create a creative dialogue between designers and users and help formulate inventive ideas that are rooted in the users' own experience.

All the standard study methods in HCI have been designed to ensure that new artifacts fit with the needs of the users, but there is another way of looking at the results of an enquiry. Rather than considering it as a specification that has to be followed, we can choose to see it as *inspiration* for the innovation process. As an example, *cultural probes* is a method developed to collect input from users that is intended to inspire new ideas, by letting users perform a series of evocative tasks – for instance, taking pictures of their environment (Gaver et al. 1999). This represents a departure for HCI, a field that has always stressed the needs of users as the foundation for new technologies (participatory design being one obvious example of this, among many others). What if we acknowledge the fact that most users are not designers, have no particular technical skills and do not have the time or inclination to come up with new product ideas? Then, instead of being specifications, user studies become one of many inputs into the process of inventing new artifacts. This is closer to an artistic process, where the goal is to create something that is unique and has a certain effect, rather than something that fulfills certain set criteria. On the surface, this seems very different from the scientific process, where every result has to come from an a priori articulated hypothesis. Yet, despite the demands for formal rigor, the fact is that

successful scientists will also often take inspiration from unexpected sources and come up with solutions based on impulses that may seemingly have nothing to do with the work at hand (Johnson 2010).

If we map the axes of innovation and enquiry in a two-dimensional field, we can start identifying where different approaches and methods appear (Figure 30.1). The user-centered design methods typical of HCI all fall in the center of the diagram; they include a modicum of both invention and enquiry. Pure idea generation, such as brainstorming and conceptual art projects, strive for a very high degree of originality and invention but little in the form of enquiry. Pure studies, however, whether it is engineering feasibility to test out a new technology or long-term ethnographic studies to investigate a setting, score high on the enquiry axis but have little or no ambition for invention.

It would seem that to produce great innovations we would need to have both a high level of originality and invention (to come up with something genuinely new) and a very strong grounding base of enquiry in the real world (to know what the possibilities and opportunities really are). If we sketch out this 'ideal' method, it would fall in the top right-hand corner of our diagram, where both axes are at or close to their maximum. This is what I call *grounded innovation.*

So how can we achieve grounded innovation, where we maximize the impact of enquiry and invention? In reality, even with the best methods this is not a state that can be realistically achieved – if so, everybody would already be producing world-class leading products! Even if we could formulate a method that captured this, the results of any such method will only be as good as the people who are carrying it out. However, by being aware of the potential that is there, and by consciously adapting our methods to get closer to the desired state, I argue it is entirely possible to systematically work in a way that has a higher chance of producing innovations than other methods.

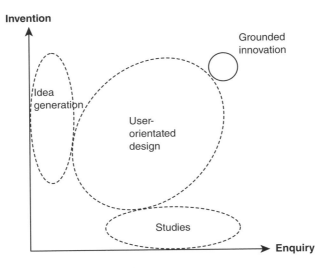

Figure 30.1 Grounded Innovation attempts to maximize the degrees of both invention and enquiry in an innovation process (© Future Applications Lab/Viktoria Institute).

To clarify what we mean by grounded innovation methods, let us take a concrete example of how my group has strived for achieving this. The *transfer scenarios* method provides a way to transfer grounding from one domain to inspire inventions in another (Ljungblad and Holmquist 2007; Ljungblad 2008). As part of a large European project, we were given the task of coming up with ideas for what future robot technology might look like. Robots are a complicated class of digital artifacts. By definition, a robot is any machine that has the possibility of autonomously acting on the physical world, but in the public mind, robots are much more narrowly defined. Ask anybody to draw a robot and most will come up with a mechanical humanoid, perhaps a somewhat menacing science-fiction-inspired image. If you ask what they would like a robot to do if they had one, it will include typical human tasks such as 'do the dishes'. In our early work with robots, we tried various ways to break this image, including guided brainstorming sessions. However, it was clear that the ideas we came up with were not well grounded in reality. They were both too far out to be implemented technically and they did not provide any actual use or fill any human need that would make them attractive in the long run.

We first decided to attack the latter problem, of envisioning compelling functions and roles for robots beyond the science-fiction scenarios. However, the problem with finding any kind of use for robots is that people really have no idea what to do with them! It does not really help to ask prospective users what they want, because the answers will be just as unrealistic and ill-founded as whatever researchers can come up with, if not more. Also there are few situations to study where robots are part of everyday activities, outside of highly controlled factory environments and clearly defined tasks such as vacuuming (where iRobot's Roomba has one of the few successful household robot products). It was obvious we needed a new approach to our methods of enquiry, one that would give inspiration to design without fettering us to existing preconceptions.

One of the opportunities for innovation identified above is *marginal practices*. By looking at user groups outside of your own general experience, it might be possible to come up with solutions to problems that you would otherwise not think of. A potential marginal practice for robots might be robotics researchers, who are already living with various robot prototypes of some sort. However, we realized that the grounding in an innovation process does not necessarily have to come from the same domain as that of the technology – it might even help if they are different. Instead it could be useful to find some activity or group that has properties similar to the one we are innovating for – we can call that analog practices. We set out to find an analog practice that was suitable for robots, basing the search on the properties we had identified for robots, including:

- autonomous
- emerging behaviors
- taking advantage of the physical world.

Several previous robot designs have been based on common household pets, such as dogs (e.g. Sony's AIBO). However, compared to current digital technology, dogs are in fact extremely complex and even the most basic dog behavior is almost impossible to replicate with today's technology. Instead, we found another group of animals also matched the set criteria, but with less complex behaviors. It turns out that exotic pets such as insects, spiders and reptiles have very devout owners, but their overall complexity in behavior is significantly less than that of a dog. Furthermore, most people do not have any experience with these types of animals and would have fewer preconceptions about what they could do. Finally, as with many marginal practices, it turns out, although the size of the group of people who own this kind of animal is quite small compared to more of people who mainstream pets, they are also often highly articulate and happy to share their reasons for owning these animals with other people.

The researchers set out to gather information about owners of exotic pets. Through various organizations, they found ten owners of lizards, spiders, snakes and other animals, who were willing to sharing their experiences (Figure 30.2). The data gathering was performed as semi-structured interviews, where the interviewer has a set of guiding questions to keep the conversation on track, but can also allow for detours on the way if

Figure 30.2 To get inspiration for new robotic agents, we interviewed owners of exotic pets, such as this spider (© Future Applications Lab/Viktoria Institute).

something is of particular interest. Examples of the questions asked were as follows.

- What do they consider important qualities of their pets?
- Why are they interested in this kind of animal?
- What do they do with them?
- What are the pets doing?
- Do the have any social interaction with other pet owners?

What came back from these interviews was a rich set of statements about the reasons for keeping these unusual pets, what kinds of activities they performed, how the owners interacted with others around their interest, and so on. In other words, this was real *grounding* in actual experiences. What made it useful for us was that these people owned something that had at least some analogous qualities to the robot technology that we were ultimately interested in.

The subjects had many different motivations for owning and enjoying their pets. For instance, one saw them not so much as pets but as objects of a more distanced interest, acknowledging that you cannot interact with a snake in the same way as with, for instance, a dog:

> … I mean the snakes are constructed in a specific way and if you get them you have to accept that they aren't any cozy pets or alike, you have to have them as your interest.

Another had a more instrumental and activity-centered way of approaching them, being interested in the challenge of breeding:

> Yes, well, it's mostly that it is exciting and a challenge to develop certain colors and things like that.

There were also many stories about the personal relationships that would develop between owner and pet, even though the interviewee was of course fully aware that in reality a lizard does not really form actual relationships with people:

> Eh … here is a leopard gecko, it is partially sighted, so I have fed it with tweezers since it was small, and now it is a bit over a year, so it is pretty special to me.

The next step was to organize all this material in some form to make it useful for design purposes. First, all references to particular types of pets were removed and replaced with the neutral word 'agent' (we did not want to use 'robot' because of the connotations mentioned previously). The idea was that the object of interaction would be neutralized, effectively leaving a hole, which could be filled with

technology of our own design. With this neutralized data, where a leopard gecko lizard was on an equal footing with a black widow spider or any other kind of exotic pet, it turned out that many of the statements fell in the same general area. For instance, there was a group that had to do with how owners cared for and related to their pets in their environments:

> … partly it is fun to build these environments, and partly it is that I can spend hours to just sit and look at them when I have fed them or something like this.
>
> Well, it should be … be like a furniture preferably, nice to look at and at the same time easy to care for.
>
> Well, I have had it [the terrarium] in the living room, and then … well it's like a little extra furniture piece with a jungle theme.

What came out of this process were four rough groups of different ways to relate to the agents (Figure 30.3). The next step was to turn these groups into more concrete cases. For this, we used a technique called *personas*. This was originally created to make it easier to envision use-cases for products such as Web pages (Cooper 2004). A persona is a fictions person, created to illustrate a particular kind of user. A persona can be quite elaborate, complete with full name, work history, family background, hobbies, personality traits and so on. The idea is that by using the persona, it is easier for a designer to imagine different functions and obstacles that might occur. For instance, on a shopping website, it might be very useful to distinguish between

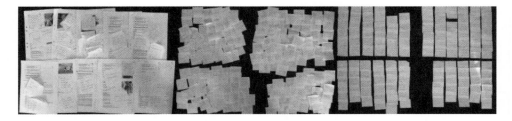

Figure 30.3   The results from the interviews were organized into four clusters, denoting four different ways of relating to the agents (© Future Applications Lab/Viktoria Institute).

what is needed for the persona 'Single father with two kids shopping for sports clothes' versus 'Fashion-conscious teenager looking for something to wear over the weekend' – the two different cases may have very different needs and this can expose flaws in a design without needing a complete user study.

The difference in our case was that we had no pre-existing use cases we wanted to apply the personas to – instead, these fictitious characters and their activities were created to inspire new technology that does not yet exist. In the personas, although specific traits such as name and background were invented, the activities they performed with the agents were inspired by the interactions we had seen with real-world exotic pets. The persona descriptions were quite elaborate and included a variety of ways in which each person related to some as-yet unspecified technology.

There were four personas in total, each based on one of the four groups that were found in the data. The four personas that emerged were *Nadim*, who considered his agents a hobby; *Anne*, who considered them as a part of her interior design; *Magda*, for whom they were an extension of her identity; and *Christopher* who used them as a contact mediator. To give a flavor of what the personas looked like, here are some excerpts from two of them (Ljungblad et al. 2006: 575–580).

Anne is a 41-year-old physiotherapist. When she is not working she enjoys getting together with friends and family. She lives with her boyfriend in a one-bedroom apartment in the suburbs of a small city. She is interested in interior design and a wall of the living room is occupied with agents. Her fiancé is not so fond of the agents, but he is of Anne, so the agents can stay. Anne had had the agents a long time before they got together and she is never going to get rid of them. Anne is fascinated by the feeling the agents give the interior. She believes they create pleasant surroundings to live in, as the room feels more alive and dynamic.

Nadim is 32 years old and works as a network engineer, living alone in a two-bedroom flat in a small town. One of the rooms is Nadim's hobby room and this is where he keeps his agents. Most of the people in Nadim's home town do not know he owns agents; it is not something he goes around talking about. He has always had a great interest in collecting and exploring various things, and as he got older he became fascinated in having agents as a hobby. Nadim finds it exciting to try to understand their behavior and sees them as a research area where there is always something more to learn. He has specialized in a type of agent that communicates through colors. He enjoys watching them communicating to each other and changing their patterns.

The next step in this process was to fill in the 'holes' in the personas with actual technology. It is important to consider that, at this stage, the scenarios were still very open-ended; there were many ways in which we could imagine technologies that could provide the kinds of interactions that the personas suggested. However, at this point we also introduced some technological constraints. We wanted the final designs to be something that could actually be built and demonstrated with the resources available in the project, rather than some science-fiction vision that might never see the light of day. This meant working with technologies the team was already fairly familiar with, such as mobile phones, projected displays and existing research robots that could be easily modified.

The final designs were the result of a tight interplay between the interactions described in the personas and the technology we had to work with. The first scenario, with Nadim and his hobby agents, resulted in a system we called *GlowBots* (Figure 30.4; Jacobsson, Fernaeus et al. 2008). Inspired by the idea of agents that could 'breed' to somehow create new patterns, we imagined a collection of small autonomous robots that could change appearance by means of a colorful display. By putting the robots next to each other, each one would be inspired by the pattern shown on their neighbors and create a new pattern that was a mix of the existing ones. For the actual implementation, we used a small educational robot platform called the *E-Puck*, which was developed by one of the other partners in the project. This coffee-cup sized

Figure 30.4 GlowBots are small autonomous robots that communicate using a colorful light display (© Future Applications Lab/Viktoria Institute).

robot provided most of the necessary technology to realize the scenario, including wheels for moving around and a set of infrared receivers and transmitters to communicate between the agents. However, the E-Puck did not have any visual display, so we needed to design and produce a round LED display that fitted snugly on top of the existing robot (Jacobsson, Bodin et al. 2008). With that in place, we wrote custom software to control the E-Pucks to exhibit the new behavior.

In the second scenario, Anne uses her agents to make her living room feel more alive. To achieve this, we wanted a large display that could cover an entire wall. In the future, it might be possible to realize this with technologies such as electronic ink or textiles, but for this implementation we settled on a data projector. To make the display dynamic, we used algorithms that can create life-like simulations of plants that grow and wilt away over time. Since we wanted the display to somehow reflect Anne's everyday experiences, the system can receive images

taken by a mobile phone camera; the colors and shapes extracted from these images become the basis for the appearance of the plants. *The Flower Wall* (Figure 30.5) allows users to send pictures from their phones (using Bluetooth or e-mail) to the system, and every new picture will be turned into an attractive flower that grows and wilts away over time (Petersen et al 2009). For demonstration purposes, this process is quite fast, taking place in the order of minutes. However, the intention is that if this system were installed more permanently, the flowers would take weeks or months to grow and eventually disappear, creating the impression of a true 'living' wall in the user's home.

We do not yet know if these particular examples are innovations that could be turned into successful products, because the technology that they require (robust autonomous robots and large, unobtrusive displays) is still too expensive for any mass-market application. However, both systems have been demonstrated to many thousands of people in various

Figure 30.5 *The Flower Wall* **lets users create a living tapestry from their own mobile phone pictures (© Future Applications Lab/Viktoria Institute).**

settings, with a lot of positive feedback. Both of them display unique properties of autonomous agents that could very well appear in products when the technology is ready. As we learned earlier, inventions like this are only a first step towards successful innovation – but by using a process grounded in technology as well as users, we may at least be closer to the goal than if we had been working on truly blue-sky ideas.

The transfer scenarios method shows how it is possible to have a creative interplay between enquiry and invention, which is necessary to achieve grounded innovation. The enquiry was both user-orientated and technical; the interviews with owners of exotic pets gave us a grounding in real-world practices, whereas a careful consideration of available technology ensured that it would be possible to implement the results. While the user studies defined the outlines of the interaction, the technology also directly inspired the results – the GlowBots

might have been entirely different if we had used another robot platform, and the Flower Wall was affected by our experience with camera phones as well as experiments with other forms of dynamic decoration. Most importantly, the fact that user studies and technology from totally different domains were combined provided a creative spark that could not have been achieved otherwise.

Of course, transfer scenarios have their caveats, as do all methods. It is not always easy to find a suitable practice to match a particular technology and it is hard to provide any guidelines except to rely on intuition and trial-and-error. The final outcome is very much dependent on the skills and technical knowledge of the designers; given another set of people, the results could be completely different, for better or worse. However the idea of taking a marginal practice as inspiration and the method of turning raw data into personas is a solid process that can be easily replicated. No method can completely guarantee the quality of the results – that is ultimately up to the skills of the people who actually perform the work – but this method can at the very least provide a combination of enquiry and invention that takes us one step further to grounded innovation.

In closing, when working with industry, or more generally towards the goal of producing true innovations (in Denning's sense), it is valuable to at least be aware of the very long road from invention to product. Sometimes, it might be that researchers put too much value in the pure inventions they produce, whether procedural or technical, and too little in the complex web of other circumstances that need to be taken into account for them to actually be taken up and used. Furthermore, to be truly innovative, it is necessary to have a firm grounding in enquiry, but it is equally crucial to provide the additional spark of unexpected inventions. Finally, when it comes to the field of HCI, it is worth noting that outside of success stories such as the graphical user interface, many of the most successful inventions have to do with processes and procedures rather than actual artifacts. For instance, methods for evaluation of interfaces and for producing system specifications grounded in real-world user needs have become commonplace in software development, something that to a great extent is the result of HCI. Thus, while HCI certainly has a role to play in producing new innovations, a large part of our mission in industry can be to contribute understanding and methodology to innovation processes in a larger context – instead of simply coming up with the next shiny gadget!

NOTE

This chapter is adapted from the book *Grounded Innovation: Strategies for Creating Digital Products,* published by Morgan Kaufmann. For more information, visit http://groundedinnovation.net

REFERENCES

Cooper, A. (2004) *The Inmates Are Running the Asylum.* Harlow: Sams Publishing.

Denning, P.J. (2004) 'The Social life of Innovation', *Communications of the ACM*, 47(4): 15–19.

Drucker, P.F. (2006) *Innovation and Entrepreneurship.* New York: Harper.

Gaver, W.W., Dunne, A. and Pacenti, E. (1999) 'Cultural Probes', *Interactions*, 6(1): 21–29.

Jacobsson, M., Bodin, J. and Holmquist, L.E. (2008) 'The See-Puck: A Platform for Exploring Human–Robot Relationships', *Proceedings of the Twenty-sixth Annual SIGCHI Conference on Human Factors in Computing Systems*, ACM, 141–144.

Jacobsson, M., Fernaeus, Y. and Holmquist, L.E. (2008) 'GlowBots: Designing and Implementing Engaging Human Robot Interaction', *Journal of Physical Agents*, 2(2).

Johnson, S. (2010) *Where Good Ideas Come From: The Natural History of Innovation.* New York: Riverhead Books.

Ljungblad, S. (2008) 'Beyond Users: Grounding Technology in Experience'. PhD thesis, Department of Computer and Systems Sciences, Stockholm University, No. 08-004.

Ljungblad, S. and Holmquist, L.E. (2007) 'Transfer Scenarios: Grounding Innovation with Marginal Practices', *Proceedings of the SIGCHI Conference on Human Factors in Computing Systems*, ACM, 737–746.

Ljungblad, S., Walter, K., Jacobsson, M. and Holmquist, L.E. (2006) 'Designing Personal Embodied Agents with Personas', *Proceedings of the 15th IEEE International Symposium on Robot and Human Interactive Communication*, 575–580.

Muller, M.J. and Kuhn, S. (1993) 'Participatory Design', *Communications of the ACM*, 36(6): 24–28.

Petersen, M.G., Ljungblad, S. and Håkansson, M. (2009) 'Designing for Playful Photography', *New Review of Hypermedia and Multimedia*, 15(2).

Rogers, Y. and Belotti, V. (1997) 'Grounding Blue-sky Research: How Can Ethnography Help?', *Interactions*, 4(3): 58–63.

Simon, H. ([1969]1996) *The Sciences of the Artificial*, (3rd edn). Cambridge, MA: MIT Press.

Von Hippel, E. (1986) 'Lead Users: A Source of Novel Product Concepts', *Management Science*, 32(7): 791–805.

Afterword: Looking to the Future

Sara Price, Carey Jewitt and Barry Brown

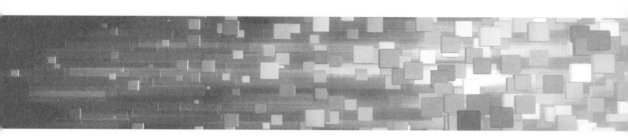

This *Sage Handbook of Digital Technology Research* offers a wide-ranging foray into important and contemporary perspectives of digital technology research. It has situated the research field historically, culturally and politically; highlighted key characteristics and pertinent issues for digital technology research for the twenty-first century and their implications for research; introduced central theoretical and analytical approaches; and illustrated exemplary research across a number of different technologies and sites of practice.

As this book shows, a broad range of methodological approaches can legitimately be used in digital technology research, ranging from exploratory approaches (which offer valuable opportunities for novel discoveries and new directions in this nascent field), to situated approaches (which offer contextualized research), to more experimental approaches to technology-based research and in many cases a 'mixed methods' approach is taken. One of the challenges facing researchers

undertaking digital technology research is choosing the appropriate methodology and we hope that the chapters in this book offer a valuable foundation for supporting readers' selection through understanding the challenges of different methodological approaches and ways to use them effectively.

The specific challenges of researching digital technologies and undertaking research with digital technology are discussed across the parts and chapters in this book. Some of these challenges are focused and specific to methods or sites of research – for example, defining and managing 'context', managing online information, contrasting lab-based versus in-the-wild studies or fostering innovation. Other methodological challenges are, however, more overarching. For example, as many of the chapters in this handbook have shown, the rapid development of technology and co-evolution of social–cultural practices presents a number of challenges for research. One consequence of new technology designs, new interfaces and new infrastructures is that technologies and devices are often novel to users, are not embedded in existing practices and routines and in many cases demand different kinds of practices. While this is exciting and compelling and, as seen in some chapters in this book, it can be central to the provocative nature of the research, it also brings significant challenges to some areas of digital technology research.

As an example let's take education, where research is exploring innovative technology use in teaching and learning. In this context, the impact of the 'novelty' of a technology on the research process and findings needs to be considered and accounted for. Cutting-edge technologies are often functionally unstable, interfering with the smooth running of digital environments and either requiring on site technical support or the running of studies 'off site' or in a lab-based context. New technologies and devices are generally not embedded into the practice, curriculum or policy context of education and, as such, often require participants, or users, to engage in new practices. This not only creates a complex relationship between the innovative applications and established practices but also highlights the need for fostering appropriation by the target user group. For example, in education contexts, not only does research need to address the learning value of new technology learning tools, but also the need for teachers to be able to design, adapt or even build technologies for them to become valuable teaching tools. One issue arising here is that the hype and users' expectations are not matched to the way the new technology can be embedded or appropriated into current practice.

Other key challenges for researching digital technologies addressed across the handbook include developing methods and procedures to effectively make use of automatically generated data – for example, computer-logged data stored by gaming systems. As several chapters highlight, in terms of analysis, the use of digital technologies in itself generates more disparate means of storing records – for example, data collection may include video data, audio data, as well as data relating to digital activity (computer-logged data), which may include GPS data, body sensor data, as well as system activity. Such data collection results in diverse forms of data records – for example, video with audio, separate audio and computer-logged data, which can prove problematic for analysis when one needs to cross-reference the status of events at any one time, thus highlighting the need for recorded data to be synchronized to precisely, to the second. Such issues are important when considering methods of data capture that span media and are used to support post hoc analyses. Digital technologies also give rise to large amounts of automatically generated data, both digital video data and computer-logged data. This provides exciting opportunities for social science research, opening up access to rich data sets, while also generating large amounts of data to manage and analyse, and raising questions of how to get good 'quality' data. For example, the lengthy duration of digital video data may be streamed in real time, providing richness in terms of quantity, but not necessarily in terms of quality content for the research questions. A significant challenge, therefore, is to develop more sophisticated and effective methods for 'quality' data collection and analytics – particularly ways to automatically capture relevant data and facilitate pertinent analytical procedures.

Linked to both these challenges are the methodological issues raised across a number of chapters concerning scale and scalability of research – moving from micro analytical approaches to macro concerns and generalizing from the particular. Much digital technology research, as the chapters in this handbook suggest, is also conducted over short timescales, often in the form of one-off interventions, and the challenges of longitudinal studies with technologies is exasperated by the rapidly changing technology environment and the impracticality for work and educational environments, for instance, of investing heavily in new technologies to be researched over time. Extensive deployment and implementation is, therefore, scarce. Another associated challenge is that of researching technologies in authentic contexts, as people's lived experiences of technology tends to occur over several different sites and unfolds over long stretches of time. This means there is often no reliable test bed for researching either extensively or longitudinally. Consequently, large numbers of disparate case studies emerge, making coherence of findings problematic.

These and other challenges associated with the continued refinement and development of methodologies need to be addressed to ensure research responds to the particularity of digital technologies and the full potential of undertaking research with digital technology can be realized.

Extending this issue of methods for effective capture and analysis of automatic data, discussed above, raises the question of how to make good use of data that can be sourced differently through digital technologies. For example, the ubiquitous, worldwide pervasive nature of networking and online technologies gives rise to the potential for accessing a wider number of users. The point made across this handbook is that digital tools generate and support new forms of data and, perhaps more importantly, they offer new ways to represent knowledge, engender new practices and create and connect local and global communities in new ways. Central to this is the fostering of innovative and transformative ways of interacting, expressing and forms of thinking. As the chapters in this handbook illustrate, such changes mean that non-verbal forms of interaction, and thus non-verbal forms of data, are becoming more prominent within social research on digital

technologies across a range of theoretical and methodological approaches and research contexts – for instance, in relation to theories of embodiment, space and place. As these and other chapters indicate, this is pushing the development and use of digital technologies for research purposes. For example, gesture, both through interaction with touchscreen interfaces and bigger body movements, can be automatically captured through sensor technologies, in real time and across different spaces. As many authors in this handbook note, further work is needed to extend existing methods and to develop new methods that look beyond people's talk and take account of ways of expression through bodily interaction, including gesture and action. Contemporary digital environments are therefore inherently multimodal, both in terms of representation modalities (e.g. texture, colour and shape of objects as well as digital representation) and interaction modes (e.g. talk, gaze, manipulation, gesture). This explores the value of taking a multimodality approach as a new way of describing and classifying 'embodied' forms of interaction and examining how embodied action can be played out differently in digital environments, such as learning, or in medical practice contexts.

In summary, this handbook provides an understanding of the 'space' and 'scope' for researching digital technologies, pointing to some of the important ways in which people use technology and the current status of research that uses digital technologies, as well as the main challenges of identifying and engaging with appropriate theoretical and methodological approaches. We hope that it provides a comprehensive 'toolkit' of theoretical perspectives to further develop the field of digital technology research.

Index

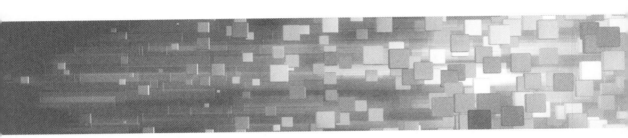

Page references to Figures or Tables will be in *italics*. Where there are Notes, the letter 'n' will follow the page number. HCI stands for Human–Computer Interaction.